Ground-Based Wireless Positioning

Ground-Based Wireless Positioning

Kegen Yu, Ian Sharp and Y. Jay Guo

CSIRO ICT Centre Australia

John B. Anderson, *Series Editor*

A John Wiley and Sons, Ltd, Publication

This edition first published 2009
© 2009 John Wiley & Sons Ltd.,

Registered office
John Wiley & Sons Ltd, The Atrium, Southern Gate, Chichester, West Sussex, PO19 8SQ, United Kingdom

For details of our global editorial offices, for customer services and for information about how to apply for
permission to reuse the copyright material in this book please see our website at www.wiley.com.

Library of Congress Cataloging-in-Publication Data

Yu, Kegen.
 Ground-Based wireless positioning / authors, Kegen Yu, Ian Sharp, and Y. Jay Guo.
  p. cm.
 Includes bibliographical references and index.
 ISBN 978-0-470-74704-9 (cloth)
 1. Location-based services. 2. Radio direction finders. 3. Mobile geographic information systems.
I. Sharp, Ian. II. Guo, Y. Jay. III. Title.
 TK5105.65.Y8 2009
 621.384'191–dc22

                                                                              2009007115

A catalogue record for this book is available from the British Library.

ISBN 978-0-470-74704-9(H/B)

Set in 10/12 pt. Times by Thomson Digital, Noida, India.
Printed in Great Britain by CPI Antony Rowe, Chippenham, Wiltshire

Contents

About the Authors

Dr **Kegen Yu** received a BEng degree from Jilin University, China, a MEng degree from the Australian National University, Australia, and a PhD degree from the University of Sydney, Australia, in 1983, 1999 and 2003 respectively. From 1983 to 1997 he worked as a practicing engineer and later a lecturer at Nanchang University, China. From 2003 to 2005 he was employed as a Postdoctoral Fellow at the Centre for Wireless Communications, University of Oulu, Finland, researching on wireless positioning and communications theory. Since November 2005 he has served as a Research Scientist at CSIRO working on ad hoc wireless positioning systems, wireless sensor networks and reconfigurable radio. Kegen has published three book chapters, and over 40 refereed journal and conference papers.

Ian Sharp is a Senior Consultant on wireless positioning systems. He has over 30 years of engineering experience in radio systems. His initial involvement in positioning technology was in aviation and later, in the 1980s, with the Interscan microwave landing system (MLS). In the later 1980s to the early 1990s, Ian was the R&D manager for the Quiktrak covert vehicle tracking system. This system is now commercially operating worldwide. From the mid 1990s to 2007 Ian worked at the CSIRO mainly on developing experimental radio systems. He was the inventor and architect designer of CSIRO's precision location system (PLS) for sports applications. The PLS has been successfully trialed in Australia and the USA. Ian holds a number of patents relating to positioning technology.

Professor **Y Jay Guo** is the Director of the Wireless Technologies Laboratory and Theme Leader of Broadband for Australia in the CSIRO ICT Centre, Australia. Prior to this appointment in August 2005, Jay held various senior positions in the European wireless industry managing the development of advanced technologies for the third-generation (3G) mobile communications systems. Jay has over 20 years of industrial and academic experience in antennas, signal processing and wireless networks. He has published three technical books and over 100 scientific papers in top-tier research journals and at international conferences. He holds 12 patents in wireless technologies. He is a Fellow of IET, Adjunct Professor at Macquarie University, Australia, and Guest Professor at the Chinese Academy of Science (CAS).

Preface

With the advent of the global positioning system (GPS) in the 1990s, civilian applications for location-based services have been increasing steadily. As the cost of GPS chipsets has decreased, integration into mobile/cellular phones has become feasible, further driving the rapid development in the consumer market. The GPS was originally developed for military applications with global reach, but with the increasing urbanization of the world population the main location-based services growth area is likely to be in cities. Further, as city dwellers spend most of their time indoors, where GPS is not effective, it appears that future developments of positioning systems are likely to require good performance within buildings. Thus, the focus of this book is on short- to medium-range positioning systems, with a particular emphasis on indoor performance with a projected accuracy of 1–2 m. Such systems, particularly those using cheap hardware, are not currently available, but developments in the large-scale integration of both analog radios and digital signal processing chips suggests that the projected performance requirements will soon be met. The applications of such systems include people tracking within buildings, inventory management, security applications and positioning in wireless sensor networks.

The nature of indoor positioning systems is somewhat different from outdoor systems such as GPS. In the literature on GPS, much of the emphasis is on the performance of the receiver, as the satellite infrastructure is the responsibility of the US Department of Defense. In contrast, an indoor positioning system must involve all aspects, including timing and frequency synchronization, the effects of internal delays in the hardware and communications between nodes of the network. All these aspects are particularly challenging, as to achieve the 1–2 m accuracy typically requires very precise timing measurements to the order of a nanosecond. Because of the nature of indoor applications, another difference in the systems' architectures is the probable adoption of ad hoc networking, rather than using the traditional "fixed" base stations and mobile nodes typical of long-range positioning systems. While position determination based on traditional hyperbolic navigation principles is likely to remain important, we may see a widespread employment of other position determination techniques, including signal-strength methods, use of angle-of-arrival data, hybrid radio–ultrasonics techniques and even position determination simply based on the detection of neighboring nodes in a network. Because of the diversity of location-based services and applications, this book covers a wide range of positioning techniques and processing algorithms currently dispersed in the technical literature, rather than focusing on particular positioning systems. The systematic treatment of these techniques and algorithms in the book should prove beneficial to system designers, researchers and graduate students.

This book is broadly organized into two parts. The first part is on the background aspects affecting positioning performance, such as indoor radio propagation and signal processing. The second part is on the details of various position-determination algorithms and their associated positional accuracy performance.

Chapter 1 provides an introduction to indoor positioning and provides a summary of various possible technologies and the desirable characteristics of future systems. Chapters 2–5 provide an introduction to aspects which have an important influence on the performance of indoor positioning systems. Chapter 2 provides an overview of the characteristics of indoor radio propagation, particularly related to the important topics of excess propagation losses and delays. Using this material, designers can estimate the likely range and accuracy of a system. Chapter 3 is an overview of signal processing, in particular as it relates to spread-spectrum signals. The concentration on spread-spectrum modulation is attributed to the desire for accurate position determination, which in turn necessitates wide spectral bandwidths. The largest available bandwidths with reasonably high transmitter power necessary for adequate indoor range are limited to the industrial, scientific and medical bands, which restrict transmissions to spread-spectrum signal. Chapter 4 provides analysis of time-of-arrival detection accuracy in the presence of Gaussian noise and multipath signals. Chapter 5 provides an overview of indoor tracking systems and introduces important topics described in more detail in subsequent chapters.

The second part of the book, Chapters 6 to 14, focuses on the details of position-determination algorithms. Chapters 6 and 7 provide details of position determination based on the traditional base station architecture and the measurement of range or time-of-arrival data. Chapters 8 and 9 provide statistical analysis of the accuracy of position determination for the methods described in Chapters 6 and 7. Chapter 10 describes multipath mitigation techniques, which are particularly important in the indoor environment. Chapters 11–13 focus on methods applicable to ad hoc networks and low-accuracy position determination, as in wireless sensor networks. Finally, Chapter 14 provides a wide range of techniques for determining whether the received signal is line-of-sight or non-line-of-sight path; the determination of the path type can be useful in improving position accuracy.

Acknowledgments

We would like to acknowledge the contributions of our colleagues at the Wireless Technologies Laboratory in CSIRO ICT Centre (previously CSIRO Radiophysics Laboratory) over the many years of work in the area of radiolocation. From 1995 to the present (2009), various radio-positioning systems and radio-propagation measurement equipment was developed that has formed the basis for much of the material in this book. We are particularly in debt to Alija Kajan for computer programming, propagation measurements and systems' testing in Australia and the USA, to Alex Grancea, John Olip and Robert Shaw for the design and building of the radio equipment, and to Jayasri Joseph and Joseph Pathikugara for the design and building of digital hardware. We are also grateful to Dr Mark Hedley, Dr David Humphrey and Dr Phil Ho for many stimulating discussions we enjoyed.

1

Introduction

It is now over 100 years since Marconi first demonstrated a practical radio system at transcontinental distances, and in the intervening years radio technology has become increasingly important to modern life. This book is focused on an emerging application of radio technology, namely short to medium-range positioning, particularly for indoor applications. The history of the development of radio technology shows that the initial applications tend to be for business and military use, but the most rapid development occurs when the technology becomes a consumer product.

As the short to medium-range indoor positioning technology is in its infancy, it is interesting to review the adoption of radio technology in the twentieth century as a rough indication of its likely development trajectory. Initial applications of radio technology were concentrated in the area of simple text communications, but analog modulation soon allowed the transmission of voice and later vision. In the general consumer area, 'wireless' initially came to mean analog AM radio, with the 'wireless' term being used to distinguish it from wired voice communications using the telephone. Over time, the term 'wireless' in this context became out of fashion, being replaced by the more generic term 'radio', only to become a popular term again for the description of short-range radio systems. Developments during World War II expanded radio technology into new areas, such as radar and radiolocation, but with no extension into consumer products. The development of mobile phone technology commenced as early as the late 1940s, but the uptake of the technology [1] was much slower than other consumer products, such as the telephone, automobiles, radio and television. The development of cellular telephone technology initially had little impact on this situation. Indeed, as late as 1987 the growth of customers in the USA appeared to have stalled [1], but with the advent of digital cellular telephone technology the growth exploded. However, it took some 50 years for the penetration of the mobile/cellular phone to reach saturation, far longer than other consumer electronics.

The development of radiolocation technology can be traced back to the late 1930s with the invention of radar, and with the development of navigation aids for aircraft during World War II. Early systems included the instrument landing system (ILS) and wide-area navigation with LORAN [2]. Such systems were restricted to aircraft and to a lesser extent shipping. While the theory of hyperbolic radiolocation was developed in the 1940s [2], little further occurred

Ground-Based Wireless Positioning   Kegen Yu, Ian Sharp and Jay Guo
© 2009 John Wiley & Sons, Ltd

until the advent of the global positioning system (GPS) by the US Department of Defense in the late 1980s. Although GPS is based on satellites rather than terrestrial stations, the general principles are the same as those used in LORAN, and indeed much of what is described in this book. While initially GPS was mainly intended for military use, the utility of global position determination was soon realized, so that civilian applications from surveying to in-vehicle navigation were developed in the 1990s. However, explosive growth in GPS is now occurring with the marriage of the cellular/mobile phone technology with GPS, which was made feasible by the increasing availability of low-cost GPS chipsets. The integration of location databases, the Internet, digital radio communications and radio positioning into mobile devices is poised to expand into the future. Perhaps the status of radio positioning technology today could be summed up by the English mathematical physicist Sir Oliver Heaviside, who in 1891 said[1] 'Three years ago, electromagnetic waves were nowhere. Shortly afterward, they were everywhere.'

GPS and other similar technologies such as the European Galileo system are mainly intended for navigation outside of buildings, whereas many of the potential applications are indoors, including security applications, health care, emergency services and inventory management [3]. These indoor applications require a tracking system rather than a navigation one, whereby knowledge of the location of an object is required remotely. Further, the indoor radio propagation environment is much more complex than the outdoor counterpart. Consequently, accurate but economical position location systems over large indoor areas is current not commercially feasible. Nevertheless, with the availability of ever more powerful radio and digital signal-processing chips, and more sophisticated position determination algorithms, such short to medium-range positioning systems are on the threshold of commercial viability.

The key difficulty in implementing short to medium-range radiolocation is the complexity of radio propagation, particularly in an indoor environment. It is interesting from a historical perspective that the very first experimentation [4,5] with radio waves in the late 1880s by Heinrich Hertz was indoors at frequencies close to those of modern indoor radiolocation technology. In the time before electronics, Hertz performed a surprisingly comprehensive investigation into radio propagation, including the speed of propagation, polarization, reflections, diffraction and mutual interference between paths. From the details of the equipment (a tuned dipole of length 26 cm, with a parabolic reflector for extra gain), the inferred frequency used in the experiments was about 500 MHz.[2] Using spark excitation and the natural resonance characteristics of the antenna, Hertz obtained line-of-sight (LOS) ranges up to 20 m within his laboratory by using a similar receiving antenna and observing small sparks. Hertz's original observations remain valid today. In particular, Hertz observed that radio waves propagate at the same speed as light (30 cm per nanosecond) and are reflected by surfaces (particularly metallic), but will pass through other materials such as those in walls. Thus, the indoor environment is dominated by multipath propagation, and the path length is likely to be longer than the straight-line path assumed in position determination. A more thorough understanding of these propagation effects is of key importance in the design of positioning systems. This book provides an introduction to radio propagation and signal processing, with particular emphasis on theoretical and algorithmic aspects of position determination, and design of

---

[1] Heinrich Rudolf Hertz website: chem.ch.huji.ac.il/history/hertz.htm.
[2] It need hardly be pointed out that hertz (abbreviated Hz) has now replaced 'cycles per second' as the standard term for frequency.

ground-based (non-GPS) radio positioning and tracking systems. In particular, it covers some key topics in the field which have not been discussed in other books. These include the time-of-arrival measurements in multipath environments, positioning performance analysis and accuracy limitations, positioning techniques for wireless sensor networks, identification of non-LOS (NLOS) radio propagation, NLOS mitigation techniques and practical issues in designing wireless positioning systems.

1.1 Introduction to Radio Positioning

Radio positioning can be broadly defined as a method of determining the geographic position of a radio device using the properties of radio waves. Various methods have been developed over the years, including the measurements of time-of-flight (TOF), signal phase, signal strength and angle of arrival. Traditionally, the architecture of positioning systems was based on the concept of 'fixed' nodes and 'mobile' nodes whose position is required. Even GPS with satellites can be considered in this category, as a satellite's position at any time, even though moving relative to the Earth, is accurately known. For short to medium-range applications, a logical development would be to enhance the existing wireless local-area network (WLAN) technology to incorporate positioning capability. Many other technological solutions are also possible, but it appears that the general thrust will be towards position determination incorporated into small, cheap nodes which are organized as an ad hoc network. Such networks are not centrally organized, but using radio transmissions to neighboring nodes allows both data communications and position determination.

The main difficulty with developing short to medium-range positioning systems is related to the characteristics of radio propagation indoors and the requirement for accurate position determination. Because of the smaller scale of the indoor environment, most applications require an accuracy of 1–2 m or better; this is in contrast, for example, to the typical GPS accuracy of around 10 m. Indoor radio propagation is usually NLOS with multiple scattering of the signal from the transmitter to the receiver. This multipath propagation causes both additional loss of signal strength and additional propagation time above the straight-line path assumed for radio position determination. While in theory the solution could be a high-powered and wide-bandwidth system, limitations by regulatory authorities on the availability of suitable spectrum and transmitter power mean that other more complex solutions must be sorted.

Radio positioning can be based on many properties of radio waves, but most accurate systems have been based on determining the TOF, or the closely similar concept of the time-of-arrival (TOA). In principle, if the time of the start of a transmission and the time of reception are both known, then the TOF can be used to determine the range from the transmitting node to the receiving node. If two or more of these measurements are made using three or more nodes, then a two- (2D) or three-dimensional (3D) position can be calculated from basic geometry, assuming straight-line propagation. As radio waves propagate at about 0.3 m per nanosecond, and given the above-defined accuracy of an indoor positioning system of (say) 1–2 m, timing measurements must be made to the order of 1 ns. This precision of time measurement is very challenging, particularly when using small, cheap radios.

Although position determination based on TOF is straightforward in principle, there are implementation issues that make it impractical in most situations. First, the TOF method requires the clocks in all the nodes to be synchronized; this is particularly difficult in mobile nodes and requires special design features in fixed nodes. Second, the elapsed time from the transmitter to the receiver includes delays in the radio equipment. Again, for fixed nodes, calibration procedures can be devised to determine these delays, but this is difficult in mobile nodes. Thus, many positioning systems, including GPS, use TOA data for position determination and do not require time synchronization in mobile nodes. In this case, the receiver effectively measures a pseudo-range, which is the true range (estimate) plus an unknown range offset. This principle also applies to other positioning methods. For example, if the round-trip time (RTT) delay is measured between two nodes, one fixed and one mobile, then the time delay in the mobile node radio (transmitter and receiver) should be considered as an unknown parameter. While these delays are approximately known, the requirement for sub-nanosecond accuracy in specifying these delays means that even very small variations in the equipment delays cannot be tolerated. In fact, equipment delays in short-range systems represent the vast majority of the measured round-trip delay, so variations as small as a few parts per thousand, which could be due to effects such as temperature fluctuations, cannot be tolerated. If received signal levels are used for range estimation, then uncertainties in the effective transmitted power again result in the measurement of pseudo-ranges. In this case, the uncertainty is due mainly to antenna effects when the transmitter is close to an object, such as the ground or a body.

Because of the common occurrence in positioning systems of measuring pseudo-ranges rather than ranges, techniques for determining positions based on pseudo-range measurements at a number of fixed nodes is of prime interest. In the 2D case, the position determination has a simple geometric interpretation. If two pseudo-range measurements at two fixed nodes are subtracted, the common offset is eliminated, so that the locus of points with the differential TOA (DTOA) constant is a hyperbola. Similarly, with another pair of fixed nodes the DTOA defines another hyperbola, and the position of the mobile node is at the intersection of these hyperbolas. For this reason, positioning systems based on pseudo-range measurements are sometimes called 'hyperbolic navigation' systems. The analysis of position determination based on the intersection of hyperbolas is given in Appendix A, together with a summary of important characteristics of hyperbolas. However, this method is mainly of historic interest only, as more sophisticated methods are used in actual systems, as described in later chapters in this book.

While classical positioning systems are based on fixed and mobile nodes, many of the developments in short to medium-range systems are associated with ad hoc networks where all the nodes are essentially the same. In such systems, nodes exchange radio messages with data and timing information. If a node can communicate with sufficient neighboring nodes, thus sharing information, the relative positions of all the nodes can be determined. Further, if a few of these nodes have absolute positional data by some independent surveying process, then the positions of all the nodes can be determined in absolute coordinates. In some simple networks, such as wireless sensor networks (WSNs), only rather crude positional information may be required. In such cases, simple positioning strategies can be adopted. For example, the simplest methods are based on estimating position using knowledge of the surrounding neighbors' positions, and (say) defining the position at the centroid of the neighboring nodes; the known positions of neighbors may be from 'anchor' nodes with accurate positional data, or from other nodes which have previously determined their positions. More accurate positions can be obtained if the signal strength data, available in even the simplest of radio receivers, are used to estimate range. For more sophisticated ad hoc networks, TOA data can be used in a manner akin to the hyperbolic method with fixed receiving nodes, thus allowing more accurate position fixes.

The above brief introduction to radiolocation can only touch on the various aspects and methods used in positioning systems. Because there are a wide variety of applications and requirements, no one method is applicable; thus, this book attempts to cover a wide range of the major techniques. These are outlined in Section 1.3 as a guide to using this book. However, before summarizing the chapters in this book, it is useful to review the types of short to medium-range radiolocation technologies that have been used in the past and the possible future developments.

1.2 Short and Medium-range Radiolocation Technologies

While much of the theory and analysis in this book can be applied generally to all radio positioning systems, the main focus is on ground-based short (say less than 100 m) to medium-range (say up to 2000 m) systems, and more particularly indoor positioning systems. Within this definition there are many existing radio systems which currently are mainly used for data transmission, but these systems could also be used for position determination. When considering short to medium-range positioning, there is no one technology which suits all applications and situations, so it is useful to review the weaknesses and strengths of various technologies. From a technical performance point of view there are two main characteristics of practical interest, namely the positioning accuracy and the range of the radio link. However, for practical implementation, other factors are also of major importance, including the cost and size of the mobile units, the associated costs of the fixed components, such as base stations, and infrastructure costs for cabling and installation. As one might expect, there is no one technology that has all the desirable characteristics; a long-range, accurate positioning system is likely to require sophisticated and expensive hardware. Nevertheless, the processing power of the integrated circuits for both the analog radio and the digital signal-processing parts are increasing over time, while their cost is decreasing simultaneously. Thus, while an ideal indoor system may not be available at the time of writing (2009), one can define the characteristics of such a system and provide some performance estimates. However, developments in the technology are constrained by fundamental physics and other constraints, such as the availability of radio spectrum, so the performance of a positioning system is bounded, and these bounds will not change over time. Thus, while this book is not aimed at any one particular technology and implementation, the overall thrust is aimed at next-generation short to medium-range positioning systems which are comparatively cheap and which have good positioning accuracy and range.

The starting point in defining potential future systems is current technology. The following list is not comprehensive, but represents the range of possible positioning technologies.

1. *WLANs* WLAN technology currently is based almost exclusively on the IEEE 802.11 specification,[3] which defines a number of different physical layers for data transmission operating in the 2.4 GHz industrial, scientific and medical (ISM)[4] band and at 5.2 GHz.[5]

---

[3] http://standards.ieee.org/getieee802/802.11.html

[4] The ISM bands are unlicensed, but all users must share the band. The potential of mutual interference limits the radio transmissions to spread-spectrum types (frequency hopping or direct sequence) which have an ability to resist the effects of interference.

[5] The 802.11a WLAN operates in a specially allocated band at 5.2 GHz, but is restricted to indoor data transmission applications. The 5.8 GHz ISM band is not currently used for 802.11 WLAN systems, so this band would be an ideal candidate for a positioning system.

Position determination is not part of the standard hardware; but, as the transmissions are based on spread-spectrum modulation with bandwidths of 1–20 MHz, depending on the specific version of 802.11, positioning capability could be added. The simplest implementation currently is based on signal strength measurements, but techniques using TOA could also be implemented. The range of current WLANs is typically limited to about 50 m indoors, and a similar range could be expected for a positioning application. While the cost of the hardware, particularly the mobile unit, is modest, the system is based on fixed base stations installed in a building, with typically connections to a wired local-area network (LAN). For a positioning system, many more base stations would be required, thus leading to higher infrastructure cost.

2. *Ultra-wideband (UWB)*   UWB technology is a response to the limitation in available spectrum, which in turn limits the data transmission rates and the accuracy of positioning systems. UWB uses very wide bandwidths (at least 550 MHz), but is restricted to a very low transmitter power density (watts per hertz) to minimize the interference to other existing radio systems which use part of the same frequency band (3–10.7 GHz) [6]. UWB technology is mainly aimed at short-range, high data-rates links. However, the large bandwidths are ideal for indoor positioning systems, as the large bandwidths mitigate the effects of multipath propagation by allowing very fine (sub-nanosecond) time resolution. The indoor positioning accuracy for UWB is of the order of 20 cm, but typically the range is limited to about 10 m. As a consequence, the number of base stations required to cover an area will be large, so the wired infrastructure costs would be high, at least for the first-generation systems. However, the hardware costs in future-generation UWB systems will be lower, but the short range and high base station spatial density means that other, cheaper technologies are likely to be used in preference unless the superior positioning accuracy is required.

3. *Field strength systems*   The simplest method of estimation radio propagation range is by signal strength measurements. Such methods have been implements using WLAN hardware [7] and more recently in WSNs and ad hoc networks. While signal strength measurements are available in even the simplest of single-chip radios, the accuracy of such measurements is limited, partly due to the simple implementation of the radio, but more importantly due to effects such as interaction with the human body when worn by a person and indoor propagation effects. For indoor applications the signal strength varies in a complex fashion, so there is no simple function between loss and range. One method to improve the accuracy is by generating a radio-loss map of the building, which allows a database-matching algorithm to improve the accuracy of the position fix. However, even with these techniques, and particularly when using cheap body-worn mobile units, the positioning accuracy is likely to be of the order of 5 m. Thus, signal strength methods are likely to be limited to applications where only crude positional accuracy (such as in WSN applications) is sufficient.

4. *Radio frequency identification (RFID)*   RFID is a simple radio technology mainly aimed at providing identification of objects based on a unique multidigit identifier, usually associated with inventory management. RFID tags are small and very cheap, but their range is limited to about 1 m. Normally, a special tag reader, often hand held, is used to identify the tag. Such technology is not particularly aimed at position determination, but a simple adaptation would allow its use for position determination to a limited extent. For example, tag readers at key locations, such as doorways, can be used to identify people moving through a building.

While the tags are cheap, the infrastructure costs would be high if a large number of readers are employed to provide a good positioning coverage. For example, in an office building, one reader located at the entrance of each office may be necessary. In such a case the positioning accuracy would be typically a few meters (the size of a room), but the infrastructure costs to interconnect the readers would be high.

5. *Next-generation indoor positioning system*   To provide some idea of the next generation of indoor positioning systems, a general specification can be devised, in part based on the strengths and weaknesses of the existing technology solutions summarized above. Thus, the system radio should have a typical indoor link ranges of at least 30 m, and a ranging accuracy of 1–2 m. Ideally, the positioning system should be 'piggybacked' onto an existing data system and the analog radio should be based on a single chip. To minimize infrastructure costs, the system architecture should be based on the ad hoc concept, whereby nodes communicate with their neighbors for position determination, and there is no central organization of the network. In such a system there is no distinction between base stations and mobile units. To minimize infrastructure costs, wired communications should be limited to a few interconnections to a wired LAN. Ideally, the hardware is battery powered with a long battery life.

To further investigate the relationships between these various radio technologies, Figures 1.1–1.3 show the 'operational region' of each technology plotted in three different domains. Note that these diagrams represent the various technologies when employed for radio positioning, which typically is not their main use currently. Note also that these diagrams are indicative of the technologies rather than being an accurate representation.

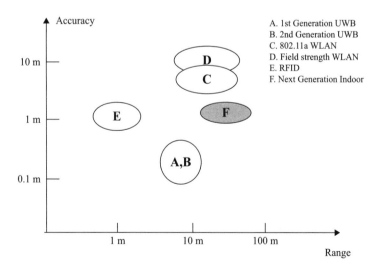

Figure 1.1   Representation of the range and positioning accuracy of various indoor radiolocation technologies. The ideal future technology (F) will provide a good compromise between range and accuracy.

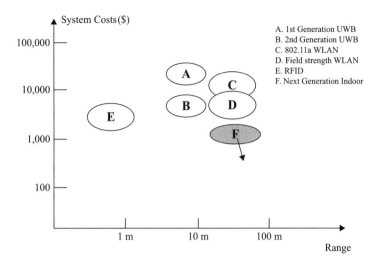

Figure 1.2 Representation of the range and system costs of various indoor radiolocation technologies. The ideal future technology (F) promises to provide the lowest system costs, with significant improvements over time.

Figure 1.1 is a plot of the positioning accuracy versus the link range. To some extent, existing technologies either have good positioning accuracy or long link ranges, but not both. Thus, these technologies would tend to be limited either to applications where good positioning accuracy is required over a relatively small area, or alternatively a large coverage area is required but with lower positioning accuracy. The suggested area of operation of the

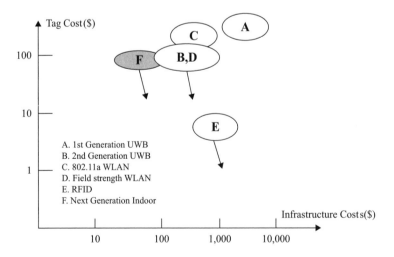

Figure 1.3 Representation of the cost of a tag and infrastructure costs of various indoor radiolocation technologies. The arrows show the expected trend over time.

next-generation location systems has similar or better range than current WLAN systems and an accuracy intermediate between those of UWB and WLAN systems. Thus, from an operational point of view, the next-generation positioning system should cover a wider range of applications.

Figure 1.2 shows a plot of system cost versus the positioning accuracy. The system cost is defined as the cost of the 'fixed' components (base stations) and any central network controller, but not the cost associated with the deployment. The system cost is based on installing nodes in a building to cover an area of $2000\,\mathrm{m}^2$ to an accuracy broadly defined in Figure 1.1. Current system costs tend to be too high to provide economical coverage in large buildings. Unlike coverage for data systems, for correct operation the mobile unit must communicate simultaneously with typically four or more base stations (or neighboring nodes in an ad hoc network), so the spatial density needs would be much greater, typically 3–10 times of the data system. As a consequence, the system costs are much higher for a location system than a comparable wireless data system covering the same area. The short range of UWB (particularly first-generation systems) makes it expensive to cover a large area. For WLAN-type systems, particularly those based on field strength measurements, the cost is significantly less than UWB systems, but systems costs would still be too high to cover large areas economically. RFID technology can be cheap, but the costs are critically dependent on the required density and cost of tag readers. The suggested next-generation system costs should be quite low due to the 'fixed' and mobile units being essentially the same, and costing no more than $100. The arrow in Figure 1.2 shows the suggested trend over time, with costs decreasing significantly, and improved hardware and signal processing increasing the operational range. Thus, next-generation systems have the potential of covering large areas economically, resulting from a combination of lower unit cost and comparatively long operational range. Note that as the number of units required to cover a given area is proportional to the square of the link range, so link range has a big impact on system costs.

Figure 1.3 shows another representation of the performance of the various systems, this time the tag costs versus the infrastructure costs. In this context the infrastructure costs are defined as the cost of installing the 'fixed' nodes (base stations). As infrastructure costs are largely associated with the cost of labor, which is expected to rise over time, there is no benefit in infrastructure costs from the reducing costs of hardware, as is appropriate with the tags, whose cost is expected to fall rapidly over time. The number of tags required for a particular application would vary widely, so it is not possible to provide estimates of the total costs (total tags plus system plus infrastructure). However, for large systems with a large number of tags, overall costs would probably fall over time.

When comparing the various technologies in Figure 1.3 it is clear that technologies based on cheap tags operating in an ad hoc network are by far the most economical, as the infrastructure costs are low or nonexistent as there are only a few fixed base stations. Indeed, an ad hoc positioning system is possible with only two fixed nodes, so that position determination is largely based on communications between mobile nodes. Such a design can cover large areas at minimal costs. On this basis, the suggested next-generation system is clearly superior to the other technologies.

Thus, in summary, from Figures 1.1–1.3 it can be observed that next-generation indoor positioning systems promise good accuracy (1–2 m) and the ability to cover large areas at low cost. Given these parameters, such systems should find many applications in the future.

1.3 Overview of the Book

The book covers a wide range of topics associated with position determination in short to medium-range systems, with particular emphasis on indoor positioning using networks of simple nodes. Because of the diversity of topics, it is not expected that the reader would progress serially through the book, but would access areas of particular interest. Each chapter is written such that it can be read in a standalone fashion, although the material in a general area of interest may involve more than one chapter. The book is broadly divided into two groups of chapters, with Chapters 2–5 providing an overview of radio propagation, signal processing (particularly relating to spread-spectrum signals) and systems' aspects of positioning systems. Chapters 6–13 are focused on methods of position determination, with mathematical development, details of specific algorithms and performance charts. These chapters are loosely grouped into sub-topics, according to positioning accuracy and type of positioning system architecture. Finally, Chapter 14 provides methods of identifying LOS and NLOS propagation, mainly based on statistical analysis of the received radio signal.

Thus, to provide some guidance to using the book, the following subsections provide a brief summary of the content of each chapter.

1.3.1 Radio Propagation (Chapter 2)

The performance of any radiolocation system is ultimately based on the propagation characteristics of radio waves. Radio propagation for outdoor large-scale systems is close to free space, so that the performance of such systems is relatively easy to determine. However, radio propagation associated with short-range systems in an urban environment, particularly indoors, is very complex and places challenges on the design of positioning systems. Chapter 2 provides an overview of indoor radio propagation, with particular emphasis on loss and delay excess relative to free-space propagation. The material is presented without deep mathematical analysis, but rather in the form of charts based on measured data. The information provided allows the radio propagation performance to be estimated for indoor situations, based on the properties of materials in the building and the architectural layout. In addition, Appendix B provides an overview of measurement techniques for determining the radio propagation characteristics in a multipath environment.

1.3.2 Signal Detection by Correlation (Chapter 3)

The most common method of radio positioning is based on the measurement of the TOA of the radio signals. However, because of the requirements for multiple access to the radio channel, radio pulse methods are restricted to very low powered UWB short-range systems, so that most common technology is based on spread-spectrum signal modulation. The key to this method is the pseudo-random modulating signal, which spreads the radio energy in a manner akin to the truly random noise in nature. The detection of such a signal requires the received signal to be correlated with a copy of the code used for modulating the transmitted signal. Thus, the properties of this correlation process are vitally important in determining the performance of spread-spectrum-based positioning systems. Chapter 3 analyses the relevant characteristics of the correlation process in relation to direct-sequence spread-spectrum signals, including the effective generation of a narrow time-domain pulse (signal despreading) used for TOA estimation, multiple access properties, resistance to interference and the multipath mitigation properties. The topic of spread-spectrum is described in a vast amount of literature [8,9], so the specific topics discussed in this book are limited to the important practical areas of correlators, accumulators, interference performance and the concept of process gain.

1.3.3 Bandlimited Time-of-Arrival Estimation (Chapter 4)

The key technique used in positioning systems is based on the estimation of the TOA of a radio signal at a receiver. Because of the spectral limitations by regulatory authorities on radio transmissions and the need for wide bandwidths for ranging accuracy in a multipath environment, the most accurate positioning systems are based on direct-sequence spread-spectrum signals. Chapter 4 provides algorithms for estimating the TOA from these bandlimited signals and estimates their performance in the presence of both Gaussian noise and multipath signals. Simple analytical formulae are developed to allow designers to estimate the accuracy of TOA measurements.

1.3.4 Fundamentals of Positioning Systems (Chapter 5)

Chapter 5 provides an overview of the most important aspects of radio positioning, particularly in an indoor environment. The chapter briefly describes the effects of radio propagation on the performance (propagation range and TOA measurement accuracy) of a radio positioning system. Two broad types of positioning system are described: navigation systems and, the focus of this book, tracking systems. An introduction to important design aspects is described, including such topics as time synchronization of nodes and how the internal delays in the radio equipment are measured or compensated for by the design of the system. The chapter also briefly reviews the various methods of position determination, including TOA methods, TOF methods, received signal strength (RSS) methods, and hybrid radio/ultrasonics systems.

1.3.5 Position Determination

Chapters 6 and 7 describe methods of position determination based on using TOA, angle-of-arrival (AOA) or signal strength measurements. However, the main emphasis is on the processing of TOA data, as these provide the most accurate position determination. As these measurements are the arrival time at the receiver rather than the TOF, the effective measurement data are pseudo-ranges (range plus an unknown offset), rather than ranges. The algorithms in these chapters are based on processing pseudo-range data in a classical system architecture with fixed base stations used to determine the location of a mobile node.

1.3.5.1 Noniterative Position Determination (Chapter 6)

Chapter 6 focuses on noniterative position determination methods, where the raw measured data are processed using analytical formulae to provide the position of a mobile node directly. The particular algorithms comprehensive analyzed are spherical interpolation (SI), quasi-least-squares (QLS), and linear correction least-squares (LC-LS). From simulations with various arrangements of base stations, and ranging errors based on models introduced in Chapters 2 and 5, it is shown that there is no one ideal algorithm. While these methods directly produce a position fix, the analytical equations can be computationally intensive and, thus, may not be suitable for simple network nodes with limited computational capability.

1.3.5.2 Iterative Position Determination (Chapter 7)

Chapter 7 provides numerical computational alternatives to the analytical methods of Chapter 6. In particular, iterative methods are used to solve the nonlinear equations which describe the position determination problem. The algorithms described include the Taylor series least-squares (TS-LS) method, the iterative optimization method and the maximum likelihood (ML) method. Iterative methods based on filtering the data (particularly the use of Kalman filters) are also discussed. These iterative methods require an initial 'guess' of the position, and in some circumstances, particularly when large measurement errors are present, the algorithm may not converge. However, in applications where continuous periodic updates are required, the previous position can be used as the initial position, which usually results in rapid convergence and position determination with less computation than the noniterative methods.

1.3.6 Positioning Accuracy

Perhaps the most important performance aspect of a positioning system is the accuracy of the position determination. The positioning accuracy has two broad aspects, namely geometric factors relating the position of the fixed and mobile nodes and the statistical performance due to random measurement errors. Chapters 8 and 9 describe various statistical metrics for determining the accuracy of a positioning system.

1.3.6.1 Positioning Accuracy Evaluation (Chapter 8)

Chapter 8 introduces various accuracy metrics, including Cramer–Rao lower bound (CRLB), geometric dilution of precision (GDOP), root-mean-squared error (RMSE) and cumulative distribution probability (CDP) of the location errors. The CRLB method is then used for the analysis of LOS and NLOS scenarios and the performance of the least-squares (LS) estimator; the iterative optimization-based location algorithm is evaluated against the CRLB. Also, the impact of the anchor location errors (see also Chapters 11 and 12) is investigated. The results of these methods are illustrated through simulations.

1.3.6.2 Geometric Dilution of Precision Analysis (Chapter 9)

One important aspect that affects positioning accuracy is the relative geometric relationship of the fixed nodes to the mobile node whose location is to be determined. The most common method of describing these geometric effects is GDOP, which is the ratio of the statistical position error to the standard deviation (STD) of the measurement error. Chapter 9 provides the statistical background to GDOP and provides relatively simple analytical formulae for its calculation for geometries typical of short-/medium-range positioning systems. The chapter also describes the limitations in the use of GDOP in short-range systems, where the measurement errors are a substantial fraction of the propagation range.

1.3.7 Multipath Mitigation (Chapter 10)

For radio positioning systems, particularly those operating indoors, the main degradation in performance is due to multipath propagation of the radio waves. Because of the scattering of the signal, the TOA and other measurements are corrupted, resulting in positioning errors. Chapter 10 considers various methods of mitigating these effects, including the residual weighting of the data, constrained optimization, Kalman filtering, smoothing data to exclude 'bad' measurements, error statistics to perform ML estimation under a number of scenarios and database-based pattern-matching position determination methods. The various methods are demonstrated through simulations, with graphical representation allowing comparison between the techniques.

1.3.8 Anchor-based Localization

Unlike classical positioning systems, which have 'fixed' (base stations) and 'mobile' nodes, ad hoc and wireless sensor networks have nodes with similar characteristics. By transmitting and then receiving messages from its immediate neighbors, each node can determine its relative position within the network. However, if at least two of the nodes have known absolute positions (for example, from an independent survey), then all the nodes with relative positions can also determine their absolute positions. The nodes whose absolute positions are known thus play an important part in any positioning scheme in ad hoc networks; such nodes are referred to as 'anchor' nodes. Chapters 11 and 12 investigate the characteristics of such ad hoc networks which include a few anchor nodes.

1.3.8.1 Anchor-based Localization for Wireless Sensor Networks (Chapter 11)

Chapter 11 describes position determination techniques in ad hoc networks with simple nodes, such as WSNs. The general characteristics of WSNs are first described and then some basic position determination methods are discussed, which at the simplest only require the identity of nearby anchor nodes. Other techniques are also described, including multihop localization algorithms and TOA-based localization. A number of practical parameters are also briefly considered, including the clock frequency offsets, internal delays in the equipment and clock time offsets.

1.3.8.2 Anchor Position Accuracy Enhancement (Chapter 12)

The localization accuracy in a WSN depends not only on the accuracy of ranging measurements, but also on the positioning accuracy of the anchors. While independent surveying of anchor nodes is possible, ideally the WSN infrastructure itself should perform the surveying of the anchor points. Chapter 12 considers various possible methods of determining anchor node absolute positions, including low-cost GPS receivers and wideband and UWB ranging. However, the main focus of this chapter is on more accurate anchor-to-anchor parameter estimates to enhance the anchor location accuracy. This accurate anchor location information is also helpful in localizing ordinary sensor nodes. The particular topics discussed include modeling LOS and NLOS propagation, accuracy improvement algorithms based on both distance and AOA measurements and CRLB as a performance benchmark. The techniques are illustrated through simulations.

1.3.9 Anchor-Free Localization (Chapter 13)

In Chapter 11, node localization was studied for scenarios where some nodes are anchors whose absolute positions are known *a priori*. In some circumstances there are no anchor nodes in the network, so that no absolute location information is available at any node. For such anchor-free networks the key issue of localization is how to determine the relative positions of the nodes accurately and to determine the graph of the node configuration accurately. Chapter 13 considers techniques for determining relative positions based on exchange of radio transmissions with neighboring nodes. Both single-hop and multihop techniques are considered, as well as localization accuracy measures, including the CRLB and the approximate distance error lower bound.

1.3.10 Non-Line-of-Sight Identification (Chapter 14)

In radio positioning, one of the dominant factors that affects the positioning accuracy is the NLOS radio propagation which happens when the direct, straight radio path between the transmitter and receiver is blocked. Compared with the LOS condition, the signal travels an extra distance and time under the NLOS condition. The NLOS propagation also results in an extra power loss and an AOA bias. To achieve improved location estimation accuracy, it is desirable to determine whether the measurements come from LOS propagation or NLOS propagation. The methods investigated in Chapter 14 include identification based on calculating the error variance through data smoothing, a number of well-known statistical distribution tests, level crossing rate and fade duration of the received signal envelope, nonparametric methods, a joint TOA and RSS-based method and AOA-based methods.

References

[1] G. Calhoun, *Digital Cellular Radio*, Artech House, 1988.
[2] J.A. Pierce, A.A. McKenzie and R.H. Woodward, *Loran*, MIT Radiation Laboratory Series, volume 4, McGraw-Hill, New York, 1948.
[3] K. Pahlavan, X. Li and J.-P. Makela, 'Indoor geolocation science and technology', *IEEE Communications Magazine*, **40**, 2002, 112–118.
[4] M.H. Shamos, *Great Experiments in Physics*, Courier Dover Publications, 1987.
[5] R.W. Burns, *Communications: An International History of the Formative Years*, IET, 2004.
[6] I. Oppermann, M. Hamalainen and J. Iinatti, *UWB Theory and Applications*. John Wiley & Sons, Ltd, Chichester, 2004.
[7] M.A. Youssef, A. Agrawala and A.U. Shankar, 'WLAN location determination via clustering and probability distributions', in *Proceedings of IEEE International Conference on Pervasive and Communications*, pp. 143–150, March 2003.
[8] A.J. Viterbi, *CDMA Principles of Spread Spectrum Communication*, Addison-Wesley, 1995.
[9] R.E. Ziemer, R.L. Peterson and D.E. Borth, *Introduction to Spread Spectrum Communications*, Prentice Hall, 1995.

2

Radio Propagation

A radio-based positioning system exploits the propagation of radio signals from transmitters to receivers. Many properties of radio propagation can be used for positioning, including the propagation delay (the most commonly used property), the signal strength, the signal phase and the AOA. While radio propagation in free space is well defined and allows these propagation characteristics to be accurately defined mathematically using Maxwell's equations, the details of propagation in other environments, particularly indoors, is too complex for such an approach. Nevertheless, the general characteristics of radio propagation are important for designing all types of radio positioning system.

For the design of radio positioning systems, two characteristics of radio propagation are particularly important. This first characteristic is the propagation loss, or the loss excess above the free-space loss in a multipath-dominated environment. For systems based on the signal strength this characteristic is clearly of prime importance, as the loss directly determines the range. More typically for systems based on the measurement of the TOA, the loss characteristics determine the range of propagation and, hence, the coverage area per node. As at least three and preferably five or more receiving nodes are required for an accurate position fix, the propagation loss characteristics determine the density of the deployed nodes. If the propagation loss is too severe, then the deployment of a positioning system over a large area may become infeasible.

The second characteristic of radio propagation which is particularly important for TOA-based systems is the delay excess. As the position determination assumes straight-line propagation, any delay excess due to scattering from obstacles along the path translates directly into errors in the position estimates. While direct-sequence spread-spectrum signals have multipath mitigation characteristics, bandwidth restrictions will almost always result in errors in the TOA measurement.

This chapter provides an overview of radio propagation in the context of radio-based positioning, particularly at microwave frequencies used in ad hoc wireless LANs, and indoor positioning systems. Section 2.1 presents the theory of radio propagation with particular reference to modeling multipath propagation, the associate signal spectrum and the statistics of the RSS. Section 2.2 discusses radio propagation at different scales, including the large-scale variation with distance greater than 100 wavelengths, the medium-scale with distances of the order of 10–100 wavelengths and the small-scale with distance less than 10 wavelengths. Section 2.3 describes the measurement of radio propagation characteristics. Section 2.4 is

Ground-Based Wireless Positioning   Kegen Yu, Ian Sharp and Jay Guo
© 2009 John Wiley & Sons, Ltd

focused on the characteristics of large-scale propagation. Section 2.5 studies the characteristics of received signals, particularly in an indoor environment. Specific issues discussed include the spectrum, amplitude statistics, environmental impulse response, propagation through walls and the effect of bandwidth. Finally, the effects of the interaction between antennas and the local environments are presented in Section 2.6, which include the scattering of the human body for body-worn antennas and the reflections of the ground.

2.1 Statistical Multipath Theory

For outdoor systems it is usually safe to assume that the direct path is present and dominates over other scattered signals. As a consequence, determining the TOF from the transmitter to the receiver is largely based on the straight-line path of the dominant signal. The situation is similar for indoor systems located in large areas such as sporting stadiums, except that the multipath signals tend to be larger than the outdoor counterparts. However, most indoor environments will have no clear LOS; thus, the received signal is dominated by multiple scattered signals. There are two main consequences of this NLOS propagation. First, the signal scattering and absorption results in added propagation losses, which in turn limits the operational range, particularly for low-powered devices typical of WSNs and ad hoc networks. Second, and more importantly, the strongest signal at the receiver usually propagates in a non-straight line, so that measurement of the TOF will incur a delay excess which affects the positional accuracy. Further, even when the straight-line path (or close to it) is present, the time difference between the direct and scattered paths is rather small in time (typically only a few nanoseconds), so that very wide bandwidths are required to provide accurate position determination.

Thus, to determine the performance of a positioning system, particularly indoors, an analysis of multipath propagation is essential. The analysis should include the amplitude characteristics and the spectral characteristics; the latter are particularly important, as most systems are either very wideband or are based on spread-spectrum technology. The topic of radio propagation in a scattering environment is too vast to be covered in detail, so only an overview of the main topics is covered here. For more details the reader should refer to specialist books [1–3].

2.1.1 Multipath Model

We start with a simplified mathematical model of indoor multipath propagation. The model is based on multiple scattering (reflections, diffractions, absorptions) of the radio-frequency (RF) signal by walls/floors/ceilings and other smaller objects in a building. Further, the model is based on the scattering from a finite number of objects whose sizes are typically greater than one wavelength; for systems operating in the range of 2.5–6 GHz this means that the objects would be greater than about 10 cm in size, but the major scattering objects would typically be much larger in size. With these restrictions, the geometric theory of diffraction (GTD) [4] can be applied to simplify the calculations of the scattered signals. GTD approximates the scattered signal by straight-line propagation paths similar to geometric optics, with the source amplitude (and phase) related to the geometry of the incident ray and the geometry of the scatterer. This scattering environment is assumed in the analysis illustrated in Figure 2.1. The transmission of the signal from the transmitter to the receiver occurs via scattering from objects in the transmission path. Typically, multiple scattering will occur from multiple objects.

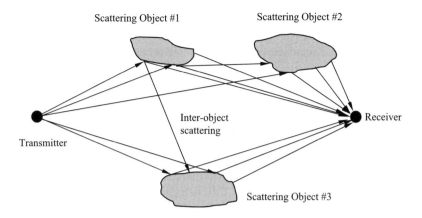

Figure 2.1   Geometry of the scattering environment.

For the multipath model the exact nature of the scattering along the propagation path is not required, as only the signal incident onto the receiver is of interest. Thus, the received signal can be considered as comprising a moderately large number of components from sources located in the vicinity of the receiver. Applying this approximation, the received signal can be expressed by

$$S = \sum_{i=1}^{N} \left[a_i(f) e^{j\phi_i(f)} \right] \frac{e^{-j2\pi R_i/\lambda}}{R_i/\lambda} \tag{2.1}$$

where $a_i e^{j\phi_i}$ is the complex amplitude of the scattering from point i before the signal is incident onto the receiving antenna, R_i is the propagation distance and λ is the wavelength.

2.1.2 Spectrum of Received Signal

Because of the requirement for high time resolution to achieve good positioning accuracy, the transmitted signal is assumed to be wideband, so the spectrum of the signal is of interest. Although the signals used in positioning systems tend to have wider bandwidths than RF communications systems, the signal bandwidth is normally still a small fraction of the carrier frequency,[1] so it is appropriate to express the RSS as a function of the frequency relative to the centre frequency of the band. Thus, with the signal frequency measured relative to the carrier frequency, the received signal from (2.1) can be expressed in the form

$$S(\Delta f) = \sum_{i=1}^{N} A_i(\Delta f) e^{j[2\pi(\Delta f/c)R_i + \phi_i^*(\Delta f)]} \tag{2.2}$$

where Δf is the frequency offset from a nominal center frequency and A_i is the amplitude obtained by combining all the constant terms in Equation (2.1), and can be considered to be a

---

[1] The exception to this may be UWB systems, which by definition must have a bandwidths of at least 550 MHz. Current Federal Communications Commission (FCC) rules for the UWB systems restrict transmissions to 3.1–10.6 GHz; so, apart from UWB pulse systems, the bandwidth to carrier frequency ratio tends to be much less than unity.

slowly varying (with frequency) random variable. Similarly, ϕ_i^* is a random phase variable which varies slowly with frequency.

Consider now the real component of the received signal as a function of frequency across the band:

$$
\begin{aligned}
\mathrm{Re}[S(\Delta f)] &= \sum_{i=1}^{N} A_i(\Delta f)\cos\left[2\pi\frac{\Delta f}{c}R_i + \phi_i^*(\Delta f)\right] \\
&\approx \sum_{i=1}^{N} A_i^0\cos\left(2\pi\frac{\Delta f}{c}R_i + \phi_i^0\right) \\
&= \sum_{i=1}^{N}[A_i^0\cos(\phi^0)]\cos\left(2\pi\frac{\Delta f}{c}R_i\right) - \sum_{i=1}^{N}[A_i^0\sin(\phi^0)]\sin\left(2\pi\frac{\Delta f}{c}R_i\right)
\end{aligned}
\tag{2.3}
$$

where the assumption has been made that the amplitude and phase of each scattered signal component are approximately constant across the frequency band. Thus, the real part of the signal can be expressed as two components, each of which is a summation of the scattering sources; each component has two independent random variables, one related to the amplitude/phase of the source and the other related to the phase associated with the propagation from the source to the receiver. Although the statistical properties of the various variables are not known, the central limit theorem (CLT) can be invoked to obtain the statistical properties of the summations. Thus, the real part of the received signal can be expressed as the sum of two Gaussian functions. Further, as the phase term is expected to have a uniform distribution in the range $[0, 2\pi]$, the cos/sin of the phase will have a symmetrical distribution with zero mean; consequently, the real part of the received signal will also have a zero mean.

A similar analysis of the imaginary part of the received signal shows that it also has a Gaussian distribution with zero mean and the same variance as the real part. Further, it can be shown that the real and imaginary components are statistically independent. Thus, as will be shown in the following section, the amplitude of the received signal is expected to have a Rayleigh distribution as a function of the frequency across the band, provided that the samples are sufficiently separated in frequency to be statistically independent. The signals will be correlated at small frequency separations, but the signals can be considered as uncorrelated at a large frequency separation. A crude measure of the required frequency separation is given by

$$
\frac{\Delta f R_i}{c} > 1 \quad \text{or} \quad \Delta f > \frac{c}{R_i} \approx \frac{f_0}{R_i/\lambda_0}
\tag{2.4}
$$

For example, with a center frequency of 5 GHz and a range of 3 m, typical of a room in an office building, the necessary separation in frequency for statistical independence will be 100 MHz. Therefore, across a 550 MHz band (minimum UWB bandwidth) the number of signal fades will be of the order of 5.

2.1.3 Rayleigh Statistics in a Non-Line-of-Sight Scattering Environment

The analysis in Section 2.1.2 showed that the amplitude statistics of the RF signal in a scattering environment where there is NLOS propagation should exhibit Rayleigh statistics. The following provides formulae associated with these statistics.

From Equation (2.2), the signal at a particular frequency can be expressed in the form

$$S = \sum_{i=1}^{N} A_i \, e^{j\phi_i} = \sum_{i=1}^{N} A_i \cos \phi_i + j \sum_{i=1}^{N} A_i \sin \phi_i = X + jY \qquad (2.5)$$

where A_i is the signal amplitude with unspecified statistics and ϕ_i is the random phase with assumed uniform statistics as the propagation paths are many wavelengths in length.

Now consider the statistics of the real and imaginary components separately, namely the evaluation of the mean and variance of X_i and Y_i. These components of the signal are the product of two random variables. As they are statistically independent, the expected values are

$$\begin{aligned} \mathrm{E}[X_i] &= \mathrm{E}[A]\mathrm{E}[\cos \phi] = \mu_a * 0 = 0 \\ \mathrm{E}[Y_i] &= \mathrm{E}[A]\mathrm{E}[\sin \phi] = \mu_a * 0 = 0 \end{aligned} \qquad (2.6)$$

The expected values of the cosine and sine of the random phase are zero, as shown in the following details. The probability density function (PDF) of the cosine of the random variable ϕ $(c = \cos\phi)$ is

$$p(c) = 2U(\phi)\left|\frac{d\phi}{dc}\right| = \frac{1}{\pi\sqrt{1 - c^2}} \qquad (2.7)$$

The factor of 2 is due to the fact that there are two values of each ϕ for each value of c. The expected value (mean) and variance of the random variable c are

$$\mathrm{E}[c] = \mu_c = \int_{-1}^{1} \frac{c}{\pi\sqrt{1 - c^2}} \, dc = 0$$

$$\mathrm{var}[c] = \sigma_c^2 = \int_{-1}^{1} \frac{c^2}{\pi\sqrt{1 - c^2}} \, dc = 1/2 \qquad (2.8)$$

Similar expressions apply to the random variable $s = \sin\phi$. The variance of X_i (the product of two random statistically independent variables) can be evaluated as follows:

$$\mathrm{var}[X_i] = \sigma_x^2 = \mu_a^2\sigma_c^2 + \mu_c^2\sigma_a^2 + \sigma_a^2\sigma_c^2 = 1/2(\mu_a^2 + \sigma_a^2) \qquad (2.9)$$

(This is strictly only true for large N.) A similar expression applies to the variance of Y_i, so that $\sigma_x^2 = \sigma_y^2 = \sigma^2$.

Another property of X_i and Y_i is that they are statistically independent. This can be shown by proving that the correlation coefficient ρ is zero. From the definition of the correlation coefficient:

$$\rho = \frac{\mathrm{E}[(X_i - \mu_x)(Y_i - \mu_y)]}{\sigma_x\sigma_y} = \frac{\mathrm{E}[X_iY_i]}{\sigma_x\sigma_y} = \frac{\mathrm{E}[A_i^2\cos\phi_i\sin\phi_i]}{\sigma_x\sigma_y} = \frac{\mathrm{E}[A_i^2]\mathrm{E}[\sin(2\phi_i)]}{2\sigma_x\sigma_y} = \frac{\mathrm{E}[A_i^2] * 0}{2\sigma_x\sigma_y} = 0$$

$$(2.10)$$

In summary, X_i and Y_i are two statistically independent variables with zero mean and a nonzero variance defined by (2.9). The real and imaginary components of the signal are the summation of N of these random variables, all with the same statistics. Thus, the statistics of X and Y are given by

$$E[X] = E[Y] = 0$$
$$\text{var}[X] = \text{var}[Y] = N\sigma^2 = \sigma_t^2 \qquad (2.11)$$

Equation (2.11) shows that the real and imaginary components of the signal have zero mean and a variance which only depends on the total power σ_t^2 in the scattered signals. Further, as X and Y are the summation of random variables, then from the CLT the PDF of each approaches Gaussian as the number of scattering rays becomes large. The effect of the number of scattering rays on the PDF is considered in more detail in Section 2.1.4. Thus, the PDFs of the X and Y components are given by

$$p_x(X) = p_g(X) \qquad p_y(Y) = p_g(Y)$$
$$p_g(u) = \frac{1}{\sqrt{2\pi}\sigma_t} \exp(-u^2/2\sigma_t^2) \qquad (2.12)$$

The statistics of the magnitude of the signal ($R = (X^2 + Y^2)^{0.5}$) are now computed. The PDF of the signal is the joint probability of the X and Y random variables, so in rectilinear coordinates the PDF is

$$p_R(X, Y) = p_g(X)p_g(Y) = \frac{1}{2\pi\sigma_t^2} \exp\left[-(X^2 + Y^2)/2\sigma_t^2\right] \qquad (2.13)$$

In (2.13), the joint probability is the product of the individual probabilities, as X and Y are statistically independent. The joint probability is more conveniently expressed in polar coordinates (R, θ). The relationship between the two PDFs is given by

$$q_s(R, \theta) = p_R(X, Y)\left|\frac{\partial(X, Y)}{\partial(R, \theta)}\right| \qquad (2.14)$$

By noting that $X = R\cos\theta$ and $Y = R\sin\theta$, the determinant of the matrix of the partial differentials is given by

$$\left|\frac{\partial(X, Y)}{\partial(R, \theta)}\right| = \left| \begin{matrix} \cos\theta & \sin\theta \\ -R\sin\theta & R\cos\theta \end{matrix} \right| = R \qquad (2.15)$$

Using equations (2.13) and (2.15) in (2.14), the required joint distribution in polar coordinates is

$$q_s(R, \theta) = \frac{R}{2\pi\sigma_t^2} \exp\left(-R^2/2\sigma_t^2\right) \qquad (2.16)$$

Noting that $q_s(R, \theta)$ does not explicitly include θ, the angular dependence can be integrated out:

$$q_R(R) = \int_{-\pi}^{\pi} q_s(R, \theta)d\theta = \frac{R}{\sigma_t^2} \exp\left(-R^2/2\sigma_t^2\right) \qquad (2.17)$$

Equation (2.17) gives the PDF of the signal amplitude and is the required Rayleigh statistical distribution. In deriving this statistical distribution, only the following assumptions were made:

1. The phase has uniform statistical distribution over range 0 to 2π.
2. The amplitude and phase of each signal component are statistically independent.
3. The number of (significant) signal components N is large.

Of these assumptions, only the last assumption may be called into question in some cases. However, even when the number of significant signal components is not large, the general statistical characteristics of the individual components (zero mean, finite variance) mean that the overall statistics will exhibit statistics similar to the Rayleigh distribution. Also note that no particular statistical distribution on the amplitude of the scattered signals is required in the derivation. Thus, in conclusion, the Rayleigh scattering statistics should apply to most environments where there are a significant number of scattering objects near the receiver.

Finally, the statistical properties of the Rayleigh distribution can be computed:

$$
\begin{aligned}
\mu_R &= \int_0^\infty R q_R(R)\mathrm{d}R = \sqrt{\pi/2}\sigma_t \\
\sigma_R^2 &= \int_0^\infty R^2 q_R(R)\mathrm{d}R = (2 - \pi/2)\sigma_t^2
\end{aligned}
\tag{2.18}
$$

The associated signal power ($P = R^2$) statistics can be computed from the amplitude statistics:

$$
q_P(P) = q_R(R)\frac{\mathrm{d}R}{\mathrm{d}P} = \frac{\exp(-P/2\sigma_t^2)}{2\sigma_t^2} = \frac{\exp(-P/\bar{P})}{\bar{P}}
\tag{2.19}
$$

Curiously, (2.19) shows that the PDF of the signal power is maximum at $P = 0$, while (2.17) shows the PDF of the signal amplitude is a minimum zero at $R = 0$.

The median of the signal is given by the 50% probability of the cumulative distribution. Thus, the median is obtained from the solution of the equation

$$
\int_0^R q_R(x)\mathrm{d}x = 0.5
\tag{2.20}
$$

Solving for R gives the median signal as $[2\ln(2)]^{0.5}\sigma_t$.

The RMS signal is given by

$$
R_{\mathrm{rms}} = \left[\int_0^\infty P q_P(P)\mathrm{d}P\right]^{1/2} = \sqrt{2}\sigma_t = \bar{P}
\tag{2.21}
$$

As can be observed from (2.17), there is a finite probability of a small signal level, much less than the mean signal level given in (2.18). The physical reason for this is that the individual signal vectors can sum to a small value if the phases of the vectors are appropriate; such small

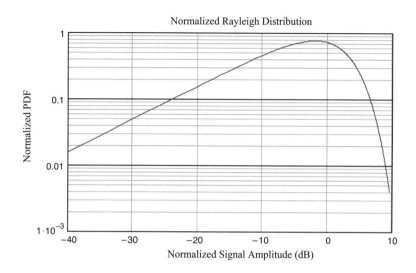

Figure 2.2 Normalized Rayleigh distribution plotted with the normalized signal level in decibels.

total signals are referred to as Rayleigh fading. Normalizing the signal relative to the mean signal results in the normalized PDF

$$\hat{R} = R/\mu_R$$

$$\hat{q}_R(\hat{R})d\hat{R} = \frac{\pi}{2}\hat{R}\exp\left(-\frac{\pi}{4}\hat{R}\right)d\hat{R} \qquad (2.22)$$

The normalized Rayleigh distribution is shown plotted in Figure 2.2, where the normalized signal amplitude is expressed in decibels and the PDF is plotted on a log scale to enable small probabilities to be observed. Thus, from Figure 2.2 it can be observed that there is a probability of about 0.01 that the signal fades by 40 dB relative to the mean signal. Also observe that the most probable signal level is about 2 dB less than the mean signal and that the maximum possible signal is about 10 dB greater than the mean signal.

From (2.3) it is observed that the vector phase is a function of both the frequency and the propagation distance of individual components, so Rayleigh fades occur both as a function of the frequency across the signal bandwidth and spatially as a function of position. As the phase is related to the propagation distance between the scattering source and the receiver measured in wavelengths, even small movements of less than a wavelength can change the individual vector phases significantly and, thus, dramatically change the combined resultant received signal amplitude. Note that in such a small distance the amplitudes of each scattered signal will change little, so the effect is purely associated with the signal phase.

In the frequency domain, signal fades across the signal bandwidth have a major effect on the performance of radio systems operating in a severe multipath environment. For narrow-band systems Rayleigh fades can result in such a large loss in signal that reception can be lost. However, positioning systems require wide bandwidths to provide adequate time resolution and, thus, are less susceptible to Rayleigh fading. In particular, if the system is based on spread-spectrum signals, the effective timing information associated with the signal is spread across the bandwidth of the signal, effectively ensuring great redundancy. Hence, a loss of a small part of the signal across the signal bandwidth due to Rayleigh fading results in comparatively little loss in performance; the analysis of these effects on the measurement of the TOA of the signal is best analyzed in the time domain, as shown in Chapters 3 and 4.

2.1.4 Distribution Convergence

The theory in Section 2.1.3 shows that the PDF will approach the Rayleigh distribution provided the number of scattering sources is large. However, the number of scatters required for the distribution to reasonably closely approach a Gaussian distribution in the real and imaginary components of the signal remains unresolved. In this subsection, simulations of the received signal are used to estimate the number of significant signals required, and then a more rigorous mathematical analysis is used to examine the convergence behavior as the number of scatters becomes large.

For the numerical evaluation, the signal amplitudes of the scattered signals are assumed to have a uniform distribution in the range from 0 to 1. The distribution of amplitudes is a function of the scattering environment, but all environments will have a wide range of amplitudes. The uniform distribution is thus only a rough guide to a realistic situation. However, as shown in Section 2.1.3, the resulting PDF does not depend on the amplitude distribution, and so a uniform distribution is satisfactory for the investigation of the behavior.

The number of scattering sources N is varied in the numerical simulation. The output distribution is compared with the nominal Gaussian distribution. Clearly, with $N = 1$ there is no interference, so that the signal amplitude is simply A.

The first example is with $N = 2$, as shown in Figure 2.3a. The distribution is not closely matched to the Gaussian distribution, but shares the properties of having zero mean and symmetry about the origin. The peaks in distribution are related to the PDF of individual components. The second examples are with $N = 3$. Two samples are shown in Figure 2.3b. The shapes are now closer to the Gaussian curve, but there is still considerable variability in the shape. The third examples are with $N = 5$. Two samples are shown in Figure 2.3c. The shapes are generally Gaussian in shape, but some variability remains. The final example is with $N = 10$. In this case there is quite close agreement with the theoretical Gaussian curve.

From the numerical simulations it can be concluded that with approximately ten scatters or more, the total real and imaginary components will closely approach Gaussian statistics, and as a consequence the combined signal amplitude will closely approach Rayleigh statistics. As it seems probable that, in a typical indoor environment, the number of significant scattering sources (greater than a few wavelengths in size) is at least 10, the conclusion is that virtually all indoor NLOS environments will exhibit Rayleigh statistics quite closely.

The fact that the signal amplitude components approach a Gaussian distribution is a consequence of the CLT. The CLT states that if a variable is the sum of a large number of independent random variables, then the statistics will approach Gaussian independent of the statistics of the underlying random processes. The numerical example above suggests in the signal scattering case the number of scatters required is of the order of 5–10. To more formally consider the limiting process, an extension to the proof of the CLT is required. As the proof is given in advanced statistical analysis books, such as [5], only an outline analysis is given.

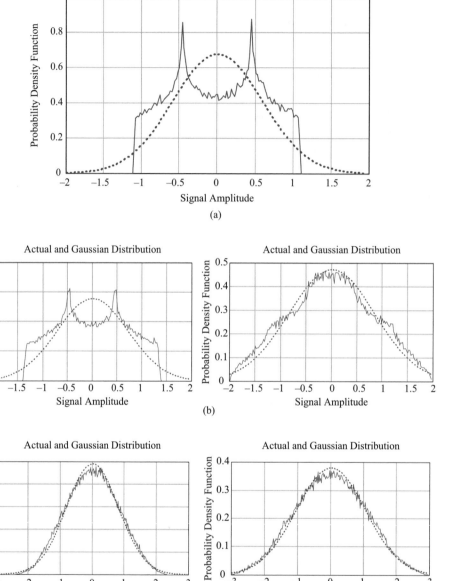

Figure 2.3 (a) PDF with $N = 2$, $A = [0.32, 0.77]$. (b) Two examples with $N = 3$; the shapes vary, but they correspond approximately with the Gaussian shape. (c) Two examples with $N = 5$; the shapes are approximately Gaussian, but there remains some variability. (d) Example with $N = 10$; the simulation curve is quite close to the theoretical Gaussian shape.

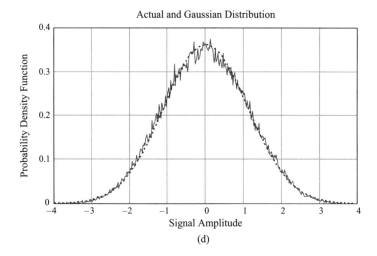

Figure 2.3 *(Continued)*

For the analysis it is convenient to normalize the random variables X_n, such that the sum of N such variables S_N has a distribution with zero mean and unit STD. Thus:

$$\frac{S_N - E[S_N]}{\sigma(S_N)} = \frac{Z_N}{\sqrt{N}} = \frac{\sqrt{N}(\bar{X}_N - \mu)}{\sigma} \approx N(0,1) = e^{-t^2/2} \tag{2.23}$$

where the sample mean is μ and the STD is σ. Thus, the required probability function is

$$G_N(x) = P\left[\frac{\sqrt{N}(\bar{X}_N - \mu)}{\sigma} \leq x\right] \tag{2.24}$$

The CLT is then $G_N(x) \to \Phi(x)$ as $N \to \infty$, where $\Phi(x) = \int_{-\infty}^{x} e^{-t^2/2}$. However, to proceed with the analysis, the characteristic functions[2] of these two probability functions are used to analyze the convergence behavior. Thus, the required characteristic function is

$$g_N(t) = g_{Z_N/\sqrt{N}}(t) = g_{Z_N}\left(\frac{t}{\sqrt{N}}\right) = \left[g_{Z_1}\left(\frac{t}{\sqrt{N}}\right)\right]^N \tag{2.25}$$

where the multiplication property of characteristic functions has been invoked. Expanding (2.25) as a Taylor series about $t = 0$ and using the derivative properties of characteristic functions and the normalization in (2.23) results in the series

$$g_1(t) = 1 - \frac{t^2}{2N} - \left[j\frac{t^3}{6N\sqrt{N}}E[Z_1^3] - \frac{t^4}{24N^2}E[Z_1^4] + O(t^5)\right] \tag{2.26}$$

---

[2] The characteristic function is the Fourier transform of the PDF. Thus, the properties of Fourier transforms can be directly applied to characteristic functions. For example, the PDF of the sum of two random variables is the convolution of their individual PDFs. Thus, from the properties of Fourier transforms, the characteristic function of the sum is the product of their individual characteristic functions.

For large N the term in square brackets can be ignored; so, inserting (2.26) into (2.25) results in

$$g_N(t) = \left(1 - \frac{t^2}{2N}\right)^N \xrightarrow[N \to \infty]{} e^{-t^2/2} \tag{2.27}$$

which is the limiting proof of the CLT. To examine this convergence behavior expressed by (2.26), a particular distribution of the random variable X_n is required. Thus, consider the case of scattering as given in the numerical case above, where it is assumed the scattering coefficients are uniformly distributed $U[0, 1]$. The moment generation function for a general uniform distribution $U[a, b]$ is given by

$$m_n = E\left[Z_U^n\right] = \frac{1}{n+1} \sum_{i=0}^{n} a^i b^{n-i} \tag{2.28}$$

Applying (2.28), it is easily shown that $m_3 = 0$ and $m_4 = 9/5$, so the characteristic function (2.25) for this uniform distribution case becomes

$$\hat{g}_N(t) = \left(1 - \frac{t^2}{2N} + \frac{3}{40} \frac{t^4}{N^2}\right)^N \tag{2.29}$$

Although this calculation is based on a uniform distribution, a similar result can be obtained for other distributions by applying the appropriate moment generation function to (2.26).

By comparing (2.27) and (2.29), a measure of the convergence as a function of N can be determined. In particular, an error function can be defined as

$$\xi(t, N) = \frac{g_N(t) - \hat{g}_N(t)}{g_N(t)} = 1 - e^{t^2/2} \hat{g}_N(t) \tag{2.30}$$

Figure 2.4 shows the convergence behavior of equation (2.30). The convergence is a function of both the number of scattering sources N, and the deviation distance from the mean. Thus, for large distances from the mean, say more than two STDs, the convergence is slow. Figure 2.4 shows that at two STDs the convergence is within about 16% for five scatters, and 7% for 10 scatters. However, at one STD the convergence is good, even for a relatively small number of scatters.

2.1.5 Rician Statistics in a Line-of-Sight Scattering Environment

For outdoor or at short-range indoor propagation conditions, a direct path from the transmitter to the receiver may exist, so the NLOS assumptions associated with Rayleigh statistics are no longer valid. However, even when an LOS path exists, other scattering paths similar to that assumed for the NLOS analysis will still be present, so the NLOS analysis can be performed by adding a nonrandom signal S. Thus, from (2.13), the distribution joint probability distribution becomes

$$p_r(X, Y) = \frac{1}{2\pi\sigma^2} \exp\left[-\frac{(X - S)^2 + Y^2}{2\sigma^2}\right] \tag{2.31}$$

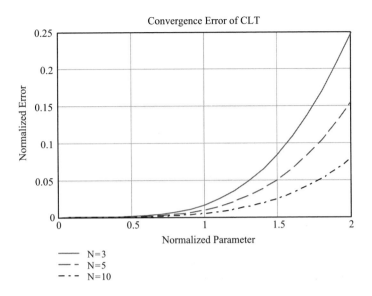

Figure 2.4 Convergence behavior of the CLT. The x-axis is normalized parameter t, so the data is limited to a maximum of two STDs.

These statistics are referred to as Rician statistics after Rice [6], who first analyzed problems with these statistics. Proceeding as for Rayleigh statistics by converting to polar coordinates results in the equivalent form

$$q_r(R, \theta) = \frac{R}{2\pi\sigma^2} \exp\left[-\frac{R^2 + S^2 - 2RS\cos\theta}{2\sigma^2}\right] \tag{2.32}$$

The distribution as a function of R only can be obtained by integration with respect to θ:

$$p_r(R) = \int_{-\pi}^{\pi} q_r(R, \theta)d\theta = \frac{R}{\sigma^2} \exp\left(-\frac{R^2 + S^2}{2\sigma^2}\right)\left[\frac{1}{2\pi}\int_{-\pi}^{\pi} \exp\left(\frac{RS\cos\theta}{\sigma^2}\right)\right]d\theta \tag{2.33}$$

Note that the integral can be expressed as a zeroth-order modified Bessel function; namely:

$$I_0(x) = \frac{1}{2\pi}\int_{-\pi}^{\pi} \exp(x\cos\theta)d\theta \tag{2.34}$$

Thus, using (2.34) in (2.33), the final expression for the Rician distribution is

$$p_r(R) = \frac{R}{\sigma^2} \exp\left(-\frac{R^2 + S^2}{2\sigma^2}\right) I_0\left(\frac{RS}{\sigma^2}\right) \tag{2.35}$$

It is convenient to normalize this basic equation so that there is one universal curve. The signal amplitude is normalized by the scattered signal parameter σ, so that $r = R/\sigma$. Also, the

Rician factor K is introduced, which is the ratio of the direct signal power to the scattered signal power. Thus, the Rician factor is given by

$$K = \frac{S^2}{2\sigma^2} \tag{2.36}$$

and the normalized Rician statistics equation becomes

$$p_r(r) = r \exp\left[-\left(K + \frac{r^2}{2}\right)\right] I_0(\sqrt{2Kr}) \tag{2.37}$$

The characteristics of the normalized Rician statistics are shown in Figure 2.5, with the Rician factor K as a parameter. For $K = 0$, the Rician statistics reduce to Rayleigh statistics. As the Rician factor becomes large (say >3), the Rician statistics approach the normal distribution. The mean and variance of the Rician distribution are given by

$$\mu_r(K) = \int_0^\infty r p_r(r) dr = \sqrt{\frac{\pi}{2}} \, e^{-K/2} [(1+K) I_0(K/2) + K I_1(K/2)] \tag{2.38}$$

$$var_r(K) = \int_0^\infty r^2 p_r(r) dr - \mu_r^2 = 2(K+1) - \mu_r(K)^2 \tag{2.39}$$

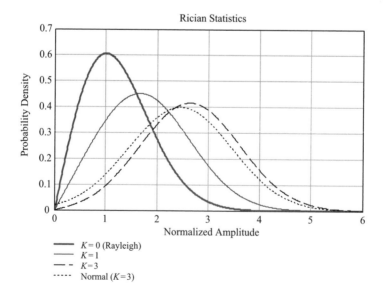

Figure 2.5   Normalized Rician statistics, with K as a parameter.

Because of the complexity of these expressions, approximate expressions are useful. For the mean, two expressions are appropriate, one for small K and one for large K:

$$\mu_r = \sqrt{\frac{\pi}{2} + \sum_{n=1}^{\infty} \left(\frac{2}{\pi}\right)^n \frac{(-1)^n}{(2n-1)!} K^n} \qquad (K < 2)$$

$$= \sqrt{2K}\left(1 - \frac{1}{4K} + \frac{1}{K^2}\right) \qquad (K \geq 2)$$

(2.40)

Similarly, the asymptotic expressions for the variance are

$$\text{var}_r(K) = 1 - \left(\frac{\pi}{2} - 1\right)\exp\left(-\frac{K}{\sqrt{2}}\right) \qquad (K < 2)$$

$$= 1 - \frac{1}{4K}\left(1 + \frac{1}{K} - \frac{1}{K^2}\right) \qquad (K \geq 2)$$

(2.41)

As the Rician statistics approach normal statistics for moderate Rician factors with $K > 3$, calculations can normally be based on the normal distribution in most practical situations.

2.2 Radio Propagation Characteristics at Different Distance Scales

The statistical analysis in Section 2.1 provides information of the small-scale variation of the signal in, say, a particular room of a building. These signal characteristics are important in designing the TOA detection subsystem, but they provide no information on the large-scale variation within the total coverage area of a transmitting node. To estimate the range of transmission, the large-scale variation of the signal is required. While in free space a simple inverse-square law can be used to determine the signal strength as a function of range, in an NLOS environment the complexity of the environment makes calculations of the large-scale signal attenuation with distance difficult. Further, while the average variation with distance is important, the medium-scale variation of the signal strength is also an important parameter in the design of indoor positioning systems. While the small-scale variation in signal strength is associated with complex (amplitude and phase) constructive and destructive interference between the signal components, medium-scale effects are associated with signal diffraction and the absorption of RF energy by obstacles such as walls along the propagation path. These medium-scale effects are typically on a scale of 5–50 wavelengths (0.5–5 m at 3 GHz) and, thus, are associated with major structural elements of a building. By extending this argument further, the large-scale effects are at scales of greater than 50 wavelengths, while the Rayleigh and Rician statistical effects are appropriate at scales of less than five wavelengths. These dividing lines are only rough guidelines which are useful for engineering analysis of the complex propagation environment of indoor radio propagation at microwave frequencies.

2.2.1 Large-scale Propagation Characteristics

Large-scale propagation effects are important in determining the range of a transmitting node. Given system performance parameters such as the transmitter power, the receiver sensitivity and antenna gains, a link budget can be devised to estimate the available loss for satisfactory operation. From this estimate and the large-scale attenuation characteristics, an estimate of the mean maximum operating range can be determined and, hence, the node spatial density required for the positioning system.

In this section, the propagation loss characteristics are presented. As the propagation characteristics are complex, the approach taken is one of empirical generic equations to represent the propagation loss, with the parameters determined by appropriate fitting of these equations to measured data.

The basic stating point for loss determination is the free-space loss, as this represents the minimum possible loss. From Maxwell's equations it is known that the RF energy reduces as the square of the distance from a point radiating source, so it can be shown [1,6] that the free-space path loss is given by

$$l_{\text{free\_space}} = \left(4\pi \frac{R}{\lambda}\right)^2 \tag{2.42}$$

Note that the attenuation can be expressed solely as a function of one parameter, namely the propagation distance expressed in wavelengths. The only other term in this simple expression is a proportionality constant.

An example of the propagation loss as a function of distance for an indoor case is plotted in Figure 2.6. The figure also shows the free-space loss for comparison, as well as the LS fit based

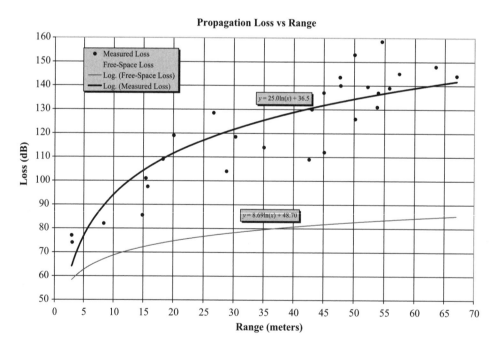

Figure 2.6   Scatter diagram of the measured path loss from an indoor transmitter to points in the adjacent buildings. Also shown are the free-space loss and the LS fit of the measured loss data to a logarithmic curve (when plotted in decibels).

on a power-law model. As can be observed, the measured loss is considerably greater than the free-space loss and it exhibits some random (medium-scale) variation. In determining the large-scale variation, a smooth curve can be plotted based on an appropriate empirical model. One possible indoor propagation model based on the analytically derived free-space attenuation model is of the form

$$l = c\left(\frac{R}{\lambda}\right)^{\gamma} \tag{2.43}$$

In this power-law model, the attenuation as a function of distance is controlled by the parameter γ, with $\gamma > 2$ as the loss is greater than the free-space case. The loss is assumed to be a function of the propagation distance measured in wavelengths, as with the free-space case. The proportionality constant c is determined by fitting the model to the data, usually based on a QLS manner. Instead of performing an LS fit to the loss data directly, the usual technique is to perform the LS fit to the loss data expressed in decibels. Thus, the model equation is

$$L(\text{dB}) = a + b\log(R/\lambda)$$
$$a = 10\log(c) \qquad b = 10\gamma \tag{2.44}$$

Thus, the model loss in decibels[3] plotted against the logarithm of the range measured in wavelengths is a straight line, with the slope related to the power parameter γ and the y-intercept related to the constant parameter c. These parameters are determined by performing a linear LS fit to the data. Note that this result is not a true LS fit to the original data, but the results are usually close to the true LS-fit solution.

The LS fit to the data are shown in Figure 2.6. The STD between the model curve and the scattered measured data is 11.5 dB. The measurement carrier frequency is 6.5 GHz. Note in this case that the propagation model is plotted with logarithmic loss but with linear range; the reason for this is that the x-axis plotted linearly more clearly displays the loss characteristics over the range interval of interest, namely 5–70 m. From the LS fit to this data the power-law parameter $\gamma = 6.5$ (compared with $\gamma = 2$ for free space), and the constant parameter is $c = 4.74 \times 10^{-7}$. As the power-law exponent $\gamma = 6.5$ is much greater than the free-space exponent, the losses increase much more rapidly with propagation distance. While this model is perhaps a natural extension to the free-space law, the model defined by (2.36) is not totally satisfactory. For example, LOS conditions will exist when the propagation range is short, implying that $\gamma = 2$ at short range, but a larger value applies at long ranges, with presumably a transitionary region at intermediate ranges. This effect cannot be captured with a constant power-law exponent. Another drawback is the difficulty in relating the power-law parameters to the physical nature of the building. In practice, the power-law exponent γ is determined by measurements in specific buildings, with reported values in the range of $\gamma = 3–6$ for indoor propagation [7–9]. However, the exact value appears quite sensitive to the specific loss data, with, for example, the data in Figure 2.6 having a power-law parameter $\gamma = 6.5$. While the value of the power-law parameter γ can be roughly estimated based on the scattering and absorption of the RF energy, the constant parameter c is even more difficult to specify in particular cases.

---

[3] The convention used is that lower case text l is used for the loss in the standard form and upper case L when specified in decibels.

For these reasons, an alternative model is proposed that is more closely related to the physical situation.

An alternate approach to the analysis of the propagation loss data is to consider the total loss as the sum of the free-space loss plus the additional loss associated with the propagation processes, such as scattering from surfaces, diffraction around corners and the transmission losses through the walls of buildings. While a detailed analysis is too complex, the general loss mechanisms can be analyzed mathematically or determined from measurements. For example, the transmission losses through various wall types can be both measured and estimated by theoretical analysis [1,10]. Similarly, diffraction losses around right-angle bends typical in buildings can be determined. Other losses, such as scattering from objects (such as furniture) in buildings, are more difficult to estimate, so some correction using measurements would still be necessary. Nevertheless, at least in principle, the loss excess can be estimated based on the known characteristics of buildings.

The propagation loss excess data from Figure 2.6 is plotted in Figure 2.7, together with a linear fit to the data. Although there is a considerable STD of 10 dB, slightly less than the 11.5 dB for the power-law scatter in Figure 2.6, there appears to be a linear trend with propagation distance. This linear relationship can be explained by assuming the scattering/diffraction/wall transmission losses are related to the number of such occurrences along the propagation path. Thus, the loss excess for an approximately homogeneous environment will be proportional to the propagation range at least in a statistical sense. The excess loss of about 13 dB at short range in this case can be explained by the particular circumstances of the measurement points where the propagation to these points is through a wall separating the transmitter from the receiving points, so there are no LOS points even at close range. Thus, this short-range loss can be interpreted as the transmission loss through a wall. As a first-order approximation, this loss can be assumed to be representative of all walls and, thus, can be used to estimate loss excesses for other propagation paths.

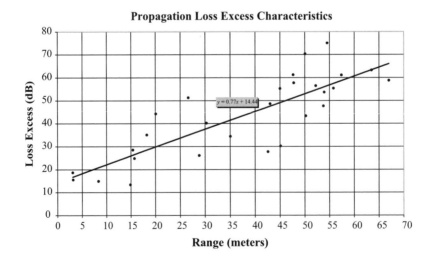

Figure 2.7 Scatter diagram of the measured path loss excess above the free-space loss. The excess loss appears to be linearly related with the propagation range. The STD of the fit error is 10 dB.

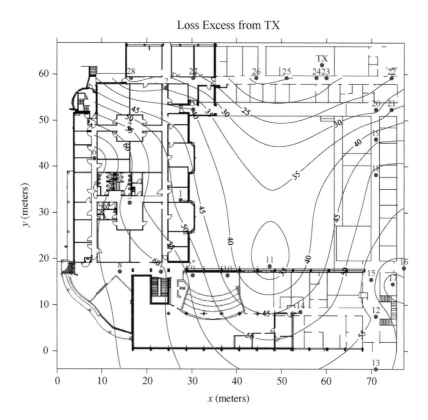

Figure 2.8   Loss excess map. The loss excess varies from about 15 dB close to the transmitter to 65 dB at the far end of the building from the transmitter. The contour step is 5 dB. The buildings surround a central outdoor courtyard, which is largely open space except for some trees.

The linear relation slope parameter is about 0.8 dB per meter of propagation (see Figure 2.7). As the loss excesses are closely related to the architecture of the building (particularly the spatial frequency of walls along the propagation path), this parameter can be related to the average number of walls per unit length along the path. Thus, a simple model for the loss excess (in decibels) is given by

$$\Delta L = \alpha \eta R = \beta R \tag{2.45}$$

where α is the average attenuation per obstruction along the propagation path and η is the average number of obstructions per unit length (meter) along the path. From Figure 2.7, $\beta = 0.77$ dB/m, and for this building (see Figure 2.8 for a map of the building) the average number walls per meter η is estimated at about 0.25 obstacles per meter, so that the attenuation parameter is $\alpha = 3.1$ dB per obstacle. The estimate above of 13 dB transmission loss through walls seems to suggest that the α parameter is more associated with diffraction/scattering losses rather than wall transmission losses. While the above analysis is for a particular case, the general principles can be used on other buildings with the parameters adjusted as appropriate.

Thus, (2.45) can be used to estimate the propagation losses in a statistical fashion if the attenuation and architectural parameters of the building are known. Further, for a given propagation path a deterministic estimate can be made from the number obstacles M along the path, so that the loss excess is

$$\Delta L = \alpha M \tag{2.46}$$

A simple model of propagation would assume a straight-line path between the transmitter and the receiver, but a more appropriate concept for indoor propagation is the 'path of least resistance'. For example, buildings such as typical offices (not open plan) will have corridors connecting the smaller rooms, so that the least-loss path between rooms would be along corridors, even though this path is longer than the straight-line path through walls.

In the example, the average loss per unit distance along the corridor was measured at 0.3 dB/m, which is considerably less than the 0.8 dB/m for all the data, as shown in Figure 2.7. Although this data set is limited in extent, the propagation losses along corridors are considerably less than those for a general path through the building. This example shows the importance of defining the propagation path from the transmitter to the receiver, as the straight-line path may not be the path of least propagation loss.

Based on the above observations, the loss characteristics are now examined in more detail. The loss excess data shown in Figure 2.7 is shown plotted as a contour map in Figure 2.8. The contour plot is based on a limited number of measurement points, so that the map is rather coarse in details. Nevertheless, some of the characteristics described above can be observed in the map.

First, observe that the contour lines in the buildings are approximately normal to the propagation path from the transmitter. This observation is consistent with the loss excess being proportional to the propagation distance. Thus, for example, the 45 dB contour near the building on the left runs approximately parallel to the outer wall. This shows that most of the loss excess in this region is due to losses associated with the walls close to the transmitter.

Second, the contours near the top of the map are approximately parallel to the x-axis. This observation is consistent with the propagation path being along the corridor (at the top of the contour map). As the propagation losses are much lower along the corridor than through walls, the propagation loss with distance in this case is different from paths through walls. Thus, the propagation path to points adjacent to the corridor must include the corridor rather than the straight-line path through walls, which have larger losses.

Finally, consider an example relevant to the above observations. The large loss excesses near point 6 (70 dB) are expected due to the large number of walls the straight-line path must pass through. However, by drawing the straight-line path, the number of walls along the path can be counted to be 10, so that the loss along this path would be estimated by the above method at $10 \times 13 = 130$ dB. As the actual loss is much less, the 'path of least resistance' must be some path other than the straight line. In contrast, a path along two corridors (total length 70 m) requires the transmission through only three walls plus two diffraction losses. Thus, the total losses for this path can be estimated using the above-defined method as $(70 \times 0.3) + (3 \times 13) = 60$ dB plus two diffraction losses. As the measured loss excess is 70 dB, the above analysis is consistent if the diffraction losses are each 5 dB. As this estimate appears to be reasonable, the conclusion is that the actual path is as described. Note that the extra path length for the 'path of least resistance' will have important consequences for the performance of indoor TOF positioning systems.

Although the simple method of estimating propagation losses indoors was illustrated with a particular example of a building, the method should be widely applicable because the construction of buildings follows universal principles. Thus, it is suggested that the loss excess can be estimated by the following method. The straight-line path is plotted and the number of walls used to estimate the loss using (2.46). A second estimate of the propagation loss excess is made by defining a path that minimizes the number of walls along the path and uses corridors to the maximum extent. In this case, the loss per meter along corridors and the diffractions losses around corners are used to estimate the loss excess. The minimum loss excess from these two estimates is chosen as the excess loss. This method was used to estimate the loss excess in the building on the left in Figure 2.8; the results are illustrated in Figure 2.9. The model loss excess is based on estimating the losses from wall penetrations loss, corridor loss per meter and diffraction losses. The LS fits to the data are wall penetration loss 1.3 dB, corridor loss 0.5 dB/m and diffraction loss 4.1 dB. The STD of the difference between the measured and model data is 5.8 dB, which is considerably less than the variation in Figures 2.6 and 2.7. The frequency is 5.2 GHz, with ranges up to 40 m. It is concluded that the above-described method can provide moderately accurate correlation between measured and estimated loss excesses.

The data in Figure 2.7 provides measured excess loss data at 6.5 GHz in an indoor environment up to a range of 70 m. These measurements were made using a system which has a relatively high performance in terms of range, as it has a high process gain (about 87 dB). Other systems (such as WSNs) are likely to be inferior in performance with respect to range; thus, the data suggest that indoor systems will have maximum ranges of about 40 m in an office-type environment, based on more typical link budgets. Open-plan offices and open indoor areas (such as sporting arenas or theatres) will have larger ranges, limited by the LOS. As the only significant available bandwidth for high-power (up to 4 W) transmissions is at 2.4 and 5.8 GHz (which has higher propagation losses and, hence, shorter ranges than at 2.4 GHz), the suggested

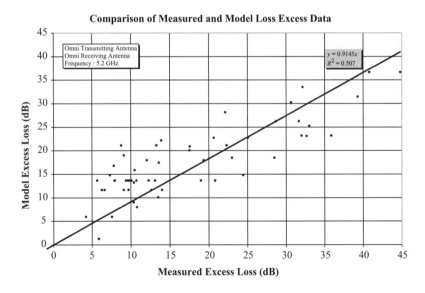

Figure 2.9 Comparison of the measured loss excess with the model loss excess for an indoor environment.

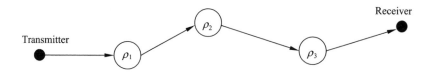

Figure 2.10   Simplified diagram showing the scattering of RF energy along a path from the transmitter to the receiver. Note there are four sub-paths but three scattering obstacles along the path.

40 m range limit appears to be valid for all conceivable RF systems operating in a cluttered indoor environment. Note the UWB systems will have greater positional accuracy, but typically are limited to about 10 m in range, due to the low transmitter power.

2.2.2 Medium-scale Propagation Characteristics

From the introductory comments to Section 2.2, medium-scale variation is defined as being associated with *energy* scattering and absorption from large objects along the propagation path from the transmitter to the receiver. To measure the medium-scale energy scattering, the small-scale signal variation must be averaged of an area of the order of 5×5 wavelengths. The STD between the measured loss and large-scale model loss described in Section 2.2.1 can be explained by the medium-scale random variations; hence, all large-scale models will exhibit random errors when compared with measured data.

Because of the complex nature of the propagation environment, direct deterministic calculation of the signal path attenuation variation on a medium scale is usually not feasible, so that statistical methods must be adopted. Consider the simplified propagation diagram shown in Figure 2.10, which shows the scattering/reflection coefficient ρ for each leg of a path from a transmitter to a receiver. Although only one such path is shown, the received signal will be composed of many such paths p in practice, but only the dominant (smallest loss) path is considered initially. From the transmitter power and the receiver power, the propagation loss[4] is given by

$$\frac{P_{tx}}{P_{rx}} = l_p = \prod_{n=1}^{N} \rho_n \tag{2.47}$$

where N is the number of *significant* obstacles along the path. For indoor propagation, the significant obstacles along a propagation path are usually walls or associated with diffraction around corners. The nature of the path attenuation and scattering losses is unknown, but these are assumed to be random with undefined statistics. As the loss measurements are in decibels, it is more useful to consider the log of l_p, so that

$$L_p(\text{dB}) = 10 \log(l_p) = 10 \sum_{n=1}^{N} \log(\rho_n) \tag{2.48}$$

---

[4] Here, by definition, the *path loss* is a number greater than unity, or in decibels the loss is a positive number. Alternatively, the *path gain* is the reciprocal of the loss and will be less than unity and a negative number in decibels.

Thus, when expressed in decibels, the loss is the sum of the scattering coefficients expressed in decibels. As the scattering coefficients are presumably independent random variables, their logarithms will also be random, so the path loss in decibels will also be a random variable. However, as the path loss in decibels is expressed as the sum of a moderately large number of independent random variables, invoking the CLT implies that the statistics will approach the normal distribution as the number of scattering paths becomes large. The numerical example in Section 2.1.5 shows that the number of variables required is quite modest, with $N = 5$ being sufficient to quite closely approach the normal distribution. Thus, when the loss is measured in decibels, the expected distribution will be log-normal, at least for a simple dominant path. Note that the measurement of the medium-scale propagation loss requires the averaging out of the small-scale statistical fluctuations described in Section 2.1, where the area for averaging is over a few wavelengths. The smoothing of the medium-scale fluctuations over a large area, say 50×50 wavelengths, results in the large-scale variation in the propagation loss with distance, as described in Section 2.2.1.

The above simplified analysis shows that, with scattering losses from obstacles along a single propagation path, the expected received signal power will be a random variable exhibiting log-normal statistics. However, the real-world propagation environment will have many such propagation paths from the transmitter to the receiver. The dominant path may have the fewest number of obstacles and typically would be closest to the straight-line path, while the other paths tend to be longer and encounter more obstacles along the path. In the simple model, where the scattering losses from each obstacle results in a defined decibel loss and additional path range, the received signal power from a path will decrease exponentially with the excess path length. These expectations are confirmed, at least approximately, in the measurements described in Section 2.3.

Consider the multiple path case by extending the path loss model of (2.47) to include the main dominant (zeroth) path plus other longer paths P, so that the path-loss equation becomes

$$\frac{P_{tx}}{P_{rx}} = l_p = \prod_{n=1}^{N} \rho_n^0 + \sum_{p=1}^{P} \left(\prod_{n=1}^{N_p} \rho_n \right) = T_0 + \sum_{p=1}^{P} T_p = T_0(1 + 1/K) \qquad (2.49)$$

Note that in the typical indoor NLOS situation the dominant path replaces the LOS path assumed in the classical Rician analysis, so that in this case the Rician factor $K = T_0 / \sum_{p=1}^{P} T_p$. To further the analysis it is now assumed that the power in the dominant path exceeds the sum of the powers in the other paths; or, in terms of the analysis in Section 2.1.5, the Rician factor is greater than unity. Now, as before, the loss excess in decibels is given by

$$L_p = 10 \log[T_0(1 + 1/K)] = 10 \log T_0 + \frac{10 \ln(1 + 1/K)}{\ln(10)} \approx 10 \log T_0 + \frac{10}{\ln(10)T_0} \sum_{p=1}^{P} T_p$$

$$(2.50)$$

where the approximation $\ln(1 + x) \approx x$ for small x has been invoked. The first term in the final expression in (2.50) is the same as in (2.49), and this was shown to be a Gaussian random variable. The second term in the final expression is the sum of random variables, and by the

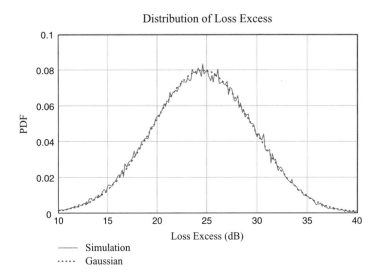

Figure 2.11 Simulated and theoretical Gaussian distribution of the loss excess measured in decibels. The mean excess loss is 25 dB and the STD is 5 dB. The results shown are based on a range of 40 m.

same argument regarding the CLT the statistical distribution is also a Gaussian random variable. As the sum of two Gaussian random variables is also a Gaussian random variable, the conclusion is that for the case $K \gg 1$ the resulting overall propagation loss (in decibels) is a Gaussian random variable. While this result is strictly only true for $K \gg 1$, the log-normal distribution can be expected to be approximately true for smaller values of K.

To test this theoretical analysis, a simulation of the multiple path case can be used. For the simulation, the propagation loss data are based on the measured data described in Section 2.2.1, and in particular the loss-distance parameter $\beta = 0.77$ dB/m (see (2.45)), and the obstacle density factor $\eta = 0.25$ obstacles per meter. Based on these parameters, the loss per obstacle is 3.1 dB and there will be nine obstacles in the 40 m range of the dominant path used in the simulation. For the other paths it was assumed that $P = 10$, and these 10 paths had 10, 11, ..., 19 obstacles. As the extra path loss is about 3 dB per obstacle, the signal is attenuated by a little more than a factor of 2 for each additional obstacle. Thus, the power in these additional paths is a geometric series, and the total power should sum (on average) to a little less than the power in the dominant path, thus ensuring on average $K > 1$.

The results of the simulation in Figure 2.11 show the statistical distribution of the loss excess based on 100 000 Monte Carlo simulation runs. Also shown is the Gaussian (log-normal) distribution with the same mean and STD as the simulation data. The mean K for the simulation is 4.3, although K varies widely from near zero to about 200. Evidently, the distribution is close to log-normal, even though the assumption $K \gg 1$ is not strictly true in all cases.

2.3 Measurements

There are fundamentally two approaches to measuring the radio propagation characteristics: in the time domain or in the frequency domain. In theory, as measurement in one domain can be converted to the other domain using a Fourier transform (or its inverse), the choice of method is more related to practical consideration associated with the equipment. The most flexible method is to use a network analyzer to measure the frequency response of the environment. This method can determine the transfer function at each frequency across a wide band of frequencies and, thus, determine the propagation losses and the propagation delays by the inverse Fourier transform of the spectral data. However, as will be explained, the method is somewhat limited by the need for a cable from the transmitter to the receiver, and the sensitivity can be less than required for long propagation paths. The alternative time-domain approach is more sensitive and does not need a cable for relative delay measurements, but absolute delays with this method cannot be measured. Thus, there is no one perfect method for all situations, so both techniques are used in practice, depending on the requirements.

Because of the requirement to measure the scattering from major objects along the path from the transmitter to the receiver, and the desire for high measurement resolution, very wide bandwidths are necessary. If a bandwidth of 1 GHz is used, then the time resolution is about 2 ns, equivalent to 60 cm. Although these bandwidths are associated with UWB systems, the performance of narrower band systems can be derived from the wideband data by the appropriate mathematical processing of the raw data.

The details of the measurement techniques are given in Appendix B. These techniques were used to generate the data described in the following sections.

2.3.1 Measurements in a Room

In this section, some measurement results using the frequency-domain method are presented, thus illustrating the measurement techniques presented in Appendix B and providing some typical propagation data inside buildings. The measurements were made in an L-shaped laboratory, as shown in Figure 2.12. The laboratory comprises work benches and shelves with equipment. The dimensions are shown in Figure 2.12, and the ceiling height is 2.6 m. The

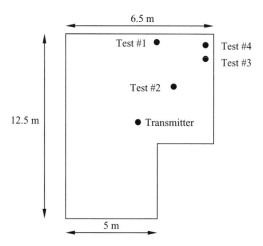

Figure 2.12 Diagram of the room in which the measurements were taken. The positions of the receiving antenna (four positions) and the transmitter position are also shown.

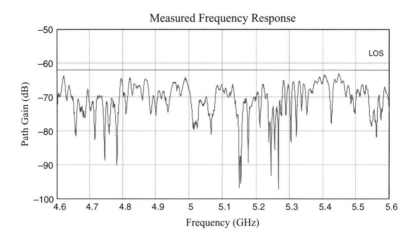

Figure 2.13   Measured amplitude frequency response for Test #4. The propagation distance is 5.8 m, corresponding to 61 dB LOS loss, and the mean loss is 70 dB.

transmitter was located near the center of the room with clear LOS in all directions. The antenna is omnidirectional, so that the transmitted signal is expected to be scattered by the walls, roof, floor and objects in the room. The data presented in the following subsections relate to the NLOS measurements (Test #4).

2.3.1.1 Spectrum Characteristics

Figure 2.13 shows the spectrum as measured by the network analyzer in the frequency band 4.6–5.6 GHz. This test is NLOS, so the propagation loss is expected to be greater than in free space, which in this case is 70 dB (mean), or 9 dB more than the LOS loss. The signal level relative to the mean varies from $+7$ dB to -27 dB, with the characteristic sharp Rayleigh fades particularly noticeable. This characteristic is expected from the theory presented in Section 2.2. The number of fades greater than 10 dB is 15, compared with the rough estimate given by (2.4) of 20 fades.

Because the received signal is delayed by the propagation delay between the antennas, the received signal will have a phase slope across the band. This effect is shown in Figure 2.14. From Appendix B, equation (B.1), the expected characteristic without multipath is a straight line, with the slope defining the propagation delay (distance). The actual measured phase characteristic shows some variation from the straight line, but is generally in agreement with expectations. By performing a LS fit, an estimate can be made of the phase slope, which in this case is equivalent to a 25 ns delay. As the actual delay is about 19 ns, the phase slope is not a very good method of estimating the propagation delay accurately.

From the theory in Section 2.2, the spectrum amplitude statistics in a scattering environment should exhibit Rayleigh statistics. The measured amplitude statistics extracted from the spectral characteristics in Figure 2.13 are shown in Figure 2.15, together with the Rayleigh distribution normalized to the mean signal amplitude. As can be observed, the match is quite good, signifying that theoretical Rayleigh statistics are a good match to measured NLOS data.

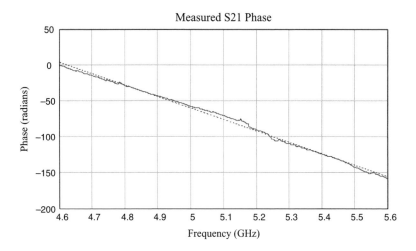

Figure 2.14   Measured phase frequency response for Test #4. The mean slope (dotted) is equivalent to a 25 ns delay. The LOS delay is 19 ns.

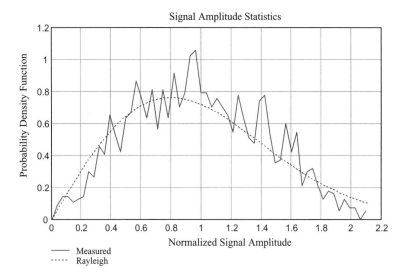

Figure 2.15   Computed signal amplitude statistics across the frequency band.

2.3.1.2 The Impulse Function

The impulse response function can be generated from the measured frequency response data presented in Section 2.3.1.1. The data presented are the amplitude response and the RAKE[5] receiver gain. A RAKE receiver is one that combines all the various delayed impulse responses in a coherent fashion, thus improving the receiver sensitivity. With such a receiver, the

---

[5] The term RAKE derives from the individual, delayed impulse responses being reminiscent of the prongs of a rake.

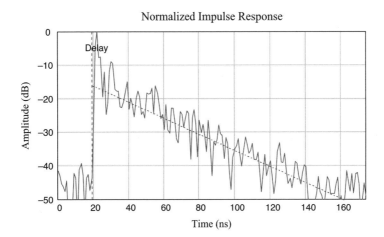

Figure 2.16 Computed impulse response. The LOS distance is 5.8 m, equivalent to 19.3 ns. The measured delay is in close agreement. The 'Delay' parameter is the LOS propagation delay. The linear decay slope (dotted line) is 0.24 dB/ns.

multipath signals can potentially be used to improve the performance of a receiver operating in free space.

The impulse response calculated from the spectral data in Figure 2.13 is shown plotted in Figure 2.16 with logarithmic amplitude. Note that, apart from the initial pulse associated with the first significant signal to arrive at the receiver, the scattered signals decrease linearly with the delay excess (or exponentially using a linear amplitude scale). The extent of the delays is large at about 150 ns at the limit of the measurements. As this delay is equivalent to a propagation path excess of 45 m, much greater than the size of the room shown in Figure 2.12, the received signal must have undergone multiple scattering, losing energy on each scattering incident.

Another useful measure of the signal scattering is the RAKE receiver gain, as shown in Figure 2.17. Each 'finger' of the rake is separated by 1 ns (the reciprocal of the bandwidth) and, hence, largely statistically independent. In this case the RAKE gain is less than 3 dB, which shows that the power in all the delayed scattered signals is less than that in the dominant path. However, from Table 2.1, summarizing the results of four tests, it can be observed that the RAKE receiver gain can be much greater, with the maximum value being 8 dB for one NLOS case.

2.3.1.3 Summary of Results

Although the measurement set is rather limited, the results from the four test measurements shown in Figure 2.12 are summarized in Table 2.1. The test results include two LOS and two NLOS, summarized as follows:

Test #1 Transmitter antenna is 30 cm from wall. Receiver antenna is about 5.4 m (18 ns) distant, approximately in the center of the room, at 1 m height. LOS propagation.
Test #2 Transmitter antenna is near centre of room. Antenna is about 3.2 m (10.6 ns) distant, approximately in the center of the room, at 1 m height. LOS propagation.
Test #3 Transmitter antenna is near wall on the floor behind a wooden cabinet. Antenna about 5.5 m (18.3 ns) distant, approximately in the center of the room, at 1 m height. No LOS.

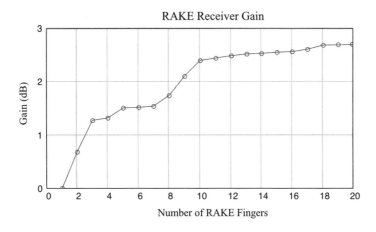

Figure 2.17 RAKE receiver gain characteristics.

Test #4 Transmitter antenna is near wall on the floor behind a metal cabinet. Antenna about 5.8 m (19.3 ns) distant, approximately in the center of the room, at 1 m height. No LOS.

Most of the parameters of the propagation have been described previously. The delay spread is a measure of how the impulse is spread relative to the peak signal power, usually at or near the leading edge. The mean delay is defined by

$$\mu_d = \frac{\sum_{n=1}^{N} a_n^2(\tau_n - \tau_0)}{\sum_{n=0}^{N} a_n^2} \tag{2.51}$$

and the delay spread is defined as

$$\tau_d = \sqrt{\left[\sum_{n=1}^{N} a_n^2(\tau_n - \tau_0)^2 / \sum_{n=0}^{N} a_n^2\right] - \mu_d^2} \tag{2.52}$$

Table 2.1 Summary of measurements made in the RF laboratory room.

Parameter	Test #1	Test #2	Test #3	Test #4
Type of propagation	LOS	LOS	NLOS	NLOS
LOS range (m)	5.4	3.2	5.5	5.8
LOS delay (ns)	18.0	10.6	18.7	19.3
LOS loss (dB)	61	57	61	62
Loss excess (mean loss minus LOS) (dB)	−0.6	−1.4	9.0	7.9
Power-delay slope (dB/ns)	−0.29	−0.28	−0.21	−0.27
Mean delay (ns)	10.8	8.9	16.3	6.9
Delay spread (ns)	15.0	13.2	19.1	13.2
RAKE receiver gain (dB)	5.2	4.8	7.9	2.8

where $n=0$ is the peak signal, a_n is the amplitude of the nth impulse signal and τ_n is the corresponding relative delay. The delay spread is an indication of how spread out the impulse response is, relative to the time of the signal peak. If the signal impulse *amplitude* decays exponentially with time constant τ_p, as in Figure 2.16 for example, then the mean delay and the delay spread are both given by $\tau_p/2$.

An alternative expression for the delay spread can be derived which interestingly links the delay excess and loss excess parameters. From (2.45), the loss excess parameter β (dB/m) defines the excess loss per unit distance. Alternatively, assuming an exponential model for the impulse response, the power loss per unit time is given by $10/\ln(10)\tau_d$, expressed in decibels per nanosecond. Thus, using the speed of propagation c (0.3 m/ns), these two measures can be equated, resulting in the expression to the delay spread:

$$\tau_d = \frac{10}{\ln(10)c\beta} = \frac{10}{\ln(10)c\alpha\eta} \qquad (2.53)$$

Thus, measurements of the loss excess can be used to estimate the delay spread, and vice versa. Further, as the loss excess parameter β is related to two other parameters (the obstacle density η and the loss per obstacle α), then, at least in principle, the delay spread can be estimated from parameters which are directly related to the architecture and construction materials of the building. For example, from Section 2.2.1, the loss excess parameter β is 0.77 dB/m; so, using (2.53), the estimated delay spread is 19 ns, which is close to the measured values given in Table 2.1.

2.4 Excess Delays in Radio Propagation

In position location systems, the main parameter that affects the positional accuracy most is the accuracy in measuring the TOF from the transmitter to the receiver. For an indoor system where the conditions are largely NLOS, the propagation delays will be in excess of the straight-line path delay assumed in the position determination process. The determination of the delay excess is more complex than the estimation of the corresponding loss excess discussed previously, even though the propagation mechanisms are the same. The reason for this is that the delay excess as measured results from a combination of the propagation conditions and the measurement technique. As a consequence, there is no one 'correct' delay excess for a given propagation path; rather, the delay excess as measured will depend on other factors, such as the signal bandwidth and the signal-to-noise ratio (SNR) at the output of the receiver. Indeed, as it can be argued that the indoor straight-line NLOS path will always be present, albeit severely attenuated, then, given sufficient receiver sensitivity and time resolution, the delay excess should approach zero with the appropriate measurement technique. However, these observations of theoretically near-zero delay excesses are of little value when considering real positioning systems with limited bandwidth and signal sensitivity.

Given that delay excesses are difficult to estimate theoretically, reliable data are generally only available from actual measurements using the techniques described in Section 2.3. The weakness of such an approach is that these measurements relate to particular buildings or building types; thus, a general approach for an arbitrary building is more difficult for delay excess than for loss excess, as described in Section 2.2. Nevertheless, some general principles can be established, with two characteristics being particularly important, namely the general characteristic that the delay excess tends to increase with range and that the accuracy improves with bandwidth. These two effects are considered in more detail in the following subsections. The general aim is to provide information that allows some mathematical models to be developed to estimate the range excess in practical situations.

2.4.1 Effect of Range

In an indoor environment, the propagation delay excess is a function of the propagation distance. The measured TOA will be based on the first significant signal detected above the receiver noise floor. As the direct path is often severely attenuated and the first significant signal results from a zigzag path from multiple scattering from obstacles, the measured delay excess will tend to increase with the propagation distance. Owing to the complexity of the indoor scattering environment, the delay excess data is usually based on measurement rather than theoretical analysis. For the following analysis and mathematical modeling we will use the results of the delay excesses using UWB transmissions in [9]. As noted previously, UWB will have superior ranging accuracy than the longer range but narrower band systems, so the measured data should represent an upper limit on the performance of all systems. As the performance is a function of the transmission bandwidth (as discussed in detail in Section 2.4.2), the results are initially presented based on 500 MHz bandwidth only, but later extended to other bandwidths. This 500 MHz bandwidth is typical of first-generation UWB systems, but much greater than the bandwidth available in the ISM bands.

The delay excess performance in a typical indoor environment is shown in Figure 2.18, plotted as a function of propagation distance. The data show that ranging errors tend to increase approximately linearly with distance, but with random errors approximately symmetrical about the trend line. Figure 2.18 also shows that the scatter tends to increase with distance, although this is not a dominant effect, and there also appears to be a slight positive bias to the

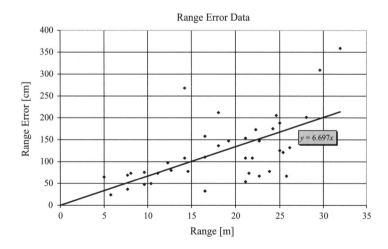

Figure 2.18   An example of the ranging error for an NLOS example with a 500 MHz bandwidth and an RF frequency of 5 GHz. The mean error varies from about 0.5 m at short range (5 m) to about 3.5 meters at 32 m. The trend line is the linear LS fit to the data, with zero error at zero range.

random scatter, with a few points with significantly greater delay excesses above the trend line. The modeling of these random variations can be based on various standard statistical distributions, including Gaussian, exponential, log-normal, and Weibull. However, to keep the analysis relatively simple, the random component will be modeled as a Gaussian random variable. Such a model underestimates the low-probability large ranging errors. Based on the above example and explanation, the ranging errors for NLOS propagation in buildings is proposed to be modeled by the expression

$$\Delta R = mR + N(0, \sigma) \tag{2.54}$$

where m is the slope of the linear bias error, R is the range and the Gaussian (or normal distribution N) has zero mean and a STD σ. The parameters in (2.54) can be related directly to the characteristics of the propagation environment. In summary, the description is as follows.

Slope parameter m   This parameter defines the dominant characteristics of the delay excess; namely, on average, the delay excess increases with distance (assumed to be linear). This effect is due to the scattering from obstacles near the straight-line path, which increases the delay due to the zigzag path. In a quasi-homogeneous environment, the number of obstacles along the path will be approximately proportional to the propagation distance; thus $\Delta R \propto R$. The constant of proportionality m can be related to the number of scatters per unit length N_s and the average extra path length per scatter δR_s, so that $m = N_s \delta R_s$. Thus, at least in principle, the parameter m can be related to the building characteristics. In practice, this parameter is difficult to determine without measurements, so that the parameter m in a model will be a random variable over an interval defined by typical values in the particular type of building.

Gaussian parameter σ   The total delay excess is the sum of the delay excesses from each scatter. Because of the variation in the scattering from each object, the total delay for each path between scatters will have a random component. Thus, an added random component needs to be included in the modeling of the delay excess, in addition to the component defined by the parameter m. While the statistics of each individual scatter are not known, the statistics of the sum of the delays will be approximately Gaussian due to the CLT, provided the number of scatters is not too small. Thus, to a first order of approximation, the dominant random component in the statistical variation would be expected to be Gaussian.[6] However, this model does not adequately describe the low-probability extraordinary delays associated with large objects and Rayleigh fading. In this simple model, these effects are ignored.

The statistical performance of the model and actual data from Figure 2.18 are shown in Figure 2.19. The model statistical data are obtained from a Monte Carlo simulation based on the model in (2.46). The ranges in the simulation were randomly chosen with uniform distribution between 5 m and 32 m, which approximately matches the ranges in the measured data. The LS fit of the cumulative distribution shows fair agreement over the span of the range error (0 to 4 m), with the RMS fit error to the cumulative distribution of 0.043. However, as noted above, the Gaussian model underestimates the probability of large errors, but also overestimates small

---

[6] A similar argument on the loss excess due to scattering predicts a statistical variation of the loss excess to be lognormal. This statistical distribution is confirmed by measurements.

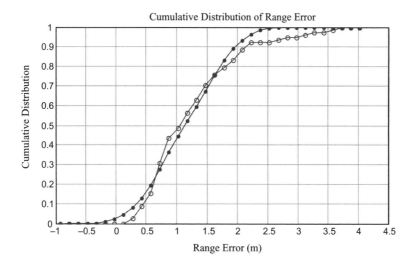

Figure 2.19 Cumulative distribution comparison between the measured data (circles) and the model data (dots) based on the parameters $m = 6.2$ cm/m and $\sigma = 0.34$ m for the Gaussian distribution. The RMS error in the cumulative distribution between the model and the measured data is 0.038.

range errors. Further, in practice, the model parameters for a particular operating environment will not be known; this uncertainty in the model parameters typically will be greater than those associated with the statistical nature of the random errors.

The characteristics of the delay excess are similar in other examples given in the referenced paper, but the parameters vary somewhat depending of the particular propagation environment. Table 2.2 summarizes the parameters associated with four cases of NLOS propagation.

In general, the parameters for a particular environment will not be known. In such cases, modeling can be based on selecting the parameters closest to one of those listed in Table 2.2, using the general description of the building type as a guide. Alternatively, the model can be generalized, so that the model parameters themselves are random variables. While the statistics

Table 2.2 Summary of some NLOS delay excess modeling data based on simulating the measured data in the cited reference.

Name	Slope parameter m (cm/m)	STD σ (m)	CDF[a] RMS error	Comment
Child Care	6.2	0.34	0.038	Plaster/wooden studs. Maximum of 7 walls in 33 m
NIST North	5.9	0.21	0.048	Sheet rock and aluminum studs. Maximum of 12 walls in 41 m
Sound	8.5	0.37	0.073	Cinder block. Maximum 7 of walls in 41 m
Plant	30.4	3.25	0.050	Steel construction. Very hostile multipath environment

[a] CDF: cumulative distribution function.

of these variables can be determined only by measurements in a large number of environments, such data are not available. Thus, as a fallback, the parameters could be selected based on the range of the parameters in Table 2.2. For example, the parameters could be assumed to have a uniform distribution between the upper and lower limits defined in the table.

Based on the above suggestions, the performance of a (500 MHz bandwidth) system could be estimated through simulations. Note that this would be an upper estimate in performance for systems with less bandwidth, so that lower bandwidth systems will have greater measurement errors. The effect of the signal bandwidth is considered in the next subsection.

2.4.2 Effect of Signal Bandwidth

As noted in the introduction to this section, the delay excess is a function of the signal bandwidth and the propagation environment. Thus, while the bandwidth effect is not strictly a radio propagation effect, it is appropriate to consider such effects as part of the propagation analysis. For more details on the effect of bandwidth on TOA, refer to Chapter 4.

The experiments described previously in Section 2.3.3 can be performed to determine the delay excess as a function of bandwidth. If the network analyzer bandwidth is set to a large value, then the recorded data can be processed with various bandwidths up to the maximum, using the one set of measurements. In the following example, the RF frequency is 7.5 GHz and the measurement bandwidth used is 3 GHz, which has a nominal resolution of 10 cm.[7] As indoor propagation effects are expected to result in ranging errors in excess of this value, the measurements are expected to provide the 'true' delay excesses. However, as the measurement bandwidth is reduced, the ranging accuracy will eventually be dominated by the measurement resolution rather than the propagation characteristics. In any practical system the design bandwidth should not be greater than that dictated by the errors induced by the propagation environment; thus, it is important to estimate the bandwidth required for a specified positional accuracy.

Table 2.3 shows the delay excess for different bandwidths, with the bandwidth reduced by factors of 2, 4, 8 and 16. The graphical representation is given in Figures 2.20 and 2.21.

The mean range error (shown in Figure 2.20) varies as the measurement resolution in a nonlinear fashion. The model of this behavior is based on a simple interpretation of the delay measurement process. The raw impulse response at high resolution is complex, but this complexity has two main features that are important in measuring the arrival time of the impulse. First, the first significant signal is delayed relative to the straight-line arrival time. This excess delay is a largely a function of the layout of the building and is essentially independent of frequency (in the UWB band). At sufficiently high measurement resolution, the first significant measured signal will be independent of the bandwidth; further, this delay will also be present in lower bandwidth measurements. The second significant characteristic of the impulse response is the observation that the peak signal is delayed (often significantly) relative to the first significant detectable signal. As the bandwidth reduces, the effective measured impulse response is the high-resolution impulse convolved by the measurement weighting function. Thus, at lower resolution, the impulse response shape is dominated by the peak of the signal rather than the smaller components preceding the peak. Thus, as the bandwidth is progressively decreased, the measured arrival time moves from the first significant signal delay to the delay

---

[7] The nominal resolution is defined as the reciprocal of the bandwidth and converted to the equivalent propagation distance using the speed of propagation (30 cm per nanosecond).

Table 2.3   Summary of the performance at different bandwidths.

Reduction factor	1	2	4	8	16
Time resolution (ns)	0.33	0.66	1.33	2.66	5.33
Distance resolution (cm)	10	20	40	80	160
Mean range error (cm)	37	49	63	97	115
STD in range error (cm)	17	26	31	94	106
Peak range error (cm)	76	111	137	470	545

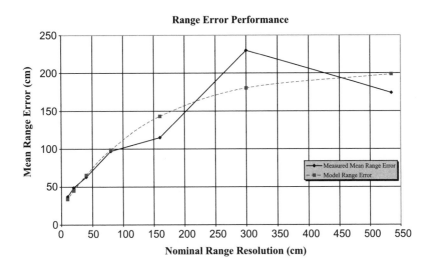

Figure 2.20   Measured and model data on the mean ranging errors.

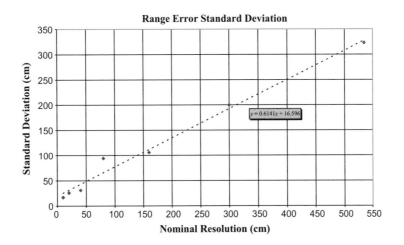

Figure 2.21   Measured and model data on the STD of ranging errors.

associated with the peak of the impulse response. As a consequence, the proposed model for the (mean) measured arrival time of the impulse is given by the sum of the minimum detectable signal delay plus a component which varies from the minimum measurable delay to the delay of the peak. If this variation is assumed to be exponential with the measurement resolution, then the model delay is given by

$$\tau = \tau_0 + \Delta\tau(1 - e^{-\rho/\rho_0}) \tag{2.55}$$

where τ_0 is the minimum detectable delay, $\Delta\tau$ is the delay from the first detectable signal to the signal peak, ρ is the measurement resolution and ρ_0 is the exponential parameter of the model. The model parameters can be estimated by fitting the model to the measured data defined in Table 2.3, resulting in the following estimated parameters:

τ_0 22 cm (0.73 ns)
$\Delta\tau$ 178 cm (5.93 ns)
ρ_0 182 cm (5.46 ns)

As can be observed from Figure 2.20, there is a reasonable fit between the model and the measured range errors (24 cm RMS error). Note that these estimated parameters are strictly only applicable to the particular measurements performed, but it is probably reasonable to suggest that these data would approximately apply to similar indoor operating environments. A consequence of the model is the prediction that at low bandwidths the measurement delay error will have a maximum, in this particular case 200 cm. This limiting delay will be reached when the resolution is much greater than the parameter ρ_0.

While the mean ranging error is an important parameter for indoor position location, the STD of the range error is the main parameter of interest for hyperbolic position determination; this concept is discusses in detail in Chapter 5. The results for the STD in the range excess are shown plotted in Figure 2.21, which shows that the STD in the range error is essentially a linear function of the resolution, and is given by the equation

$$\sigma_\tau(\text{cm}) = 17 + 0.61\rho \tag{2.56}$$

The characteristics of this model are quite different from the corresponding mean delay model, as the STD continues to increase (within the limits of the available data) as the nominal resolution becomes larger (smaller bandwidths). At bandwidths greater than about 200 MHz the STD is somewhat less than the mean, but the STD at smaller bandwidths rapidly becomes much greater than the mean delay. As the STD is the important parameter in hyperbolic navigation systems, a bandwidth of the order of 250 MHz is required to achieve positional accuracy of the order of 1–2 m. For a system with a bandwidth of 60 MHz, the predicted positional accuracy is of the order of 3–5 m.

The above models of the effect of signal bandwidth can be used for predictions. Table 2.4 shows some examples of how the (model) delay error varies with the signal bandwidth. To achieve the delay error of the order of 1–2 m, the signal bandwidth must be of the order of 250 MHz. Alternatively, increasing the bandwidth from 1 GHz to 8 GHz (UWB bandwidths) results in little further improvement. Very low bandwidths of the order of 10 MHz are included in the table, although the model has not been validated below 60 MHz; thus, this prediction should be used with caution.

Table 2.4 Estimated mean and STD of the delay error as a function of bandwidth.

Bandwidth (MHz)	Resolution ρ (cm)	Mean delay error (cm)	STD delay error (cm)
10	3000	203	1850
50	600	200	385
250	120	120	90
1000	30	56	35
8000	3.75	26	19

The conclusion from these observations is that, for practical indoor position systems using TOA data for position determination, a minimum bandwidth of 250 MHz is required to achieve 1–2 m accuracy. The accuracy slowly improves at greater bandwidths, but the residual errors even at high bandwidths are of the order of 20 cm. Thus, bandwidths beyond 1 GHz have little benefit in improving positional accuracy in situations where NLOS propagation conditions predominate.

2.5 Antenna Effects

The performance of a radio system depends on many factors, including the propagation environment and the performance of the radio transmitter and the receiver. One of the often neglected components of the radio transmitters and receivers, particularly in small cheap components as found in WSNs and ad hoc networks, is the antennas. In the analysis of the performance of the system, it is often assumed that the antenna performance is as defined in isolation in free space, whereas in reality the antennas typically interact closely with the local environment, often with detrimental results on the overall system performance.

2.5.1 Antennas on the Human Body

One common application for positioning systems is the tracking of people, and in such cases the mobile tag is typically mounted on the body, say on a belt around the waist. In such cases it is clear that the antenna performance will be severely affected by the presence of the human body. This effect is in two areas. The first is the close proximity of the body will affect the matching of the antenna, thus reducing the effective gain of the antenna. However, the dominant effect is the loss of signal due to diffraction around the human body. As the typical frequencies of operation are in the range 2–6 GHz, the wavelength is much smaller than the dimensions of the trunk of a human and, thus, the diffraction losses can be significant. Thus, in computing the performance of a system, an allowance must be made for these diffraction losses.

The typical performance of a system operating at 2.4 GHz is shown in Figure 2.22. These tests were performed using a radio system designed to monitor the health of elderly people in their home [11]. In this test, the mobile unit was located on the chest and hip, with the antenna beam pointed at the Home Station for 0° of rotation. When the angle of rotation is 180°, the body is directly blocking the signal path to the Home Station. The antenna in the mobile unit is a small centimeter-sized antenna typically used in Bluetooth applications. Because of the small size of the antenna, the radiated signal is not very polarized, so the orientation of the antenna is not very important. The small size also affects the antenna gain, so that these small antennas are typically quite lossy, with peak gain of around −3 dBi. Detuning due to the nearby body is not very

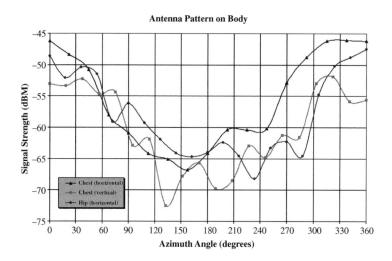

Figure 2.22 Effect of rotating the body with the mobile unit at three different locations on the body. The angle of 0° corresponds to a direction pointing towards the Home Station.

significant, as the matching is dominated by an internal groundplane. As the body width is of the order of three wavelengths, considerable diffraction losses can be expected. This loss can be specified as a front-to-back ratio. From the measured data in Figure 2.22, the front-to-back ratio averaged over the three locations is 20 ± 1 dB. Thus, the effect of diffraction around the human body is considerable at 2.4 GHz and can be expected to be even larger at high frequencies. The effect on the performance of an indoor positioning system is thus considerable. For optimum performance (accuracy) the base stations of a positioning system should 'surround' the mobile device. However, with the signal diffraction losses in some directions, the effective range to some nearby base stations will be severely restricted. For example, suppose the system is designed for a nominal range of 50 m. Then, from Figure 2.6, and assuming a loss of 20 dB, the range-loss data indicate that the range will be reduced to about 20 m. As a consequence, the density of base stations will have to be increased to compensate for the reduced range.

2.5.2 *Effect of Ground Reflections*

The effect of the environment on antenna performance described earlier is related to the close interaction of the antenna with the structure on which it is mounted. However, another effect on performance related to antennas is associated with reflections from large nearby structures which in effect create an 'image' antenna; this image antenna can be considered as an additional radiation source which interferes with the true antenna radiation, resulting in loss of performance. While such reflective interference can occur from building structures, the most common source of interference is from ground reflections. This effect is briefly analyzed and an example of measured performance in a sporting application (rowing) is presented. However, the same problem is often present in many other applications, such as in WSNs with nodes deployed at ground level or on other large, flat surfaces.

To illustrate the impact of ground reflections on the performance of a radio positioning system, consider the application of tracking rowing boats. In this application there are no obstacles preventing LOS propagation, but the water surface represents a good reflector of radio waves. Further, the mobile device will be located very close to the water surface, typically at a height of only about 25 cm.[8] While this height is several wavelengths above the water and, thus, does not directly affect the radiating performance of the antenna, the very low height in relation to the propagation distances greatly reduces the combined received signals at the remote base stations around the rowing course. With such shallow reflection angles, the reflection coefficient for radio waves of all polarizations will be close to -1. In this case it can be shown [1,12] that the extra loss over the free-space loss in signal strength is given by

$$\Delta L(\text{dB}) = 20 \log \left(\frac{\lambda r}{4 \pi h_1 h_2} \right) \tag{2.57}$$

where h_1 and h_2 are the heights of the base station and mobile antennas, r is the range and λ is the wavelength. Note that the loss increases as the height decreases, so that in the rowing application with the mobile close to the water the extra loss will be large. With typical deployment the base station heights are about 5 m above water level, so applying (2.57) at a range of 2000 m[9] and a frequency of 2.4 GHz shows that the extra loss is 24 dB. The free-space loss at that range is 106 dB, so that the total loss is 130 dB. The total loss is given by

$$L(\text{dB}) = 20 \log \left(\frac{r^2}{h_1 h_2} \right) \tag{2.58}$$

From (2.58) it can be observed that the loss increases by 40 dB per decade in range, compared with 20 dB per decade for free-space propagation. Thus, the effect of the ground reflections is to reduce greatly the effective range of the propagation from mobile nodes, even though the radiated power remains unaffected by the nearby ground reflections.

The measured loss from a mobile node to a base station is shown in Figure 2.23 for a 2000 m race, with the base station being at about the 1000 m point (or midway). The mobile is mounted on a plastic lane number plate on the bow of the boat at a height of about 25 cm above the water and, thus, is relatively free from blockage from the boat and the crew, except when the signal path is directly along the axis of the boat towards the rear. The measured loss at the start is 110 dB (no blockage) and at the finish 115 dB, where some blockage of signal along the path to the base station can be expected. The simple two-ray model above predicts a loss of 118 dB at a range of 1000 m, while the free-space loss is 100 dB, so the actual losses are closest to the two-ray model predictions. The base station is located at about the 1000 m point, so for the first half of the race the boat is approaching the base station, while in the second half the boat is receding from the base station. Note that the loss is greater when receding, as the radio path includes diffraction losses around the rowers. In this case the diffraction losses are about 5 dB. The smooth curve in the figure is the free-space loss, which is in good agreement with the measurements at short range.

The above rowing example shows the importance of considering the effect of ground reflections in estimating the propagation range when nodes are deployed close to the ground.

---

[8] The best location for attaching the mobile device is at the bow on the plate for attaching the number used for identification at rowing regattas. This location also allows direct measurement of the positions of boats along the rowing course, as the tip of the bow is used to determine race order.
[9] The standard course length for Olympic-type rowing is 2000 m.

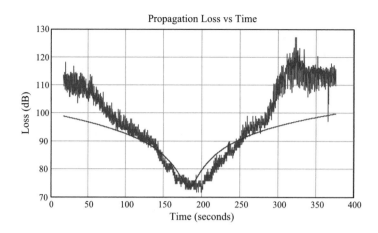

Figure 2.23 Propagation loss to base station as a function of time during a 2000 m race. The smooth curve is the free-space loss. The boat speed is about 5.5 m/s, from which distances can be estimated.

For example, nodes in WSNs and ad hoc networks often can be located at ground level or placed on top of large, flat objects. In such cases it is important that the antenna is mounted normal to the planar surface, similar to that described for the mounting of antennas on the roof of vehicles. However, it is common in small self-contained nodes to incorporate the antenna internally, with the beam directed normal to the plane of the flat, thin node enclosure. In such cases, when the node is placed on a flat surface, the groundplane effects described above become operative, resulting in a severe loss in range performance. Thus, when deploying WSNs and ad hoc network nodes, careful consideration needs to be taken regarding the type and orientation of the antenna.

References

[1] H.L. Bertoni, *Radio Propagation for Modern Wireless Systems*, Prentice Hall PTR, 2000.
[2] J.D. Parsons and J. David, *The Mobile Radio Propagation Channel*, John Wiley & Sons, Ltd, 2002.
[3] T.S. Rappaport, *Wireless Communications: Principles and Practice*, Prentice Hall, 2002.
[4] D.A. McNamara, C.W. Pistorius and J.A. Malherbe, *Introduction to the Uniform Geometrical Theory of Diffraction*, Artech House, 1990.
[5] G. Roussas, *A Course in Mathematical Statistics*, second edition, Academic Press, 1997.
[6] S.O. Rice, 'Mathematical analysis of random noise', *Bell System Technical Journal* **24**, 1945, 46–156.
[7] N. Alsindi, B. Alavi and K. Pahlavan, 'Measurement and modeling of ultrawideband TOA-based ranging in indoor multipath environments', *IEEE Transactions on Vehicular Technology*, **58**, 2009, 1046–1058.
[8] B. Alavi and K. Pahlavan, 'Modeling of the TOA-based distance measurements error using UWB indoor radio measurements', *IEEE Communications Letters*, **10**(4), 2006, 275–277.
[9] C. Gentile and A. Kik, 'An evaluation of ultra wideband technology for indoor ranging', in *Proceedings of IEEE Globecom*, pp. 1–6, 2006.
[10] S.Y. Seidel and T.S. Rappaport, '914 MHz path loss prediction models for indoor wireless communications in multifloored buildings', *IEEE Transactions on Antenna and Propagations*, **40**(2), 1992, 207–217.
[11] L.S. Wilson, P. Ho, K.J. Bengston, M.J. Dadd, C.F. Chen, C. Huynh and R.W. Gill, 'The CSIRO hospital without walls home telecare system', in *Proceedings of the Seventh Australian and New Zealand Intelligent Information Systems Conference*, pp. 43–46, 2001.
[12] E.C. Jordan and K.G. Balmain, *Electromagnetic Waves and Radiating Systems*, Prentice Hall, 1968.

3

Signal Detection by Correlation

A basic requirement of a wireless location system is to estimate the TOA of a 'pulse' of radio signal at the receiver. In theory, many techniques are possible, including UWB pulse detection, correlation techniques using spread-spectrum transmissions and the processing of orthogonal frequency division modulation (OFDM) signals to estimate the impulse response of the environment from the frequency response. In all such techniques, the accuracy of the estimation of the TOA is directly related to the bandwidth of the RF signal; thus, for systems which require high positional accuracy, a bandwidth of 100 MHz or greater is desirable, particularly for indoor applications, where severe multipath propagation can cause large timing measurement errors. Although large bandwidths are available in ISM bands at RF frequencies greater than 10 GHz, high propagation losses and the cost of implementing radios (particularly at the transmitters) tend to limit the practical implementation to lower frequencies. Conversely, while the radio propagation characteristics at 900 MHz are favorable, the available bandwidths are less than 10 MHz, thus limiting the accuracy of the system. Therefore, from a practical stand point, the only suitable spectrum is in the 2.4 GHz and 5.8 GHz ISM bands, where bandwidths of up to 80 MHz and 150 MHz respectively are available. However, the operation in these bands is limited to spread-spectrum modulation, either frequency-hopping or direct-sequence. For radiolocation systems, the latter modulation technique is particularly useful, as it simultaneously provides a direct method of estimation of the TOA and provides protection from other uses of these unlicensed radio bands.

In this chapter, an overview of the principles of the signal processing used in position location systems is provided, with particular emphasis on detection of direct-sequence spread-spectrum modulated signals. In Section 3.1 the characteristics of digitally generated transmitter signals are briefly reviewed. Section 3.2 provides an overview of the receiver signal processing associated with despreading of a spread-spectrum signal. The topics covered include correlators, accumulators, interference performance and the concept of process gain.

3.1 Transmitter Signal

Although the transmitter signal is analog, typical modern implementation is based on digitally generated signals. Thus, we consider a system that generates the spread-spectrum

Ground-Based Wireless Positioning   Kegen Yu, Ian Sharp and Jay Guo
© 2009 John Wiley & Sons, Ltd

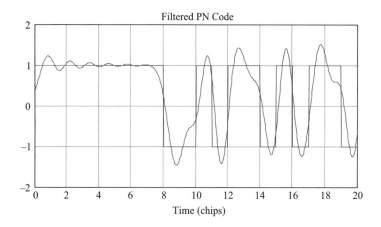

Figure 3.1 Ideal (square) and filtered pn-code. The filter 3 dB cutoff is at $0.875/T_c$ and is 33 dB at $1/T_c$.

signal digitally. Also, it is assumed that the bipolar analog pseudo-random noise (pn)-code is filtered so that the spectrum is confined to the main lobe within $\pm 1/T_c$, where T_c is the chip period of the pn-code. This filtering is required because the radio spectrum is limited by the spectral rules defined by national regulatory bodies such as the FCC in the USA. The filtering of the binary pn-code can be effectively performed by a fast Fourier transform (FFT) with oversampling of the binary signal and digitally truncating the spectrum at the required bandwidth. The filtered time-domain signal can then be reconstructed by performing an inverse FFT and taking the real part. The resulting pn-code in the time domain will have ripples, as shown in Figure 3.1.

The analog signal illustrated in Figure 3.1 is bandlimited to $\pm 1/T_c$ and, thus, can be represented digitally without error if sampled at the Nyquist rate, or $2/T_c$ (or greater). The transmitter signal is thus represented by signal samples which can be converted to the required analog baseband signal by a digital-to-analog converter. As the output signal from the converter is constant between samples when a non-return-to-zero converter is used, the reconstructed signal is not an accurate representation of the original time-domain signal. In particular, the reconstructed signal can be represented mathematically as

$$s_{tx}(t) = [s(t)\mathrm{comb}_T(t)] * \Pi t_T(t) \qquad (3.1)$$

where T is the sample period, the 'comb' function is an infinite sequence of delta functions separated by time T and the rectangular function $\Pi_T(t)$ is unity for $|t| \leq T/2$ and zero otherwise. This expression can be understood as follows. The analog signal $s(t)$ is sampled with a period T, and this is represented by the term in the bracket. However, the effect of the converter is to hold the output constant between samples, and this is represented mathematically by the convolution with the rectangular function $\Pi_T(t)$. The spectrum of this signal is given by

$$S_{tx}(f) = [S(f) * \mathrm{comb}_{1/T}(f)]\mathrm{sinc}(fT) \qquad (3.2)$$

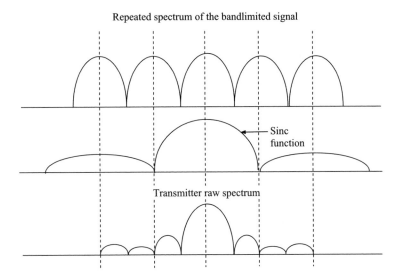

Repeated spectrum of the bandlimited signal

Sinc function

Transmitter raw spectrum

Figure 3.2 Diagram showing the form of the transmitted signal spectrum.

Figure 3.2 shows the mathematical operations implied by (3.2) in diagrammatic form. Since the transmitter spectrum is bandlimited to $\pm 1/2T$, the convolution process does not involve any spectral overlapping. Thus, the spectrum can be expressed as

$$S_{tx}(f) = \sum_{n=-\infty}^{\infty} S(f - n/T)\operatorname{sinc}(fT) \tag{3.3}$$

Equation (3.3) shows that the transmitted signal has a spectrum which has multiple copies of the spectrum of the original signal, but multiplied by a (low-pass) filter with a sinc distribution. Thus, the central part of the spectrum in the range of $f = \pm 1/2T$ approximates the required spectrum, but frequencies outside this range represent unwanted distortion. However, these outlying spectral components can be removed by an appropriate low-pass filter, with only minimal distortion to the in-band signal.

The distortion caused by the sinc filter as defined in (3.3) can be eliminated by pre-emphasis of the digital signal. Thus, the spectrum of the digital signal $S_{tx}(f)$ is defined by

$$S_{tx}(f) = \frac{S(f)}{\operatorname{sinc}(ft)} \qquad (-1/2T \leq f \leq 1/2T) \tag{3.4}$$

where $S(f)$ is the digitally truncated spectrum associated with the signal in Figure 3.1. Thus, with a combination of a low-pass filter and spectral pre-emphasis, essentially the exact desired bandlimited signal can be reconstructed from the digitally sampled signal with two samples per chip. Finally, this baseband signal is used to modulate the RF carrier, resulting in the required RF spread-spectrum signal.

(Optional)

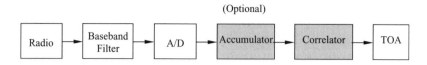

Figure 3.3   Generic block diagram in a modern radio receiver, with analog components at the front end and digital components at the back end.

3.2 Receiver Signal Processing

This section provides an analysis of the performance of correlation-based receivers, with particular emphasis on digital signal processing. A generic block diagram of the main functional components of a modern digital receiver is shown in Figure 3.3, with the components analyzed in this chapter shown in gray.

The radio system receives the RF signal and translates the signal to baseband. In a typical implementation, the RF signal is first amplified and converted to an intermediate frequency (IF). As the local oscillator phase is not synchronized to the phase of the RF signal, the baseband output is typically in the form of in-phase and quadrature components, so the output has two components to be filtered and then converted to digital signals. To ensure no aliasing in the conversion process, the baseband signal is filtered to a bandwidth less than twice the sampling rate. The optional accumulator provides additional improvement in the input SNR for the correlator. The correlator processes the complex signal to detect and track the spread-spectrum signal, which is fed to the TOA detection subsystem.

The analysis of the performance of the signal processing in direct-sequence correlator receivers is based on fundamental theorems of mathematical statistics. The form of the input signal consists of a known (pseudo-random) code plus random noise which may have more power than the signal itself. The key to processing such a signal is the correlation properties of the pseudo-random code when compared with the truly random noise. In most modern implementations, the analog input signal is low-pass filtered and sampled, so the actual data are discrete samples which can be analyzed by classical statistical methods. In particular, the signal and the random noise can be described in terms of the signal mean and the signal variance.

A key characteristic of the assumed random noise samples is that they are statistically independent. As the radio receiver baseband output is low-pass filtered to a bandwidth B, the signal must be sampled at least twice this rate to avoid aliasing effects. However, the filtering also has an effect on the noise, so the correlation coefficient of the noise is approximately given by

$$\rho(\tau) = \text{sinc}(2B\tau) \tag{3.5}$$

where the low-pass filter characteristics are assumed to be an ideal rectangular function. In this case, when the time separation τ is equal to $1/2B$, the correlation is zero, so the samples are uncorrelated. As the sample rate is $2B$, the samples are uncorrelated, as required. Figure 3.4 shows the autocorrelation function of a receiver with low-pass filters compared with the ideal function given by (3.5). The case shown is for a filter with a 3 dB cutoff frequency of $B = 35$ MHz and attenuation of 22 dB at 40 MHz. This filter is designed for a 40 Mchips/s spread-spectrum signal. As can be observed, the ideal and actual functions are essentially identical.

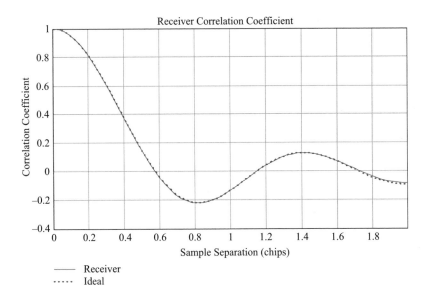

Figure 3.4 Computed actual and ideal correlation coefficient for random noise with typical low-pass filtering in spread-spectrum radios.

Having established that the noise in the sampled data is almost statistically independent, the statistical properties of the sum of these samples can be derived. To this end, two theorems are important, namely:

1. The mean of the sum of statistically independent random variables is equal to the sum of the means.
2. The variance of the sum of statistically independent random variables is equal to the sum of the variances.

In the case where all N random samples have the same mean μ and variance σ^2, the result can be expressed as

$$\mu_\Sigma = N\mu$$
$$\sigma_\Sigma^2 = N\sigma^2 \tag{3.6}$$

The key consequence of these statistical relationships is that defining the SNR as the mean signal squared to the signal variance, the output SNR is related to the input SNR by

$$\gamma_{\text{out}} = \frac{\mu_\Sigma^2}{\sigma_\Sigma^2} = \frac{(N\mu)^2}{N\sigma^2} = \frac{N\mu^2}{\sigma^2} = N\gamma_{\text{in}} \tag{3.7}$$

Thus, by accumulating statistically independent random samples, the SNR is improved by a factor equal to the number of samples summed. This principle forms the basis for the performance analysis of the correlation receivers; further performance details are given in the following subsections.

3.2.1 *Performance of Ideal Correlator*

Before considering the details of correlation receivers, some signal-processing theory appropriate to receiver performance analysis is introduced first. Although the following presentation is tutorial in nature, the general results will be applied later to the performance analysis of actual receivers.

The basic definition of correlation is given by

$$R(\tau) = \frac{1}{T} \int_T x(t) y(t + \tau) dt \qquad (3.8)$$

In (3.8), two time-domain signals $x(t)$ and $y(t)$ are correlated over a time interval T. A closely related concept to correlation is convolution. The convolution integral plays an important role in signal spectral analysis, so the relationship between these two concepts is useful in signal-processing analysis. Convolution is defined as

$$C(\tau) = \int_{-\infty}^{\infty} x(t) y(\tau - t) dt \qquad (3.9)$$

The importance of the convolution integral is the relationship between multiplying in the time domain and convolution in the frequency domain. Thus, the transform pairs are defined as

$$x(t)y(t) \Leftrightarrow X(f) * Y(f) \qquad (3.10)$$

where '$*$' represents the convolution operation in (3.10). The relationship between the convolution and the correlation operation can be seen more clearly by substituting $t' = -t$ into (3.8). Then, the correlation integral is

$$R'(\tau) = \frac{1}{T} \int_T -x(-t') y(\tau - t') dt' \qquad (3.11)$$

Apart from the $1/T$ scaling, (3.11) can be recognized as the convolution of $-x(-t)$ and $y(t)$. The Fourier transform of $-x(-t)$ can be directly calculated from the integral definition of the Fourier transform, namely

$$\mathcal{F}[-x(-t)] = \int_{-\infty}^{\infty} -x(-t') e^{-j2\pi ft'} dt' = \int_{-\infty}^{\infty} x(t) e^{+j2\pi ft} dt = X^*(f) \qquad (3.12)$$

Thus, from Equations (3.10) and (3.12), the Fourier transform of the correlation operation defined in (3.8) results in the transform pair

$$R(\tau) \Leftrightarrow \frac{1}{T} [X^*(f) Y(f)] \qquad (3.13)$$

where the superscript asterisk represents the complex conjugate. Equation (3.13) provides a convenient method of computing the correlation function in the frequency domain. If FFT is used, then it can be shown that the amount of computations required in the time-domain analysis is significantly less than the equivalent time-domain processing [1].

To decode the spread-spectrum signal, the received signal must be correlated with the pn-code known by the receiver. The decoding process essentially requires the computation of the autocorrelation function of the pn-code. For the autocorrelation function, one puts $y(t) = x(t)$; so, (3.13) yields

$$\mathcal{F}[R(\tau)] = \Phi(f) = \frac{X^*(f)X(f)}{T} = \frac{|X(f)|^2}{T} \tag{3.14}$$

Thus, the spectrum of the autocorrelation of a signal is identical to the power spectrum of the signal, normalized by the integration period. This identity can be used to determine the output spectrum without the complications of performing the correlation processing. As will be shown in later sections, computations in the frequency domain often result in more convenient analysis of the correlator performance.

The output power of the correlator receiver can be determined by integrating over all frequencies. From (3.14), one obtains

$$P_{\text{out}} = \int_{-\infty}^{\infty} \Phi(f)df = \frac{1}{T} \int_{-\infty}^{\infty} |X(f)|^2 \, df \tag{3.15}$$

The input signal power is given by

$$P_{\text{in}} = \frac{1}{T} \int_{T} x^2(t)dt = \frac{1}{T} \int_{-\infty}^{\infty} |X(f)|^2 \, df \tag{3.16}$$

Thus, if the correlator output power is normalized by the input signal power, the correlator output signal is always unity, independent of the input signal $x(t)$.

3.2.2 Correlator Performance with Sampled Data

The analysis in Section 3.2.1 is based on analog signals for the correlation process, but actual systems are based on filtered sampled signals. To analyze the performance of such systems, a different approach is required.

A technique for the generation of an arbitrary pn-code is based on passing a random series of delta functions through a filter with the appropriate spectral characteristics, as shown in Figure 3.5. The input random function $r(t)$ has evenly spaced delta functions of amplitude A_n at the chip period T_c, and can be expressed as

$$r(t) = \sum_{n=0}^{N-1} A_n \delta(t - nT_c) \tag{3.17}$$

If the signal represented by (3.17) is passed through a filter with an impulse response $p(t)$, then the output signal $s(t)$ is given by the convolution of the input signal and the impulse

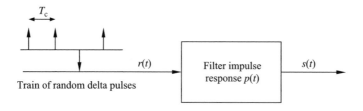

Figure 3.5 Generation of filtered pn-code using a train of random delta functions and a filter.

response $p(t)$ of the filter:

$$
\begin{aligned}
s(t) &= r(t) * p(t) \\
\Phi_s(f) &= \Phi_r(f)|P(f)|^2
\end{aligned}
\tag{3.18}
$$

The power spectral function of the random sequence $r(t)$ can be determined by taking the Fourier transform of the autocorrelation function $R(\tau)$. For the case of $\tau = 0$, the delta functions all align, so that $R(0) = N$. For all offsets that are not an integral multiple of the chip period T_c, the delta functions will not align, so the autocorrelation function will be zero. When $\tau = nT_c$, the delta function amplitude will be the summation of random variables $-A_n$ and $+A_n$ with equal probability. As $N \to \infty$, the expected value of summation will be zero, with a variance $A^2 N$. Thus, the autocorrelation function of $r(t)$ and the associated power spectrum are given by

$$
\begin{aligned}
R_r(\tau) &= \frac{A^2 N \delta(\tau)}{T} = \frac{A^2 \delta(\tau)}{T_c} \\
\Phi_r(f) &= \frac{A^2}{T_c}
\end{aligned}
\tag{3.19}
$$

Thus, the power spectrum of the filtered pn-code is obtained by substituting (3.19) into (3.18), resulting in

$$
\Phi_s(f) = \frac{A^2|P(f)|^2}{T_c}
\tag{3.20}
$$

This expression is true for any shaping filter; the following subsections apply this equation to various filter types.

3.2.2.1 Correlation with Ideal Pn-Code

While (3.20) is generally applicable to any pulse shape, it is instructive initially to analyze the unfiltered case, namely with a rectangular pulse of period T_c. Applying (3.20) with a unit amplitude signal ($A = 1$) to the pulse $p(t) = \Pi_T(t)$ gives

$$
\begin{aligned}
\Phi_{s0}(f) &= T_c \, \text{sinc}^2 f T_c \\
P_{s0} &= \int_{-\infty}^{\infty} \Phi_{s0}(f) df = 1
\end{aligned}
\tag{3.21}
$$

Thus, as expected the signal power for a binary pn-code is unity.

Of particular interest for the TOA determination is the shape of the autocorrelation function, which is obtained by taking the inverse Fourier transform of the autocorrelation function in (3.21), resulting in

$$R_{s0}(\tau) = \mathcal{F}^{-1}[\Phi_{s0}(f)] = \Lambda_{T_c}(t) \tag{3.22}$$

where the function $\Lambda_T(t)$ is of triangular shape centered at $t = 0$, of amplitude unity and width $\pm T$. If the received signal is in the form $s(t) = Apn(t)$, where A is the complex signal amplitude and the local pn-code relative to this signal is $pn(t - t_{rx})$ then the cross-correlation output will be $A\Lambda_T(t - t_{rx})$. Thus, the cross-correlation is in the form of a triangular pulse whose position in time can be used to determine the phase of the local pn-code relative to the pn-code in a received signal.

Now consider the output signal from the correlator, as shown in Figure 3.6a. The correlator can be replaced by the equivalent matched filter [2], as shown in Figure 3.5b, where the integration time is $T = NT_c$.

Now consider the effect of the above signal processing on random noise. The random noise is assumed to have a constant power spectral density of η. The radio receiver bandwidth B_{radio} is approximately equal to the reciprocal of the chip rate, so the output noise (and input to the correlator) is given by

$$N_{radio} = \eta \int_{-\infty}^{\infty} |H(f)|^2 \, df \approx 2\eta B_{radio} \tag{3.23}$$

Note that the integral includes the negative frequencies, so the effective bandwidth is twice the baseband bandwidth.

The correlator can be represented by the matched filter equivalent, so the correlator output noise is given by

$$N_{corr} = \eta \int_{-\infty}^{\infty} |H(f)|^2 \, df = \eta \int_{-\infty}^{\infty} \text{sinc}^2 fT \, df = \frac{\eta}{T} = \frac{\eta}{NT_c} \tag{3.24}$$

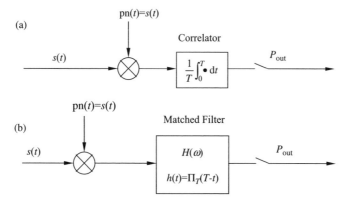

Figure 3.6 (a) Block diagram of the correlator and the associated signals. (b) Matched filter equivalent to the integrator in (a). In the figure the switch is open until the integration is complete, when it closes. This output is latched using a sample-and-hold circuit (not shown).

As the signal power from (3.21) is unity, the input and output SNRs are given by

$$\gamma_{in} = \frac{1}{N_{radio}} \approx \frac{1}{2\eta B_{radio}}$$

$$\gamma_{out} = \frac{1}{N_{corr}} = \frac{NT_c}{\eta} \tag{3.25}$$

$$\frac{\gamma_{out}}{\gamma_{in}} = (2B_{radio}T_c)N = A_{rx}N$$

Thus, the processing by the correlator has resulted in an improvement in the SNR by a factor equal to the number of chips in the pn-code multiplied by a factor (A_{rx}) which is related to the receiver bandwidth and the chip period. This improvement factor is referred to as the 'process gain'. Normally, the baseband bandwidth is set to $1/T_c$, then the process gain becomes $2N$. For accurate tracking of the signal epoch, the receiver gain factor A_{rx} should be of the order of 2. If accurate tracking is not required, then the baseband bandwidth can be reduced to about half the chip rate, and the process gain is equal to the number of chips in the pn-code. However, the minimum practical limit for A_{rx} is about unity, as the correlation properties degrade rapidly for lower values.

3.2.2.2 Correlation with Filtered Pn-code

The characteristics of the ideal pn-code, with the binary signal being ± 1 with sharp rectangular transitions between the binary values, results in a wideband spectrum, as given by (3.21). For real systems, the signal spectrum must be limited to a finite bandwidth. This filtering of the ideal pn-code results in distortions in the corresponding autocorrelation function and, hence, the output of the receiver correlator. As the time resolution in aligning the local code to that of the received signal is directly related to the chip rate, which in turn defines the bandwidth, the chip rate should be kept as high as possible in the design of a real system. A reasonable compromise between the requirement to bandlimit the signal and simultaneously to have as high a chip rate as possible is to restrict the signal to the frequency band $[-1/T_c, 1/T_c]$. The effect of this bandlimiting on the shape of the autocorrelation function is analyzed in the following.

The bandlimited pulse shape $p'(t)$ can be calculated by truncating the spectrum of the ideal pn-code and then taking the inverse Fourier transform. The truncated spectrum is given by

$$P'(f) = \Pi_{2/T_c}[T_c \, \text{sinc}(T_c f)] \tag{3.26}$$

and the corresponding pulse normalized by the chip period is given by

$$p'(\hat{t}) = 2 \int_{-1/2}^{1/2} \text{sinc}[2(\hat{t} - x)] \, dx \qquad (\hat{t} = t/T_c) \tag{3.27}$$

The bandlimited pulse shape $p'(\hat{t})$, together with the rectangular wideband pulse, is shown in Figure 3.7a. The bandlimiting results in the pulse spreading beyond the nominal width of $\pm T_c/2$, and the amplitude is increased. Note that the time integral of the normalized pulse remains unity. The corresponding autocorrelation function is shown in Figure 3.7b, together

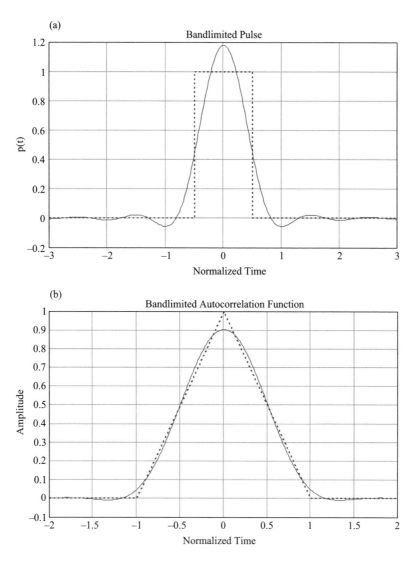

Figure 3.7 (a) Normalized pulse shape with the spectrum limited to $\pm 1/T_c$. (b) Autocorrelation function with the spectrum limited to $\pm 1/T_c$.

with the nominal triangular function. It is seen that bandlimiting reduces the amplitude to 0.9 and the sharp corners of the triangular pulse are smoothed out. The bandlimited pulse also has small-amplitude sidelobes before and after the main pulse. However, the general shape remains unaltered, with approximately the same linearly rising and falling edges. It is important to note that the detection of the position of the pulse peak becomes more difficult due to the flat pulse shape of the autocorrelation function near the peak.

Further restricting the bandwidth results in more severe distortions in the autocorrelation function. These effects and the performance of various detection methods for determining the timing of the received signal relative to a local clock are considered in detail in Chapter 5.

3.2.3 Performance of Accumulators

A correlator is the main method of detecting a direct-sequence spread-spectrum signal by enhancing the output SNR. However, an auxiliary method of also improving the SNR is by the use of an accumulator in conjunction with the correlator. As the process gain is essentially a function of the time of signal integration, increasing the process gain using a correlator alone requires increasing the length of the pn-code. However, such an approach has consequences which may be undesirable.

The main disadvantage of increasing the code length is the extra signal processing required to decode the signal. Even when using efficient techniques such as the FFTs, the processing complexity to acquire the signal is of the order of $N \log_2(N)$, where N is the code length, while correlation after acquisition is proportional to N for each point in the correlogram (a minimum of at least two points is required). Further, systems which use the pn-code for both data transmission and position determination place further restrictions on the length of the pn-code. If the pn-code length is increased to improve the output SNR for position determination, then the data rate of the data system is correspondingly reduced. Further, the system aspects of data reception can be quite different from position determination. For example, the position determination requires simultaneous reception of the signal at multiple anchor nodes; to ensure the reception at the most distant base station, a high process gain is required. In contrast, the data reception is required only at the closest base station, so a lower process gain is adequate; this implies a shorter pn-code and a consequential higher data rate. Thus, with a correlator-only system, there is a conflict in requirements between these two aspects of the system requirements. This conflict can be resolved by the use of an accumulator. A system can have a short pn-code for data transmission with consequential high data rate, while the accumulator can be used for improving the SNR for the position determination function. The accumulator is usually located before the correlator, although post-correlation accumulation is also possible. In the following the performance of both analog and digital accumulators is discussed.

3.2.3.1 Analog Accumulator

The first case considered is an analog signal, or a digital signal with, say, eight or more bits, so that the digital signal is essentially equivalent to the analog signal. In both cases the performance of the accumulator can be analyzed by considering the effect on the signal and the noise by the accumulation process.

Consider an accumulator located before the correlator. The requirement is to accumulate successive pn-codes input to the correlator. Because the samples accumulated are separated in time by the length of the pn-code, the noise components will be statistically independent. Further, if the frequency error is not too large,[1] then the individual complex data samples will have essentially the same phase and, thus, the samples can be accumulated coherently. In such a case, the mean signal is simply the sum of the means of the individual signals, and the variance of the accumulated signal is the sum of the individual variances. While the actual implementation does not divide the accumulation by the number of samples accumulated, it is convenient

---

[1] In the case of a moving device, the Doppler frequency offset also needs to be considered. This effect ultimately limits the extent of any accumulation of pn-codes.

in the analysis to normalize the resulting accumulation so that the output statistics of the accumulator can be directly compared with the input statistics. Hence, if M samples are accumulated, the expected mean output is the input signal amplitude and the output noise power is $1/M$ the input noise power, where M is the number of pn-codes accumulated. As a consequence, the resulting output SNR is M times greater than the input SNR. Further, owing to the CLT, the output statistics will be approximately Gaussian, regardless of the input statistics. Thus, the effect of the accumulator for the analog input signal is simply to increase the SNR by a factor equal to the number of samples accumulated. As this signal is used as input to the correlator, the output from the correlator will be increased by the same factor. Note that no time synchronization to the pn-code is required; the only requirement is that the length of the pn-code is known by the receiver.

3.2.3.2 Accumulator with 1-bit Data Input

The analysis of the performance of an analog accumulator also can be used for a digital accumulator, provided the number of bits used to digitize the signal is sufficiently large to represent the analog signal adequately. As the number of bits representing the signal is reduced, the performance will reduce. Of practical interest is when the data is represented by just 1 bit; then, a simple comparator can be used for digitization and a 1-bit summer used in the accumulator. This simplification greatly reduces the complexity of the digital hardware required and, thus, is important for simple nodes as used in WSNs. In such a case, the reduction in the performance of the accumulator is required for performance calculations.

The input signal to the sampling comparator is assumed to consist of a signal of amplitude A plus random Gaussian noise of STD σ, so that the sampled data have a normal distribution given by $N(A, \sigma^2)$. In the following analysis, it is convenient to normalize the signals so that the noise STD is unity and the signal PDF is $N(A/\sigma, 1)$. With this definition, the input SNR is $\gamma_{in} = a^2 = A^2/\sigma^2$, where a is the normalized signal amplitude.

Consider the comparator output statistics. If p is the probability that the signal is positive, then the probability density function is

$$f(x) = (1-p)\delta(-1) + p\delta(1) \qquad (3.28)$$

where the binary outputs are ± 1. For Gaussian noise, the probability p is given by

$$p = \Phi(a)$$

$$\Phi(x) = \frac{1}{\sqrt{2\pi}} \int_{-\infty}^{x} e^{-u^2/2} \, du = \frac{1}{2}\left[1 + \mathrm{erf}\left(\frac{x}{\sqrt{2}}\right)\right] \qquad (3.29)$$

The mean signal at the output of the comparator is given by

$$\mu = \int_{-\infty}^{\infty} xf(x)\,dx = (-1)(1-p) + (+1)p = 2p - 1 = 2\Phi(a) - 1 \qquad (3.30)$$

The variance of the output signal is given by

$$\sigma^2 = \int_{-\infty}^{\infty} x^2 f(x) dx - \mu^2 = (-1)^2 (1-p) + (+1)^2 p = 1 - \mu^2 = 4p(1-p) \qquad (3.31)$$

Now consider the accumulation of M samples with the statistical parameters as defined above and normalized by a factor $1/M$. Then, the mean and variance of the output are given by

$$\mu_\Sigma = \sum_{m=1}^{M} \left(\frac{1}{M}\right) \mu = \mu = 2\Phi(a) - 1$$

$$\sigma_\Sigma^2 = \sum_{m=1}^{M} \left(\frac{1}{M}\right)^2 \sigma^2 = \frac{4p(1-p)}{M} = \frac{4\Phi(A)[1 - \Phi(a)]}{M} \qquad (3.32)$$

Thus, the accumulator output SNR for 1-bit input data is given by

$$\gamma_{1bit_{acc}} = \frac{\mu_\Sigma^2}{\sigma_\Sigma^2} = \frac{M[2\Phi(a) - 1]^2}{4\Phi(a)[1 - \Phi(a)]} = \frac{M[2\Phi\sqrt{\gamma_{in}} - 1]^2}{4\Phi\sqrt{\gamma_{in}}(1 - \Phi\sqrt{\gamma_{in}})} \qquad (3.33)$$

Of particular interest is the performance at low input SNR (SNR $\ll 1)^2$. In this case, the function $\Phi(x)$ can be approximated by

$$\Phi(x) \approx \frac{1}{2} + \frac{x}{\sqrt{2\pi}} \qquad (x \ll 1) \qquad (3.34)$$

Applying (3.34) to (3.33), the SNR approaches the limit

$$\gamma_{1bit\_acc}\big|_{\gamma_{in} \to 0} = \left(\frac{2}{\pi}\right) M \gamma_{in} \qquad (3.35)$$

As discussed in Section 3.2.3.1, the corresponding output SNR for an analog accumulator is given by $\gamma_{analog\_acc} = M\gamma_{in}$, so the 1-bit accumulator has a reduced output SNR relative to the ideal accumulator by $10\log(2/\pi) = -1.96$ dB.

Thus, the 1-bit accumulator has a similar performance to the analog accumulator for low input SNR, except that the output SNR is reduced by about 2 dB. At high input SNR the 1-bit accumulator is actually superior to the analog accumulator. For more details, refer to Section 3.2.5 on the 1-bit correlator.

In typical implementation, the accumulator is followed by a correlator, so the output of the accumulator is the input to the correlator. Because of the accumulation process in both the analog and the 1-bit cases, the signal PDF will be approximately Gaussian according to the CLT. Further, in the 1-bit case, the accumulator output will be over the range $[-M, +M]$, so the number of bits in the output is $1 + \log_2(M)$. For example, if 16 samples are accumulated, then the number of bits in the output is 5, compared with only one bit at the input. As a consequence,

---

[2] The performance at high SNR is not particularly relevant, as the accumulator effect of improving the SNR is not required in this case.

the accumulator process has increased the number of bits in the correlator input signal; so, for all practical purposes, the signal is equivalent to the analog signal output from an analog accumulator.

3.2.3.3 Coherent Accumulation

The accumulation process described above is based on the assumption that the signal is coherent across the period of the accumulation. This assumption is generally valid, but the signal phase can vary significantly over the accumulation period at high mobile node speeds. For example, consider a mobile travelling at 15 m/s (or 120 wavelengths per second at an RF frequency of 2.4 GHz) relative to a base station. For a 255-chip pn-code, 10 Mchips/s and 16 codes accumulation, the period of the accumulation is about 400 μs and the distance moved in this time is about 0.05 wavelengths (or 18°). Thus, the phase of the last sample accumulated will be 18° rotated relative to the first. In fact, the actual rotation will be twice this figure for the case of the signal transmitted from a mobile whose local oscillator is locked to the base station signal, as the Doppler frequency adjustment applies to the mobile local oscillator and the same Doppler frequency shift applies on the return path to the base station. Thus, the formula for the phase error (in radians) is

$$\Delta\phi = 2\pi\Delta f\Delta t = 2\pi\left(\frac{v_{\text{mobile}}}{\lambda_{\text{RF}}}\right)[\cos(\psi_{\text{ref}}) + \cos(\psi_{\text{bs}})](MNT_{\text{c}}) \qquad (3.36)$$

where N is the pn-code length, M is the number of pn-codes accumulated, T_{c} is the chip period and ψ_{ref} and ψ_{bs} are the angles to the timing reference site and the receiving base station respectively. The accumulation of the samples (in phase and quadrature) can be interpreted as a vector summation rather than the summation of sample magnitudes. In this case, the locus of the tip of the summation vectors lie on a circle rather than a straight line, and it can be shown that the magnitude of the accumulated vector relative to the coherent summation magnitude is given approximately by

$$\frac{S_{\text{vector}}}{S_{\text{coherent}}} \approx \frac{\sin(\Delta\phi)}{\Delta\phi} \qquad (3.37)$$

For the case given above $\Delta\phi = 36°$, so the summation is reduced by a factor of 0.935 or 0.6 dB, which is negligible. However, as the speed increases and the phase error $\Delta\phi$ approaches 180°, the accumulator output approaches zero. For design purposes it is recommended that the worst-case phase error is set at 45°. Given the maximum speed of the mobile device and the RF wavelength, the maximum integration period can be determined using (3.36). This integration time is divided between the number of accumulations M and the pn-code period N. As previously discussed, the choice of M and N depends on other factors, such as the data rate requirement and the signal-processing capabilities of the equipment.

3.2.4 Interference

Typically, positioning systems use ISM bands, usually either at 2.4 GHz or 5.8 GHz. However, these bands can be used by other systems which cause interference. The interference in the ISM band may include frequency hopping. The interference associated with frequency hopping is essentially carrier wave (CW), or a sine wave with random frequency. As the accumulation or correlation process is usually over a relatively short period of time compared with the hop period, it can be assumed that the interfering signal is coherent over the period of the signal processing in the receiver.

Let us consider two particular cases, namely a multi-bit (or analog) accumulator and a 1-bit accumulator. The analysis of the corresponding correlators is similar. However, if the architecture of the receiver is such that the accumulator is before the correlator, then, as shown previously, the input signal to the correlator is multi-bit; thus, the analysis of a 1-bit correlator is not required.

3.2.4.1 Performance of Analog Accumulator with Carrier Wave Interference

For the analysis of the performance of an analog accumulator, it will be assumed that the interferer is a coherent CW signal; so, after mixing down to baseband, the interference signal can be represented by

$$
\begin{aligned}
S_{\text{interfer}}(t) &= a\cos(\Delta\omega t + \phi) = a\cos\phi\cos(\Delta\omega t) - a\sin\phi\sin(\Delta\omega t) \\
&= a_1\cos(\Delta\omega t) - a_2\sin(\Delta\omega t) \\
\Delta\omega &= 2\pi(f_{\text{interferer}} - f_0)
\end{aligned}
\tag{3.38}
$$

where $f_{\text{interferer}}$ is the frequency of the interferer and f_0 is the frequency of the desired signal. The average power of the interferer is $a^2/2$, while the power in the pn-code (± 1) is unity. The signal processing in an accumulator is linear, so the effect of the interference signal can be analyzed separately from the desired signal. The accumulator sums samples at a constant spacing in time of one pn-code period (N chips with a chip period of T_c). The accumulator output for the interfering signal is given by

$$
\begin{aligned}
S_{\text{interferer}}(t) &= a_1\sum_{m=1}^{M}\cos(\Delta\omega m T_s) - a_2\sum_{m=1}^{M}\sin(\Delta\omega m T_s) \\
&= a_1\sum_{m=1}^{M}\cos(m\theta) - a_2\sum_{m=1}^{M}\sin(m\theta) \\
T_{\text{pn}} &= NT_c \qquad \theta = \Delta\omega T_{\text{pn}}
\end{aligned}
\tag{3.39}
$$

Equation (3.39) can be simplified as [3]

$$
\begin{aligned}
S_{\text{interferer}}(\theta,\phi) &= \left[\frac{\sin(M\theta/2)}{\sin(\theta/2)}\right]\left[a_1\cos\left(\frac{M+1}{2}\right)\theta - a_2\sin\left(\frac{M+1}{2}\right)\theta\right] \\
&= a\left[\frac{\sin(M\theta/2)}{\sin(\theta/2)}\right]\cos\left[\left(\frac{M+1}{2}\right)\theta + \phi\right]
\end{aligned}
\tag{3.40}
$$

where ϕ is the constant offset angle between the interference signal and the receiver local oscillator, which can be considered generally as a random number in the range $[0, 2\pi]$. As M is an integer, the accumulator output repeats itself with a period of 2π. An example of the

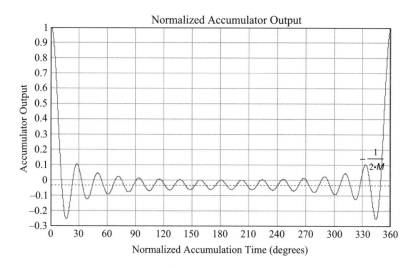

Figure 3.8 Accumulator output with $M = 16$ and normalized by a factor $1/M$ so that the nominal output of the accumulator for the signal is unity. The phase $\phi = 0$ and the interference amplitude is unity.

amplitude function is shown in Figure 3.8. The amplitude characteristics are largely determined by the first term in brackets. Assuming the signal amplitude is unity, then the signal-to-interference ratio (SIR) at the input and output of the accumulator and the associated process gain are given by

$$\gamma_{in} = \frac{1}{a^2/2} = \frac{2}{a^2}$$

$$\gamma_{out} = \frac{1}{S_{interferer}(\theta, \phi)^2} \tag{3.41}$$

$$G_{proc} = \frac{\gamma_{out}}{\gamma_{in}}$$

The SIR at the output is a function of the accumulator amplitude function (see Figure 3.8). For example, when $\theta = \phi = 0$, the output SIR is $1/a^2$, so that the process gain is 1/2. This is the worst-case result, where the interference signal is at maximum during the sampling; such a circumstance has very low probability. A more appropriate estimate is to average over the random phase offset ϕ, in which case the average process gain with $\theta = 0$ is unity. Therefore, in the case where $\theta \rightarrow 0$, the accumulator has, on average, no effect on reducing the effect of the interference. However, when the normalized time parameter θ is large ($M\theta$ is not near 0 or 2π), the normalized power is of the order of $a^2/2M^2$, so the process gain is of the order of M^2. Another measure is to estimate the process gain averaged over all values of the parameters θ and ϕ. In particular:

$$\overline{S_{interferer}} = \frac{1}{2\pi} \int_0^{2\pi} S_{interferer}(\theta, \phi)^2 \, d\theta = \frac{a^2}{2M} \tag{3.42}$$

Note that the mean interference power at the output is independent of the phase ϕ, so the mean output SIR is given by $\overline{\gamma_{\text{out}}} = 2M/a^2$. Thus, from the first equation in (3.41), the mean process gain is simply M, or the same process gain as for a random interference signal (see Section 3.2.3.1).

Thus, the conclusion is that for a CW interferer the mean process gain is the same as for random interference signals, but in the worst case there is a processing loss of 3 dB. The worst case occurs when the interference signal frequency is offset from the carrier frequency by an integer multiple of the pn-code repetition rate. However, as the interference in an ISM band will be from a frequency-hopping source, the interference will only occur infrequently when the interference source 'hops' into the band of the positioning system.

3.2.4.2 Performance of 1-bit Accumulator with Carrier Wave Interference

The first task in determining the performance of a 1-bit accumulator is to determine the statistical properties of a randomly phased sine wave. Assume $x = \sin \theta$, where θ is a random phase variable with a uniform distribution $p(\theta)$ over the range $[-\pi, \pi]$, then $q(x)$, the PDF of x, can be expressed as

$$q(x) = 2p(\theta)\left|\frac{\mathrm{d}\theta}{\mathrm{d}x}\right| = \frac{1}{\pi\sqrt{1 - x^2}} \qquad (x \le 1) \tag{3.43}$$

From (3.43), the variance of x can be calculated to be 1/2.

The performance of the 1-bit accumulator is calculated using the same method as described for Gaussian noise, except that the $\Phi(x)$ function is replaced by

$$\Gamma(\alpha) = \int_{-1}^{\alpha} q(x)\,\mathrm{d}x = \frac{1}{2} + \frac{\sin^{-1}(\alpha)}{\pi} \qquad (\alpha \le 1)$$

$$\approx \frac{1}{2} + \frac{\alpha}{\pi} \qquad (\alpha \ll 1) \tag{3.44}$$

where α is the amplitude of the pn-code. The input SIR is $\gamma_{\text{in}} = 2\alpha^2$.

From (3.44), the output SNR of a 1-bit accumulator normalized to the nominal output SNR of an analog accumulator can be directly inferred by replacing Φ with Γ in (3.30), resulting in

$$\gamma_{1\,\text{bit}} = \frac{M[2\Gamma(\alpha) - 1]^2}{4\Gamma(\alpha)[1 - \Gamma(\alpha)]} = \frac{M[2\Gamma(\sqrt{\gamma_{\text{in}}/2}) - 1]^2}{4\gamma_{\text{in}}\Gamma(\sqrt{\gamma_{\text{in}}/2})[1 - \Gamma(\sqrt{\gamma_{\text{in}}/2})]} \tag{3.45}$$

The main interest is at low SIR, so the approximation in (3.44) can be applied to (3.45), resulting in

$$\frac{\gamma_{1\,\text{bit}}}{\gamma_{\text{in}}} = \frac{(2/\pi^2)M}{1 - (2/\pi^2)\gamma_{\text{in}}} \qquad (\gamma_{\text{in}} \ll 1) \tag{3.46}$$

Finally, the limiting process gain performance as the input SIR approaches zero relative to the nominal process gain $G_0 = M$ is

$$\left.\frac{G_{1\,\text{bit}}}{G_0}\right|_{\gamma_\text{in}\,\to\,0} = \frac{2}{\pi^2} = -7\,\text{dB} \qquad (3.47)$$

Thus, the 1-bit accumulator is 7 dB worse than the average performance of an ideal analog accumulator in the presence of sine-wave interference. This 7 dB reduction compares with only 2 dB reduction when the interference is Gaussian noise.

3.2.4.3 Multiple Access Using an Accumulator

The main purpose of an accumulator before the correlator is to provide the additional process gain, but an accumulator can also be used to implement a form of multiple access to the radio channel. As the radio spectrum is a finite and limited resource, some form of multiple access to the radio channel is necessary to implement radiolocation functionality servicing many nodes. Common types of multiple access include frequency-division multiple access (FDMA), time-division multiple access (TDMA), and code-division multiple access (CDMA) [4]. A less well known method of multiple access which is particularly applicable to radiolocation systems will now be described. This technique uses the characteristics of the accumulated signal described previously in Section 3.2.4.1 and the spectral characteristics of direct-sequence spread-spectrum signals.

A direct-sequence spread-spectrum signal used to measure TOA usually has a relatively short pn-code which is transmitted repeatedly. For a periodic signal, the spectrum is broken up into spectral lines separated in frequency by the repetition rate. This spectral characteristic can be determined by considering an arbitrary signal $p(t)$ time constrained to $0 \leq t \leq T$, so that the repeated transmission of this signal can be represented by the expression

$$s(t) = p(t) * \text{comb}_T(t) \qquad (3.48)$$

where the $\text{comb}_T(t)$ function is an infinite series of delta functions separated in time by a period T. Taking the Fourier transform of (3.48) results in the spectrum

$$S(f) = \frac{1}{T}P(f)\text{comb}_{1/T}(f) \qquad (3.49)$$

Thus, the comb function effectively converts the continuous spectrum $P(f)$ into a line spectrum as stated above. For example, consider a spread-spectrum signal with a 10 Mchips/s rate and an $N = 255$ chip pn-code. In this case the pn-code period is $T = 25.5\,\mu s$ and the repetition rate is about 40 kHz. As the transmitted spread-spectrum is typically limited to the main lobe of $\pm 1/T_c$ (or $\pm 10\,\text{MHz}$), the spectrum is rather finely subdivided into about 500 individual lines.

Now consider the spectral characteristics of the output of the accumulator, as defined in (3.40). The output signal is zero at regular points corresponding to

$$\frac{M\theta}{2} = 2\pi(f_\text{interferer} - f_0)\frac{MT_\text{pn}}{2} = m\pi \qquad (m = 1, 2, \ldots) \qquad (3.50)$$

Thus, if the interference spectral frequency occurs at

$$\Delta f(m) = f_m - f_0 = \frac{m}{MT_\text{pn}} \qquad (3.51)$$

then the effect of the accumulator is to nullify the interference signal associated with that spectral line. In the case of M multiple simultaneous transmissions of the pn-codes, $M - 1$ of these transmissions can be considered as interference signals. However, if these transmissions have an RF carrier frequency offset by an amount given by (3.51), then the accumulator output from these $M - 1$ transmissions will be zero. As this process is mutually orthogonal, the accumulator has effectively allowed M simultaneous spread-spectrum signals without causing mutual interference.

Consider the case of the above numerical example, with $M = 10$. In this case the RF carrier offset frequency is $\Delta f = 4\,\text{kHz}$ and the carrier offsets are 0, 4, 8, ..., 36 kHz. In effect, the 10 signals interleave in the frequency domain; as such, this multiple access method is called interlaced FDMA. Observing that the offsets in frequency are very small compared with the total spectral bandwidth, the total spectral requirement is essentially the same as for a single transmission, so the method is very spectrally efficient. This technique of multiplexing spread-spectrum signals is subject to a US patent [5].

The number of simultaneous signals M in the above multiplexing technique is unconstrained in theory, but practical considerations limit the number. First, the method requires very accurate control of the RF carrier frequency to ensure the orthogonality of the signals. For example, if in the above example the offset frequency parameter Δf is constrained to with 1% of its nominal value (40 Hz), then if the RF frequency is in the 5.8 GHz ISM band the frequency accuracy of the RF frequency is about 0.1 ppm. Such high-frequency accuracy is well beyond that of local oscillators in typical radios, so some form of frequency control is required; such methods are discussed in Chapter 5. Second, if the node is mobile, then even with frequency control the RF frequency will suffer from an offset error due to the Doppler effect. However, if the speeds are limited, which is typical in wireless sensor networks, then the Doppler frequency offset is small. For example, at a walking speed of 1 m/s and an RF frequency of 5.8 GHz, the Doppler offset frequency is about 20 Hz, which is within the above requirement for frequency accuracy.

Another limitation of the method is that the use of the accumulator increases the effective symbol period and, hence, reduces the effective data rate. In the above numerical example, the cumulative symbol period is 255 μs (with $M = 10$), so the symbol rate is about 4000/s. If, for example, 2 bits per symbol (quadrature phase-shift keying modulation) is used, then the data rate is 8000 bits/s. Such data rates are rather low, but may be acceptable for WSNs. Other modulation techniques can be used to further enhance the data rate, if required.

3.2.5 Performance of 1-bit Correlator

In the case where an accumulator is not used before the correlator, the performance of a 1-bit correlator is of practical interest, particularly in a mobile device, where the simpler implementation of 1-bit signal processing is attractive.

When the correlator is preceded by an accumulator, it was shown in the previous section that the accumulator output is essentially analog, regardless of the type of input signal. When there is no proceeding accumulator, the input signal may be either 1-bit or multi-bits. Since the multi-bit correlator performs essentially in the same way as an analog correlator as discussed in Section 3.2.2, only the performance of a correlator with 1-bit input data is analysed.

The analysis of the 1-bit correlator is very similar to that described previously for the 1-bit accumulator. The correlator sums N such samples, so the mean and STD of the output of the correlator become

$$\mu_{1\,\text{bit}} = N\mu = N[2\Phi(a) - 1]$$
$$\sigma_{1\,\text{bit}}^2 = N\sigma^2 = 4N\Phi(a)[1 - \Phi(a)] \tag{3.52}$$

where the noise is assumed to be Gaussian. Thus, the process gain, namely the output SNR $\gamma_{1\,\text{bit}}$ normalized by the input SNR γ_{in}, of the 1-bit correlator is given by

$$\frac{\gamma_{1\,\text{bit}}}{\gamma_{\text{in}}} = G_{\text{proc}}(\gamma_{\text{in}}) = Nf(\gamma_{\text{in}}) = \frac{N[2\Phi(a) - 1]^2}{4A^2\Phi(a)[1 - \Phi(a)]} = \frac{N[2\Phi(\sqrt{\gamma_{\text{in}}}) - 1]^2}{4\gamma_{\text{in}}\Phi(\sqrt{\gamma_{\text{in}}})[1 - \Phi(\sqrt{\gamma_{\text{in}}})]} \tag{3.53}$$

Of particular interest is the performance at low input SNR. Applying the approximation defined by (3.34)–(3.53), the process gain reduction factor approaches the limit

$$f(\gamma_{\text{in}})|_{\gamma_{\text{in}} \to 0} = \frac{2}{\pi} \tag{3.54}$$

Thus, the 1-bit correlator has a reduced output SNR relative to the ideal correlator by $10\log(2/\pi) = -1.96$ dB. Note that this result is the same as for the accumulator.

For comparison purposes, it is convenient to normalize the output correlogram peak to unity and again compare the 1-bit correlator with an ideal correlator with an N-chip input pn-code. The output noise STDs of the ideal and 1-bit correlators are given by

$$\sigma_0 = \frac{1}{\sqrt{N\gamma_{\text{in}}}}$$
$$\sigma_{1\,\text{bit}} = \frac{\sigma_0}{\sqrt{f(\gamma_{\text{in}})}} \tag{3.55}$$

The ideal and 1-bit correlator output noise STDs are plotted as a function of the input SNR for both the ideal and 1-bit correlators in Figure 3.9. In both cases, the mean correlogram amplitude

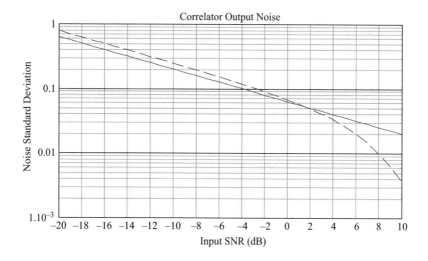

Figure 3.9   The normalized correlator output noise STD for both an ideal correlator (solid line) and the 1-bit correlator (dotted line) with $N = 255$. The correlogram amplitude in both cases is normalized to unity.

is normalized to unity. As can be observed, for input SNRs less than about 2 dB, the output noise of the 1-bit correlator is worse than an ideal correlator, approaching the performance given by (3.54). Further, for input SNR greater than 2 dB, which is generally of no practical interest, the 1-bit correlator is superior to the 'ideal' correlator. The reason for this somewhat surprising result is that, at high input SNR, the 1-bit digitization effectively reproduces the binary pn-code and, hence, there are fewer errors.

References

[1] A.V. Oppenheim, *Discrete-Time Signal Processing*, 2nd edition, Prentice Hall, 1999.
[2] A.J. Viterbi, *CDMA Principles of Spread Spectrum Communication*, Addison-Wesley, 1995.
[3] I.S. Gradshteyn and I.M. Ryzhik, *Table of Integrals, Series, and Products*, Academic Press, 7th edition, 2007.
[4] J.G. Proakis, *Digital Communications*, 4th edition, McGraw-Hill, New York, 2001.
[5] M.J. Yerbury and G.C. Hurst, 'Spread-spectrum multiplexed transmission system', US patent 5,063,560, November 1991.

4

Bandlimited Time-of-Arrival Measurements

In this chapter we will consider the performance of TOA measurements using bandlimited radio signals. Because all radio systems must operate in an environment of regulation of the radio transmissions, particularly the transmitter power and spectral bandwidth, the design of positioning systems must consider these imposed constraints. The transmitter power limits, together with the radio propagation losses, place limits on the receiver SNR, while the bandlimiting places limits on the effective time resolution of the receiver. Thus, the algorithms used to measure the TOA should be designed to provide good performance, both regarding the receiver noise and multipath interference. As will be shown, to some extent a compromise must be reached between these two requirements, so that the optimum solution depends on the particular operating environment of the positioning system. However, the main focus will be on multipath mitigation techniques rather than good noise performance, as the designer has at least some control over the receiver noise, but the operating environment is an imposed constraint.

Before analyzing the performance of various receiver signal-processing algorithms with a limited bandwidth signal, it is useful to consider the case where the bandwidth is not a constraint. Thus, it will be shown that if the SNR is sufficiently high, and the signal bandwidth is large (essentially infinite), then algorithms can be designed which have zero TOA measurement errors in a multipath environment where the LOS direct signal is stronger than the combined multipath signals. While this is mainly a theoretical construct which is not of use in severe multipath environments, the concept is useful in guiding the design of the signal-processing algorithms. For example, the above-defined operating constraints approximate the operating conditions for the GPS in outdoor environments. In particular, the paths to the GPS satellites can be clear of obstruction and, importantly, the GPS C/A mode of operation (civilian mode) has an operating bandwidth much greater than the chip rate of the pseudo-random codes used to modulate the radio signal. Thus, the GPS signal-processing algorithms can utilize the theoretical framework which largely eliminates multipath tracking errors (but this is ultimately constrained by the finite signal bandwidth). In contrast, in terrestrial-based positioning systems, particularly those operating indoors or where multipath signals are dominant, the

Ground-Based Wireless Positioning   Kegen Yu, Ian Sharp and Jay Guo
© 2009 John Wiley & Sons, Ltd

bandwidth limitations mean that GPS-type signal-processing algorithms are not appropriate. Thus, this chapter will introduce algorithms that are specifically designed for bandlimited signals in severe multipath environments.

Because the signal is bandlimited, the number of independent samples in the correlation diagram (correlogram) is limited. In particular, if the spectral bandwidth is limited to twice the chip rate (the most common form of bandlimiting the signal), then the nominal triangular-shaped correlogram has at most four samples: two on the leading edge and two on the trailing edge. For a peak-detection algorithm, two samples either a 0.25 or 0.5 chips apart are typically used in estimating the location of the peak. By appropriate curve fitting, the location of the peak can be estimated. As the shape of the correlogram is flat near the peak, if the two samples are too close together the accuracy will reduce due to Gaussian noise and multipath signals. An alternative approach is to use the leading edge of the correlogram, where the slope is greater and less susceptible to multipath. As there are only two independent samples on the leading edge, there are only a limited number of feasible algorithms that can be used. Two such algorithms are considered and their performance compared with the peak-detection algorithm. In general, these leading-edge algorithms have superior performance in the presence of multipath, but inferior Gaussian noise performance. However, in typical operating environments (indoors in particular), the leading-edge algorithms are preferred, with a reduction in TOA measurement errors of typically 50% or more.

The following is a summary of the contents of the subsections of this chapter.

1. Section 4.1 introduces the wideband multipath theorem, showing that with sufficient bandwidth the effects of multipath interference can be mitigated essentially entirely, provided the direct signal is greater that the combined multipath signals.
2. Section 4.2 analyzes theoretically the effect that bandlimiting has on the process gain of the receiver.
3. Section 4.3 introduces simplified models of the correlation diagram (correlogram) that are used in the performance analysis of the TOA algorithms.
4. Section 4.4 provides an analysis of the peak-tracking algorithm, including the degrading effects of Gaussian noise and a single multipath interference source.
5. Section 4.5 provides an analysis of the leading-edge projection tracking algorithm, including the degrading effects of Gaussian noise and a single multipath interference source.
6. Section 4.6 provides an analysis of the leading-edge ratio tracking algorithm, including the degrading effects of Gaussian noise and a single multipath interference source.
7. Section 4.7 briefly analyzes the effect that the multipath signal phase has on the performance of the algorithms.
8. Section 4.8 provides a summary of the performance of the various algorithms, indicating their weaknesses and strengths.

4.1 Wideband Multipath Theorem

The performance of real radio systems is limited by their operating environment and constraints on the bandwidth. For data systems, the bandwidth (together with the SNR) governs the potential maximum data rates, while for position-location systems based on measuring the TOA of the radio signal the main impact on performance is limiting the accuracy of the time measurement and, hence, the position. In the simple case of a baseband pulse, this limited bandwidth determines the rise time of the pulse and, hence, the accuracy of the TOA measurement. For direct-sequence spread-spectrum, the limited bandwidth places an upper limit on the chip rate of the baseband pseudo-random modulating signal and, hence, the effective rise time of the reconstructed correlation diagram (or correlogram) 'pulse'. Thus, in both cases the bandwidth of the system has a dominant effect on performance. However, before considering the effect of limiting the bandwidth, it is constructive to consider the case of a large bandwidth and, more particularly, the degrading effect on performance due to multipath transmissions of the radio signal. This analysis is particularly relevant, as the civilian mode of operation of the GPS has a bandwidth 20 times the chip rate and, thus, can be considered as a wideband system in this context.

Before considering the effect of limiting the bandwidth, it is useful to review the general approach taken by GPS receivers to minimize the effects of multipath signals. A common technique used in GPS receivers is to use two correlogram samples near the peak, say ±0.1 chips relative to the nominal peak. Figure 4.1 shows the correlogram signal appropriate for the method. If the bandwidth is infinite, then the ideal correlogram shape is triangular, while the multipath signal will be delayed with a shape determined by the characteristics of the multipath signals. In this simplified analysis, a direct signal will be assumed and the multipath signal is assumed to be smaller in amplitude than the direct signal. What will be shown is that if the signal is not bandlimited and the SNR is large, then, theoretically, appropriate signal processing can result in zero TOA errors even in a multipath environment.

To understand this effect, consider determining the TOA by detecting the time of the peak of the signal. Without multipath interference and noise, the peak of the triangular correlogram

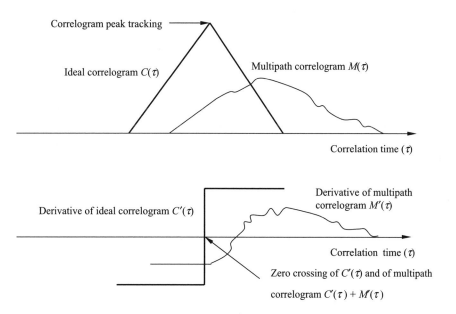

Figure 4.1   Geometry associated with the peak-tracking algorithm, showing tolerance to multipath.

clearly gives the correct TOA, but with a multipath case it is more appropriate to locate the peak where the derivative of the correlogram $C(\tau)$ is zero. For the ideal correlogram $C(\tau)$ is triangular in shape, so that the derivative has a discontinuity at the peak; hence, the TOA point is located where the derivative crosses the zero amplitude axis, as shown in Figure 4.1. Now consider the case where there is additionally an arbitrary multipath interference function $M(\tau)$[1] disturbing the derivative function. Provided the derivative of the multipath signal is less than the corresponding derivative of the direct signal, the position of the peak will be unaffected by the multipath interference. This analysis suggests that if the direct signal is present, and the multipath signal is not too strong, then with an infinite bandwidth the determination of the location of the peak of the correlogram should result in an accurate measurement of the TOA.

The above discussion suggests the strategy for determining the TOA to a wideband spread-spectrum signal. Practical tracking loops estimate the correlogram at two closely spaced points near the peak and the tracking loop attempts to equalize these signals. Current-generation GPS receivers typically use two correlator outputs separated by 0.1 chips. This technique is only an approximate method of tracking the peak, so that a real receiver will have some residual errors due to multipath. However, it should be stressed that this technique is only valid for wideband systems.[2]

The general effect of bandlimiting the signal can also be inferred from the above analysis approach. As the signal becomes bandlimited, the derivative of the direct signal correlogram more slowly changes from negative to positive, so that with multipath the position of crossing the origin becomes less defined. As the bandwidth reduces, the width of this area of uncertainty increases, so that the TOA estimate becomes increasingly inaccurate. Thus, in a severely bandlimited correlogram, the location of the peak of the signal in a multipath environment becomes increasingly unsatisfactory. Also, if the direct signal is blocked, then the peak of the correlogram may be considerably delayed relative to smaller but more direct signals, so that a more appropriate approach is to consider the leading edge of the correlogram, regardless of the bandwidth.

4.2 Bandlimited Correlogram Characteristics

The analysis in Section 4.1 shows that for wideband spread-spectrum systems with a high SNR the TOA of the spread-spectrum signal can be accurately determined, even in multipath conditions, provided there is a direct signal. However, most practical systems do not comply with these constraints, particularly those operating in severe multipath conditions indoors and limited radio bandwidth. This section computes the effect of bandlimiting the correlogram.

While actual systems constrain the bandwidth of operation with appropriate filters with a transfer function $H(f)$, the computational technique used in the following analysis is to compute the spectrum of the infinite bandwidth correlogram, and then null the spectrum above the required bandwidth. While this technique is not based on realisable physical filters, the

---

[1] In this simplified analysis, only the magnitude of the correlogram is considered, ignoring the complex (amplitude and phase) of actual signals. Nevertheless, the overall conclusions remain valid.

[2] In this context, a wideband system is defined by the ratio of the RF bandwidth to the chip rate. For a wideband system this ratio should be much greater than unity. For example, the GPS C/A mode has a ratio of 20:1.

analysis captures the essential features of bandlimiting. Further, typical implementations are often based on the digital generation of the pn-code where spectrum nulling can be applied, and thus the analysis technique is appropriate.

The correlogram (or auto-correlation function of the pn-code) can be computed using the formula

$$c(\tau) = \int s(t)s(t+\tau)\mathrm{d}t = \mathcal{F}^{-1}[|S(f)|^2] \tag{4.1}$$

where $\mathcal{F}^{-1}[*]$ is the inverse Fourier transform and $S(f)$ is the Fourier transform of the signal $s(t)$. For the case of a bandlimited pn-code, the signal can be expressed as

$$|S(f)|^2 = |\mathrm{PN}(f)|^2 \Pi_{2B}(f) \tag{4.2}$$

where $\Pi_{2B}(f)$ is a rectangular weighting function of bandwidth $\pm B$ and $\mathrm{PN}(f)$ is the spectrum of the pn-code. By taking the inverse transform of (4.2), the bandlimited correlogram is given by

$$c(\tau) = c_0(\tau) * [2B\mathrm{sinc}(2B\tau)] \tag{4.3}$$

where $c_0(\tau)$ is the infinite bandwidth correlogram, and the asterisk represents the convolution operation. The nominal infinite-bandwidth correlogram is triangular in shape, with unit amplitude and a width of ± 1 chip.

It is convenient to normalize the infinite-bandwidth correlogram to unit amplitude and the correlation time to the chip period. The convolution can then be expressed as

$$c(\tau) = 2\beta \int_{-1}^{1} c_0(t)\mathrm{sinc}[2\beta(t-\tau)]\mathrm{d}t \tag{4.4}$$

where β is the bandwidth normalized to the chip rate, so that $\beta = B\tau_0$, where τ_0 is the chip period. For example, typically $\beta = 1$ for spread-spectrum systems which limit the signal to the main lobe. For GPS C/A receivers, typically $\beta = 10$. Note in this context that the 'bandwidth' is defined at -3 dB, while the actual RF channel is defined over a wider band. Commonly, the radio signal is limited at the first spectral null of the ideal infinite bandwidth signal, so the radio bandwidth is twice the bandwidth defined by the β parameter.

The effect of bandlimiting can be observed from Figure 4.2. First, the sharp intersections of the straight-line segments associated with the triangular shape are smoothed out, with the shape becoming progressively smoother as the bandwidth decreases. In particular, the peak of the correlogram becomes a broad smooth curve with zero derivative at the peak. Thus, detecting the peak by the derivative is not appropriate for the bandlimited case. In fact, as the accuracy of determining position on a curve is proportional to the derivative of the function, bandlimiting results in poor performance of peak detection methods. Informally, as the peak 'width' near the peak increases as the bandwidth decreases, the accuracy reduces with bandwidth. Second, the amplitude progressively decreases as the bandwidth is reduced. Third, the shape of the correlogram beyond ± 1 chip becomes distorted, with the exact shape dependent on the exact nature of the filtering. For the case where $\beta = 1$, the magnitude of the sidelobes is small, and the sidelobes are only slightly larger for $\beta = 0.75$. However, the sidelobes are considerable when $\beta = 0.5$. The presence of sidelobes when peak-detection algorithms are used is of little

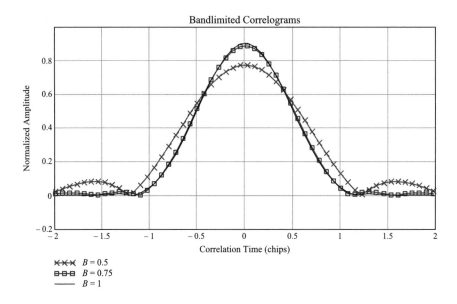

Figure 4.2   Effect of bandlimiting the correlogram. Note that the B parameter on the figure is the β parameter defined in the above text, and that the magnitude of the correlogram is plotted. The case of $\beta = 1$ is used in the analysis of the performance of the TOA algorithms in the following sections.

importance, but the presence of sidelobes has a significant impact on performance when leading-edge algorithms are used. The reason for this is that for optimum performance the first significant signal on the leading edge should be used to determine the TOA. If there are sidelobes due to bandlimiting, then these artifacts can be incorrectly inferred to be signal, resulting in inaccurate TOA determination. To minimize these effects, the detection threshold must be increased; as will be shown, this reduces the accuracy when multipath signals are present.

The β parameter has a significant effect on the performance of the TOA. Generally, the β parameter should be selected to be as low as possible, as this maximizes the chip rate for a given bandwidth. However, as discussed above, if the β parameter is made too small, then the resulting distortions in the correlogram result in loss of performance, particularly in a multipath environment. In the following subsection the analysis is based on a conservative choice of $\beta = 1$. However, the choice $\beta = 0.75$ appears close to the optimum choice, as this increases the chip rate be 33% (and, hence, correspondingly reduces the potential effects of multipath interference), but with only minimal extra distortion in the shape of the correlogram.

By noting that the peak of the correlation corresponds to the signal amplitude, the bandlimited peak amplitude can be expressed as

$$P(\beta) = 2 \int_0^\beta \text{sinc}^2 x \, dx \qquad (4.5)$$

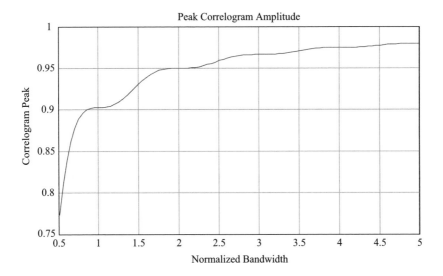

Figure 4.3   Correlogram peak amplitude as a function of the bandwidth parameter.

Another useful parameter (which will be used later in a model in Section 4.3) is the reduction in the relative amplitude, given by

$$\delta = 1 - P(\beta) = 1 - 2\int_0^\beta \mathrm{sinc}^2 x\, \mathrm{d}x \qquad (4.6)$$

It will be shown that the multipath errors are closely related to this parameter.

The variation of the correlogram peak as a function of the bandwidth parameter is shown in Figure 4.3. Observe that the amplitude only slowly reduces for bandwidth parameters $\beta \geq 1$, but the amplitude decreases rapidly for lower bandwidths. The minimum bandwidth parameter for a practical system is $\beta = 0.5$. In this case the radio bandwidth is equal to the chip rate, but this degree of bandlimiting is still satisfactory for data transmissions.[3] However, a common design for TOA measurements is to select a bandwidth to chip-rate ratio of $\beta = 1$. This choice results in the correlation gain reduced by $20\log(0.9)$, or about 1 dB, but otherwise the correlogram shape has minimal other distortions, as shown in Figure 4.3.

4.3 Model of Bandlimited Correlogram

Section 4.2 provided a rigorous derivation of a bandlimited correlogram, but the complexity of the expressions make it difficult to apply in practice. This section introduces approximate models for the bandlimited correlogram, based on the amplitude reduction parameter δ of

---

[3] The second-generation CDMA cellular phone system IS-95 has a radio bandwidth of about 1 MHz and a chip rate of about 1 Mchips/s, so $\beta = 0.5$ in this case, as the null-to-null bandwidth of the unfiltered signal is ± 1 MHz.

Section 4.2. The main purpose of the models is to allow computer modeling, theoretical analyses and real-time computation of the epoch of the signal, based on determining the location of the peak of the correlogram. The approach proposed is to use a least-squares fit of the measured data to obtain the model parameters and, hence, determine the location of the correlogram peak (which by definition is the pn-code epoch location). Thus, the model should be based on approximating the shape of the correlogram near the peak; the shape of the correlogram a long way from the peak is of no interest in this TOA determination method; thus, modeling the shape near the peak should give accurate results even if the overall shape of the model function is not a good match.

Two models are suggested: one is based on a hyperbola and one on a Gaussian curve. Both these curves have a smooth curved peak with an amplitude defined by the δ parameter given by (4.6). The hyperbola asymptotically approaches the straight lines of the triangular shape of the wideband correlogram, whereas the Gaussian curve has a shape at low amplitude that more closely matches the true bandlimited correlogram.

4.3.1 Hyperbolic Model

From the shape of the correlogram computed in Section 4.2, it is clear that the required function is a rounded triangle. A suitable function that meets this criterion is a hyperbola. With correlation time normalized to the chip period, the correlation function can be approximated by the hyperbolic function

$$H(\tau) = A \left[1 - \sqrt{\left(\frac{\tau - \tau_{\mathrm{p}}}{\tau_0}\right)^2 + \delta^2} \right] \qquad H(\tau) \geq 0 \qquad (4.7)$$

where τ_0, A and τ_{p} are parameters to be determined from the least-squares fitting procedure and δ is the known amplitude reduction parameter (a function of the spread-spectrum normalized bandwidth – see (4.6)). In particular, the computed parameter τ_{p} is the required pn-code epoch. Note that, with a zero offset peak, the magnitude of the correlogram at $\tau = 0$ is $A[1 - \delta]$ as required, whereas far from the peak the function becomes $A[1 \pm \tau/\tau_0]$, or the required triangular shape, with τ_0 the chip period.

4.3.2 Gaussian Model

The Gaussian model is defined by the equation

$$G(\tau) = A(1 - \delta)e^{-[(\tau - \tau_p)/a]^2} \qquad (4.8)$$

where a, A and τ_p are parameters to be determined from the least-squares fitting procedure and δ is the known amplitude reduction. The Gaussian shape parameter a is related to the chip period. The value of a for a given bandlimiting parameter β can be determined by a least-squares fit to the 'exact' shape.

Figure 4.4 shows a comparison between the 'exact' correlogram shape and the two models for the bandwidth parameter $\beta = 1$. The model parameters are determined by a least-squares fit procedure. In general, the match is good in both cases, but the Gaussian fit is the closest fit overall.

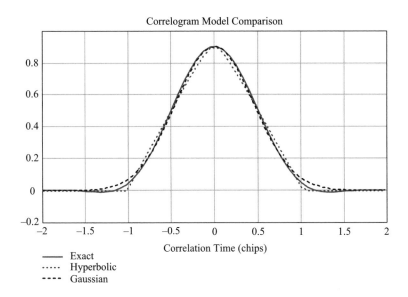

Figure 4.4 Comparison between model and exact correlogram for bandwidth parameter $\beta = 1$. The model parameters are: $\delta = 0.097$, $\tau_0 = 1.022$ (hyperbolic model), $a = 0.634$ (Gaussian model).

The model parameters as a function of the bandwidth parameter β are shown in Figure 4.5. The parameters are normalized so that correlation time is in chips. Thus, the hyperbolic parameter approaches 1 chip as the bandwidth increases, as the shape approaches the wideband triangular form.

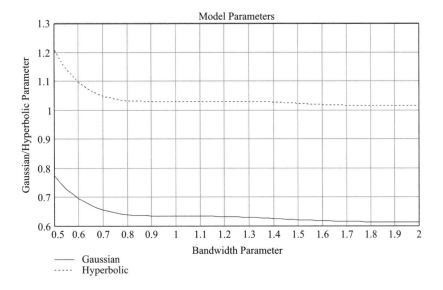

Figure 4.5 Gaussian parameter a and hyperbolic model parameter τ_0 as a function of the bandwidth parameter β. The parameters are normalized so that the correlation time is in chips.

4.4 Peak-Tracking Algorithm Performance

Peak tracking is one of the fundamental and practical TOA estimation methods. Since the TOA tracking algorithm and the early–late gate device were developed [1], it has been widely used in wireless communication, positioning and tracking [2–4]. This is mainly due to its implementation simplicity compared with the more advanced methods such as the super-resolution MUSIC algorithm and ESPRIT method [5–7].

The determination of the TOA of the signal based on detection of the peak of the bandlimited correlogram is analyzed in this section. The analysis is based on the model of the correlogram shape near the peak, as established in Section 4.3. While the correlogram shape is a continuous function of correlation time, because the function is bandlimited the continuous function can be replaced by a sampled version without loss of accuracy, provided the sampling rate is greater than the Nyquist rate. Thus, in the case where the bandwidth of the signal is twice the chip rate, the Nyquist rate is two samples per chip.[4] While sampling at a higher rate is possible, it is important to understand that no further information about the shape of the correlogram is possible. As the correlogram width (without multipath) is ± 1 chip, at most four samples fully describe the shape of the bandlimited correlogram. However, in the case of a peak-detection algorithm, only the two largest samples are used.

The peak-tracking algorithm is based of calculating the correlation at two points (termed P_1 and P_2), one leading and one lagging relative to the peak by τ chips. If the sampling clock in the receiver is misaligned by $\Delta\tau$ from the nominal zero position, the difference between the correlogram amplitudes using the hyperbolic model at these two points is

$$P_1 - P_2 = A\left[1 - \sqrt{\left(\frac{\tau + \Delta\tau}{\tau_0}\right)^2 + \delta^2}\right] - A\left[1 - \sqrt{\left(\frac{-\tau + \Delta\tau}{\tau_0}\right)^2 + \delta^2}\right] \qquad (4.9)$$

If the sampling points are chosen such that they are well removed from the flat region near the peak (or mathematically $\tau \gg \tau_0\delta$ or $\tau \gg 0.1$ chips with $\beta = 1$), then the differential amplitude can be approximated to

$$P_1 - P_2 = -2A\left[\frac{\tau/\tau_0}{\sqrt{(\tau/\tau_0)^2 + \delta^2}}\right]\left(\frac{\Delta\tau}{\tau_0}\right) \approx -2A\left(\frac{\Delta\tau}{\tau_0}\right) \qquad \left(\frac{\tau}{\tau_0} \gg \delta\right) \qquad (4.10)$$

Thus, the differential amplitude is approximately linearly related to the timing misalignment; thus, this measurement can be used to correct for the misalignment. Alternatively, if the sampling points are close together, as used in GPS (or mathematically $\tau \ll \tau_0\delta$ or $\tau \ll 0.1$ with $\beta = 1$), then the differential amplitude is approximated by

$$P_1 - P_2 \approx -2A\left(\frac{\tau}{\tau_0\delta}\right)\left(\frac{\Delta\tau}{\tau_0}\right) \qquad \left(\frac{\tau}{\tau_0} \ll \delta\right) \qquad (4.11)$$

---

[4] In the case of GPS C/A signal, the corresponding Nyquist rate is 20 samples per chip, so that detection of the signal near the peak is possible.

However, the linear relationship in (4.11) is only valid over a very limited range of timing offsets (about $\pm\delta/2$ or ±0.05 chips for the $\beta = 1$ case), so that for larger offsets the differential amplitude asymptotically approaches

$$P_1 - P_2 = -2A\left(\frac{\tau}{\tau_0}\right)\text{signum}(\Delta\tau) \qquad \left(\frac{\tau}{\tau_0} \ll \delta, \quad \Delta\tau \gg \tau\right) \qquad (4.12)$$

Thus, is this latter case, the differential amplitude is independent of the timing misalignment; so, apart from the sign of the differential amplitude, no information on the timing misalignment is obtained from the measurement. The behavior of the differential amplitude as a function of the misalignment and the separation of the measurement points is shown graphically in Figure 4.6.

The differential signal can be used in a feedback loop to maintain tracking with a nominal zero displacement. The effects of random noise and multipath interference on the tracking performance are considered in the following subsections.

4.4.1 Peak-Tracking Algorithm Performance with Random Noise

The performance of the peak-tracking algorithm is now considered when the correlogram amplitude measurements are corrupted by noise. The corrupting noise is assumed to be Gaussian with zero mean, but the combined statistics of the measured amplitude will exhibit

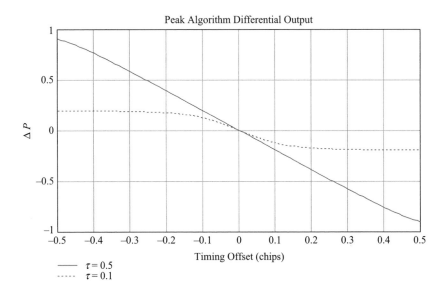

Figure 4.6 Peak-tracking algorithm output as a function of timing misalignment with the points separated by ±0.5 chips and ±0.1 chips. In the former case the delta amplitude is very near linearly related to the offset over the range ±0.5 chips, while in the latter case the function is linear over a small range of about ±0.05 chips and then asymptotically approaches limits of ±0.2. In this asymptotic case, the differential measurement is unrelated to the offset, except for the sign of the offset.

Rician statistics. However, with a large SNR for this application, the statistics closely approximate Gaussian with a zero mean.

Consider the amplitudes at P_1 and P_2 are corrupted with noise n_1 and n_2. Thus, the differential amplitude becomes

$$P_1 - P_2 = -2A\left(\frac{\Delta\tau}{\tau_0}\right) + (n_1 - n_2) \tag{4.13}$$

The tracking loop will try to null the differential amplitude, so that the tracking error will be

$$\frac{\Delta\tau}{\tau_0} = \varepsilon = \frac{1}{2}\left(\frac{n_1 - n_2}{A}\right) \tag{4.14}$$

The tracking error ε will be a random variable; the statistics of the tracking error are now calculated. As the noise is assumed to have zero mean, the mean tracking error will be zero. The variance of the tracking error is given by

$$\text{var}(\varepsilon) = E\left[\varepsilon^2\right] = \frac{1}{4}\left[\frac{n_1^2 + n_2^2 - 2n_1 n_2}{A^2}\right] \tag{4.15}$$

Noting that the SNR is $\gamma = A^2/\overline{n^2}$ and the correlation coefficient between the two noise components is ρ, the STD in the tracking error becomes

$$\sigma_\varepsilon(\text{chips}) = \frac{1}{\sqrt{2}}\sqrt{\frac{1-\rho}{\gamma}} \tag{4.16}$$

The correlation coefficient between the noise at P_1 and P_2 has as yet not been defined. However, an estimate of the coefficient can be made, based on the properties of the receiver correlator. Clearly, if the separation between P_1 and P_2 is zero, then the correlation coefficient of the noise will be one. Further, if noise samples are separated in time by more than about the reciprocal of the bandwidth of the signal, then they will be largely uncorrelated ($\rho = 0$). As in this case the bandwidth is $\pm 1/\tau_0$, the noise will be uncorrelated if the samples are separated by more than half a chip. Analysis shows that the correlation coefficient will vary approximately linearly between these two extremes, so that the correlation coefficient with a separation of x chips ($x \leq 0.5$) can be approximated by $\rho_c = 1 - 2x$. For a typical design with the correlogram sample points separated by more than half a chip the correlation coefficient is zero; thus, in this case, for Gaussian noise the STD in the epoch tracking error is

$$\sigma_\varepsilon = \frac{1}{\sqrt{2}}\sqrt{\frac{1}{\gamma}} = \frac{K}{\sqrt{\gamma}} \qquad (K = 1/\sqrt{2}) \tag{4.17}$$

Note that the tracking error can be expressed as a constant divided by the square root of the SNR; this form is general to all tracking algorithms, but the constant K is particular to each algorithm. Thus, the Gaussian noise tracking performance of various algorithms can be compared by the associated algorithm constant K.

When the separation between the two sample points is small the correlation becomes larger, approaching unity; thus, according to (4.15), the STD in the tracking error will decrease. However, according to (4.11), the tracking error will increase as the two sampling points become closer together, so the two effects are in opposite directions. If the separation of the two measurement points is s chips apart, and using (4.10) to model the nonlinear effects, then the STD in the tracking error due to the two effects will result in modifying (4.17) by the factor F:

$$F = \sqrt{[1 - \rho_c(s)]\left[1 + \left(\frac{2\delta}{s}\right)^2\right]}$$

$$\sigma_\varepsilon = \frac{F}{\sqrt{2}}\sqrt{\frac{1}{\gamma}}$$

(4.18)

The modifying factor F for the STD in the tracking error is shown plotted in Figure 4.7 as a function of the point separation. For separations less than 0.2 chips the error rapidly increases. Thus, the point separation should be greater than 0.2 chips and should be based on the multipath performance, rather than the Gaussian noise performance.

Consider an example with a chip rate of 10 Mchips/s (chip period 100 ns). In this case, with an SNR of 20 dB (typically the minimum useful operating SNR) the STD in the tracking error according to (4.17) is 7 ns (about 2 m). By increasing the SNR this error can be further reduced. Thus, for example, if the SNR is increased to 40 dB, the error reduces to 20 cm. In general, errors associated with receiver noise will not be the limiting effect on performance; rather, the effects of multipath are dominant. The effects are considered in the next section.

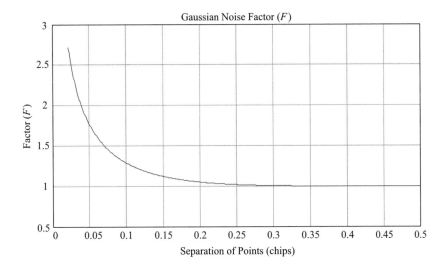

Figure 4.7 Variation in tracking error factor F due to Gaussian noise as a function of the separation of the sampling points. The assumed correlation coefficient is $\rho_c = e^{-4s}$.

4.4.2 Peak-Tracking Multipath Performance

The calculation of the multipath performance requires a model of the multipath interference signal. While this signal will in practice be complex, consisting of many interference sources, a common approach is to calculate the tracking error due to a single in-phase interfering source. Thus, the calculations only provide a rough guide to the real situation, but importantly provide a method for comparing various tracking algorithms in the presence of multipath signals.

The two-ray multipath interference is defined by two parameters, namely the relative amplitude m and the multipath delay τ_m. As the amplitude is defined relative to the direct signal, the direct signal amplitude can be set to unity without loss of generality. Further, for initial analytical calculations, the tracking error will be defined as the error in the position of the signal peak, mathematically defined by the derivative of the combined signal function being zero.

The shape of the bandlimited correlogram was established in Section 4.3, but for analytical calculations in this section the shape near the peak will be approximated by a Gaussian function with a shape parameter a, as established in Section 4.3. The hyperbolic approximation defined in Section 4.3 is not used here due to the complexity of the expressions. The combined correlogram is thus defined by

$$C(t) = e^{-(t/a)^2} + m\,e^{-[(t-\tau_m)/a]^2} \tag{4.19}$$

The peak is determined by differentiating (4.19) and equating to zero. This results in

$$t = m(\tau_m - t)e^{(2t\tau_m - \tau_m^2)/a^2} = f(t) \tag{4.20}$$

The solution of (4.20) for t determines the location of the peak of the correlogram relative to the peak without the multipath interference. No closed-form solution exists in general, so that approximation methods must be employed. In particular, noting that the equation can be expressed in the form $t = f(t)$, if an initial approximated solution is t_0 ($n = 0$), then a better solution is given by $t_n = f(t_{n-1})$. This form of iterative solution can be used to determine the solution to (4.20) to any degree of accuracy required.

The overall tracking error performance, based on the iterative solution described above, is shown in Figure 4.8 for various values of the multipath parameter. For more specific details, refer to the following subsections.

4.4.2.1 Solution with $m = 1$

For the particular case where the multipath signal is equal to the direct signal (very unlikely in outdoor environments, but quite feasible in indoor environments[5]), symmetry considerations mean that the solution is $t = \tau_m/2$. More formally, with $t = \tau_m/2$ the exponent in (4.20) is zero, so that the equation reduces to

$$t = m(\tau_m - t) \Rightarrow t = \tau_m/2 \tag{4.21}$$

Thus, the worst case ($m = 1$) tracking error is equal to half the multipath delay. However, (4.20) has three solutions when the multipath delay becomes large, as the correlogram will have two

---

[5] In indoor environments, the direct signal may be totally absent. In this case, the 'direct' signal is interpreted as the dominant signal which could be delayed relative to the straight-line path between the transmitter and the receiver.

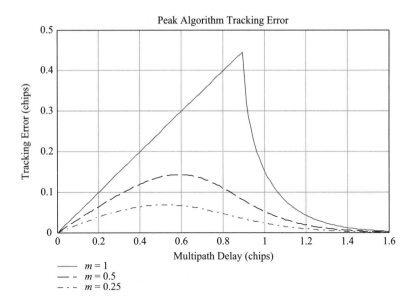

Figure 4.8 Numerical computation of the tracking error as a function of the multipath amplitude and delay parameters.

peaks and a minimum point between the peaks. In this case, the solution given by (4.21) is the position of the minimum point, rather than the first peak, which is the required solution. This effect is shown in Figure 4.8, where with $m = 1$ and multipath delays greater than about 0.9 chips the solution given by (4.21) is no longer true and a numerical solution is required. In fact, analysis shows that the transition from a single to a triple solution for a zero derivative occurs when $\tau_m = \sqrt{2}a$, which for the case in Figure 4.8 is $\tau_m = 0.897$.

4.4.2.2 Solution with Large Delay

Consider an approximate solution with large delay τ_m. As the delay becomes large, the correlogram will evolve from a single peak to dual peaks, of which only the first is of interest. Thus, as the delay becomes large (say greater than 1 chip), the tracking error (namely t in (4.20)) becomes small. Thus, using $t = 0$ on the right-hand side of (4.20), the zeroth-order solution is

$$t_0 = m\tau_m e^{-(\tau_m/a)^2} \tag{4.22}$$

Thus, for large multipath delays, the error (in chips) is proportional to the multipath signal and decreases exponentially with the delay. The maximum epoch tracking error for the zeroth-order solution is given by

$$\varepsilon_{max}\big|_{t_0} = \frac{ma}{\sqrt{2e}}$$
$$\frac{\tau_m^0}{a} = \frac{1}{\sqrt{2}} \tag{4.23}$$

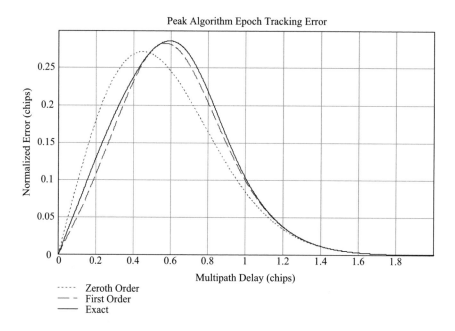

Figure 4.9 Normalized solution (error divided by m) for the tracking error with $m = 0.5$. The zeroth- and first-order solutions are shown, together with the exact numerical solution.

For the particular case considered previously (bandwidth parameter $\beta = 1$), $a = 0.634$ and the maximum tracking error (in chips) of $0.27m$, this occurs with a multipath delay of 0.45 chips. For normalized presentation of results it is convenient to divide by the multipath amplitude, so that the normalized performance curves are solely a function $f(\tau_m)$ of the multipath delay.

Applying the iterative approach described above, higher order solutions can be derived. However, as the analytical expressions become increasingly complex, the method is better suited to numerical calculations. Figure 4.9 shows the normalized solution (error divided by m) for the tracking error with $m = 0.5$. The zeroth- and first-order solutions are shown, together with the exact numerical solution. For $\tau_m > 0.5$, the first-order solution provides a good estimate of the tracking errors. The peak error is about $0.28m$ chips (slightly more than the above zeroth-order solution). Thus, for example, with $m = 0.5$ and a 10 Mchips/s rate the peak error is 4.2 m. This maximum error occurs at $\tau_m \approx 0.59$ chips or 17.5 m delay. For all multipath delays, the first-order solution provides a reasonably good estimate of the tracking errors. All the solutions are asymptotically the same for large multipath delays.

4.4.2.3 Solution with Small Delay

An approximate solution to (4.20) can be determined with small multipath delay, $\tau_m \ll a$. First, the equation can be expressed in normalized form:

$$t = M(ax - t)e^{2xt/a}$$

$$M = me^{-x^2} \qquad x = \frac{\tau_m}{a} \tag{4.24}$$

In this case, where x and hence t are small, the exponential function can be approximated by a series expansion, so that the equation becomes

$$t = M\left[ax + (2x^2 - 1)t + \frac{2}{a}x(x^2 - 1)t^2\right] \tag{4.25}$$

where only the first three terms of the exponential expansion have been included.

The quadratic of (4.25) can be solved for t. When x is small the t^2 term is much smaller than the t term. (This is also true when $x \to 1$.) However, when $x^2 \to 0.5$, the t^2 term is much greater than the t term. Thus, provided $x^2 \ll 0.5$, the zeroth-order solution can be based on only the first two terms (linear equation), resulting in

$$t_0 = \frac{m(ax)}{m(1 - 2x^2) + e^{x^2}} \tag{4.26}$$

For small x, the zeroth-order solution becomes

$$t_0 \approx \frac{m(ax)}{m+1} = \frac{m}{m+1}\tau_m \qquad (\tau_m \ll a) \tag{4.27}$$

Thus, the initial slope of the error function is $m/(m+1)$. For large x (even though this violates the original assumption that $x \ll 0.5$), the zeroth-order solution becomes

$$t_0 \approx \frac{m(ax)}{e^{x^2}} = m\tau_m\, e^{-(\tau_m/a)^2} \qquad (\tau_m \gg a) \tag{4.28}$$

Note that the solution for large x in (4.28) is the same as obtained in (4.22). Thus, the zeroth-order solution for small multipath delays as given by (4.26) is useful over wider ranges of multipath parameters than originally assumed in the above derivation. Figure 4.10 shows the zeroth- and first-order solutions based on the small multipath delay approximation for solving (4.20). When compared with the large multipath delay approximations in Figure 4.9, these small multipath delay approximations are superior, as the solutions are asymptotically correct at both small and large delays.

From Figure 4.10 it can be observed that the zeroth-order solution given by (4.26) provides a reasonable estimate of the tracking error for all multipath delays. Thus, this equation can be used to derive an analytical estimate of the peak tracking error and the delay at which this occurs. Differentiating (4.26) with respect to x and equating to zero results in

$$y = \frac{1}{2} + \frac{m}{2}[(1 + 2y)e^{-y}] \qquad y = x^2 \tag{4.29}$$

A closed-form solution to (4.29) is not possible, but as the equation is in the form $y = f(y)$ the solution can be expressed in the form of a series of solutions converging on the exact solution. Equation (4.23) can be used for an initial estimate, resulting in the first-order estimates

$$y_1 = \left(\frac{\tau_m^1}{a}\right)^2 = \frac{1}{2} + \frac{m}{\sqrt{e}}$$

$$\tau_m^1 = a\sqrt{\frac{1}{2} + \frac{m}{\sqrt{e}}} \approx \frac{a}{\sqrt{2}}\left[1 + \frac{m}{\sqrt{e}}\right] \tag{4.30}$$

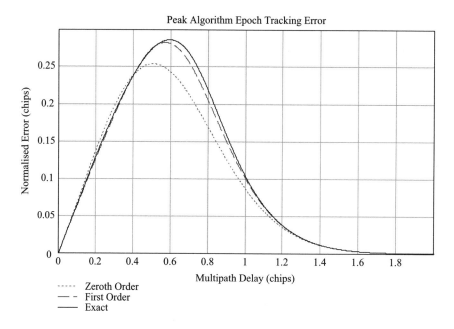

Peak Algorithm Epoch Tracking Error

- - - - - Zeroth Order
— - First Order
——— Exact

Figure 4.10 Normalized solution (error divided by m) for the tracking error with $m = 0.5$, based on the small multipath delay approximation. The zeroth- and first-order solutions are shown, together with the exact numerical solution. For all multipath delays, the first-order solution provides a reasonably good estimate of the tracking errors. All the solutions are asymptotically the same for both small and large multipath delays.

Thus, the multipath delay for maximum error is approximately a linear function of the multipath ratio m. For the case $\beta = 1$ with $a = 0.634$, the approximate solution becomes $0.45 + 0.272m$ chips. For the case of $m = 0.5$, the multipath delay estimate is 0.57 chips, compared with 0.59 chips based on the exact solution (see Figure 4.10). The corresponding maximum error can be computed by solving (4.20) for t with the above estimate of the multipath delay parameter; alternatively, Figure 4.10 can be used.

The analysis in this section has been based on the tracking error of the peak of the correlogram. However, the actual algorithm must use the correlogram amplitude as measured at points P_1 and P_2 on either side of the peak, with the tracking point defined when the two amplitudes are equal. The actual algorithm performance is expected to be very close to that determined by the derived analytical expressions, but the separation parameter τ will have some effect on the performance.

Because the signal is bandlimited, the number of independent samples on the correlogram is limited. In particular, if the spectral bandwidth is limited to twice the chip rate (the most common form of bandlimiting the signal), then the nominal triangular-shaped correlogram has at most four samples: two on the leading edge and two on the trailing edge. For the peak-detection algorithm, two samples either 0.25 or 0.5 chips apart are typically used in estimating the location of the peak. The location of the peak can be estimated by appropriate curve fitting. As the shape of the correlogram is flat near the peak, if the two samples are too close together then the accuracy will reduce due to Gaussian noise and multipath signals. An alternative approach is to use the leading edge of the correlogram, where the slope is greater and less susceptible to multipath interference. Two such algorithms, the leading-edge projection algorithm and the leading-edge ratio algorithm, are discussed in the following sections.

4.5 Leading-edge Projection Tracking Algorithm

The leading-edge detection method is another widely considered TOA estimation method in practice. The least-squares curve fitting can be applied to the rising edge of the correlation spike to reduce the effect of multipath fading [8]. Also, the ratio of pulse samples on the leading edge can be employed as a detection statistic for TOA estimation [9]. Another leading-edge detection method is using a threshold [10]. The main drawback of the threshold-based approach is the difficulty of determining the threshold. Although there are other TOA methods, such as the sign-correlation algorithm [11], the energy-collection approach [12], the maximum likelihood method [13] and the generalized likelihood ratio test algorithm [14], the peak tracking described in the preceding section and the leading-edge detection are more attractive in practice due to their good accuracy and low complexity. In this section, the leading-edge projection method is studied, whereas the leading-edge ratio algorithm will be discussed in the following section.

Because the multipath signal will always be delayed relative to the direct signal, one technique for minimizing the tracking error is to base the algorithm on tracking the leading edge of the correlogram. Clearly, if the section of the correlogram which is uncorrupted by the multipath signals is used for estimating the pn-code epoch, then in theory no tracking errors will result. This ideal situation has to be modified because of two factors, namely the signal bandwidth and the SNR. If the multipath delay is small, then only a small section of the leading edge will be uncorrupted by the multipath. However, this section of the leading edge is severely affected by bandlimiting and system noise, so that a practical algorithm must use a section of the leading edge which is only somewhat corrupted by the multipath signals. Nevertheless, the expectations are that the performance will be superior to the correlogram peak algorithm.

With reference to Figure 4.11, the projection algorithm is again based on two points (P_1 and P_2), but in this case both points are on the leading edge of the correlogram. The positions of these two points are algorithm parameters. The first point P_1 is at the leading edge at an

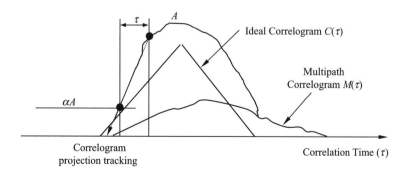

Figure 4.11 Geometry associated with the leading-edge projection tracking algorithm, showing tolerance to multipath.

amplitude defined by a threshold parameter α, with αA being the threshold amplitude and A being the correlogram amplitude. The α parameter is selected so that the P_1 point is sufficiently out of the noise and above the curvature distortion associated with bandlimiting. (See Figure 4.4 for the effects on the correlogram shape by bandlimiting the spread-spectrum signal.) A typical value is $\alpha = 0.15$, but this point can be adjusted for the SNR. The second point P_2 is defined by a separation parameter τ from point P_1; typical values are in the range of 0.25 to 0.5 chips. The basis of the algorithm is then to define the location of the signal epoch by projecting P_2 through P_1 until it intercepts the x-axis of the correlogram. The intersection of this straight line with the x-axis is defined as the signal epoch.

The performance of the algorithm with both random noise and multipath interference is analyzed in the following subsections. However, to simplify the analysis, the unfiltered (triangular) correlogram shape for the ideal correlogram is assumed to simplify the mathematical analysis greatly. This is justified, as the middle section on the correlogram (say, in the amplitude ranging from 0.2 to 0.8 with bandlimiting parameter $\beta = 1$) is essentially undistorted – namely a straight-line segment. Thus, provided the positions of P_1 and P_2 are within this range, the use of the triangular-shape correlogram will also be representative of the bandlimited case.

4.5.1 Gaussian Noise Performance

The analysis under Gaussian noise conditions assumes a simplified theoretical correlogram, namely a triangular shape for the leading edge with a rise-time of 1 chip. As shown previously, the effect of bandlimiting is to round off the sharp intersections of the triangular shape, but away from these areas the basic triangular shape remains largely unaltered. As the two points are located within this minimally distorted region, the assumption of using the ideal triangular shape has a minimal effect on the derived performance.

The following analysis assumes that the correlogram is corrupted by random Gaussian noise, with zero mean and known noise power. The noise at the two points P_1 and P_2 is not assumed to be statistically independent, but the correlation coefficient is assumed to be a function of the separation τ. The simplified geometry of the algorithm is shown in Figure 4.12. The diagram shows a triangular correlogram of amplitude A and rise time of 1 chip. The algorithm defines two points. The first point is nominally located on the leading edge of the correlogram where it

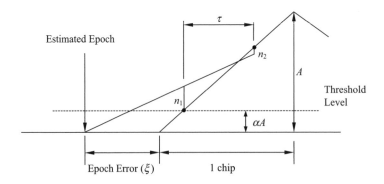

Figure 4.12 Geometry showing main correlogram, noise n_1 and n_2 and the estimated epoch.

crosses a threshold level of amplitude αA. The second point is defined as lagging the first point by τ chips. Both points are corrupted by Gaussian noise (n_1 and n_2 respectively), which is assumed to have identical Gaussian statistics with zero mean and variance σ_n^2. The epoch is estimated by extrapolating the straight line joining the points to intersect the x-axis (correlation time in chips). The error in the estimated epoch is then ξ chips.

The statistical analysis is required to determine the expected epoch error (or $E(\xi)$) and the variance of the error (or var(ξ)). The first step in the analysis is to obtain an expression for the epoch error in terms of the algorithm parameters α and τ and the SNR γ, defined as A^2/σ_n^2. Simple geometric calculations show that the epoch error is given by

$$\xi = \alpha - \tau\left(\frac{\alpha A + n_1}{A\tau + n_2 - n_1}\right) = -\frac{\left[\dfrac{\alpha}{\tau}\left(\dfrac{\Delta n}{A}\right) + \left(\dfrac{n_1}{A}\right)\right]}{1 - \dfrac{1}{\tau}\left(\dfrac{\Delta n}{A}\right)} \tag{4.31}$$

where $\Delta n = n_1 - n_2$. Note also that the final expression is in normalized form, with the noise terms divided by the signal amplitude. From the denominator, it clearly can be observed that the error has a singularity when

$$\frac{\Delta n}{A} = \tau \tag{4.32}$$

Thus, for the algorithm to provide reasonably accurate results, the SNR must obey the condition $\gamma \gg 2/\tau^2$, otherwise large epoch measurement errors can occur when the noise is sufficiently large that the singularity condition occurs (or nearly occurs). As, typically, the algorithm parameter τ is about 0.5 chips, this condition implies that the SNR should be much greater than 8 (or 9 dB). As the typical SNR will exceed 20 dB, this restriction is not a practical problem. However, reducing τ to (say) 0.1 chips results in the unsatisfactory requirement that the SNR must be much greater than 23 dB. Further, increasing the parameter τ beyond about 0.7 is not possible, as the pulse width is 1 chip. Thus, a practical range for τ is about 0.3 to 0.7 chips.

Providing the SNR is sufficiently large, (4.31) can be simplified using the first few terms of the series expansion to

$$\xi = -\left[\left(\frac{n_1}{A}\right) + \frac{\alpha}{\tau}\left(\frac{\Delta n}{A}\right) + \frac{1}{\tau^2}\left(\frac{n_1}{A}\right)\left(\frac{\Delta n}{A}\right)\right] \tag{4.33}$$

The mean of the epoch error is now obtained from the expectation $E(\xi)$. The expectation of the first two terms is zero (the noise is assumed to have zero mean) and the expectation $E(n_1 n_2)$ is related to the noise variance by the correlation parameter ρ. The only nonzero term results in the estimate of the mean epoch error as

$$E(\xi) = -\frac{1}{\tau^2}\left[\frac{\sigma_n^2(1-\rho)}{A^2}\right] = -\frac{1-\rho}{\tau^2\gamma} \tag{4.34}$$

Thus, the algorithm results in a biased estimate of the epoch location (provided the correlation coefficient is not unity). However, for an SNR of (say) 20 dB and a spacing parameter τ of 0.5 chips, the worst-case mean error is only 0.04 chips, which is small but significant. However, as the SNR can be measured, the bias error can be corrected, if this is considered desirable for maximum accuracy. For most practical situations, the mean error can be considered zero.

The variance of the epoch error can be computed with the aid of the expression $\text{var}(\xi) = E(\xi^2) - E(\xi)^2$. The expectation of ξ^2 can be calculated directly from the square of (4.33) and taking the expectation of the individual terms, resulting in the expression

$$E(\xi^2) = \frac{1}{\gamma}\left[1 + 2\left(\frac{\alpha}{\tau}\right) + 2\left(\frac{\alpha}{\tau}\right)^2\right] \tag{4.35}$$

where higher order terms of the SNR have been dropped. Noting that the square of (4.34) only results in higher order terms of the SNR (and are thus also ignored), the first-order estimate of the variance of the epoch error is simply given by (4.35). From (4.35), it can be observed that the variance in the epoch error is a weak function of the ratio α/τ, as the constraints on α (always small, in the range 0.1–0.2) and τ (in the range 0.3–0.7 – see above) means that $\text{var}(\xi) \approx 1/\gamma$. For example, if $\alpha = 0.15$ and $\tau = 0.5$ chips, then $\text{var}(\xi) = 1.69/\gamma$. For comparison, the corresponding expression for the peak algorithm (see (4.35)) is $\text{var}(\xi) = 0.5/\gamma$. Thus, the leading-edge algorithm has the worst noise performance, but the aim of the leading-edge algorithm is to improve the multipath performance. As the system designer can control the receiver noise, but not the multipath, this performance tradeoff is appropriate.

Thus, the requirement to minimize the error in the epoch estimate in the presence of Gaussian noise is to make the threshold level as small as possible and the separation of the two points as large as possible; these constraints will minimize both the bias error and the variance of the error. In practice, however, the STD of the epoch error is only weakly dependent on these parameters, so that the STD is approximately equal to

$$\sigma_\xi \approx \sqrt{\frac{1}{\gamma}} \tag{4.36}$$

Thus, for an SNR of (say) 30 dB, $\alpha = 0.15$ and $\tau = 0.5$ chips, the STD in the estimated epoch position is about 0.041 chips, or about 4 ns for a 10 MHz chip rate. This variation can be reduced by averaging many such estimates.

4.5.2 Multipath Performance

The analysis of the effect of multipath errors is again based on the simplified triangular correlogram used for the Gaussian analysis, plus a single multipath interference path. Further, the phase of the multipath signal is assumed to be the same as the direct signal (this restriction will be eased later). The multipath signal is assumed to be m times the amplitude of the direct signal ($m \le 1$). While it is recognized that real multipath conditions will involve many interfering signals, the single-interference analysis provides the basis for the superposition of multiple signals.

The geometry of the multipath correlogram is shown in Figures 4.13 and 4.14. There are two cases, depending on the multipath delay. The correlogram is measured at two points separated by τ chips (typically $\tau = 0.5$). The first (or threshold) point is τ_t chips from the leading edge. The multipath signal is delayed by τ_m chips relative to the direct signal. The direct signal has

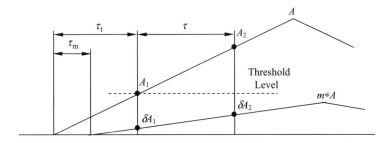

Figure 4.13   Case 1. Multipath delay τ_m less than the threshold delay τ_t.

amplitude A and the multipath signal has an amplitude mA. The epoch is determined by the extrapolation of the line joining points P_2 and P_1 to the time axis.

The epoch error due to the multipath signals δA_1 and δA_2 respectively at points 1 and 2 is given by

$$\varepsilon = \left[\frac{A_1}{A_2 - A_1} - \frac{A_1 + \delta A_1}{(A_2 + \delta A_2) - (A_1 + \delta A_1)}\right]\tau \tag{4.37}$$

Three cases can be identified, namely when the multipath delay is less than the threshold delay, when the multipath delay is between the delays of points 1 and 2 and when the delay is greater than the delay of point 2. In the latter case, there is clearly no corruption of the data at the measurement points, so that there is no epoch measurement error. For the case where the multipath delay is less than the threshold delay, (4.37) can be evaluated to give

$$\Delta\tau_e = \frac{m}{m+1}\tau_m \qquad (\tau_m < \tau_t) \tag{4.38}$$

If the multipath ratio parameter m is small, then the measurement error (in chips) is approximately m times the multipath delay. Note that in this case the algorithm parameters (such as the threshold delay and the separation delay) do not affect the measurement error; the error is solely a function of the multipath signal characteristics (delay and multipath ratio). If (4.38) is compared with (4.27), for the peak-detection algorithm it is observed that the expressions are the same. However, the constraint in (4.38) limits the maximum error by a parameter in the leading-edge algorithm, while there is no such constraint for the peak algorithm.

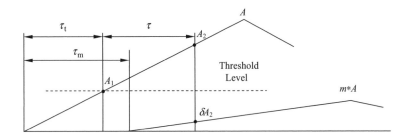

Figure 4.14   Case 2. Multipath delay τ_m lies between the threshold delay τ_t and the delay of the second measurement point.

For the case where the multipath delay lies between the delays of the measurement points, the epoch measurement error of (4.37) can be evaluated to give

$$\Delta\tau_e = \frac{m\tau_r\tau_t}{\tau + m\tau_r} \qquad (\tau_r = \tau_t + \tau - \tau_m > 0) \tag{4.39}$$

Thus, the delay is again proportional to the multipath ratio m. However, the error is also proportional to the threshold delay τ_t and to τ_r, the multipath delay relative to P_2. Note that τ_r is constrained to less than the algorithm parameter τ; so, again, the maximum error is constrained by a parameter of the algorithm, rather than the multipath delay.

Figure 4.15 shows the epoch error as a function of the multipath delay. Observe that the error peaks when the multipath delay is equal to the threshold delay ($\tau_m = \tau_t$). In particular:

$$\Delta\tau_e \leq \frac{m}{m+1}\tau_t \approx m\tau_t \tag{4.40}$$

Again, comparing (4.40) with the corresponding expression (4.27) for the peak algorithm shows the maximum error is proportional to the multipath ratio m, but the maximum error for the leading-edge algorithm is constrained by an algorithmic parameter, while for the peak algorithm the maximum error is determined by the shape of the correlogram (and hence the bandwidth). Thus, unlike the peak algorithm, where the epoch errors due to multipath signals are related directly to the multipath signal parameters (multipath delay and multipath ratio), the leading-edge projection algorithm is more constrained, so that the maximum error is essentially given by the threshold delay parameter. To minimize the epoch error for a given multipath

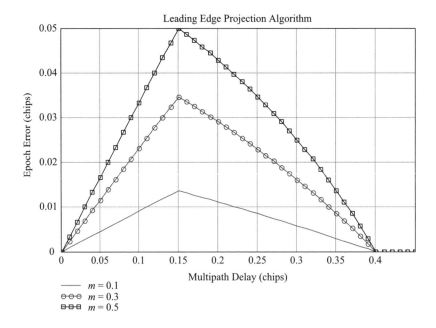

Figure 4.15 Epoch errors, as a function of the multipath delay for three levels of multipath. Threshold delay is 0.15 chips and the separation 0.25 chips.

signal, the threshold delay should be as small as possible. In practice, the threshold value is related to the SNR and the signal bandwidth. With bandlimiting of the signal, the correlogram near the threshold curves significantly, making projection measurements inaccurate. Further, if the threshold is set too low, then noise causes 'false' peaks, resulting in large epoch errors. Thus, in practice, the threshold typically is limited such that $\tau_t \geq 0.15$ for the bandlimiting parameter $\beta = 1$. The threshold must be adjusted accordingly for other band-widths. For example, if the bandwidth parameter is reduced to $\beta = 0.5$ (the minimum acceptable value), then the threshold should be increased to 0.2, as the bandlimited correlogram has significant sidelobes which can be confused for the (small) direct signal if the threshold is not increased.

Now consider a comparison with the performance of the peak-tracking algorithm. With a typical threshold delay of 0.15 chips (or $\alpha = 0.15$) and a multipath parameter $m = 0.5$, the peak error is 0.05 chips. For comparison, the corresponding peak algorithm error is 0.135 chips, or 2.7 times greater. Thus, the leading-edge algorithm has much reduced errors due to multipath; but, as shown previously in Section 4.5.1, the Gaussian noise errors will be 1.8 times greater. Thus, there is a tradeoff of worse noise performance for improved multipath performance.

4.6 Leading-edge Ratio Algorithm

This section describes another epoch tracking algorithm, namely the leading-edge ratio algorithm. Like the projection algorithm described in Section 4.5, the ratio algorithm uses two points (P_1 and P_2) on the leading edge of the correlogram to determine the location of the epoch. However, as the name implies, the algorithm uses the ratio of the correlogram amplitudes as a measure of correlation time on the leading edge. As a ratio is used, the calculated epoch is independent of the absolute amplitude of the correlogram. Further, unlike the projection algorithm, the leading-edge ratio algorithm does not require the correlogram peak for normalization of the tracking parameters, and thus the ratio algorithm is expected to perform somewhat better than the projection algorithm. Also, unlike the projection algorithm, which can have large tracking errors (see (4.32)), the leading-edge algorithm restricts the epoch to a narrow range of at most 1 chip on the leading edge of the correlogram.

For the analysis of the performance of the ratio algorithm, the correlogram shape used is not the idealized triangular shape used for the projection algorithm; rather, a Gaussian shape is assumed. For example, the correlogram shape with a bandlimiting parameter $\beta = 1$ is shown in Figure 4.16, which shows that the bandlimited correlogram is close to a Gaussian shape. The Gaussian function is defined as

$$G(t) = e^{-(t/a)^2} \tag{4.41}$$

Based on the Gaussian approximation to the correlogram, the ratio of the P_1 and P_2 amplitudes is

$$\rho(t) = \frac{G(-t)}{G(-t+\tau)} = e^{(\tau^2 - 2t\tau)/a^2} \qquad (t > \tau) \tag{4.42}$$

where τ is the separation of the points P_1 and P_2. Figure 4.16 shows the correlogram with typical tracking parameters of $\rho_0 = 0.5$ and $\tau = 5/16$ chips, where ρ_0 is the set-point

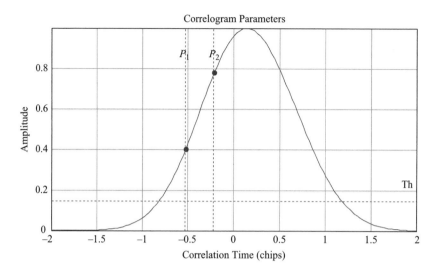

Figure 4.16 Gaussian model correlogram showing the location of P_1 and P_2 for typical tracking parameters. The bandwidth parameter is $\beta = 1$.

for the tracking loop. The threshold parameter (Th) sets the minimum acceptable amplitude for P_1.

The basis of the tracking algorithm is that the ratio is a function of the correlation time parameter t. Figure 4.17 shows this function for the above-defined parameters. Notice that the ratio is approximately a linear function of the displacement from the nominal set-point with

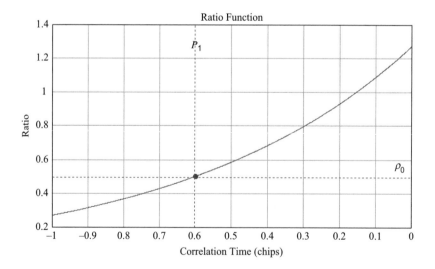

Figure 4.17 Ratio function for correlogram with a separation of P_1 and P_2 of 5/16 chip parameters. Notice that within about ± 0.3 chips of the set-point the error function is approximately linear. The correlation time is zero at the peak, so the tracking point is on the leading edge.

$\rho_0 = 0.5$ (or $t_0 = 0.523$ chips). In particular, by differentiation of the ratio function, the slope is given by

$$\frac{d\rho}{dt} = -\frac{2\tau}{a^2}\rho(t) \approx -\frac{2\tau}{a^2}\rho(t_0) \tag{4.43}$$

where t_0 is the correlation time at the set-point where $\rho = \rho_0$. Thus, by measuring the ratio, the correction of the epoch is given by

$$\Delta t = \frac{\rho - \rho_0}{d\rho/dt|_{t=t_0}} = \alpha_0 \Delta\rho \tag{4.44}$$

At the set-point, the derivative is about $-0.78\,\text{chips}^{-1}$, so the correction factor is about $\Delta t = -1.28\Delta\rho$ chips. Thus, if the error in the ratio can be computed, then the corresponding epoch error can be estimated.

4.6.1 Gaussian Noise Performance

From the above introductory analysis, the effect of Gaussian noise on the ratio function can be estimated. The basic technique is to calculate the error in the ratio function due to errors in the correlogram amplitudes at P_1 and P_2 and then use the approximate linear relationship (4.44) with the epoch tracking error to determine the epoch error These errors are then assumed to be random (Gaussian with zero mean), so that statistical properties can be estimated. The end result of the analysis is the STD of the tracking error as a function of the STD of the noise and the algorithm parameters.

Consider the corruption of the correlogram with noise, the characteristics of which are not yet stated except for the assumption that the noise amplitude is much less than the signal amplitude. If the signals at P_1 and P_2 are A_1 and A_2 respectively and the corresponding noise are n_1 and n_2, then the error in the ratio is given by

$$\Delta\rho = \frac{A_1 + n_1}{A_2 + n_2} - \frac{A_1}{A_2} = \frac{n_1 - \rho_0 n_2}{A_2 + n_2} = \frac{\lambda_1 - \rho_0 \lambda_2}{1 + \lambda_2}$$

$$\rho_0 = \frac{A_1}{A_2} \qquad \lambda = \frac{n}{A_2} \approx \frac{n}{A} \tag{4.45}$$

In (4.45), ρ_0 is the ratio set-point and A is the correlogram peak amplitude, such that $\lambda = n/A$ is the reciprocal of the *amplitude* SNR, which is assumed to be large. Equation (4.45) is exact for any noise; but, assuming the SNR is large, the expression for the ratio error can be approximated by

$$\Delta\rho \approx \lambda_1 - \rho_0 \lambda_2 - \lambda_1 \lambda_2 + \rho_0 \lambda_2^2 \tag{4.46}$$

Now consider the noise to be random with zero mean and variance σ_n^2 for both n_1 and n_2. The two noise sources are not assumed to be statistically independent, but related by the correlation coefficient ρ_c, the value of which is currently undefined. As the noises are random variables, the ratio error will also be a random variable, whose statistical properties are to be determined.

First consider the mean of the ratio error. Using the fact that the noise has zero mean, the mean (or expected value) of the ratio error is given by

$$\mu_{\Delta\rho} = E\left[\lambda_1 - \rho_0\lambda_2 - \lambda_1\lambda_2 + \rho_0\lambda_2^2\right] = -\frac{E[n_1 n_2]}{A_2^2} + \rho_0\frac{E[n_2^2]}{A_2^2} = (\rho_0 - \rho_c)\frac{\sigma_n^2}{A_2^2} \qquad (4.47)$$

Noting that the SNR γ is the ratio of the signal power A^2 to the noise variance, the final expression for the mean of the ratio error is

$$\mu_{\Delta\rho}(\text{chips}) \approx \frac{1}{\gamma}(\rho_0 - \rho_c) \qquad (4.48)$$

Note that the symbol ρ_0 is the ratio function set-point and ρ_c is the correlation coefficient for the noise in P_1 and P_2. The SNR of a receiver will be at least 20 dB, so the corresponding mean error will be less than 0.01 chips. Thus, for all practical purposes, the mean error in the ratio function can be considered zero.

The variance of the ratio error will now be estimated. From (4.46), the variance is given by

$$\sigma_{\Delta\rho}^2 = E[\Delta\rho^2] = E[(\lambda_1 - \rho_0\lambda_2 - \lambda_1\lambda_2 + \rho_0\lambda_2^2)^2] \approx E[\lambda_1^2] + \rho_0^2 E[\lambda_2^2] - 2\rho_0 E[\lambda_1\lambda_2] \qquad (4.49)$$

In (4.49), use has been made of the mean of the ratio function being very nearly zero, and third-order and higher terms are ignored as the SNR is high. Assuming the noise variances at P_1 and P_2 are the same and their means are zero, the STD of the ratio error from (4.49) becomes

$$\sigma_{\Delta\rho} = \sqrt{\frac{1 + \rho_0^2 - 2\rho_0\rho_c}{\gamma}} \qquad (4.50)$$

The STD in the epoch can now be calculated using (4.44), which provides an approximate linear translation:

$$\sigma_\varepsilon = |\alpha_0|\sqrt{\frac{1 + \rho_0^2 - 2\rho_0\rho_c}{\gamma}} \qquad (4.51)$$

The STD in the ratio depends on the set-point ρ_0 and the noise correlation coefficient. In the worst case the correlation coefficient is one and in the best case the correlation coefficient is zero. Thus, the ratio error STD is in the range

$$\frac{|\alpha_0|(1 - \rho_0)}{\sqrt{\gamma}} \leq \sigma_\varepsilon \leq \frac{|\alpha_0|\sqrt{1 + \rho_0^2}}{\sqrt{\gamma}} \qquad (4.52)$$

For a set-point of $\rho_0 = 0.5$, $\alpha_0 = 1.28$ chips and an SNR of 20 dB, the worst-case STD is in the range of 0.064 to 0.143 chips, or 7.3–16.3 ns at 10 Mchips/s.

The correlation coefficient between the noise at P_1 and P_2 has as yet not been defined. However, an estimate of the coefficient can be made based on the properties of the receiver correlator. Clearly, if the separation between P_1 and P_2 is zero, then the correlation coefficient of the noise will be one. Previous discussion in Section 4.4.1 established that the correlation coefficient with a separation of τ chips ($\tau < 0.5$) can be approximated by $\rho_c = 1 - 2\tau$ and is essentially zero for greater separations. Applying this to (4.51) results in the STD in the epoch as

$$\sigma_t = |\alpha_0| \sqrt{\frac{1 + \rho_0^2 - 2\rho_0(1 - 2\tau)}{\gamma}} \qquad (\tau \le 0.5) \qquad (4.53)$$

For example, with $\rho_0 = 0.5$ and $\tau = 5/16$ chips, the STD in the epoch due to Gaussian noise becomes $1.197/\sqrt{\gamma}$ chips. If the SNR is 20 dB, then a 10 Mchips/s system will have a STD of 0.12 chips (2 ns), or 3.6 m. If the SNR is increased to 40 dB, then this reduces to 5 cm.

Figure 4.18 shows the (normalized) STD parameter as a function of the ratio set-point ρ_0 and the P_1–P_2 separation τ. As can be observed, there is not much variation in the tracking error as a function of the separation and ratio set-point parameters; thus, these parameters will be defined by the requirement of multipath mitigation rather than the reduction of the effects of Gaussian noise.

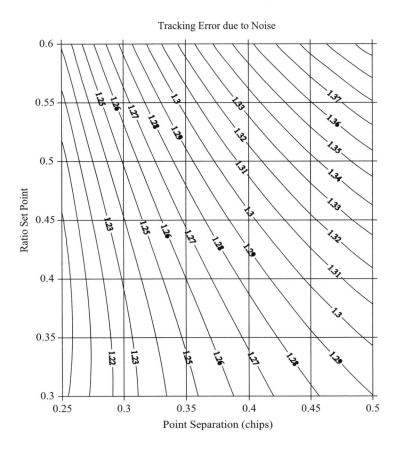

Figure 4.18 Noise STD as a function of the P_1 and P_2 separation and the ratio set-point. The STD (in chips) is the plotted parameter times the square root of the SNR.

4.6.2 Multipath Interference Performance

As with the other algorithms analyzed, the performance with multipath interference will be estimated assuming a single interference source, with the delay and amplitude as parameters. The analytical calculations for even this simple case are too complex to be useful, so that a numerical solution will be presented. The epoch tracking error is a function of the algorithm parameters ρ_0 and τ and the multipath amplitude and delay parameters m and τ_m. For the following performance graphs it is assumed that $m = 1$ (multipath signal equal to the direct signal); the performance curves are approximately proportional to the multipath amplitude, so that the epoch tracking errors for more realistic multipath amplitudes can be determined by scaling by the multipath amplitude parameter m.

The performance with a constant separation of 5/16 chip between P_1 and P_2 is shown in Figure 4.19 as a function of the ratio set-point ρ_0 as a parameter in the range 0.4–0.6 and with a constant threshold of 0.15. The maximum error occurs with a multipath delay in the range 0.35–0.45 chips. As can be observed, the tracking error decreases as the tracking set-point is reduced. Reducing the set-point has the effect of moving the points P_1 and P_2 closer to the beginning of the correlogram where the correlogram is less distorted by the multipath signal, but where it is more subjected to noise and correlogram distortions. For small multipath delays, the data in the graph shows that the epoch error is given by

$$\Delta\tau_e = \frac{m}{m+1}\tau_m \tag{4.54}$$

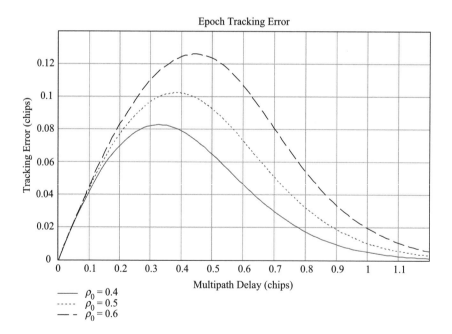

Figure 4.19 Computed epoch tracking error with a separation of 5/16 chips, a threshold of 0.15 and a multipath amplitude parameter $m = 1$. The ratio set-point ρ_0 ranges from 0.4 to 0.6.

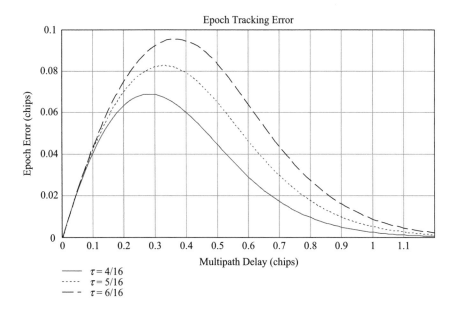

Figure 4.20   Computed epoch tracking error with a ratio set-point of $\rho_0 = 0.4$, a threshold of 0.15 and a multipath amplitude parameter $m = 1$. The separation τ ranges from 4/16 to 6/16 chips. For smaller multipath ratios ($m < 1$) the errors can be obtained by appropriate scaling by m.

This expression is inferred from the numerical data, but is the same as obtained analytically for the leading-edge projection algorithm. The peak errors for ρ_0 of 0.4, 0.5 and 0.6 are respectively 0.083 chips, 0.102 chips and 0.126 chips. The corresponding peak error for the leading-edge projection algorithm is 0.075 chips with the same threshold of 0.15, but the ratio algorithm has an option of reducing the set-point parameter ρ_0 to reduce the peak errors. However, the minimum practical value for ρ_0 is about 0.3.

The performance with $m = 1$, a constant ratio set-point of 0.4 and with the separation varying from 4/16 to 6/16 chips is shown in Figure 4.20. As can be observed, the tracking error decreases as the separation decreases. The maximum error occurs with a multipath delay of 0.29 to 0.37 chips. The separation must be at least 4/16 chips if the minimum threshold is set to 0.15. The peak errors for the point separations of 4/16 chips, 5/16 chips and 6/16 chips are respectively 0.069 chips, 0.093 chips and 0.095 chips.

The combined effect of varying the separation and the set-point is shown in Figure 4.21, with the above lower limits applying. The maximum error is plotted for the case of $m = 1$. The minimum error occurs when $\rho_0 = 0.35$ and $\tau = 4/16$ chips, and the error increases as each of the parameters increases. However, the tracking errors with Gaussian noise (see Section 4.6.1) *decreases* as these parameters increase, so that a compromise solution is required.

The suggested parameters to be used are $\rho_0 = 0.45$ and $\tau = 4/16$ chips. These parameters result in a maximum normalized multipath tracking error of 0.073 chips. However, as the algorithm has three main parameters, the algorithm can be fine-tuned to cater for a wide range of circumstances.

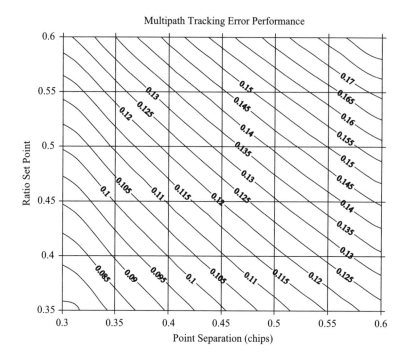

Figure 4.21 Contour plot of the maximum epoch tracking error as a function of the separation of points P_1 and P_2 and the ratio set-point (0.35 to 0.6), and with a (maximum) multipath amplitude of $m = 1$ and a threshold of 0.1. The suggested parameters are a separation of 5/16 chips and a ratio set-point of 0.45, when the tracking error is about 0.095 chips.

4.7 Multipath Phase

The analysis of the multipath performance of the epoch tracking algorithms in previous sections assumed that the multipath interference signal was in phase with the direct signal. While this assumption simplifies the mathematical analysis, the actual multipath signal will have a phase difference from the direct path which is related to the geometry of the radio signal scattering environment. While the real-world environment will have multiple scattering sources, again the analysis will be restricted to a single scattering source. However, if necessary, more complex situations can be analyzed by the summation of the sources.

Figure 4.22 shows the geometry of the multipath interference, where the differential path lengths result in a phase angle given by

$$\phi = \frac{2\pi}{\lambda}(r_1 - r_2) \tag{4.55}$$

where λ is the wavelength of the radio signal. For the types of radiolocation system considered in this book, the wavelength is typically in the range of 5–20 cm, so that the

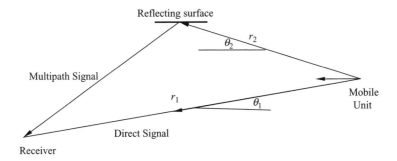

Figure 4.22 Geometry of multipath interference. The vehicle moves such that the angle between the velocity vector and the direct signal path to the receiver is θ_1. The multipath signal is scattered off an object so that the angular direction of the scattered signal at the receiver is, in general, different from the direct signal. This situation results in a variation in the received signal phase over time.

propagation ranges r will be many wavelengths. Now the magnitude[6] of the total signal S_t is then given by

$$|S_t| = |A + mAe^{j\phi}| = A\sqrt{(1 + m\cos\phi)^2 + (\sin\phi)^2}$$
$$\approx A\left(1 + m\cos\phi + \frac{m^2}{2}\sin^2\phi + \dots\right) \tag{4.56}$$

Thus, provided the multipath ratio is relatively small, the magnitude of the correlogram signal can be approximated by the first two terms in (4.56). Further, with reference to the previous epoch error analysis, the results can be extended to the arbitrary phase case by substituting $m\cos\phi$ for m.

Thus, for example, consider the epoch error for the leading-edge projection algorithm. From (4.40), the epoch error is now constrained by

$$\Delta\tau_e \leq \frac{|m|\cos\phi(t)}{1 + |m|\cos\phi(t)}\tau_t \tag{4.57}$$

The phase $\phi(t)$, in general, is a function of time. For example, with reference to Figure 4.21, for a vehicle moving with a speed v, the phase–time function is given by

$$\phi(t) = 2\pi f_s t$$
$$f_s = \frac{\partial}{\partial t}\left(\frac{r_1 - r_2}{\lambda}\right) = \frac{v}{\lambda}(|\cos\theta_1 - \cos\theta_2|) \tag{4.58}$$

For example, if the mobile is moving at 15 m/s and the radio frequency is 2.4 GHz (an ISM band frequency commonly used for spread-spectrum systems), then the spatial frequency will be up to 240 Hz (depending on the geometry, the cosine factor has a magnitude up to 2). This spatial frequency can also be interpreted as the Doppler frequency associated with a moving radio

---

[6] The epoch tracking algorithms only use the magnitude of the signal.

source. Because the receiver correlator requires phase coherence over the period of the pn-code, this phase variation places constraints on the length of the code.

Now consider the behavior of the epoch tracking error as the vehicle moves. Provided the multipath signal is not too large relative to the direct signal, the epoch error will vary in an approximately sinusoidal fashion over time; thus, the error will oscillate both positively and negatively. Thus, while the multipath signal is delayed relative to the direct signal, the measured TOA can be less than the direct straight-line path. However, in the case where there is no direct path (indoor propagation), the 'direct signal' is interpreted as the strongest path (always delayed relative to the straight-line path), so in this case there is a positive bias error in the measured TOA, with variation about this biased mean. In the case of a stationary receiver, the variation (over time) will not occur, so the nature of the epoch error cannot be determined from the static data. However, in the case of a moving transmitter, data averaging in the receiver can be used to minimize the variation, provided the measurements are performed at a sufficiently high update rate. In this case, the multipath errors can be averaged to essentially zero in the direct case (provided the multipath amplitude is not too large). In the case of NLOS signals, the averaging will reduce the variation in the measured TOA, but a positive bias will remain. These characteristics of the measurement errors have an important impact on the accuracy of the positioning system and are discussed in detail in Chapter 5.

4.8 Performance Summary of Tracking Algorithms

This chapter has provided an analysis of the performance of both a simple peak-detection algorithm and two versions of leading-edge algorithms. The performance of these algorithms was then assessed in the presence of Gaussian noise and a single multipath interference signal. The peak-detection algorithm can be considered the 'reference' algorithm, so that the performance of other algorithms can be compared relative to this reference. The results of the analysis in this chapter are summarized in Table 4.1.

The basis of the analysis is a bandlimited spread-spectrum signal, which after correlation results in a bandlimited correlogram. With typical bandlimiting of a baseband bandwidth of twice the chip rate (defined by the parameter $\beta = 1$), the correlogram can be completely defined by sampling points half a chip apart, although interpolation is possible. Thus, in all the algorithms, only a limited number of sampled data points are available, and in the algorithms analyzed the number of points is limited to two. In the case of the leading-edge algorithms, this also represents the maximum number of independent points on the leading edge.

The performance with random Gaussian noise is given in terms of the mean and STD of the tracking error. In the case of the mean error, it was shown for all the algorithms that the mean is essentially unbiased (zero mean error). In the case of the STD, all the algorithms result in a STD of the form $\sigma_\varepsilon = K/\sqrt{\gamma}$, where γ is the SNR and K is a constant unique to each algorithm. The values of K for typical parameters of each algorithm are 0.71, 1.3 and 1.37 (peak, leading-edge projection and leading-edge ratio algorithms respectively). Thus, the peak-tracking algorithm has the smallest tracking errors, but the leading-edge tracking algorithms are optimized to minimize errors in a multipath environment.

The multipath interference performance of the algorithms is based on the simple case of a single interference path of relative amplitude m and delay is τ_m. In this case there is no simple formula to summarize the performance, except all the algorithms exhibit a tracking error of the form $\varepsilon = (m/m + 1)\tau_m$ when the multipath delay is small. However, the maximum error

Table 4.1 Summary of the performance of various epoch tracking algorithms. For multipath, the multipath amplitude is m and the delay is τ_m. Note that the formulae are normalized so that tracking timing errors are in chips.

Algorithm	Gaussian Noise	Multipath
Peak	$\sigma_\varepsilon = \dfrac{1}{\sqrt{2\gamma}}$	$\varepsilon = m\tau_m e^{-(\tau_m/a)^2}$ a is the shape parameter $\varepsilon = \dfrac{m}{m+1}\tau_m \qquad (\tau_m < a)$
Leading-edge projection	$\sigma_\varepsilon = \dfrac{1}{\sqrt{\gamma}}\sqrt{1 + 2\left(\dfrac{\alpha}{\tau}\right) + 2\left(\dfrac{\alpha}{\tau}\right)^2}$ α is threshold parameter $\beta = 1: \sigma_\varepsilon = \dfrac{1.3}{\sqrt{\gamma}}$	$\varepsilon = \dfrac{m\tau_t\tau_r}{\tau + m\tau_r} \leq \dfrac{m}{m+1}\tau_t$ τ_t is the threshold delay parameter $\tau_r = \tau_t + \tau - \tau_m > 0$
Leading-edge ratio	$\dfrac{\alpha_0\sqrt{1+\rho_0{}^2}}{\sqrt{\gamma}} \geq \sigma_\varepsilon \geq \dfrac{\alpha_0(1-\rho_0)}{\sqrt{\gamma}}$ ρ_0 is ratio set point $\alpha_0 = 1.28$ chips for $\beta = 1$ $\beta = 1: \dfrac{0.77}{\sqrt{\gamma}} \leq \sigma_\varepsilon \leq \dfrac{1.38}{\sqrt{\gamma}}$ $\sigma_{\varepsilon t} = \|\alpha_0\|\sqrt{\dfrac{1+\rho_0^2 - 2\rho_0(1-2\tau)}{\gamma}} \quad (\tau \leq 0.5)$ $\sigma_\varepsilon = \dfrac{1.13}{\sqrt{\gamma}} \qquad (\tau = 5/16, \rho_0 = 0.5)$	$\varepsilon = \dfrac{m}{m+1}\tau_m \qquad (\tau_m < a)$ $\varepsilon \leq 0.083m \quad (\rho_0 = 0.4, \tau = 5/16)$ ρ_0 is the ratio set point τ is the separation parameter

characteristics of the peak and leading-edge algorithms are very different. For the peak algorithm (with a bandwidth parameter $\beta = 1$) the peak error is $0.27m$ chips, essentially independent of the only algorithm parameter (the separation of the sampling points). However, the leading-edge algorithms have the characteristics that the maximum error is limited by the parameters of the algorithms. For the leading-edge projection algorithm the error is given by

$$\varepsilon = \frac{m\tau_t\tau_r}{\tau + m\tau_r} \leq \frac{m}{m+1}\tau_t$$

so that the algorithm threshold parameter τ_t can be chosen to restrict the maximum errors significantly. While in theory choosing a very small threshold parameter can essentially eliminate the multipath errors, the effect of bandlimiting and receiver noise places a minimum useable limit on the threshold parameter. Nevertheless, with a conservative limit of $\tau_t = 0.15$, the peak error with $m = 0.5$ (the common value used for comparisons of performance) is 0.045 chips, compared with 0.135 chips for the peak algorithm. Thus, the leading-edge algorithm is about three times better than the peak algorithm *when the signal is bandlimited*. No analytical formulae are available for the leading-edge ratio algorithm, but the numerical results are very similar to the leading-edge projection algorithm. The main advantage of the leading-edge ratio algorithm is that it is generally more robust, as the projection algorithm can result in very large tracking errors at low SNR.

The relative multipath performance of tracking algorithms with a bandlimited signal is best judged by normalizing the results by the signal bandwidth. Using this method, the performance diverse systems can be compared on an equitable basis. For example, the GPS algorithms, which exploit the wide band of the GPS signal, can be compared with the algorithms in this chapter. Because the accuracy is related to the bandwidth used by the system and the relative amplitude of the multipath interference, a multipath figure of merit F_m is defined as

$$F_m = \frac{E_{max}(2B)}{m} = \frac{2\varepsilon_{max}B\tau_0}{m} = \frac{2\varepsilon_{max}\beta}{m} \tag{4.59}$$

where the RF signal bandwidth is $\pm B$, τ_0 is the chip period and ε_{max} is the maximum tracking error in chips with a multipath ratio m. (The smaller the figure of merit is, the better the performance is.) Equation (4.59) can be applied to calculate the figure of merit for the peak and leading-edge algorithms. From (4.23), the figure of merit for the peak algorithm is given by

$$F_m(\text{peak}) \approx \frac{2a\beta}{\sqrt{2e}} = 0.54 \qquad (\beta = 1, a = 0.634) \tag{4.60}$$

Note that the figure of merit is a function of the bandwidth and the correlogram shape parameter a, but is independent of the multipath parameters. Thus, it is suggested that the figure of merit should be applicable to all multipath situations, not just the particular case analyzed.

From (4.40), the figure of merit for the leading-edge projection algorithm is

$$F_m(\text{leading}) = \frac{2\beta\tau_t}{m+1} = 0.2 \tag{4.61}$$

where the standard multipath amplitude of $m = 0.5$ has been used and the threshold delay parameter is $\tau_t = 0.15$ chips. Note that for small-amplitude multipath the figure of merit is independent of the multipath parameters, and it is actually better (smaller) for larger multipath ratios m. As noted previously, the leading-edge algorithms have multipath error about 2.5 times less than the peak-detection algorithm, and this is reflected in the figure of merit.

A formula for the figure of merit for the leading-edge ratio algorithm cannot be determined, as there is not an analytical expression for the multipath error. However, from Table 4.1, the maximum error for the algorithm parameter $\rho_0 = 0.4$ is 0.083 m; so, applying (4.59), the figure of merit is 0.17, which is similar to the leading-edge projection algorithm.

It is interesting to compare these results with the performance of GPS, which has an RF bandwidth of 20 MHz, and modulated at 1 Mchips/s in the civilian mode, so $\beta = 10$. The best-case 'narrow correlator' GPS receivers have a reported performance [15] of about 9 m error for a relative multipath m of 0.5, which corresponds to $F_m = 1.2$. Thus, the GPS receivers have considerably poorer performance than the bandlimited algorithms described in this chapter. This result is a consequence of the design of the GPS C/A signal, which poorly exploits the 20 MHz bandwidth. This design is deliberate, as the US Department of Defense intended the civilian mode of operation to be much worse than the military mode of operation.

References

[1] J.J. Spilker, 'Delay-lock tracking of binary signals', *IEEE Transactions on Space Electronics and Telemetry*, **9**(31), 1963, 1–8.

[2] M.K. Simon, J.K. Omura, R.A. Scholtz and B.K. Levitt, *Spread Spectrum Communications Handbook*, McGraw-Hill, 1994.

[3] E. Sourour, G. Bottomley and R. Ramesh, 'Delay tracking for direct sequence spread spectrum systems in multipath fading channels', in *Proceedings of the IEEE Vehicular Technology Conference (VTC)*, pp. 422–426, September 1999.

[4] W. Cheol and D.S. Ha, 'An accurate ultra wideband (UWB) ranging for precision asset location', in *Proceedings of the IEEE Conference on Ultra Wideband Systems and Technologies*, pp. 389–393, November 2003.

[5] R.Q. Schmidt, 'Multiple emitter location and signal parameter estimation', *IEEE Transactions on Antennas and Propagation*, **34**(3), 1986, 276–280.

[6] R. Roy and T. Kailath, 'ESPRIT-estimation of signal parameters via rotational invariance techniques', *IEEE Transactions on Acoustics, Speech, and Signal Processing*, **37**(7), 1989, 984–995.

[7] H. Saarnisaari, 'TLS-ESPRIT in a time delay estimation', in *Proceedings of the 47th IEEE Vehicular Technology Conference (VTC)*, pp. 1619–1623, May 1997.

[8] G.G. Messier and J.S. Nielsen, 'An analysis of TOA-based location for IS-95 mobiles', in *Proceedings of the IEEE Vehicular Technology Conference (VTC)*, pp. 1067–1071, September 1999.

[9] K.C. Ho, Y.T. Chan and R.J. Inkol, 'Pulse arrival time estimation based on pulse sample ratios', *IEE Proceedings: Radar, Sonar and Navigation*, **142**(4), 1995, 153–157.

[10] R.J. Fontana, 'Recent system applications of short-pulse ultra-wideband (UWB) technology', *IEEE Transactions on Microwave Theory and Technology*, **52**(9), 2004, 2087–2104.

[11] K. Yu and I.B. Collings, 'Performance of low complexity code acquisition for direct-sequence spread spectrum systems', *IEE Proceedings: Communications*, **150**(6), 2003, 453–460.

[12] K. Yu, J.P. Montillet, A. Rabbachin, P. Cheong and I. Oppermann, 'UWB location and tracking for wireless embedded networks', *Signal Processing*, **86**(9), 2006, 2153–2171.

[13] V. Lottici, A. D'Andrea and U. Mengali, 'Channel estimation for ultra-wideband communications', *IEEE Journal on Selected Areas in Communications*, **20**(9), 2002, 1638–1645.

[14] Z. Tian and G.B. Giannakis, 'A GLRT approach to data-aided timing acquisition in UWB radios-part I: algorithms', *IEEE Transactions on Wireless Communications*, **4**(6), 2005, 2056–2067.

[15] A.J. van Dierendanck, P. Fenton and T. Ford, 'Theory and performance of narrow correlator spacing in a GPS receiver', *Journal of the Institute of Navigation*, **39**(3), 1992, 265–283.

5

Fundamentals of Positioning Systems

Much of the literature on wireless positioning is based on the GPS, but the focus of this chapter is on position determination in systems such as ad hoc wireless networks and WSNs. Further, the systems considered are short range, typically operating in indoor environments such as office buildings or sporting arenas, where GPS cannot be used. Such indoor environments are particularly challenging due to the characteristics of radio propagation.

There are two main problems with indoor radio propagation, both associated with the effect the building structure has on radio waves. First, the scattering of radio signals means that it is difficult to determine the TOF of the signal accurately [1–3], which is the fundamental basis for position determination. As the propagation speed of radio waves is 30 cm/ns, and the positioning accuracy requirement is typically 1–2 m or better, very precise timing measurements are required. Second, the indoor propagation paths are typically not LOS, so the radio signal is attenuated significantly more than for outdoor environments where LOS propagation conditions typically apply. Further, as the devices used in ad hoc networks are usually power constrained, the transmitter signal cannot be increased to compensate for the extra propagation losses. These constraints make the design of such positioning systems very challenging.

The simplest conceptual system for position determination is to measure the TOF from a mobile device to fixed measuring points such as base stations. However, TOF is difficult to measure due to a number of factors. First, TOF in principle needs synchronization of clocks in both the transmitting and receiving nodes; this is difficult, particularly in the mobile nodes. Second, the measurement of the TOF will include delays in both the transmitter and receiving nodes. As these radio delays are not accurately known and can vary over time, the measurement techniques must include compensation for these factors. Thus, TOF is not the fundamental measurement in practical systems. Instead, the system design is usually based on the TOA measurement as described in Chapter 4. The design of a positioning system based on the fundamental measurement of TOA thus needs to have important subsystems in addition to the position calculation, namely the synchronization of the clocks in the nodes and the calibration of the transmitter and receiver radio delays. This chapter provides an overview of the concepts of these two important subsystems and shows how such a system can convert the TOA

Ground-Based Wireless Positioning   Kegen Yu, Ian Sharp and Jay Guo
© 2009 John Wiley & Sons, Ltd

measurements into 'pseudo-ranges' required by the position determination process. A pseudo-range is defined as the required range plus an unknown offset which is common to all the measurements.

Given that the positioning system hardware can measure pseudo-ranges from a mobile node to the base stations, the last important step is to determine the position of the mobile node. This chapter also provides an overview of the types of algorithm used to determine the position and, importantly, the accuracy of the position fix. The discussion of the position determination process shows that there is not one technique which is optimal to all situations, so more complex position determination strategies are necessary, particularly in a multipath environment, where large TOA measurement errors can occur.

This chapter presents an overview of the principles of positioning systems, with particular emphasis on terrestrial radiolocation systems. In Section 5.1, two types of positioning system are introduced, namely navigation and tracking, and the consequences on system design are discussed. Section 5.2 presents the architecture of positioning systems, and tracking systems in particular. The design of these systems is discussed based on the measurement of TOA for transmissions from a mobile node to fixed nodes. Design issues associated with time and frequency synchronization and with the compensation for delays in the radios are also discussed. In Section 5.3, different methods of determining positions, particularly using pseudo-range data, are introduced. The concept of hyperbolic navigation is described, with a mathematical analysis in Appendix A. Finally, Section 5.4 provides an introduction to various performance issues in tracking systems, particularly related to indoor radio propagation, and the associated measurement errors.

5.1 Navigation Systems and Tracking Systems

While radiolocation principles are generally applicable to all types of system, the details vary with each particular type of application. In particular, two general classes of positioning systems can be defined, namely navigation systems and tracking systems. A navigation system is one where the position of the mobile device is required at the mobile device; GPS is a typical example. A tracking system is one where the position of the mobile device is not required at the mobile device, but is required elsewhere; radar is a typical example.

For radiolocation systems, these two categories can be further distinguished by the location of the transmitters and the receivers. For a navigation system the transmitters are typically in 'fixed' nodes. In this context, 'fixed' means that the node position is known. In terrestrial systems these nodes are usually referred to as 'base stations', which are deployed throughout the coverage area. For GPS, the 'fixed' nodes are in satellites which orbit the Earth, but as the GPS accurately specifies their location at any point in time, the concept is essentially the same. The advantage of such an arrangement is that the number of mobile devices a navigation system can support is essentially unlimited, as any number of navigation devices can receive the broadcast signals. One complication of this arrangement of broadcast signals is that multiple signals are required to determine a location. While time-division multiplexing would be possible, the movement of the node between the individual transmissions complicates the position determination, so that simultaneous transmission of the broadcast signals from the fixed node is highly desirable. However, the receiver in the mobile node must be able to decode each signal by some method to recover the individual broadcast signals. The simplest method for a radio system is to broadcast on different individual frequencies, but a more sophisticated approach is to encode each radio signal with a different code (CDMA). As was discussed in Chapter 1, spread-spectrum technology provides a method for achieving both TOA and CDMA. Such a design is used in GPS.

As a tracking system does not require positional information at the mobile node, it is more appropriate that the mobile node transmits the radio signal and the receivers are in the fixed nodes. For simple nodes, such as ones used in ad hoc networks and WSNs, having the receiver in the fixed nodes has considerable advantages, as the signal processing associated with receiving and decoding the signal is much greater than that required for transmitting. If each mobile node transmits on a different frequency, uses a different code or transmits at a different time (TDMA), then the receivers in the fixed nodes can decode each transmission and use the TOA of each signal to determine the position. Naturally, the TOA data from each node must be forwarded to a central point for the position determination. However, there is a limit to the number of such simultaneous transmissions owing to the limited frequencies, time slots or codes. Therefore, a tracking system will have a finite tracking capacity, in contrast to the unlimited capacity of the navigation system.

A subtle issue with terrestrial tracking systems is the near–far problem. If one mobile node is near a base station and a second mobile node simultaneously transmitting is far from the same receiving node, then the strong signal from the first mobile node will effectively 'jam' the reception of the signal from the second node. The reason for this effect is that the receiver automatic gain control (AGC) will reduce the overall gain of the receiver based on the stronger signal, so that the SNR for the weaker signal is severely reduced and can result in loss of detection. If only one base station receives the signal, then a fast-acting transmitter-to-receiver power control loop can largely overcome this problem. However, as a tracking system requires simultaneous reception of the signal at multiple receiving sites, such a transmitter power control loop is not appropriate. The simplest method to overcome this near–far problem is to allow only one mobile to transmit at a time (TDMA), but such a system will limit the maximum number of transmissions per second, thus limiting the overall capacity[1] of the system. However, this concept fits in well with the power-saving strategies required in typical ad hoc systems such as WSNs. In such systems, the transmitter needs only to be turned on for a very short period of time for tracking purposes, and this procedure simultaneously saves battery power. With appropriate design, the transmission times can be much less than a millisecond, so that the system capacity can be many thousands of positions per second. Owing to the short range of the radio propagation, such a capacity is more than adequate for practical systems, due to the limited spatial density of nodes. For example, the typical indoor range is 30–40 m, or a coverage area of about $1000\,\mathrm{m}^2$. If the mobile node density (say people) is one per $10\,\mathrm{m}^2$, then several hundred people can be tracked at an update period of 1 s. For slow-moving or fixed nodes, higher spatial densities can be tracked at lower update rates, so the limitation in the tracking capacity is not a practical problem. For example, if the node density were 100 per square meter and the transmission period is 0.25 ms, then the corresponding update period is 25 s, which is more than satisfactory for applications such as inventory control.

Whilst much of the material in this book is equally applicable to both navigation systems and tracking systems, the focus will be on tracking systems. Because of the dominance of GPS for position location, much of the literature and techniques are related to navigation systems rather

---

[1] The overall capacity is a product of the number of mobile nodes and the number of transmissions per second of each node.

than tracking systems. Further, as the majority of the GPS is related to the systems associated with the satellites, which is the providence of the US Department of Defense, details of design and operation of that part of the GPS is usually beyond the scope of the literature. It should be noted that the software and hardware of the GPS are much more complicated than those used in small terrestrial systems such as ad hoc networks. Thus, most of the remainder of this chapter will focus on the concepts associated with small tracking systems.

5.2 System Architecture

For the determination of position using radio technology or other related propagation technologies such as ultrasonics, the basic requirement is to estimate the distance from a mobile node to a fixed node. If there are sufficient such measurements, then a 2D or 3D position fix can be obtained by suitable processing of the measured data. The simplest conceptual accurate systems are based on measuring the TOF; then, knowing the propagation speed, the distance can be readily obtained. Some other less accurate methods are also applicable, such as estimating positions from the RSS or deriving position information (even less accurate) based on broad geometric considerations. These methods are discussed in Chapters 6, 7 and 13.

Because of the prime importance of TOF systems, this section will mainly focus on how TOF (or the associated TOA) measurements are made. Simply measuring TOF or TOA is not sufficient to allow the position to be determined, as the propagation delay includes the delays associated with the baseband and radio equipment in both the transmitter and the receiver. Thus, the design must allow for these other delays to be measured (essentially an equipment calibration process), or the design is such that these other 'nuisance' system parameters are eliminated by the measurement processes. These complications mean that even if the clocks in all nodes are synchronized (ideally to 'absolute' time), simply measuring the transmit and receive times in the baseband electronics is not sufficient to allow accurate positions to be determined. Note that if (say) an overall accuracy of 1 m is required, then the sum of all measurement errors (including the radio delay parameters) must not exceed 3 ns. This observation means that all components of a radiolocation system with 1 m accuracy need to have very accurate timing measurements.

Despite the desire to estimate the propagation range by appropriate measurements, many systems, both tracking and navigation (including GPS), cannot directly measure the actual range itself; instead, they measure what is called 'pseudo-range'. A pseudo-range is by definition the actual true range plus a range offset which is the same (unknown constant) for all the measurements used to determine a position. Mathematically, the pseudo-range is given by

$$P_n = R_n + R_c \tag{5.1}$$

where the subscript n denotes measurements at base station n for a tracking system and measurement from transmitter n for a navigation system. This measurement equation typically applies to systems based on TOA measurements, but it has wider applications as well. For TOA systems, the offset constant is usually associated with an unknown radio delay parameter common to all measurements. For systems based on measuring the RSS, the offset constant is associated with signal strength measurement errors associated with the mobile node antenna gain and receiver signal strength indicator (RSSI) calibration errors. While pseudo-range data may seem inferior to range measurement, it will be shown in Section 5.4 that there are benefits to pseudo-range systems when the measurements are corrupted by bias errors, which are typical when indoor multipath conditions exist.

5.2.1 Navigation Systems Design

While the focus of this book is on tracking systems, the design of navigation systems requires the pseudo-range determination to be discussed briefly. As stated previously, pseudo-range measurements are often assumed in the description of position determination algorithms. In this discussion, it is assumed that the clocks in the transmitting nodes are synchronized by a process which is irrelevant for this discussion. Note, however, that this time synchronization must be at the transmitting antenna (phase center) and, thus, must compensate for delays through the radio transmitting equipment. Thus, the synchronization process must include measurements of these delays. The following brief analysis also assumes that the transmitted signal is a spread-spectrum signal based on the repeated transmission of a pn-code, as for example in GPS.

The TOA measurement in the receiver can be expressed in terms of the relevant propagation and other delays, with the measured time defined by a phase ϕ of the pn-code clocks in the transmitter and the receiver. The fixed nodes (base stations) are indicated by subscript 'b' and the mobile receiving device by subscript 'm'. With this notation, the receiver TOA measurement can be expressed as

$$\text{TOA}_b = \phi_0 + \frac{R_b + \beta_b}{c} + \Delta_m - \phi_m \tag{5.2}$$

where Δ_m is the delay in the mobile radio receiver and β is the measurement (bias) error. Note that, as all the transmitter clocks are synchronized, all the transmitting base stations have the same clock phase ϕ_0. Equation (5.2) can be expressed in the form of a pseudo-range by grouping all the constant terms into one term, so that

$$P_b = c\text{TOA}_b = (R_b + \beta_b) + c(\Delta_m + \phi_0 - \phi_m) = R_b + R_c + \beta_b \tag{5.3}$$

Note that this procedure effectively eliminates the unknown mobile mode clock phase (relative to the reference clocks in the transmitters) and the radio delay, as they are common to all the measurements. Thus, in this case, the TOA measurements are in the form of pseudo-ranges (with measurement errors), with no further processing required. It is seen that, in the case of navigation systems, TOA and pseudo-ranges can be used interchangeably, but the underlying reason should be understood. It should be noted that this principle does not apply to tracking systems, as will be shown in Section 5.2.2, so applying navigation systems' analysis to tracking systems can result in fallacious results.

Finally, because of the properties of the navigation TOA data, it is clear that the time difference of arrival (TDOA) can be used to eliminate the constant terms, thus resulting in

$$\text{TDOA}_{a,r} = \text{TOA}_a - \text{TOA}_r = \frac{(R_a - R_r) + (\beta_a - \beta_r)}{c} \tag{5.4}$$

where typically node r is used as a reference for all the TDOA calculations. However, such a procedure may result in the TDOA data being infected by 'bad' bias data from the reference base station r. In general, therefore, algorithms based on TDOA data have worse performance than those based on the TOA (or pseudo-range) data directly.

5.2.2 Tracking Systems Design

As will be shown, tracking systems are more complex than navigation systems, and the TOA cannot be directly equated to the equivalent pseudo-range. For a tracking system, the mobile node transmits and the base stations receive, so the TOA equation is given by

$$\text{TOA}_b = \phi_m + \frac{R_b + \beta_b}{c} + \Delta_b - \phi_0 \tag{5.5}$$

where it has been assumed that the baseband clocks in the base stations are all synchronized ($\phi_b = \phi_0$). However, the delay parameters for each base station will be all different, even if they are all the same type of radio. Further, these delays will vary over time due to effects such as temperature and ageing, which may be different for each base station. If these delays are invalidly assumed to be constant and equal, then the TOA will include an unknown variable component which will translate into positional errors. Measurements of typical radios used in ad hoc networks and WSNs show that these delay variations can be at least 5 ns, and are often greater. Thus, for accurate position determination, the system must include some method of measuring the radio delay parameter in each base station. Assuming such measurements can be made, (5.5) can be rearranged into the form

$$c(\text{TOA}_b - \Delta_b) = P_b = (R_b + \beta_b) + c(\phi_m - \phi_0) = R_b + R_c + \beta_b \tag{5.6}$$

Therefore, provided the system can both synchronize the clocks in each base station and estimate the radio delay parameters, the receiver output again is in the form of a pseudo-range. Note, however, that there is no longer a one-to-one relationship between TOA and pseudo-range, as is the case for the navigation system. In practice, the receiver design is such that the output is not the TOA, but is the pseudo-range, with the details of its generation hidden from the position determination algorithm.

5.2.3 Time Synchronization

The above analysis of the generation of pseudo-ranges assumes that the clocks in the base stations are 'synchronized' by some undefined method. This section provides a simple example of how this synchronization process could be achieved. The intention of the example is to demonstrate the general principles of the synchronization method, rather than being intended for practical implementation.

The synchronization of two clocks, in general, means meeting two requirements. The first requirement is that the period (or frequency) of the clock 'ticks' must be made equal or confined to a very small error. The second requirement is that the phases of the clocks must be synchronized. If, say, the second ticks are aligned, then the overall time must also be aligned. Therefore, a timing reference signal must be more complex than a simple oscillator such as a sine wave or square wave. In a practical position location system, this overall time synchronization needs only to resolve the time ambiguity to a length of time about equivalent to the largest dimension of the coverage area. Thus, if (say) the coverage area is 1000 m in diameter, then the ambiguity period is around $3\,\mu s$. For spread-spectrum systems, this ambiguity period thus defines the minimum length of the pn-code.[2]

While the clock synchronization process implies that all the clocks in the 'fixed' nodes are synchronized in time, the process is not actually aimed at clock synchronization per se; rather, as will be shown, the design of the clock synchronization subsystem is intended to allow the receiver to output pseudo-range data derived from raw measured TOA associated with transmissions from the mobile nodes. The main complicating factor in designing the clock synchronization subsystem is that the base station clock is in the baseband electronics, whereas the received signals are processed by a radio before interfacing to the baseband subsystem. The radio receiver and transmitter both have considerable time delays from the input to the output, and these delays are often much greater than the radio propagation delays which are the basis for position determination. Since by definition a clock synchronization process implies processing some external clock signal, and the means of communication is via a radio, the synchronized base station clock must be delayed relative to some external reference clock. As the radio delays are not accurately known and can vary as a function of time, a simple time offset to compensate for the radio delays is not satisfactory for accurate position determination. For example, if the accuracy is to be of the order of 1–2 m, then the clock synchronization process, including compensation for the radio delays, should be accurate to a nanosecond or better. Such precise clock synchronization requires careful design of the clock synchronization subsystem of the base station.

The clock synchronization process typically involves one of two key techniques. The first method is to incorporate in the design a subsystem which continuously calibrates the TOA measurements by estimating the delays in the radio and other components. Thus, the raw TOA measurements can be corrected for these delays without the need to correct the base station clock physically. The second method is more subtle. If the clock synchronization process and the measurement of TOA from the mobile node both involve the same delays through the equipment, then the TOA measurement is effectively independent of these delays and, hence, there is no need to measure and correct for the equipment delays.

5.2.3.1 Simple Example of Time Synchronization

To illustrate the principles of the clock synchronization process, a simple design is described and analyzed below. For this illustrative example, the base station clocks are synchronized to an external master clock using a cable connecting the master clock to each base station. While a wireless solution is preferred, it is more complex; thus, in this introductory section, it is more appropriate to illustrate the principles, rather than have to deal with the extra complications of a radio-based solution.

A block diagram of the example time synchronization subsystem is shown in Figure 5.1. A reference master clock is connected to the base stations by a common cable, with each base station connected by another short cable to the antenna port of the base station. This system allows the base stations to communicate with each other via the cable, as well as receiving the clock synchronization signals from the master clock. These communications use the same RF signal as used for wireless communications to/from the mobile devices. The antenna/cable connection will require a switch (not shown) for correct operation.

---

[2] In systems such as GPS where the propagation time far exceeds the length of the pn-code, other techniques are required to resolve the time ambiguity.

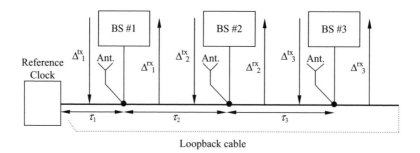

Figure 5.1   Block diagram of the clock synchronization subsystem.

The clock synchronization process is as follows. The reference subsystem transmits a short timing reference (RF) signal which is received by each base station using the same method as used to measure the TOA signals from a mobile device. The TOA at each base station is measured relative to its initially unsynchronized clock. Thus, the TOA at base station #1 (closest to the timing reference subsystem) can be expressed as

$$\text{TOA}_1 = \phi_0 + \tau_1 + \Delta_1^{\text{rx}} - \phi_1 \tag{5.7}$$

where τ_1 is the delay from the output of the timing reference subsystem to the connection point on the cable to BS #1, and Δ_1^{rx} is the receiver delay from that point through the radio to the baseband clock and ϕ_0 is the reference clock phase. The synchronization process is simply to adjust the base station local clock so that the epoch (the zero time point in the pn-code) of the local pn-code is aligned to the epoch received timing signal. Thus, from (5.7), the local clock phase is given by

$$\phi_1 = \phi_0 + \tau_1 + \Delta_1^{\text{rx}} \tag{5.8}$$

which is delayed relative to the reference clock by two unknown values. Similar expressions apply to other base stations further along the cable, so that, with reference to Figure 5.1, the clock phase of base station n is given by

$$\phi_n = \phi_0 + \sum_{i=1}^{n} \tau_i + \Delta_n^{\text{rx}} \tag{5.9}$$

Now consider the reception of signals from a mobile device m at base station n. The TOA measurement is made using the base station clock as a reference, so that from (5.9) the TOA is given by

$$\text{TOA}_{m,n} = \phi_m + \frac{R_n}{c} + \Delta_n^{\text{rx}} - \phi_n = \frac{R_n}{c} + (\phi_m - \phi_0) - \sum_{i=1}^{n} \tau_i \tag{5.10}$$

where R_n is the range from the mobile device to base station n and ϕ_m is the (unknown) phase[3] of the clock in the mobile node. Finally, assuming that the delay parameters τ_i are known

---

[3] For most systems the synchronization of the clock in the mobile is difficult, so that the design needs to consider the mobile clock phase as one of the unknown parameters of the mobile, in addition to its position.

values, (5.10) can be arranged into the form of a pseudo-range given by

$$P_{m,n} = c\left(\text{TOA}_{m,n} + \sum_{i=2}^{n} \tau_i\right) = R_n + c(\phi_m - \phi_0 - \tau_1) = R_n + R_c \qquad (5.11)$$

Note that all the unknown terms that are independent of the base stations are grouped together, as these terms are associated with the pseudo-range constant R_c. Thus, provided the cable delay parameters τ_i are measured by some technique, the TOA measurement can be converted into a pseudo-range. In this case the pseudo-range offset parameter is a combination of the mobile clock phase measured relative to the master clock phase, and the cable delay to the first base station. As the mobile clock phase is random (no synchronization), the pseudo-range offset parameter is also random and, hence, unknown to the receiver. Note also that the technique utilizes the two methods mentioned previously. First, some hardware-related parameters (in this case related to the distribution cable) need to be measured (calibrated); second, the base station receiver delay parameter Δ_n^{rx} is eliminated by the hardware design for the reception of the reference clock signal and the signal from the mobile device.

5.2.3.2 Determination of Cable Delays

As can be observed from the pseudo-range (5.11), the determination of pseudo-range by the receiver requires knowledge of the delays in the cable connecting the base stations to the reference clock. For this example system, the cable delays will be stable over time, so the simplest strategy is to measure the cable delays once during installation and then provide each base station receiver with the required information. The measurement of the delays can be performed by simply measuring the physical length and then converting to the electrical length by the known propagation speed for the cable. Alternatively, a more accurate approach would be to use an instrument, such as a network analyzer, to measure the electrical lengths of the cable directly.

While the above procedure is adequate for this example system, a more desirable approach is for the necessary delay parameters to be determined by the system itself. As the receiver is designed to measure TOA data accurately, the TOA measurements, with appropriate design, can be used to estimate the required delay parameters. However, while the required delay parameters are associated with the cable delays only, the use of the radio receiver in the measurements will involve the radio delay parameters as well. These 'nuisance' parameters complicate the design of the delay calibration subsystem.

The basis of the cable delay calibration procedure is the transmission of timing messages between the base stations using the cable for the communications. Consider a pair of base stations, a and b, where a transmits and b receives, and vice versa. By considering the propagation delays through the cables and the radio equipment, the two TOA measurements can be shown to be given by

$$TOA_{a,b} = \Delta_a \qquad (b > a)$$
$$TOA_{b,a} = \Delta_b + 2\sum_{a+1}^{b} \tau_i \qquad (5.12)$$

where the Δ parameter associated with each base station is the sum of the transmit and receive delays through the radio equipment. These equations can be solved for the cable delay parameters τ_i, as well as the 'nuisance' radio delay parameters Δ_i. If there are N base stations, then there are $N-1$ cable delay parameters (τ_1 does not need to be determined – see (5.11)) and N radio delay parameters. The first expression in (5.12) directly solves for the radio delay parameter, but only for base stations $(1, \ldots, N-1)$, as the last base station cannot transmit 'upstream'. Applying the second expression in (5.12) and using the radio delay parameters previously determined, the cable delay parameters can be determined for all the cable delays $(2, \ldots, N-1)$ except for the first and last. Thus, the above procedure has determined all the required delay parameters, except that cable delay for the last base station N.

To resolve the last cable delay parameter, the following procedure is suggested. The nonresolution of the last cable delay parameter is a consequence of the linear arrangement to the base stations along the cable and the sequential order of base stations along the cable. However, if the sequence is reversed (so that base station N becomes base station 1, and vice versa), then the same procedure described above can be used to determine the cable delay between base stations N and $N-1$ in the original configuration. Note that the loopback cable from the last base station in the original configuration (dotted in Figure 5.1) is now effectively associated with the first cable delay parameter τ_1 and can be ignored, so that the new configuration delay parameter τ_2 is equivalent to the original configuration delay parameter τ_N. Thus, this procedure has determined the last cable delay parameter necessary for determining the pseudo-ranges (5.11) for all base stations. For ease of actual implementations, it is suggested that the cable forms a loop, with the last base station N placed physically close to the timing reference module.

The above procedure requires that the distribution cable is connected to the timing reference module from either end during the calibration process. This reconfiguration could be performed manually, or ideally could be automated with an electronically controlled RF switch. If the technique were automated, then the calibration procedure could be performed regularly during normal (tracking) operation, thus ensuring the system remains calibrated at all time.

5.2.4 Technique without Time Synchronization

The design of a tracking or navigation system normally requires some form of clock synchronization in the 'fixed' nodes. Section 5.2.3 described a method that uses cables for the distribution of a reference timing signal. Other methods use a wireless distribution system to distribute the timing reference signal. However, for ad hoc networks, the need to distribute a reference timing signal, particularly using cables, is not desirable. Further, the distinction between 'fixed' and 'mobile' nodes tends to become fuzzy in ad hoc networks. Therefore, a method that does not require any (accurate) clock synchronization[4] in the nodes of the network is highly desirable. This section describes one such technique. As will be shown, with appropriate design, again only using TOA measurements in each node, the receiver output can be manipulated into the form of a pseudo-range. Thus, while the method of operation is different, the output format is the same, hence allowing for the same position determination algorithms to be used in all cases.

---

[4] While accurate time synchronization for position determination may not be available, lower accuracy time synchronization, appropriate for determining time slots in a TDMA system, is typically implemented in ad hoc networks.

The basis of the technique is to measure the RTT for a pair of nodes. As position determination requires many such links, the procedure is as follows. The nodes whose position is required transmit a signal, which is received by nearby nodes. These nodes then reply, one at a time, so that the original node can measure the RTT to all the nearby nodes. However, absolute time is not measured. As the transmission is a repeating pn-code, code phase rather than time is measured. Further, all transmissions commence at the epoch of the pn-code as determined by the local clock in the node. Thus, while the response transmissions are spread out over time (much longer than the pn-code), the receiver in the node only measures the TOA of the responses as a phase of its pn-code. Note that all the clocks in the network are unsynchronized, but as a RTT involves the time from the start of the transmission to the time of reception of the reply, all using the same local clock in a node, synchronization of clocks is not required.

Another requirement for this method is that there is no frequency offset between the local oscillators. However, this restriction will be removed later, so that the free-running oscillators in the nodes are completely asynchronous.

The round-trip TOA measurement encompasses the total transmission delays and includes the delays in the baseband hardware, the radios and the propagation delays. A block diagram of the transmission path and the associated parameters is shown in Figure 5.2. As this pn-code repeats with a fixed period T_{pn}, it is appropriate to consider the time measurements as a phase ($0 \leq \phi \leq T_{pn}$), rather than time, which increases without limit. Further, if the clocks are assumed to accurately have the same frequency, then the local time in both nodes can be defined by a single phase offset determined at $t = 0$ as measured by their respective local clock. Thus, the TOA measurement from node m to n can be represented by the expression

$$\text{TOA}_{m,n} = \phi_m + \Delta_m^{\text{tx}} + \frac{R_{mn}}{c} + \Delta_n^{\text{rx}} - \phi_n \tag{5.13}$$

This TOA measurement involves two parameters in the transmitter and two in the receiver; but, as each module has both transmit and receive functions, each node has three parameters associated with it (the clock phase, the transmitter delay and the receiver delay).

While (5.13) is perfectly satisfactory for determining the TOA and further data processing, some simplification is possible by adopting a different time reference in each module. In

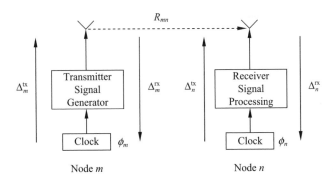

Figure 5.2 Block diagram showing the major components in the transmission path from the transmitter to the receiver.

particular, if the clock reference is altered so that the reference point is at the transmitter antenna (rather than the baseband clock), then the TOA expression becomes

$$\begin{aligned}
\text{TOA}_{m,n} &= (\phi_m - \Delta_m^{\text{tx}}) + \Delta_m^{\text{tx}} + \frac{R_{mn}}{c} + \Delta_n^{\text{rx}} - (\phi_n - \Delta_n^{\text{tx}}) \\
&= \phi_m + \frac{R_{mn}}{c} + (\Delta_n^{\text{rx}} + \Delta_n^{\text{tx}}) - \phi_n = \phi_m + \frac{R_{mn}}{c} + \Delta_n - \phi_n
\end{aligned} \tag{5.14}$$

Thus, the TOA measurement can be expressed in terms of the clock phases in the transmitter and the receiver, the propagation delay and the combined transmitter and receiver delays in the *receiving* unit only. As a consequence, only two parameters in each unit need to be defined, namely the local clock phase and the combined radio delays. As the clock phases and radio delays are unknowns to be determined as part of the pseudo-range determination process, the processing is somewhat simplified, as only two (nuisance) parameters need to be measured instead of four as in the original TOA expression.

Now consider the TOA measurements of a pair of units, namely m and n, where node m transmits and then mode n replies. Thus, applying (5.14) twice, the resulting two TOA measurements can be combined to form a pseudo-range:

$$\begin{aligned}
\text{TOA}_{m,n} &= \phi_m + \frac{R_{mn}}{c} + \Delta_n - \phi_n \\
\text{TOA}_{n,m} &= \phi_n + \frac{R_{mn}}{c} + \Delta_m - \phi_m \\
P_{m,n} &= c\left(\frac{\text{TOA}_{m,n} + \text{TOA}_{n,m} - \Delta_n}{2}\right) = R_{mn} + c\frac{\Delta_m}{2} = R_{mn} + R_c
\end{aligned} \tag{5.15}$$

Thus, by combining the two TOA measurements into a single round-trip measurement, the clock phase parameters can be eliminated, and the receiver output is in the form of a pseudo-range. However, the method requires that the radio delay parameter in each base station is determined by some calibration process, but the corresponding radio delay parameter in the mobile node does not need to be known.

The determination of the base station radio delay parameter can be based on the inverse of the above procedure for measuring pseudo-range. For 'fixed' nodes whose position is known, the range between nodes is also known. Thus, by assuming the TOA measurement is based on the straight-line radio propagation path, the TOA data can be organized as follows:

$$\text{TOA}_{m,n} + \text{TOA}_{n,m} - 2R_{mn} = \Delta_m + \Delta_n \tag{5.16}$$

If there are three nodes which mutually communicate, then a set of three equations in the three radio delay parameters can be generated; solving these simple linear equations results in the three radio delays. If there are more than three mutually communicating nodes, then the equations are overdefined, but they can be solved in a least-squares manner, thus providing a more robust estimate of the radio delay parameters due to the redundancy in the data set.

The above procedure is particularly useful in an ad hoc network. Consider a network that has three core nodes whose positions are known by some external surveying process and the remainder of the nodes are initially at unknown positions. If one such 'mobile' node can communicate with these three core nodes, then its position and radio delay parameter can be determined using the pseudo-ranges defined by (5.15) and an appropriate position determination technique to be discussed in Section 5.3. This procedure has effectively increased the number of 'fixed' nodes whose positions and radio delay parameters are known. These newly determined data can be transmitted in any response to another 'mobile' node transmission, thus allowing this new node to determine its position and radio delay parameter. Thus, by this procedure the position and radio delay calibration data can spread throughout the network. This process is dynamic; so, if a node moves, its positional data are updated and this updated data are used by other nodes in its vicinity. Similarly, a new node can enter the ad hoc network. By transmitting a request and receiving replies from at least three other nodes which have already been calibrated, the new node can successfully join the ad hoc network.

The above procedure requires no special time synchronization subsystem; thus, it is particularly useful in ad hoc networks where all the nodes are essentially the same – there is no distinction between base stations and mobile devices. However, the disadvantage is that the number of transmissions required for position determination is greatly increased over a system with a dedicated timing synchronization subsystem. In the latter case, a single transmission from a mobile node can be used for position determination, while the ad hoc signal protocol described above requires all nodes within range to response to the original transmission. As a TDMA system has a finite number of time slots, the number of positions per second that can be determined is significantly reduced. Thus, this technique is typically restricted to applications that require a low update rate, while systems with a dedicated timing synchronization subsystem are appropriate for applications where a high update rate is required.

5.2.5 Frequency Offset Determination

The technique of using round-trip pn-code phase time for measuring pseudo-ranges assumes that the frequencies in the two nodes are the same, even though the clocks are not synchronized to absolute time. This subsection describes a simple procedure to achieve the required frequency synchronization using the same TOA data used for pseudo-range measurements.

Accurate frequency synchronization is required to ensure that clock drifts between measurements do not result in TOA measurement errors. For example, if two TOA measurements are 500 ms apart in time, then a drift of 2 ppb represents a round-trip phase time error of 1 ns. As typical oscillators used in low-cost equipment have frequency errors of 1–10 ppm, a system must provide a frequency synchronization correction to reduce this initial error by at least a factor of 1000:1.

The basic concept for estimating the relative drift between two oscillators, one in the transmitter and other in the receiver, is to determine the rate of change in the TOA between measurements separated in time of the order of a few seconds. For this purpose, accurate TOA[5] data are not necessary; rather, the method requires repeatability in measurements for accurate determination of the frequency offsets. Thus, if the measurements of TOA have a repeatability of 1 ns due to system noise, and the transmissions from the first node are 1 s apart, then the clock drift can be estimated to better than 1 ppb. Further, as the frequency typically varies slowly over time (on a time-scale of at least tens of seconds), averaging of data with, say, a Kalman filter can result in significantly better estimates of the frequency drift parameter. However, generally, it

---

[5] Thus, TOA errors due to multipath propagation do not affect the accuracy of the frequency synchronization process.

can be assumed that over the measurement period the differential frequency variation is a linear function of time.

The TOA estimate from (5.14) can be modified to account for a linear drift in the clock over a short time period t, resulting in the TOA estimate as a function of time given by

$$\text{TOA}_{m,n}(t) = [\phi_m(0) + \alpha_m t] + R_{mn}/c + \Delta_n - [\phi_n(0) + \alpha_n t] \qquad (5.17)$$

where at $t = 0$ the clock phases are the nominal values, but the clock phases slowly change over subsequent time. Now consider two such measurements at two times t_a and t_b. Then, the differential change in TOA is given by

$$\begin{aligned} \delta T_{ab} &= \text{TOA}_{m,n}(t_b) - \text{TOA}_{m,n}(t_a) = \alpha_m(t_b - t_a) - \alpha_n(t_b - t_a) \qquad (t_b > t_a) \\ \alpha_{mn} &= \alpha_m - \alpha_n = \frac{\delta T_{ab}}{t_b - t_a} \end{aligned} \qquad (5.18)$$

The time differential $t_a - t_b$ is determined by the uncorrected clock in the node m; but as the measurement of the frequency offset parameter is required to be accurate only to about 1:1000, and typical clock errors will not exceed 10 ppm, the resulting drift rate measurements will be of the necessary accuracy.

Finally, if the RTT for the TOA measurement is T_r,[6], then, using the differential drift rate parameter derived from (5.18), the corrected pseudo-range is given by

$$P_{m,n} = c\left(\frac{\text{TOA}_{m,n} + \text{TOA}_{n,m} - \Delta_n - \alpha_{mn} T_r}{2}\right) = R_{mn} + c\frac{\Delta_m}{2} = R_{mn} + R_c \qquad (5.19)$$

Thus, provided there is a reasonably accurate estimate of the differential clock drift parameter, the modified expression (5.19) can be used to measure the pseudo-range between the two nodes. While the expression is strictly valid only for stationary nodes, it can be shown for relatively low speeds associated with (for example) walking that the measurement remains accurate to within 1 ns if the RTT T_r is less than about 100 ms.

5.2.6 Other Radiolocation Systems

While the main focus in this book is on radiolocation based on the TOF of radio waves, two other methods are briefly discussed in this section. The first method is to use the signal strength as an estimate of range. This method is particularly useful in WSNs, as the RSS is usually available from the radio receiver. However, as the indoor propagation characteristics are not accurately known, this method is only useful for low-accuracy position determination. The second method described is a hybrid radio–ultrasonics system. Because the radio propagation speed and the speed of sound are very different, such a hybrid system can directly measure the TOF without the complications of a radio-only system. While good accuracy can be obtained, the method is typically applicable only to short-range LOS systems due to the propagation characteristics of ultrasound.

---

[6] The real-time delay between the original transmission and the reply should not be confused with the round-trip delay measured from the TOA data. In the latter case, the basis for the measurement is the phase of the pn-code, while the former is based on the real elapsed time and will be much greater than the pn-code period.

5.2.6.1 Signal Strength Range Measurements

For low-accuracy positioning applications, particularly in systems with simple nodes, such as in WSNs, signal-strength-based range determination is a technique which is easy to implement. The following provides a summary of the method and indicates the problems with the technique which limits its effectiveness.

The basis of the technique is the variation in the RSS as a function of range between the transmitter and the receiver [4,5]. In free space, the propagation loss is a simple function of the range measured in wavelengths; thus, the range can be derived easily and with reasonable accuracy. However, in a multipath environment, particularly indoors, the RSS is a complex, partly random function of range from the transmitter to the receiver; thus, the accuracy of position fixes using the data will be relatively poor. Even in an open environment, ground reflections will be present, and these reflections significantly alter the signal strength compared with free-space propagation. In all cases, the signal strength needs to be averaged to remove the effect of rapid signal strength variations (Rayleigh fading) over a distance of the order of a wavelength. This approach is only effective for a moving device, so that the accuracy of the measurement for stationary nodes can be considerably worse. The multipath propagation also causes signal strength variation over the medium range, say of the order of 10–100 wavelengths, and these random variations typically exhibit a log-normal statistical variation with an STD of typically 6–8 dB in a severe multipath environment. As averaging is not possible in this case, such random variations result in random variations in the estimated range.

A further complication with the signal strength technique occurs when the mobile device is attached to a person, as placing a radio near a body can greatly affect the performance of the equipment. In a tracking system, the mobile device transmits, so that the effective radiated power is the important parameter in signal strength range determination. The actual effective radiated power is difficult to determine due to several factors. First, placing the device on a body has the tendency to cause an antenna mismatch, which affects the radiated power. Second, with the transmitter close to the body, the radiated signal varies considerably with the polar angle, as there can be considerable diffraction losses around the body (see Section 2.7 for details). For example, if the system operates in the 2.4 GHz ISM band, then the wavelength of the radiation is about 125 mm, which is somewhat smaller than the diameter of the trunk of the human body. As a consequence, diffraction losses of 10 dB or more can occur. This effect reduces with the frequency, so that operation in (say) the 900 MHz ISM band would be preferable. Third, the actual transmitter power in the simple radios used in WSN equipment tends to have a relatively wide variation about the nominal value, with a variation of 3 dB typical. The combination of all these effects means that there is considerable uncertainty in the effective transmitter power in typical applications. Similar arguments also apply to the performance of the receiver in a WSN, adding further uncertainty to the measurements.

For the following analysis, a tracking system is assumed with fixed base stations. These base stations are assumed to have accurately calibrated receivers and antennas which are assumed to be omnidirectional in a horizontal plane; this is an appropriate assumption for 2D applications. The basis for the signal strength range determination is the link gain/loss budget from the transmitter to the receiver. The link budget is easiest to express in decibel form, so that the budget can be expressed as the sum of individual components as follows:

$$L(R)(\text{dB}) = P_{\text{tx}} - P_{\text{rx}} + G_{\text{tx}} + G_{\text{rx}} + (\delta P_{\text{tx}} + \delta G_{\text{tx}}) \tag{5.20}$$

where $L(R)$ is the loss expressed as a function of the propagation range, P_{tx} is the transmitter power, P_{rx} is the measured received power and the G parameters are the transmitting and receiving antenna gains. Additionally, the uncertainty in the transmitter power and antenna gain is grouped at the end of the equation.

The ranging performance depends on the nature of the range-loss function, or more specifically its inverse $R(L)$. A detailed analysis is not appropriate for this introductory section, but some general observations can be made, particularly for indoor environments. For the analysis of the range–loss function, it is appropriate to divide the analysis into LOS and NLOS. The former case will only occur at close range, say 0.3 to 3 m indoors. As free-space propagation conditions will approximately apply, the propagation loss over this range interval will be about 20 dB (inverse square law), or an average of about 7.5 dB/m. Thus, even with some considerable uncertainty as defined in (5.20), the ranging errors are unlikely to exceed 1 m. NLOS conditions will apply at longer ranges and the situation is more complex, with each building having different range–loss characteristics. However, consider the special case where, over an interval of ranges, the propagation loss (in decibels) can be approximated by a linear function of range and vice versa. Ignoring the last uncertainty term in (5.20), the range to node n from node m can be determined as

$$R_n = \alpha + \beta(P_m - P_n + G_m + G_n) \tag{5.21}$$

where the linear range–loss function is $R(L) = \alpha + \beta L$. However, noting that the last term in parentheses in (5.20) only involves uncertainty at the mobile transmitter, and is independent of the receiving nodes, the range uncertainty is given by

$$\delta R_n = R_c = \beta(\delta P_m + \delta G_m) \tag{5.22}$$

Thus, for this linear case, the measurement is actually a pseudo-range as defined by Equation (5.1) and the uncertainties in the transmitter parameters will not contribute any errors to the calculated position, provided the measurement is treated as a pseudo-range in the position determination process. However, other random components of the loss will result in positional errors.

The actual range–loss function will not be a linear function, but will be more complex. Figure 5.3 shows an example of measured data, plotted with range as a function of the propagation loss. The frequency is 2.43 GHz. The measurements were made by walking around a building with a mobile device carried by a person; thus, the loss data include the effects of diffraction losses around the body. The data on the RSS were logged every 80 ms, but 10 samples were averaged to minimize the effects of Rayleigh fading. Some of the data at the shortest range are LOS, but most are not, being blocked by walls. The base station is located in an open-area atrium measuring 5 m × 5 m, with a high ceiling. Also shown is the free-space range–loss function, as well as a second-order least-squares fit to the measured data. As can be observed, a second-order polynomial fits the measured data rather well, and a linear relationship is a reasonable approximation for ranges greater than about 12 m. Thus, in this case, the above conclusion that the position will be insensitive to uncertainty at the mobile transmitter is approximately valid if the ranges are greater than about 12 m.

The ranging accuracy as a function of the range can be estimated from the fitted curve and the random variation of the loss relative to the curve. Fitting the data to the loss versus range curve

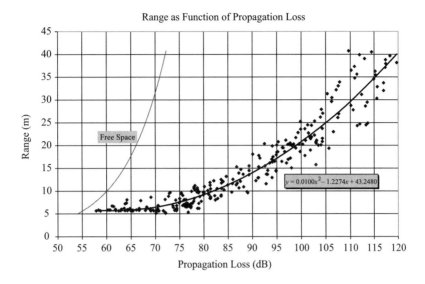

Figure 5.3 Example of the measured propagation range–loss functions for an indoor (office) environment. The curve shows the free-space range loss, and the curve associated with the dots is a second order least square fit to the measured data.

shows that the log-normal STD for this data set is 4.5 dB. Thus, if the second-order polynomial function is $R(L) = \alpha + \beta L + \gamma L^2$, then the range STD function will be

$$\sigma_R = \left|\frac{\partial R}{\partial L}\right|\sigma_{\text{log-norm}} = |\beta + 2\gamma L|\sigma_{\text{log-norm}} \tag{5.23}$$

Thus, the ranging accuracy can be expressed as a linear function of the propagation loss, but note the STD is not proportional to the loss or the range. Based on the curve fit data in Figure 5.3, Table 5.1 can be constructed of ranging STD due to random log-normal signal variation.

Thus, as indicated previously, at short range the accuracy is better than 1 m, but the accuracy reduces to worse than 5 m at the maximum loss (and range). Note that these accuracy figures are only related to the log-normal signal variation; they do not include the effects described previously associated with variations in the transmitter parameters.

While the above example is only applicable to one particular building, the general characteristics for other buildings can be conjectured to be broadly similar, so that the range–loss function also would be expected to be approximately a second-order polynomial. Thus, the general conclusions and performance characteristics from the example are likely to

Table 5.1 Propagation loss and ranging accuracy.

Propagation loss (dB)	Range (m)	Range SD (m)
70	6	0.7
100	20	3.5
120	40	5.3

be generally applicable. The best positional accuracy using the processed signal strength data would be expected if the range is interpreted as a pseudo-range.

5.2.6.2 Ultrasonics-assisted Range Measurements

Most of the positioning systems described in this book are based on the processing of radio signals. However, as the propagation speed of radio waves is so fast, the associated signal processing also must be very fast, which implies complex, power-hungry electronics. For simple applications, such as a WSN, a simpler implementation of the positioning subsystem is desirable. This subsection describes one such solution, namely a hybrid system that utilizes both radio and ultrasonics propagation [6,7].

Because the speed of propagation of sound is about 1 million times slower than that of radio waves, the signal-processing speed required is likely to be less by nearly the same factor. For a system to resolve to 30 cm in a radio system requires TOA measurement to a precision of 1 ns; in contrast, an ultrasonics system would achieve the same precision with measurement to 1 ms. As 1 ms is a long time for even the most modest of processors, the potential exists for ultrasonics signal processing to be performed entirely by software, with no specialist signal-processing hardware required. The particular method to be described utilizes both radio and ultrasonics signals, allowing the direct measurement of TOF without the need for any complex synchronization of clocks in the system nodes. However, owing to the propagation characteristics of ultrasound and the characteristics of ultrasonics transducers, the range of such a system is limited, so that position determination typically is limited to LOS applications. However, under such circumstances, the positional accuracy of such a hybrid system is likely to be superior to a radio-only system.

Consider a system where the mobile node simultaneously (say to within 0.1 ms) transmits both a radio pulse and an ultrasonics pulse. Because of the fast propagation speed of radio waves and the short distance, the radio pulse will effectively arrive 'instantaneously' at the receiver, where it is detected and starts a counter clock. Note that the delays in the radio typically will be 1 μs or less, which can be ignored even for high-precision measurements. The ultrasonics pulse propagates much more slowly and typically reaches the receiver after 1–50 ms delay. The ultrasonics signal is detected and used to stop the clock. Thus, the counter-clock is a measure of the TOF of the ultrasonics signal from the transmitting node to the receiving node.

The accuracy of the method is largely associated with the signal processing of the ultrasonics signal, so that, for example, multipath conditions for the radio wave propagation contribute a negligible error to the measurement. However, multipath propagation for the ultrasonics may result in significant errors which are similar to those associated with a radio-only system. Thus, multipath mitigation signal-processing techniques are required to process the received ultrasonics signal to obtain good accuracy. These signal-processing requirements may be similar to those used by a radio system, but the processing typically can be performed solely by software because of the slower propagation speed.

5.3 Overview of Position Determination

As was shown in previous sections, the design of the positioning systems' hardware and associated signal processing is such that either the TOF of the propagation (and, hence, the range) or, much more often, the TOA of the signal is used to estimate a pseudo-range. The final step is to use this data from a number of receivers to determine the position of the mobile node. As there are many such algorithms, which will be examined in detail in subsequent chapters, the purpose of this section is to provide an overall framework for the understanding of the process. As most systems provide pseudo-range data for the input to the position determination process, most of the discussion will be centered around the processing of pseudo-range data.

5.3.1 Hyperbolic Navigation

The design of many radiolocation systems is based on the measurement of pseudo-range data from a mobile node to fixed nodes. This concept is equally valid for both navigation and tracking systems. As the basis of pseudo-ranges is an estimate of a range plus an unknown offset which is common to all the measurements, the differential pseudo-ranges for a pair of nodes is equal to the differential ranges to the nodes. Thus, for any two nodes (a and b)

$$R_a - R_b = P_a - P_b \qquad (5.24)$$

From geometry it is known that the locus of points satisfying (5.24) in two dimensions is a hyperbola. For a review of the properties of hyperbolas, refer to Appendix A. Now consider the measurements at a second pair of nodes, say a and c, which will define another hyperbola. Clearly, the location of the mobile node will be at the intersection of these two hyperbolas. For this reason, a location system based on measured pseudo-range is sometimes referred to as 'hyperbolic navigation'. This is a generic term which applies equally to both tracking and navigation system in the context of Section 5.2.

The geometric interpretation of (5.24) is useful in obtaining the solution (position of the mobile node) to the pseudo-range measurement equations and understanding performance of such systems. Thus, if the analytical geometry associated with hyperbolas is used to solve the intersection of the hyperbolas, then analytical expressions can be obtained for the position of the mobile node. For details of one such solution to the intersection of two hyperbolas, refer to Appendix A. However, the geometric interpretation has its limitations, so that, in practice, more sophisticated algorithms are used for position determination. For example, for a tracking system with more than three measurements of pseudo-range at the 'fixed' nodes (more than four for a 3D system), the geometric solution results in multiple intersection points due to measurement errors. While it is possible to average all these solutions to obtain some sort of average best estimate of the position, nongeometric analysis avoids these difficulties. These methods are considered in detail in subsequent chapters.

The geometric interpretation of the solution is also useful in visualizing the effect of measurement errors. When there are measurement errors the hyperbolas will be shifted by (say) $\pm\delta_1$ and $\pm\delta_2$ relative to the true position, as shown in Figure 5.4. In this simple case, the error positions are located at the four vertices of a parallelogram. Further, if the angle of cut of the hyperbolas is θ, and assuming the two hyperbolas are displaced by the same amount δ, then the magnitude of the positional error is either $\delta/\sin(\theta/2)$ or $\delta/\cos(\theta/2)$. If the angle of cut becomes small, then the error parallelogram becomes long and narrow and the positional accuracy becomes poor. Thus, the accuracy depends both on the measurement errors and a geometric factor, which is related to the angle of cut between the hyperbolas. Further, the angle of cut is solely a function of the positions of the fixed nodes and the location of the mobile node relative to these fixed nodes and is independent of the measurement errors. As the mobile position becomes

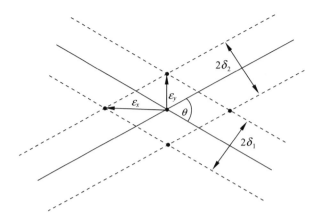

Figure 5.4 Geometry of the effect of measurement errors. The measurement errors result in the displacement of the hyperbolas, so that with errors of $\pm\delta_1$ and $\pm\delta_2$ there are four error positions on the vertices of a parallelogram. As the angle of cut decreases, the maximum error increases.

remote from the fixed nodes, simple geometric considerations show that the angle of cut between the hyperbolas becomes small; hence, the accuracy becomes increasingly poor at large distances from the fixed nodes. The best accuracy is achieved when the position is 'surrounded' by the fixed nodes, as, in this case, the angle of cut becomes large, approaching 90°.

While the error distribution is clearly 2D, it is useful to define the positional accuracy by a single number. One obvious definition for the geometric factor affecting accuracy is the radial error. With reference to Figure 5.4, the radial error, normalized by the measurement error factor δ, can be defined as

$$\text{GDOP}_{\text{hyp}} = \frac{\sqrt{\varepsilon_x^2 + \varepsilon_y^2}}{\delta} = \frac{2}{\sin\theta}. \tag{5.25}$$

As the angle of cut becomes small, the accuracy is reduced, or 'diluted' by a geometric factor, called the GDOP. A more detailed and rigorous analysis of GDOP (see Chapter 9) assumes the errors are random variables; and with the reasonable assumption that the measurement errors exhibit Gaussian statistics, it can be shown that the equiprobable contours are ellipses with a shape and orientation similar to the parallelogram in the above simple description.

5.3.2 Position Determination Summary

The determination of position using either range or pseudo-range data will be considered in detail in Chapters 6 and 7. However, before considering the details of each algorithm, it is useful to review briefly the underlying principles of the methods.

One simple approach is to interpret the solution to the problem geometrically. Using range data, the basic geometric interpretation is intersecting spheres (3D) or circles (2D), while the corresponding interpretation for pseudo-range is intersecting hyperbolic surfaces (3D) or curves (2D). However, these geometric algorithms do not easily extend as the number of nodes become large (so that there is redundant information) and also when the quality of the data varies due to factors such as SNR and multipath errors. Broadly, the position determination algorithms can be categorized as follows:

1. use of range data or use of pseudo-range data;
2. noniterative or iterative solutions;
3. differential data solutions;
4. redundancy and data weighting.

The characteristics of the various categories are briefly summarized in the following subsections.

5.3.2.1 Use of Range Data

When TOF can be directly measured (as with a hybrid radio–ultrasonics system, for example), ranges from a mobile node to fixed nodes are directly available for position determination. In this case the position is the only unknown information to be determined, so that the position is uniquely specified by two parameters for the 2D case and three parameters for the 3D case. Thus, in general, either two or three independent equations will be required to solve the problem. However, as can be recognized from a geometric approach, such equations do not result in unique results. For example, in the 2D case, the two equations can be interpreted as circles, with the position solution at the intersection of the circles. As, in general, circles intersect at two points, more information is required to provide a unique solution. Thus, in practice, the 2D case will require three equations (range measurements) and the 3D case will require four equations. In this case, there is redundant information (for example, three circles, in general, do not intersect at a single point), so that the optimum solution requires the processing of the redundant data (see Section 5.3.2.5 for more details).

While the position of the mobile node can be expressed in several ways, typically the methods are based on use of Cartesian coordinates (x, y, z) for the 3D case and (x, y) for the 2D case. However, range equations involving these variables involve a square-root operation; thus, the basic equations are nonlinear. By squaring the equations the square-root operation can be removed, but this operation introduces more nonlinear terms. Therefore, this method requires mathematical operations to remove these nonlinear terms, typically by finding the difference between the equations. The resulting equations are linear, but include a 'nuisance' term in addition to the basic positional variables. Thus, the solution is nonoptimum and requires a minimum of three range measurements for the 2D case and four range measurements for the 3D case, but without any redundancy. Further, when there are measurement errors, ranging errors result in larger positional errors than when pseudo-range data are used for position determination (see Section 5.4.3). For all these and many more reasons, the employment of range data for position determination is rarely used. For more details of the position-determination algorithms based on range measurements, refer to Chapters 6 and 7.

5.3.2.2 Use of Pseudo-range Data

Because of the difficulty of synchronization of the clock in a mobile device, and determining accurate delays through the radio equipment, most radiolocation systems measure a pseudo-range rather than a range. As a pseudo-range involves an unknown offset to the ranges to the fixed nodes, the mobile node in effect has another unknown parameter in addition to its positional parameters; this additional parameter needs to be determined by the position-fixing algorithm. Thus, the number of measurements required is three for the 2D case and four for the 3D case. However, as it was shown that the same number of measurements is required for range-based position determination, the number of measurements required is not a disadvantage of the pseudo-range-based method.

As with range-based position determination, if the position is specified in Cartesian coordinates, then the resulting equations are nonlinear. One possible technique is to use spherical coordinates rather than Cartesian, thus avoiding the square-root operation in specifying range. This approach is taken in Appendix A in solving for the intersection between two hyperbolas. Other nongeometric spherical solutions exist – see Section 6.4 for details. However, other analytical solutions to the generalized hyperbolic navigation problem also exist by using Cartesian coordinates – see Section 6.5 – without the problems associated with squaring the range equation. All these solutions provide analytical expressions for both the coordinates of the mobile node and the pseudo-range offset parameter, although the latter is usually not required and, thus, could be considered a 'nuisance' parameter. However, as the pseudo-range offset is related to the clock offset in the mobile node, the procedure effectively provides a method to synchronize its clock accurately. Thus, in an ad hoc network, the positioning procedure effectively creates another 'fixed' node, as both its position is known and its clock is synchronized. This procedure can allow positions of still further nodes to be determined in a cascading fashion.

Although the measurement of pseudo-range rather than range seems a disadvantage, the pseudo-range-based position determination is actually superior to range-based systems when there are measurement bias errors. For more details on this concept, refer to Section 5.4.3.

5.3.2.3 Types of Algorithm

Regardless of whether range or pseudo-range is used for position determination, the algorithms can be further classified as either iterative of noniterative. As previously indicated, the resulting equations require a square root when the position is specified in Cartesian coordinates, which makes analytical solutions difficult. Where analytical solutions exist, the measured data can be processed to obtain the solution directly. These expressions are often complex and involve much computation using floating-point arithmetic. For a real-time system, and more particularly in ad hoc networks, which typically have low computational capability, these analytical algorithms can be unworkable.

An alternative approach to using analytical expressions is to take a numerical approach. In particular, the nonlinear equations can be linearized, based on small displacements from the true solution. Thus, if an approximate starting position is assumed, then the linearized equations can be used to estimate a displacement which results in a new solution closer to the true solution. This procedure can be iterated until the solution converges to the required level of accuracy. Such solutions typically only require a solution to a simple set of linear equations and can be implemented efficiently even in simple processors.

The main problem with these iterative solutions is that an initial 'guess' is required. For the case where the mobile node is surrounded by fixed nodes (the usual case), and using the arithmetic mean of the positions of the fixed node as the starting point, the iterative solutions will almost always converge to the correct solution. The number of iterations required for convergence depends on how far away from the assumed initial point the true position is.

However, in practice, the number of iterations will be less than 10 in almost all cases. More typically, the starting point for the iterations will be the last known position, and if the positions are updated frequently so that the movement between updates is small, then only one iteration is required. In this case, the amount of computer processing required for the iterative update is typically much less than implementing a complete analytical solution.

Thus, a practical implementation is as follows. If no position determination has previously been performed (or the time between updates is large), then a noniterative solution is used to determine the position. Otherwise, an iterative algorithm is used, based on the last known position as the starting point. This procedure ensures that the minimal amount of processing is used to calculate the position.

One further refinement can be used when the quality of the measured data is poor. In such cases, the iterative method may not converge. In such cases, a low-accuracy analytical solution may be possible, or the analytical expressions may have no real solution (the position calculated contains complex numbers which have no physical meaning). Therefore, analytical methods are useful in extracting position information from poor data, or providing an indication that no reliable position fix is possible using the measured data.

In summary, typical implementations will use both iterative and noniterative methods to minimize the computation required and to obtain results when the iterative method fails to converge. In general, no one method is ideal for all situations.

5.3.2.4 Use of Reference Data

For a system that measures pseudo-ranges, the raw measured data from each fixed node includes an unknown offset which is the same for each node. In such cases, one approach to remove this unknown 'nuisance' parameter is to use a reference node and calculate differential measurements by subtracting the reference from all the other measurements. This simple procedure eliminates the unknown offset from all the measurements and reduces the data set for position determination by one. For a navigation system, where pseudo-ranges are equivalent to TOA data, this reduced data set is sometimes referred to as TDOA. However, such a formulation is not correct for a tracking system, where the differential data set must be based on a differential pseudo-range.

For the case where the ranging data are corrupted by random noise with zero mean with a variance that is the same for each node measurement, it can be shown [8] that, on *average*, there is no difference in performance between TOA and TDOA position determination. While this assumption may be satisfactory for outdoor applications such as GPS, these underlying assumptions are not appropriate for indoor systems, so that algorithms using differential data are generally suboptimum. Note that, while the average performance is the same, individual cases will result in different position estimates between the TOA and TDOA methods. In general, choosing as a reference the node with the smallest measurement error results in the smallest error in position. Thus, for the differential method to be effective, it is essential that the quality of the reference data is high, and ideally the best from the measured data set. The main problem with the method is that the data quality is often difficult to assess *a priori*. One possible method is to choose the node with the shortest range to the mode, as this node is likely to have the smallest measurement error. One simple method is to choose the measurement with the highest signal level, as high signal strength possibly implies LOS propagation and, hence, low ranging errors. Another approach is to use a previous position fix to determine the closest node. If no such position fix is available, then the position calculation must be performed twice: once to determine which node has the shortest range based on an arbitrary reference node and a second calculation with the presumed best choice as the reference. Finally, methods for determining if the measurement is associated with LOS or NLOS propagation, as described in Chapter 14, can be used to help in the selection of the best reference node.

Because of these complications, position determination using differential data is not recommended for indoor position determination.

5.3.2.5 Redundant Data and Weighting

Because the quality of the range or pseudo-range data can be very variable, the best positional accuracy is obtained if the low-quality data are identified and either eliminated[7] from the position determination or given a low weighting. Such procedures can be implemented only if there is redundant data, which in turn means that the number of fixed nodes must exceed the minimum necessary for a position fix. In general, the more nodes there are with range or pseudo-range data the better is the accuracy of the resulting position fix. However, the method requires some way of identifying the quality of the data. There are, in general, two methods of identifying the quality of the measured data.

The first method is for the measurement process in the radio receiver to estimate the quality of the data as part of its signal processing. The classical measure of signal quality in a system only corrupted by Gaussian noise is the SNR, which is usually easy to measure in the receiver. In this special case it can be shown that the SNR is the optimum weighting of the data. However, the more common corruption of the measured data is associated with multipath propagation, and SNR is generally only weakly correlated with multipath measurement errors. However, some receiver signal-processing algorithms are able to detect the distortions associated with multipath propagation, and these can be used to flag measurement quality.

The second method of identifying the quality of a measurement is to do it during the position determination process itself. For example, if an initial estimate is made using either equal weighting or weighting based on the quality returned by the receiver, then a fit error can be calculated if there is redundant data. A fit error is defined as the difference between the measured range[8] and the range calculated using the computed position. If this process detects a large fit error for one measurement, then a simple procedure is to eliminate this measurement (or give a low weighting) and then recompute the position. If there is sufficient data redundancy, this procedure can be repeated. However, this procedure needs to be implemented with caution. First, the computed position with 'bad' data will result in a positional error, which in turn results in errors in the fit errors. These errors can result in an incorrect identification of the 'bad' node. Second, the accuracy of a position fix is also dependent on the geometry of the nodes, as discussed in Section 5.3.1 in the context of GDOP. Therefore, eliminating a node with poor data may result in the degradation of the GDOP so that the new position fix is worse than the original position fix. Thus, ideally the (weighted) GDOP needs to be calculated as part of the process of

---

[7] An effective way of eliminating measured data from the position determination is to give the data zero weighting.
[8] In the case of pseudo-range, the position determination process will estimate the pseudo-range offset parameter; thus, the measured range can be estimated by subtracting the pseudo-range offset.

weighting the data. The estimated positional error is then computed as the product of the GDOP and the RMS fit error. Only if the estimate position accuracy improves should a node be eliminated. The main problem with this procedure is the significant extra processing required, which may be prohibitive in a real-time system.

5.4 Indoor Performance Issues

This section discusses the effect of measurement errors on the accuracy of position determination. The propagation of radio waves in an indoor environment is complex, so there will typically be no LOS propagation from a mobile node to a receiving node. Under these conditions the propagation path will be longer than the straight-line path assumed for position determination. This excess propagation delay can be considered to have a mean (bias) and a random component. It seems intuitive that the bias error is the most important factor, but the following subsections show why this is not true; in fact, the variable component has the most effect on the positional accuracy.

5.4.1 Indoor Radio Propagation Characteristics

To understand the problem of indoor position determination, some understanding of the physical nature of NLOS indoor radio propagation is necessary. As there is no unimpeded straight-line path from the transmitter to the receiver under NLOS conditions, the typical path of the strongest signal must include scattering from obstacles along the path. This scattering results in a loss in the RSS and a delay in excess of the straight-line path delay. However, a straight-line radio path still does exist, but this path typically has much greater attenuation. Thus, the path of interest for radiolocation is usually the 'path of least resistance'. Because of the problems of measuring the delay excess (as outlined in Section 5.4.2), the discussion here is based on propagation loss excess, rather than delay excess. Unlike the propagation delay excess, the loss excess is simple to measure and does not suffer from the problems described in Section 5.4.2.

An example of the loss excess measured in an office building is shown in Figure 5.5. The loss excess is defined as the difference between the actual measured loss and the loss that would have occurred if the same straight-line path were in free space. The frequency is 2.43 GHz. The measurements were made by walking around a building with a mobile device mounted on the hip of a person; thus, the loss data include the effects of diffraction losses around the body. The data on the RSS were logged every 80 ms, but 10 samples were averaged to minimize the effects of Rayleigh fading. The propagation loss excess was calculated from the known transmitter power, receiver antenna gain and the range from the mobile device to a base station. The base station is located in an open-area atrium measuring 5 m × 5 m, with a high ceiling. Some of the data at the shortest range are LOS, but most are not, being blocked by walls. The straight lines (5–12 m) show the trends at short range (propagation to offices), while the longer range lines (12–40 m) show the propagation along corridors. It is noted that the fitted loss excess is not a smooth function of distance, but has two distinct linear regions. The figure shows that there are two trend lines: one at short range and one at longer ranges. The short-range loss gradient is 2.7 dB/m in this case, while the longer range loss excess characteristics have a lower loss excess gradient of 0.55 dB/m. Also observe that the variability in the loss indicated by the dotted lines is largely independent of range.

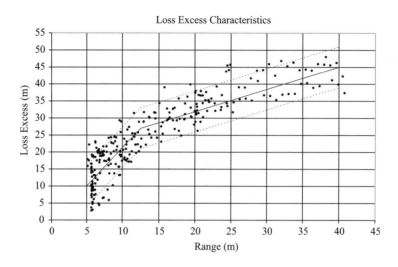

Figure 5.5   Example of the measured propagation excess losses in an indoor (office) environment.

The interpretation of these results is as follows. As the loss excess is not a linear function of distance, the propagation path cannot be a straight line, as this path will have excess losses (measured in decibels) increasing approximately linearly with distance. The straight-line path will pass through walls and other obstacles, resulting in excess losses; as the distance increases, the number of such walls increases and, hence, so does the loss. If it is assumed that the environment is quasi-homogeneous, then the excess losses are expected to increase approximately linearly with distance. The actual data show the loss (after the initial LOS conditions) increases approximately linearly with the NLOS distance and then increases more slowly with distance. The reason for this behavior is that the 'path of least resistance' is along corridors at longer ranges, not through walls. The loss along corridors tends to be only slightly greater than free-space loss; thus, the loss excess increases more slowly with distance. (The total loss is increasing more rapidly than the excess loss.) For a path between (say) two rooms in a building, the propagation path typically will be through the wall of the first room, along one or more corridors (with diffraction losses at bends, usually 90°) and finally through walls of the destination room. By counting the number of walls and the number of bends along the path, an estimate of the loss excess can be made. In the case of open-plan offices, the mechanism is similar; in this case the 'path of least resistance' is over the top of the partitions, which again has little path loss excess with distance.

As all buildings are constructed with rooms connected by corridors, this behavior is expected to be nearly universal. However, the details will vary according to the architectural structure of the building and the construction material, so that the data in Figure 5.5, while similar in the general characteristics, will differ in details.

5.4.2 Measurement Errors and their Effects

One of the most important performance parameters of a positioning system is the positional accuracy. Errors in position are a consequence of measurement errors in the range or pseudo-range. The following summarizes various aspects of this problem, particularly associated with indoor systems.

1. The measured input data for the position determination process is almost always pseudo-range, rather than the estimated range. Previous sections in this chapter show how a positioning system is designed so that the receiver output is in the form of a pseudo-range. This observation is true for many different types of system, including those based in measuring TOA, RTT delay and RSS. If there are excess delays in the radio propagation in addition to the delays through the radio equipment, then, in general, it is difficult to distinguish between these two sources of delay. Thus, any procedure designed to minimize the effect of the radio delay uncertainty will also minimize, but usually not eliminate, the effects of the propagation bias errors.

2. The concept of a bias error in a range (or pseudo-range) measurement is a statistical one associated with the analysis of a large number of measurements; a bias error has no meaning for a single measurement. The error in a measurement has components from both the bias and a random component, but these two components cannot be separated in a single measurement. If measurements are repeated at a stationary point, then at high SNR the measurements will repeat with no further information. However, if the point is moved (at least a sample per half wavelength of movement), then by definition the bias error will be (approximately) constant over (say) a meter distance. In this case, after compensating for the movement in the measurements, the variability will be associated with the random component and is specified by the STD in the measurements. Similarly, the bias error is the mean of the measurements. Thus, these concepts are only useful in experiments associated with determining the broad characteristics of the propagation environment, not for position determination. These statistical parameters are important in simulations, but are of more limited importance in position determination, both real and simulated. However, broad knowledge of the propagation environment could be used to improve (in a statistical sense only) the accuracy of the position determination.

3. As there are multiple paths, the only chance that the straight-line path can be detected by a receiver is when the time resolution of the signal processing is smaller than the differential path delays. Such fine time resolution implies a wide receiver bandwidth. Thus, the measurement accuracy depends not only on the propagation environment, but also on the nature of the transmitted signal, the signal processing and the radio hardware characteristics. The parameters that affect the measurement accuracy at a particular radio frequency include the following:

   a. *The bandwidth of the signal.* Wide signal bandwidths improve time resolution and, hence, measurement accuracy; but even very wide bandwidths cannot reduce the ranging errors to zero, as the accuracy is degraded by other factors, as discussed below. Typically, for indoor environments, about 500–1000 MHz is the maximum bandwidth required, with little benefit from greater bandwidths. However, these bandwidths are associated with UWB systems only; thus, most positioning systems with smaller bandwidths will have lower accuracy than UWB systems.

   b. *The receiver output SNR.* The SNR limits the detection of small-amplitude signals associated with straight-line but heavily attenuated signals. While straight-line signal paths will always exist, these signals may be so small that they are buried in noise and are thus of no use for determining the TOA. As the signal is weaker (and, hence, the SNR lower) at greater ranges, this implies the measured TOA will be less accurate at greater ranges, even if the signal delay characteristics are similar to those at shorter ranges. As the delay excess in fact also tends to increase with range, these two degrading effects combine and are difficult to separate in any measurement.

c. *The receiver dynamic range*. The dynamic range of the receiver limits the detection of weak signals in the presence of strong signals. The dynamic range of a receiver is typically limited to about 40–50 dB, but it can be as low as 30 dB. Increasing the dynamic range is difficult and, hence, expensive, and it is not an option in simple radios used in WSNs and other similar applications. As accurate detection of the TOA of a signal needs an SNR of at least 15 dB, the dynamic range of the receiver limits the excess loss of weak straight-line signals to about 25–35 dB.

d. *Signal-processing algorithms*. The determination of the TOA is very dependent on the signal-processing algorithms. As, in practice, the signal must be bandlimited and has a finite SNR, the processing will have a finite resolution limited by these factors. However, the best algorithms need to process the first signals that are detectable by the receiver. Thus, algorithms need to process the leading edge of the impulse signal (or equivalent) for optimum performance. Algorithms based on (for example) the peak of the signal (typical in GPS receivers, for example) will be suboptimum in an indoor propagation environment and may perform very poorly, as the first detectable signal is often not the strongest.

4. The above discussion shows that the measured delay excess characteristics of an indoor radio propagation environment are strongly influenced by the characteristics of the radio and the signal processing, so that measurements, particularly at long range, may reflect the measurement scheme more than the radio propagation environment. This observation helps to explain the great variability in reported indoor propagation delay characteristics. As the details listed in the above paragraphs are often absent in reported results, it is often difficult to assess the reliability of the results.

5. For smaller bandwidths than UWB, performance modeling needs to include the characteristics of the receiver for rejecting multipath. As a rough approximation, the performance can be modeled as an LOS signal plus an interference multipath signal with a delay and amplitude based on NLOS analysis. In a simple model, it can be assumed that the signal tracked by the receiver is either the straight-line path, possibly attenuated relative to the strongest path signal, or the first 'significant' NLOS signal. If the attenuation in the straight-line path is so great that the NLOS path is chosen as the signal path, then the delay excess should be associated with the assumed NLOS path, including diffracting around corners. However, this is the wideband case; some degradation will occur if the bandwidth is reduced. The estimation of this degradation could be on the basis of an exponential decay in the impulse response of the propagation environment, and how the receiver signal processing performs in such circumstances; the measured performance is a combination of the propagation environment, the characteristics of the signal and the receiver signal processing.

6. Now consider the process of determining a radiolocation using measurements at base stations (four or more usually) of transmissions from a mobile device. As these base stations will be roughly uniformly distributed throughout the coverage area, some of the ranges will be short, some medium length, up to the maximum range determined by the link budget of the radio system. Further, this distribution of ranges is approximately invariant with the position of the mobile within the coverage area, which is assumed to be quasi-homogeneous. As the delay excess tends to increase with range, the measurements will, on average, have

errors which increase with range. However, at longer ranges these average delay excess biases will be quite similar, so the random variation component will tend to dominate the variation in the measurement errors. Although the position-determination process is complicated, the fitting process using pseudo-ranges will include an estimate of the 'average' bias error; the estimate of this error is expected to be approximately the mean of the individual bias errors. As a consequence, the effect of the bias errors in the measurements is substantially reduced, as the mean bias is effectively incorporated into the pseudo-range constant, with the residual between the individual measurement bias and the mean bias being considered as part of the random measurement error. Assuming that the bias errors increase uniformly from zero to B_{max}, the effect of the bias errors will be an additional random component with an STD $B_{max}/2\sqrt{3} = 0.29B_{max}$. As the maximum bias and random components are typically comparable in magnitude (see Chapter 2), the combined random STD is increased by only about 4%.[9] Thus, the conclusion is that the bias errors have minimal impact on the positional accuracy; rather, the accuracy is determined by the *variation* in delay excesses. Note that this observation is restricted to pseudo-range positioning only; range-based systems will be less accurate, as the measurement bias errors feed directly into positional errors.

5.4.3 Effect of Delay Excess on Positioning Accuracy

Physical measurements indoors of radio propagation show that there are significant bias errors which increase with distance, as described in Chapter 2. However, if pseudo-range data are used for position determination, then bias errors are not as important as the variation in the bias, as will now be demonstrated. In particular, the basic measured pseudo-range data can be manipulated so that the mean of the bias measurement errors (averaged over all base stations) can be incorporated into the arbitrary pseudo-range constant, and the residual errors can be modeled by a suitable random error.

From the model of the bias errors described in Section 2.4 (and repeated here for convenience), the bias error at base station b can be expressed in the form

$$\beta_b = mR_b + N(0, \sigma_\beta) \tag{5.26}$$

where $N(\mu, \sigma)$ is the normal distribution and the other parameters are as described for the model previously in Chapter 2. As the measurements are based on pseudo-ranges which have a common constant offset for each base station measurement, (5.26) can be modified to remove the common measurement bias, so that the modified bias is

$$\beta_b = m\bar{R} + m(R_b - \bar{R}) + N(0, \sigma_\beta) \tag{5.27}$$

where the averaging (indicated by a bar symbol) is performed over all the base stations. Noting that the first term is a constant that can be incorporated into the pseudo-range constant, the pseudo-range for a base station can be expressed as

$$P_b = R_b + [R_c + m(R_b - \bar{R})] + N(0, \sigma_\beta) \tag{5.28}$$

---

[9] Assuming the random and maximum bias errors are equal and statistically uncorrelated, the combined STD will be given by $B_{max}\sqrt{1 + 0.29^2} = 1.04B_{max}$.

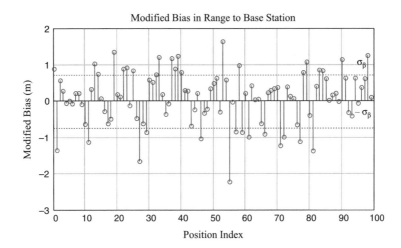

Figure 5.6 Modified pseudo-range bias errors for random positions within the coverage area with four base stations separated by 30 m. Observe that the mean bias is zero and the STD σ_β is less than 1 m. Note also that the peak bias errors are considerably less than the peak raw bias errors, which for the bias model and coverage area is 4.5 m.

If the position of the mobile node is considered a random variable, then the second bracketed term in (5.28) can also be considered a random variable in addition to the random normal distribution component. The advantage of (5.28) for describing the pseudo-range is that the effective bias errors can be considerably smaller than the raw biases given by (5.26). This property can be observed in the data in Figure 5.6, which shows the modified bias error to a base station, based on the bias model described by (5.26) and data from Section 2.4. Note in particular that the peak bias errors are about half the raw bias errors of 4.5 m (according to the model) at the maximum range of 42 m.

The advantage of using pseudo-range position-determination methods over range methods can now be understood. If the range is used for position determination, then the measurement range errors will have biases as expressed by (5.26). As no matter which position in the coverage area is considered, at least one range to a base station will be near the maximum, so that at least one measurement will have the maximum error; in the above simulation this is 4.5 m. Further, all the bias errors will be positive, so there is no averaging effect of positive and negative bias errors. In contrast, with pseudo-range measurements, the mean bias is effectively removed from the measurements, resulting in both smaller errors and the residual errors being both positive and negative, allowing for some averaging of the measurement errors in the position-fixing algorithm.

The above analysis shows that the modified pseudo-range has the form $P_b = R_b + R_c + \text{Noise}$, where the noise has zero mean. Thus, when the position-location algorithm fits the measured data, both the mobile position (and hence R_b) and the pseudo-range constant R_c will be estimated. The pseudo-range constant will be a combination of the clock offset of the transmitter and the mean bias errors as described above. The individual components of this offset cannot be determined; but as these are not individually required, this limitation is of no concern.

In conclusion, although indoor propagation conditions can result in large bias errors in the propagation delay, the use of pseudo-ranges for position determination can significantly reduce the effect of these errors on the positional accuracy.

References

[1] N. Alsindi, B. Alavi and K. Pahlavan, 'Measurement and modelling of ultra wideband TOA-based ranging in indoor multipath environments', *IEEE Transactions on Vehicular Technology*, **58**, 2009, 1046–1058.

[2] S.S. Ghassemzadeh, R. Jana, C.W. Rice, W. Turin and V. Tarokh, 'Measurement and modelling of an ultra-wide bandwidth indoor channels', *IEEE Transactions on Communications*, **52**(10), 2004, 1786–1796.

[3] C. Gentile and A. Kik, 'An evaluation of ultra wideband technology for indoor ranging', in *Proceedings of the IEEE Global Telecommunications Conference (GLOBECOM)*, pp. 1–6, November 2006.

[4] E. Elnahrawy, X. Li and R.P. Martin, 'The limits of localization using signal strength: a comparative study', in *Proceedings of the IEEE Conference on Sensor and Ad Hoc Communications and Networks (SECON)*, pp. 406–414, 2004.

[5] P. Bahl and V. Padmanabhan, 'RADAR: an in-building RF-based user location and tracking system', in *Proceedings of the IEEE Conference on Computer Communications (INFOCOM)*, pp. 775–784, 2000.

[6] N.B. Priyantha, H. Balakrishnan, E. Demaine and S. Teller, 'Anchor-free distributed localization in sensor networks', MIT LCS Technical Report, No. 892, April 2003.

[7] D. Moore, J. Leonard, D. Rus and S. Teller, 'Robust distributed network localization with noisy range measurements', in *Proceedings of the International Conference on Embedded Networked Sensor Systems*, Baltimore, MD, USA, pp. 50–61, 2004.

[8] D. Shin and T. Sung, 'Comparisons of error characteristics between TOA and TDOA positioning', *IEEE Transactions on Aerospace and Electronic Systems*, **38**(1), 2002, 307–311.

6

Noniterative Position Determination

Given the measurement results of TOA, RSS or AOA, there are two types of algorithm for calculating the actual position of a node: noniterative and iterative algorithms. A comprehensive analysis of noniterative position-estimation algorithms is presented in this chapter. In particular, it is focused on finding the position estimate using TOA measurements. A comparison of the SI, QLS, and LC-LS techniques is performed via simulations. The techniques studied throughout are suited for cellular systems and sensor networks under specific assumptions. In real systems, there always exists a clock frequency offset between each transmitter and receiver pair. In this chapter it is assumed that the clock frequency offset has been compensated or its effect can be ignored. Also, in radio transceivers, internal radio delay is inevitable. For the TDOA-based methods, it is assumed that the internal delays are the same but not necessarily known at different base stations. For other TOA-based methods it is assumed that the internal delay is known through calibration when the transceivers are built and that the variation of the internal delay is negligible.

6.1 Basic Positioning Methods

We start by studying some elementary methods for position determination. The positions are calculated directly by solving the measurement equations through the basic arithmetic operations when the system is exactly defined. A number of distance estimation techniques are first described, followed by position determination based on distance measurements. Then, position estimation based on TDOA and AOA is briefly presented.

6.1.1 Distance-based Positioning

Before discussing position determination, let us study how to measure the distance between a transmitter and receiver pair.

Ground-Based Wireless Positioning   Kegen Yu, Ian Sharp and Jay Guo
© 2009 John Wiley & Sons, Ltd

6.1.1.1 Distance Estimation Approaches

There are a number of different ways to obtain the distance measurements. The RSS-based method is probably the simplest. In this method, the path loss model is required and it can be established based on field measurements. In free space, the loss of signal strength is given by

$$P_{PL} = \left(\frac{4\pi f d}{c}\right)^2 \tag{6.1}$$

where f (Hz) is the signal frequency, d (m) is the distance between the transmitter and the receiver and $c = 2.9979 \times 10^8$ m/s is the speed of light in free space. For indoor positioning, c may be simply taken as 29.98 cm/ns for computational simplicity. Equation (6.1) is only valid under the assumption that there are no obstacles nearby to cause reflection or diffraction and that the distance between the transmitter and receiver is not too small. In most circumstance it is convenient to express the path loss in decibels and (6.1) can be written as

$$P_{PL} = 20 \log_{10}(d) + 20 \log_{10}(f) - 147.56 \tag{6.2}$$

When the transmission power is known and the received signal power is measured, the distance estimate can be obtained. For many applications, such as positioning in buildings and urban areas where rich multipath exists, the free-space path loss model cannot be employed directly. Instead, environment-specific path loss models need to be developed. A variety of path loss models exist, such as the Walfisch–Ikegami path loss model which is suited for city and urban areas [1]. This RSS-based positioning technique is simple and easy to implement; however, the distance estimation accuracy may not be satisfactory due to the unpredictable environmental variations.

Another way to obtain distance measurements is to make the RTT measurement. The basic idea of determining the RTT is illustrated in Figure 6.1. Node A at time instant t_0 sends a packet of data toward node B, which then receives the packet at time instant $t_1 = t_0 + \tau$, where τ is the propagation time from node A to node B. Owing to the TOA measurement error denoted by ε_1, the estimated TOA of t_1 becomes $\hat{t}_1 = t_0 + \tau + \varepsilon_1$. After a delay δt relative to $\hat{t}_1$ and known to node A, node B sends a packet back to node A at time $t_2 = t_0 + \tau + \varepsilon_1 + \delta t$. Then, the packet is received by node A at time $t_3 = t_0 + 2\tau + \varepsilon_1 + \delta t$. Owing to the measurement noise at node A,

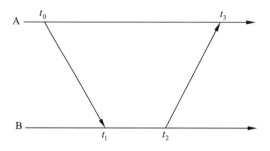

Figure 6.1   Illustration of RTT measurement.

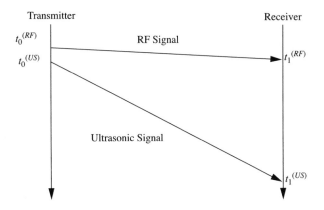

Figure 6.2 Propagation of RF signal and ultrasonic signal.

denoted by ε_2, the estimated TOA is given by $\hat{t}_3 = t_0 + 2\tau + \varepsilon_1 + \delta t + \varepsilon_2$. Consequently, the propagation time from node A to node B is estimated as

$$\hat{\tau} = \frac{1}{2}[\hat{t}_3 - (t_0 + \delta t)] = \tau + \frac{1}{2}(\varepsilon_1 + \varepsilon_2) \tag{6.3}$$

where the clock frequency difference between the two nodes is not included. The procedure of sending packets from node A to node B and then from node B to node A may continue for a certain number of rounds to improve the estimation accuracy, since the measurement errors would be averaged out. Compared with the RSS-based approach, the TOA/RTT-based method may provide more accurate distance estimates.

Alternatively, the distance measurements can be produced by jointly using an ultrasonic signal and an RF signal. The basic idea of the technique can be described with the aid of Figure 6.2. The RF signal and ultrasonic signal are sent off at the transmitter at time instants t_0^{RF} and t_0^{US} respectively. Typically, the transmission time of the RF signal and that of the ultrasonic signal are the same. The two signals arrive at the receiver at time instants t_1^{RF} and t_1^{US} respectively. In the case of simultaneous transmission, the difference between the TOA of the ultrasonic signal and that of the radio signal is virtually the propagation time of the ultrasonic signal, since the radio propagation time is extremely small comparatively. Specifically, let the TOA of the ultrasonic signal and that of the radio signal be t_1^{US} and t_1^{RF} respectively. Then, the two distance equations are

$$d = v_{US}(\hat{t}_1^{US} - t_0^{US} + n_{US})$$
$$d = c(\hat{t}_1^{RF} - t_0^{RF} + n_{RF}) \tag{6.4}$$

where v_{US} is the speed of the ultrasound signal, which is a known constant and equal to about 343 m/s in free space, depending on the medium temperature and moisture, c is the speed of light, $\hat{t}_1^{US}$ is the estimate of t_1^{US}, $\hat{t}_1^{RF}$ is the estimate of t_1^{RF}, and n_{US} and n_{RF} are the TOA estimation errors of the ultrasonic signal and the radio signal respectively. By ignoring the TOA estimation errors, it can be shown that

$$\hat{d} = v_{US}[\hat{t}_1^{US} - \hat{t}_1^{RF} + (t_0^{RF} - t_0^{US})] \tag{6.5}$$

where the fact that the speed of light is tremendously greater than the speed of sound is exploited and the difference between the two transmission time instants is known to the receiver. Typically, the transmissions are simultaneous, so that the TOF of the ultrasonic signal is the difference between the two TOAs.

6.1.1.2 Position Calculation

Figure 6.3 shows a simple example of how to determine the location of the target based on the distance measurements in a 2D environment. There are three anchors or base stations which the target can communicate with. Given the distance measurement at each anchor, the target is located somewhere on a distance circle with the anchor location as the center and the distance as the radius. In general, two distance circles are intercrossed at two points, producing two solutions to the location estimate of the target. To resolve the ambiguity of the two solutions, a third distance circle, resulting from the distance measurement at the third anchor, is required. In practice, the circles do not necessarily intersect at a point, so a method is needed to determine the position. A simple way is to compute the distance from the third anchor to each of the two solutions. Each of the computed distances is then compared with the distance measurement. We select the solution with which the computed distance to the third anchor better matches the measured one. It can be seen that the three anchors and the target should not lie in a straight line in order to produce a unique location estimate. In the following, the location estimation problem is described mathematically in a 3D environment.

Suppose that there are four anchors with known locations given by (x_i, y_i, z_i), $1 \leq i \leq 4$, and that the unknown location of the target is denoted by (x, y, z). The distance measurement equations are defined as

$$\hat{d}_i = d_i + \varepsilon_i \qquad i = 1, 2, 3, 4 \tag{6.6}$$

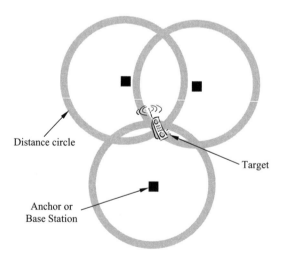

Figure 6.3 Illustration of positioning based on distance measurements.

where ε_i is the distance measurement error and

$$d_i = \sqrt{(x - x_i)^2 + (y - y_i)^2 + (z - z_i)^2} \qquad (6.7)$$

Since there are three unknown location parameters, let us first use three distance measurements, say $\hat{d}_1, \hat{d}_2$ and $\hat{d}_3$, to determine the location. Define

$$\begin{aligned}
\zeta_{i,1} &= x_i^2 + y_i^2 + z_i^2 - (x_1^2 + y_1^2 + z_1^2) \\
x_{i,j} &= x_i - x_j \\
y_{i,j} &= y_i - y_j \\
z_{i,j} &= z_i - z_j \\
f_{1,i} &= \frac{1}{2}\left\{ (\hat{d}_1^2 - \hat{d}_i^2) + \zeta_{i,1} \right\}
\end{aligned} \qquad (6.8)$$

Also define

$$\begin{aligned}
\kappa_1 &= \frac{x_{2,1}z_{3,1} - x_{3,1}z_{2,1}}{x_{3,1}y_{2,1} - x_{2,1}y_{3,1}} \\[2mm]
\kappa_2 &= \frac{x_{3,1}f_{1,2} - x_{2,1}f_{1,3}}{x_{3,1}y_{2,1} - x_{2,1}y_{3,1}} \\[2mm]
\kappa_3 &= \frac{y_{31}z_{2,1} - y_{2,1}z_{3,1}}{x_{3,1}y_{2,1} - x_{2,1}y_{3,1}} \\[2mm]
\kappa_4 &= \frac{y_{2,1}f_{1,3} - y_{3,1}f_{1,2}}{x_{3,1}y_{2,1} - x_{2,1}y_{3,1}}
\end{aligned} \qquad (6.9)$$

Then, with some mathematical manipulations, it can be shown that the estimate of the z-coordinate of the position is

$$\hat{z} = \frac{h_2}{h_1} \pm \sqrt{\left(\frac{h_2}{h_1}\right)^2 - \frac{h_3}{h_1}} \qquad (6.10)$$

where

$$\begin{aligned}
h_1 &= \kappa_1^2 + \kappa_3^2 + 1 \\
h_2 &= \kappa_1(y_1 - \kappa_2) + \kappa_3(x_1 - \kappa_4) + z_1 \\
h_3 &= (x_1 - \kappa_4)^2 + (y_1 - \kappa_2)^2 + z_1^2 - \hat{d}_1^2
\end{aligned} \qquad (6.11)$$

Accordingly, the estimates of the x-coordinate and y-coordinate can be obtained as

$$\begin{aligned}
\hat{x} &= \kappa_3\hat{z} + \kappa_4 \\
\hat{y} &= \kappa_1\hat{z} + \kappa_2
\end{aligned} \qquad (6.12)$$

Clearly, there are two solutions, but only one of them is the desired one. The ambiguity can be resolved by making use of the measurement at the fourth anchor. Specifically, the residuals of the two location estimation solutions are calculated by

$$\begin{aligned}
v_1 &= \left[\sqrt{(x_4 - \hat{x}^{(1)})^2 + (y_4 - \hat{y}^{(1)})^2 + (z_4 - \hat{z}^{(1)})^2} - \hat{d}_4 \right]^2 \\
v_2 &= \left[\sqrt{(x_4 - \hat{x}^{(2)})^2 + (y_4 - \hat{y}^{(2)})^2 + (z_4 - \hat{z}^{(2)})^2} - \hat{d}_4 \right]^2
\end{aligned} \qquad (6.13)$$

If $v_1 > v_2$, then the second solution $(\hat{x}^{(2)}, \hat{y}^{(2)}, \hat{z}^{(2)})$ is chosen. Otherwise, the first solution $(\hat{x}^{(1)}, \hat{y}^{(1)}, \hat{z}^{(1)})$ is selected. To make the residual-based judgment as reliable as possible, the fourth anchor should not be on the same plane formed by the other three anchors. Also, the distance from the fourth anchor to that plane should be greater than some threshold to make the judgment robust. The selection of the threshold will depend on the distance measurement noise. Note that this method is suboptimum, as spheres will not intersect at one point due to measurement errors.

Distance measurements are usually not very accurate, due to uncertainty in the internal delays or the power losses in the radio propagation. This applies to RSS and radio delay systems. The distance-based method is simple, but not usually used in practice. This is overcome by the time-difference methods described next.

6.1.2 Time-Difference-of-Arrival-based Positioning

When the anchors or base stations are synchronized, such as by sharing a global time reference or by cable connections, the target location may be determined by employing TDOA measurements. As shown in Figure 6.4, there are four anchors, each of which produces a TOA measurement. Supposing that the time of the start of the transmission is not known to the anchors, three TDOA equations can be obtained after eliminating the unknown transmit time. Each TDOA equation corresponds to a hyperbola, so that a unique target location estimate can be obtained as the intersection point of three hyperbolas in general. Note that this procedure also eliminates the transmit time and the internal radio delay in the transmitter, and those in the receivers if they are all the same. If the receiver delays are different, then a more complicated procedure is required.

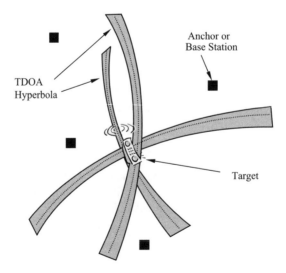

Figure 6.4   Illustration of positioning based on TDOA measurements.

The mathematical derivation of the target location in 3D environments can be described as follows. In the Cartesian system, the range (distance) between anchor node i and the mobile node (target) is given by

$$\sqrt{(x-x_i)^2 + (y-y_i)^2 + (z-z_i)^2} = c(t_i - t_0) \qquad i = 1,2,3,4 \tag{6.14}$$

where (x,y,z) and (x_i,y_i,z_i) are the coordinates of the mobile and anchor node respectively, c is the speed of propagation, t_i is the signal TOA at anchor node i and t_0 is the unknown transmit time at the mobile node. This means that the anchor clocks are synchronized, whereas the mobile node clock is not time synchronized. Let $\hat{t}_i$ be the estimate of the TOA t_i. Then, squaring both sides of (6.14) gives

$$(x-x_i)^2 + (y-y_i)^2 + (z-z_i)^2 \approx c^2(\hat{t}_i - t_0)^2 \qquad i = 1,2,3,4 \tag{6.15}$$

where the TOA estimation errors are ignored. Next, subtracting (6.15) at $i = 1$ from (6.15) at $i = 2,3,4$ produces

$$ct_0 \approx \frac{1}{2}c(\hat{t}_1 + \hat{t}_i) + \frac{1}{2c(\hat{t}_1 - \hat{t}_i)}(\beta_{i,1} - 2x_{i,1}x - 2y_{i,1}y - 2z_{i,1}z) \quad i = 2,3,4 \tag{6.16}$$

where

$$\begin{aligned}
x_{i,1} &= x_i - x_1 \\
y_{i,1} &= y_i - y_1 \\
z_{i,1} &= z_i - z_1 \\
\beta_{i,1} &= x_i^2 + y_i^2 + z_i^2 - (x_1^2 + y_1^2 + z_1^2)
\end{aligned} \tag{6.17}$$

Define the estimated TDOA between nodes i and j as

$$\delta t_{i,j} = \hat{t}_i - \hat{t}_j \tag{6.18}$$

It can be shown that eliminating t_0 from (6.16) yields

$$a_1 x + b_1 y + c_1 z \approx g_1 \tag{6.19}$$

where

$$\begin{aligned}
a_1 &= \delta t_{1,2} x_{3,1} - \delta t_{1,3} x_{2,1} \\
b_1 &= \delta t_{1,2} y_{3,1} - \delta t_{1,3} y_{2,1} \\
c_1 &= \delta t_{1,2} z_{3,1} - \delta t_{1,3} z_{2,1} \\
g_1 &= \frac{1}{2}(c^2 \delta t_{1,2} \delta t_{1,3} \delta t_{3,2} + \delta t_{1,2} \beta_{3,1} - \delta t_{1,3} \beta_{2,1})
\end{aligned} \tag{6.20}$$

and

$$a_2 x + b_2 y + c_2 z \approx g_2 \tag{6.21}$$

where

$$a_2 = \delta t_{1,2} x_{4,1} - \delta t_{1,4} x_{2,1}$$
$$b_2 = \delta t_{1,2} y_{4,1} - \delta t_{1,4} y_{2,1}$$
$$c_2 = \delta t_{1,2} z_{4,1} - \delta t_{1,4} z_{2,1} \tag{6.22}$$
$$g_2 = \frac{1}{2} \left(c^2 \delta t_{1,2} \delta t_{1,4} \delta t_{4,2} + \delta t_{1,2} \beta_{4,1} - \delta t_{1,4} \beta_{2,1} \right)$$

Combining (6.19) and (6.21), we obtain

$$x \approx \eta_1 z + \eta_2 \tag{6.23}$$

where

$$\eta_1 = \frac{b_1 c_2 - b_2 c_1}{a_1 b_2 - a_2 b_1}$$
$$\eta_2 = \frac{b_2 g_1 - b_1 g_2}{a_1 b_2 - a_2 b_1} \tag{6.24}$$

and

$$y \approx \xi_1 z + \xi_2 \tag{6.25}$$

where

$$\xi_1 = \frac{a_2 c_1 - a_1 c_2}{a_1 b_2 - a_2 b_1}$$
$$\xi_2 = \frac{a_1 g_2 - a_2 g_1}{a_1 b_2 - a_2 b_1} \tag{6.26}$$

Then, substitution of (6.23) and (6.25) into (6.16) with $i = 2$ produces

$$c(t_1 - t_0) \approx \rho_1 z + \rho_2 \tag{6.27}$$

where

$$\rho_1 = \frac{1}{c \Delta t_{1,2}} \left(x_{2,1} \eta_1 + y_{2,1} \xi_1 + z_{2,1} \right)$$
$$\rho_2 = \frac{c \Delta t_{1,2}}{2} + \frac{1}{2 c \Delta t_{1,2}} \left[2 (x_{2,1} \eta_2 + y_{2,1} \xi_2) - \beta_{2,1} \right] \tag{6.28}$$

Substituting (6.23), (6.25) and (6.27) into (6.14) for $i = 1$ and then squaring yields

$$h_1 z^2 + h_2 z + h_3 \approx 0 \tag{6.29}$$

where

$$h_1 = \eta_1^2 + \xi_1^2 - \rho_1^2 + 1$$
$$h_2 = 2 [\eta_1 (\eta_2 - x_1) + \xi_1 (\xi_2 - y_1) - z_1 - \rho_1 \rho_2] \tag{6.30}$$
$$h_3 = (\eta_2 - x_1)^2 + (\xi_2 - y_1)^2 + z_1^2 - \rho_2^2$$

The two solutions to the quadratic equation (6.29) are

$$\hat{z} = -\frac{h_2}{2h_1} \pm \sqrt{\left(\frac{h_2}{2h_1}\right)^2 - \frac{h_3}{2h_1}} \tag{6.31}$$

The two estimated z values (if both are reasonable) are then substituted into (6.23) and (6.25) to produce the estimates of the coordinates x and y respectively. To resolve the ambiguity of the two solutions, one more measurement from another anchor $\hat{t}_5$ is required. Then, using (6.13), one can resolve the ambiguity of the two solutions by comparing the residuals, which are defined as

$$
\begin{aligned}
\nu_1 &= \left[\sqrt{(x_5 - \hat{x}^{(1)})^2 + (y_5 - \hat{y}^{(1)})^2 + (z_5 - \hat{z}^{(1)})^2} - c(\hat{t}_5 - \hat{t}_0^{(1)})\right]^2 \\
\nu_2 &= \left[\sqrt{(x_5 - \hat{x}^{(2)})^2 + (y_5 - \hat{y}^{(2)})^2 + (z_5 - \hat{z}^{(2)})^2} - c(\hat{t}_5 - \hat{t}_0^{(2)})\right]^2
\end{aligned}
\tag{6.32}
$$

where $\hat{t}_0^{(1)}$ and $\hat{t}_0^{(2)}$ are the estimates of the transmission time t_0 corresponding to the respective two position estimates. The solution with the smaller residual is accepted, whereas the one with larger one is rejected.

Note that this is a suboptimum solution, since the fifth base station is only used for ambiguity resolution, rather than contributing to the solution. With five or more base stations, better accuracy would be obtained by using the LS methods to be discussed in Sections 6.3, 6.4 and 6.5.

6.1.3 Angle-of-Arrival-based Positioning

AOA can be measured by equipping the receiver nodes with an antenna array. Figure 6.5 shows an illustration of determining the location of the mobile target based on AOA measurements at

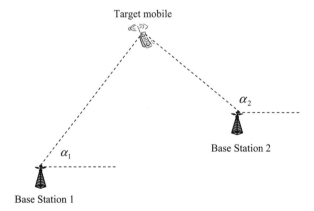

Figure 6.5   Illustration of AOA-based positioning.

two base stations in a 2D environment. Given the locations of the two base stations and the two AOA measurements $\hat{\alpha}_1$ and $\hat{\alpha}_2$, the location of the mobile target can be readily determined by solving

$$\tan \hat{\alpha}_1 x - y = x_1 \tan \hat{\alpha}_1 - y_1$$
$$\tan \hat{\alpha}_2 x - y = x_2 \tan \hat{\alpha}_2 - y_2 \tag{6.33}$$

where $\{\hat{\alpha}_i\}$ are the AOA measurements and (x_i, y_i) are the known base station coordinates. The solution to (6.33) is given by

$$\hat{x} = \frac{x_2 \tan \hat{\alpha}_2 - x_1 \tan \hat{\alpha}_1 + y_1 - y_2}{\tan \hat{\alpha}_2 - \tan \hat{\alpha}_1}$$
$$\hat{y} = \frac{\tan \hat{\alpha}_1 (x_2 \tan \hat{\alpha}_2 - y_2) - \tan \hat{\alpha}_2 (x_1 \tan \hat{\alpha}_1 - y_1)}{\tan \hat{\alpha}_2 - \tan \hat{\alpha}_1} \tag{6.34}$$

To obtain a unique solution to the unknown location, it is required that the two base stations and the target do not lie in a straight line. Also, it is worth mentioning that the two base stations must share the same orientation, which can be measured by using a compass, for instance.

When both the distance and AOA measurements can be made at each base station, the location of the target can be determined by one base station. For instance, Figure 6.6 shows an example of determining the location of the target by using the azimuth angle measurement $\hat{\phi}_i$, the elevation angle measurement $\hat{\alpha}_i$ and the distance measurement $\hat{d}_i$ in a 3D environment. The coordinates of the base station are known and set at (x_i, y_i, z_i) and the unknown location of the target is denoted by (x, y, z). The measurements can be described as

$$\hat{\phi}_i = \phi_i + n_{\hat{\phi}_i}$$
$$\hat{\alpha}_i = \alpha_i + n_{\hat{\alpha}_i} \tag{6.35}$$
$$\hat{d}_i = d_i + n_{\hat{d}_i}$$

where $n_{\hat{\phi}_i}$ is the azimuth angle measurement error, $n_{\hat{\alpha}_i}$ is the elevation angle measurement error and $n_{\hat{d}_i}$ is the distance measurement error, and

$$\phi_i = \tan^{-1} \frac{y - y_i}{x - x_i}$$
$$\alpha_i = \cos^{-1} \frac{z - z_i}{d_i} \tag{6.36}$$
$$d_i = \sqrt{(x - x_i)^2 + (y - y_i)^2 + (z - z_i)^2}$$

By using (6.36) and Figure 6.6 and dropping the measurement errors, it can be seen that the coordinate estimates are

$$\hat{x} = x_i + \hat{d}_i \sin \hat{\alpha}_i \cos \hat{\phi}_i$$
$$\hat{y} = y_i + \hat{d}_i \sin \hat{\alpha}_i \sin \hat{\phi}_i \tag{6.37}$$
$$\hat{z} = z_i + \hat{d}_i \cos \hat{\alpha}_i$$

It is worth mentioning that angle measurements are difficult to perform without large antennas, and the angle accuracy is poor in a multipath environment. Thus, AOA systems may

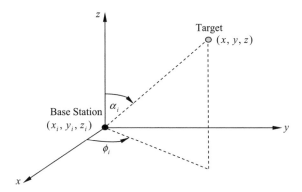

Figure 6.6 Illustration of 3D positioning based on distance and AOA measurements.

have poorer accuracy than TOA systems, especially in a rich multipath environment. Another drawback of the AOA method is that the accuracy reduces with range.

6.2 Linearization-Based Least-Squares Methods

In Section 6.1, when discussing the TDOA- and AOA-based positioning methods, the unknown positions are directly calculated by solving a group of measurement equations which are exactly determined. When the system is overdetermined (that is, the number of independent equations is greater than the number of unknown parameters), the LS technique can be applied to make use of the redundant measurements to obtain improved location estimates. In this section the simple linearization-based positioning algorithms are studied.

Consider a network in which there are N anchors whose locations are known and denoted by (x_i, y_i, z_i), $i = 1, 2, \ldots, N$. The unknown location of the target is defined as (x, y, z). When a distance measurement is made at each anchor, there are N measurement equations:

$$\hat{d}_i = d_i + v_i \qquad i = 1, 2, \ldots, N \tag{6.38}$$

where

$$d_i = \sqrt{(x_i - x)^2 + (y_i - y)^2 + (z_i - z)^2} \tag{6.39}$$

and v_i is the measurement error.

6.2.1 Linearization Based on Introducing an Intermediate Variable

Linearization of the distance measurement equations can be realized by introducing a new variable. Specifically, squaring both sides of (6.38) produces

$$2x_i x + 2y_i y + 2z_i z + (x^2 + y^2 + z^2) = (\hat{d}_i - v_i)^2 - (x_i^2 + y_i^2 + z_i^2) \tag{6.40}$$

Then, by defining

$$R = \sqrt{x^2 + y^2 + z^2}$$
$$R_i = \sqrt{x_i^2 + y_i^2 + z_i^2}$$

(6.41)

(6.40) becomes

$$-2x_i x - 2y_i y - 2z_i z + R^2 = \hat{d}_i^2 - R_i^2 + v_i^2 - 2\hat{d}_i v_i$$

(6.42)

which can be written in a compact form as

$$\mathbf{h} = \mathbf{G}\theta + \mathbf{v}$$

(6.43)

where

$$\theta = \begin{bmatrix} x & y & z & R^2 \end{bmatrix}^{\mathrm{T}}$$

$$\mathbf{h} = \begin{bmatrix} \hat{d}_1^2 - R_1^2 \\ \hat{d}_2^2 - R_2^2 \\ \vdots \\ \hat{d}_N^2 - R_N^2 \end{bmatrix}$$

$$\mathbf{G} = \begin{bmatrix} -2x_1 & -2y_1 & -2z_1 & 1 \\ -2x_2 & -2y_2 & -2z_2 & 1 \\ \vdots & \vdots & \vdots & \vdots \\ -2x_N & -2y_N & -2z_N & 1 \end{bmatrix}$$

$$\mathbf{v} = \begin{bmatrix} v_1^2 - 2\hat{d}_1 v_1 & v_2^2 - 2\hat{d}_2 v_2 & \cdots & v_N^2 - 2\hat{d}_N v_N \end{bmatrix}^{\mathrm{T}}$$

(6.44)

The LS solution of (6.43) is

$$\hat{\theta} = (\mathbf{G}^T \mathbf{W} \mathbf{G})^{-1} \mathbf{G}^T \mathbf{W} \mathbf{h}$$

(6.45)

where $\mathbf{W} = \mathrm{cov}^{-1}(\mathbf{v})$ is the weighting matrix which is set to be the inverse of the covariance matrix of $\mathbf{v}$ for Gaussian noise. In the case of unknown statistics of the error vector $\mathbf{v}$, $\mathbf{W}$ may be set at a diagonal matrix with the SNRs determined by receivers as the diagonal elements, or simply set at an identity matrix. Clearly, the solution by (6.45) is suboptimum due to the fact that the noise is amplified by squaring, and one more unknown is introduced.

6.2.2 Linearization Based on Arithmetic Operations

Another way to linearize the distance measurement equations can be described as follows. Subtracting (6.38) at $i = 1$ from (6.40) at $i = 2, 3, \ldots, N$ produces

$$2x_{1,i}x + 2y_{1,i}y + 2z_{1,i}z = \hat{d}_i^2 - \hat{d}_1^2 + R_1^2 - R_i^2 + v_i^2 - v_1^2 + 2(\hat{d}_1 v_1 - \hat{d}_i v_i) \qquad i = 2, 3, \ldots, N \tag{6.46}$$

which can be written in the same form as (6.43) but with

$$\boldsymbol{\theta} = \begin{bmatrix} x & y & z \end{bmatrix}^T$$

$$\mathbf{h} = \begin{bmatrix} \hat{d}_2^2 - \hat{d}_1^2 + R_1^2 - R_2^2 \\ \hat{d}_3^2 - \hat{d}_1^2 + R_1^2 - R_3^2 \\ \vdots \\ \hat{d}_N^2 - \hat{d}_1^2 + R_1^2 - R_N^2 \end{bmatrix}$$

$$\mathbf{G} = \begin{bmatrix} 2x_{1,2} & 2y_{1,2} & 2z_{1,2} \\ 2x_{1,3} & 2y_{1,3} & 2z_{1,3} \\ \vdots & \vdots & \vdots \\ 2x_{1,N} & 2y_{1,N} & 2z_{1,N} \end{bmatrix} \tag{6.47}$$

$$\mathbf{v} = \begin{bmatrix} v_2^2 - v_1^2 + 2(\hat{d}_1 v_1 - \hat{d}_2 v_2) & \cdots & v_N^2 - v_1^2 + 2(\hat{d}_1 v_1 - \hat{d}_N v_N) \end{bmatrix}^T$$

Accordingly, the LS estimate of the position coordinates can be obtained by using the formula given by (6.45). This solution is not optimal due to the fact that the linearization results in a loss of information. It is found through simulation that the approach in Section 6.2.1 outperforms the one based on (6.46) and (6.47) in NLOS channels. This may be due to the fact that a reference anchor is required to obtain (6.46). The accuracy of the measurements at this anchor will undoubtedly affect the accuracy of location estimation. The results provided by (6.45) are often used as the initial location estimates for more advanced estimation algorithms, such as the iterative algorithms which will be presented in Chapter 7.

6.3 Spherical Interpolation Approach

When the transmit time is not known to the receivers but the receivers are synchronized, such as in centralized networks, TDOA measurements may be employed to obtain location estimates as mentioned previously. Note that, as discussed in Chapter 5, 'differential pseudo-range' would be more appropriate than the term 'TDOA'; however, for consistency with the literature, the term 'TDOA' is still used. One such TDOA-based position estimation method is the SI method [2]. The SI method can be described with the aid of Figure 6.7. One of the receivers, say

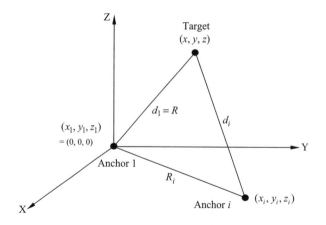

Figure 6.7 Illustration for SI approach.

anchor 1, is mapped to the spatial origin of the 3D Cartesian coordinate system. Since one anchor serves as the reference point, the accuracy of the SI algorithm will depend on the quality of the measurements observed at this reference anchor. To minimize this effect it is suggested that the shortest range base station is used as the reference, as this is likely to have the smallest measurement errors. Since the identity of this base station is not initially known, the best results are obtained by running the algorithm twice, the first time with an arbitrary reference and the second time with the chosen best reference anchor node.

With reference to Figure 6.7, define

$$
\begin{aligned}
R_i &= \sqrt{x_i^2 + y_i^2 + z_i^2} \\
\mathbf{p}_i &= [x_i \ y_i \ z_i]^{\mathrm{T}} \\
\mathbf{p} &= [x \ y \ z]^{\mathrm{T}} \\
R &= \sqrt{x^2 + y^2 + z^2} \\
d_{i,j} &= d_i - d_j = c(t_i - t_j)
\end{aligned}
\tag{6.48}
$$

where c is the speed of propagation, t_i is the TOA of the signal received at anchor i, and $t_i - t_j$ is the TDOA of the signal received at anchors i and j.

Applying Pythagoras's theorem produces

$$
(R + d_{i,1})^2 = R_i^2 - 2\mathbf{p}_i^{\mathrm{T}}\mathbf{p} + R^2 \qquad i = 2, 3, \dots, N
\tag{6.49}
$$

where use has been made of $R + d_{i,1} = d_i$. In the presence of measurement errors, (6.49) usually does not hold. To make the equation valid, the equation error ε_i is introduced and added to the left-hand side of (6.49). Then:

$$
\varepsilon_i = R_i^2 - \hat{d}_{i,1}^2 - 2R\hat{d}_{i,1} - 2\mathbf{p}_i^{\mathrm{T}}\mathbf{p}
\tag{6.50}
$$

where

$$
\hat{d}_{i,1} = c(\hat{t}_i - \hat{t}_1)
\tag{6.51}
$$

where $\hat{t}_i$ is the TOA estimate and (6.50) can be written in a compact form as

$$\varepsilon = \delta - 2R\hat{\mathbf{d}} - 2\mathbf{Ap} \tag{6.52}$$

where

$$\delta = \begin{bmatrix} R_2^2 - \hat{d}_{2,1}^2 & R_3^2 - \hat{d}_{3,1}^2 & \cdots & R_N^2 - \hat{d}_{N,1}^2 \end{bmatrix}^{\mathrm{T}}$$
$$\hat{\mathbf{d}} = \begin{bmatrix} \hat{d}_{2,1} & \hat{d}_{3,1} & \cdots & \hat{d}_{N,1} \end{bmatrix}^{\mathrm{T}}$$
$$\mathbf{A} = \begin{bmatrix} x_{2,1} & y_{2,1} & z_{2,1} \\ x_{3,1} & y_{3,1} & z_{3,1} \\ \vdots & \vdots & \vdots \\ x_{N,1} & y_{N,1} & z_{N,1} \end{bmatrix} \tag{6.53}$$

Given R, the standard LS solution for $\mathbf{p}$ is

$$\mathbf{p} = \frac{1}{2}(\mathbf{A}^{\mathrm{T}}\mathbf{WA})^{-1}\mathbf{A}^{\mathrm{T}}\mathbf{W}(\delta - 2R\hat{\mathbf{d}}) \tag{6.54}$$

where $\mathbf{W} = (1/4)E[\varepsilon\varepsilon^{\mathrm{T}}]$. Note that the parameter R in (6.54) is unknown. So, by substituting (6.54) into (6.52) produces

$$\varepsilon' = \mathbf{B}(\delta - 2R\hat{\mathbf{d}}) \tag{6.55}$$

where

$$\mathbf{B} = \mathbf{I} - \mathbf{A}(\mathbf{A}^{\mathrm{T}}\mathbf{WA})^{-1}\mathbf{A}^{\mathrm{T}}\mathbf{W} \tag{6.56}$$

Equation (6.55) can be rewritten as

$$\varepsilon'^{\mathrm{T}}\mathbf{V}\varepsilon' = (\delta - 2R\hat{\mathbf{d}})^{\mathrm{T}}\mathbf{BVB}(\delta - 2R\hat{\mathbf{d}}) \tag{6.57}$$

where $\mathbf{V}$ is a weighting matrix. Minimizing the right-hand side of (6.57) with respect to R produces

$$\hat{R} = \frac{\hat{\mathbf{d}}^{\mathrm{T}}\mathbf{BVB}\delta}{2\hat{\mathbf{d}}^{\mathrm{T}}\mathbf{BVB}\hat{\mathbf{d}}} \tag{6.58}$$

Note that the equation becomes indeterminate if $\hat{d} = 0$. For anchor nodes/base stations arranged in a circle, the error in R increases as the position approaches the center of the circle. This behavior is the opposite to the behavior of GDOP. Also note that the range estimate by this formula can be a nonphysical negative value with ranging errors. The main source of the errors is associated with the δ parameter, as this involves the squaring of the measured range data, thus magnifying the errors. To avoid the phenomenon that the estimated range becomes negative (a physical impossibility) or becomes too large, some constraints should be applied. For example, R is subject to more reasonable limits, namely to a minimum of zero and to a maximum value, say two times the radius of the circle on which the base stations lie.

Substituting (6.58) into (6.54) produces the source location estimate as

$$\hat{\mathbf{p}} = \frac{1}{2}(\mathbf{A}^{\mathrm{T}}\mathbf{W}\mathbf{A})^{-1}\mathbf{A}^{\mathrm{T}}\mathbf{W}\left(\mathbf{I} - \frac{\hat{\mathbf{d}}\hat{\mathbf{d}}^{\mathrm{T}}\mathbf{B}\mathbf{V}\mathbf{B}}{\hat{\mathbf{d}}^{\mathrm{T}}\mathbf{B}\mathbf{V}\mathbf{B}\hat{\mathbf{d}}}\right)\boldsymbol{\delta} \qquad (6.59)$$

Note that the weighting matrices $\mathbf{V}$ and $\mathbf{W}$ are employed for emphasizing the contributions of the measurements that are more reliable. In the absence of *a priori* confidences on the statistics of the errors, the weighting matrices may be simply chosen to be an identity matrix. The computational burden of (6.59) is low compared with the iterative nonlinear minimization. When using this method, the weighting matrices should be chosen properly to achieve good accuracy. Also, the solution from (6.59) can serve as an initial location estimate for more advanced iterative algorithms.

The following example is used to show how the SI method works. A 2D location area is assumed which is of a square shape with side length of 30 m. This dimensional size corresponds to an indoor environment and sensor networks. Two different numbers of anchor nodes/base stations, four and five, are considered with four different configurations, as shown in Figure 6.8. Note that, typically, a mobile node can only communicate with a rather limited number of base stations, especially in cellular networks. For each number of base stations, two different base station displacements are examined, as shown in Figure 6.8. The dashed boxes represent the location area. The location of the mobile node is randomly generated in the area and a number of 10 000 different locations are tested.

Signal propagation in indoor environments is complex in general, depending on the specific environmental structures. There are a number of references which address modeling the statistics of the TOA measurement errors in indoor conditions, but here we employ the results in [3]. The characteristics of the delay excess in this reference vary somewhat depending on the particular propagation environment. Thus, if one set of parameters is chosen, then the modeling

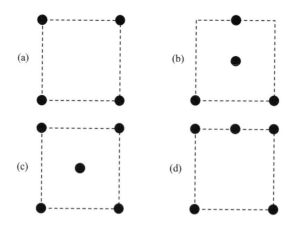

Figure 6.8 Deployment patterns of four or five base stations.

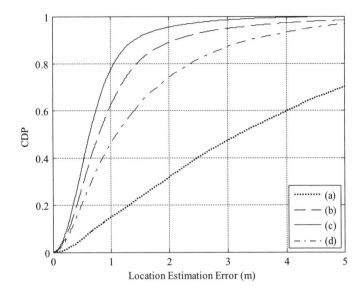

Figure 6.9 Accuracy of the SI algorithm with four and five base stations under four different configurations. (a)–(d) respectively denote the results under the base station configurations (a)–(d) in Figure 6.8.

will only be approximate. As indicated in Chapter 5, based on a broad averaging of the data, the model (500 MHz bandwidth) suggested is that the range error (in meters) is a Gaussian random variable of mean given by

$$\Delta R = c\Delta \text{TOA} = 0.25 + 0.1R \qquad (6.60)$$

where R is the true distance between the receiver and transmitter. Therefore, the range error bias varies from 0.25 m at short range to 3.25 m at 30 m, which is typically the maximum reliable indoor range with NLOS propagation. In addition to the mean bias error, there will be an additional random component. A STD of 0.5 m would be a reasonable value for simulations. Note that unlike the bias, the STD parameter is independent of the distance. At each mobile location, only one noise sample is tested. The performance measure is the CDP of the location estimation error, which is defined as the distance between the true location and the estimated one. Also, the corresponding average distance between the true location and the estimated one will be computed. In Chapter 8, more location accuracy measures will be discussed. Figure 6.9 shows the location accuracy of the SI algorithms under the four different base station configurations. One interesting observation is that the accuracy with four base stations and configuration (b) is better than that with five base stations and configuration (d). This indicates that the base station deployment is important. Without a proper deployment, more base stations may result in accuracy degradation. For the SI algorithm, it is not wise to deploy all the base stations along the boundary. Instead, one or more base stations should be located around the center of the location area.

6.4 Quasi-Least-Squares Solution

In the following, a QLS solution based on closed-form analytical expressions is described. The method was originally described in [4] for GPS applications, but it has been adapted for terrestrial applications.

The basis for the analysis is pseudo-range measurements for transmissions from a mobile node to fixed base stations indexed by i in the form as

$$S_i = R_i + R_c \tag{6.61}$$

where R_c is a constant offset which is the same for all anchor node/base station pseudo-range measurements and

$$R_i = \sqrt{(x_i - x)^2 + (y_i - y)^2 + (z_i - z)^2} \tag{6.62}$$

For TOA systems, this offset arises usually from the mobile node not having time synchronization (unlike the base stations), so that there is an arbitrary offset in time and, hence, a distance-equivalent R_c. For RSS systems, the offset is related to the uncertainty in the effective radiated power from the mobile device.[1]

For the analysis it will be assumed initially that there are just four base stations, so that there is just sufficient pseudo-range data to solve for the four unknowns, namely the mobile node position (x, y, z) and the offset parameter R_c. Later, these restrictions will be eased so that a QLS solution can be obtained [5].

The first step in the analysis is to remove the square root from (6.61) by squaring both sides of the equation. As will be shown, the analytical solution will involve dual solutions from a quadratic equation, but the invalid solution can be detected easily and, thus, ignored. Squaring (6.61) and rearranging results in

$$(R_i^2 - S_i^2) - 2(x_i x + y_i y + z_i z - S_i R_c) + (R^2 - R_c^2) = 0 \tag{6.63}$$

where $R = \sqrt{x^2 + y^2 + z^2}$. The analysis that follows is simplified by adopting a combination of matrix algebra and the Lorentz inner product, which is defined for two vectors $(\mathbf{g}, \mathbf{h} \in \mathbb{R}^4)$ as follows:

$$\langle \mathbf{g}, \mathbf{h} \rangle = \langle \mathbf{h}, \mathbf{g} \rangle = \mathbf{g}^T \mathbf{M} \mathbf{h} \tag{6.64}$$

where $\mathbf{M}$ is similar to the 4×4 identity matrix, except that the last diagonal element is -1 instead of $+1$, so that it has the following properties:

$$\mathbf{M} = \mathbf{M}^T = \mathbf{M}^{-1} \tag{6.65}$$

Based on these definitions and letting

$$\mathbf{p} = \begin{bmatrix} x & y & z \end{bmatrix}^T$$

$$\mathbf{p}_i = \begin{bmatrix} x_i & y_i & z_i \end{bmatrix}^T \tag{6.66}$$

---

[1] The variability in the radiated power in small devices such as in a WSN is partly due to the variability between the hardware in otherwise identical units, but mainly the variability is associated with the antenna location, mismatches between the antenna and the radio, and diffraction effects when body worn. This variability can exceed 10 dB.

equation (6.63) can now be written as

$$\frac{1}{2}\left\langle \begin{bmatrix} \mathbf{p}_i \\ S_i \end{bmatrix}, \begin{bmatrix} \mathbf{p}_i \\ S_i \end{bmatrix} \right\rangle - \left\langle \begin{bmatrix} \mathbf{p}_i \\ S_i \end{bmatrix}, \begin{bmatrix} \mathbf{p} \\ R_c \end{bmatrix} \right\rangle + \frac{1}{2}\left\langle \begin{bmatrix} \mathbf{p} \\ R_c \end{bmatrix}, \begin{bmatrix} \mathbf{p} \\ R_c \end{bmatrix} \right\rangle = 0 \qquad (6.67)$$

Note that there is one of these equations for each measurement (four in this case), so that combining all the equations into matrices produces

$$\mathbf{a} - \mathbf{BM\beta} + v\,\boldsymbol{\eta} = 0 \qquad (6.68)$$

where

$$\mathbf{a}_i = \frac{1}{2}\left\langle \begin{bmatrix} \mathbf{p}_i \\ S_i \end{bmatrix}, \begin{bmatrix} \mathbf{p}_i \\ S_i \end{bmatrix} \right\rangle \qquad \boldsymbol{\eta} = [1 \quad 1 \quad 1 \quad 1]^T \qquad \boldsymbol{\beta} = \begin{bmatrix} \mathbf{p} \\ R_c \end{bmatrix}$$

$$\mathbf{B} = \begin{bmatrix} x_1 & y_1 & z_1 & S_1 \\ x_2 & y_2 & z_2 & S_2 \\ x_3 & y_3 & z_3 & S_3 \\ x_4 & y_4 & z_4 & S_4 \end{bmatrix} \qquad v = \frac{1}{2}\left\langle \begin{bmatrix} \mathbf{p} \\ R_c \end{bmatrix}, \begin{bmatrix} \mathbf{p} \\ R_c \end{bmatrix} \right\rangle = \frac{1}{2}\langle \boldsymbol{\beta}, \boldsymbol{\beta} \rangle \qquad (6.69)$$

Note that a solution for the $\boldsymbol{\beta}$ vector is the required result; that is:

$$\boldsymbol{\beta} = \begin{bmatrix} \mathbf{p} \\ R_c \end{bmatrix} = [\mathbf{BM}]^{-1}(\mathbf{a} + v\,\boldsymbol{\eta}) = \mathbf{MB}^{-1}(\mathbf{a} + v\,\boldsymbol{\eta}) \qquad (6.70)$$

Finally, substitute (6.70) into the last expression in (6.69) and express the result in matrix form as

$$2v = [v\mathbf{B}^{-1}\boldsymbol{\eta} + \mathbf{B}^{-1}\mathbf{a}]^T[\mathbf{M}][v\mathbf{B}^{-1}\boldsymbol{\eta} + \mathbf{B}^{-1}\mathbf{a}] \qquad (6.71)$$

Simplifying this expression using the properties of $\mathbf{M}$ and reverting back to the Lorentz format results in

$$\langle \mathbf{B}^{-1}\boldsymbol{\eta}, \mathbf{B}^{-1}\boldsymbol{\eta}\rangle v^2 + 2(\langle \mathbf{B}^{-1}\boldsymbol{\eta}, \mathbf{B}^{-1}\mathbf{a}\rangle - 1)v + \langle \mathbf{B}^{-1}\mathbf{a}, \mathbf{B}^{-1}\mathbf{a}\rangle = 0 \qquad (6.72)$$

As can be observed, (6.72) is a quadratic equation, which can have two real roots or two complex roots. The complex roots indicate that the measured data are too poor for a fit; thus, the method can detect invalid measurements. If there are two real roots, then the task is to determine which is the correct one. The two possible solutions (v_1 and v_2) are substituted in (6.70) to obtain two estimates of the vector $\boldsymbol{\beta}$. The correct solution can be found by using these data to back calculate the expected measurement pseudo-ranges; only one will match the measurements.

The above procedure is limited to the case of four base stations, but the solution can be extended to the case of more base stations. Additionally, as the measured data can vary in quality, practical solutions require that the data are suitably weighted for an optimum solution. Further, as will be shown, the solution where the ranges to the base stations can vary widely requires the measurements to be suitably weighted, regardless of the quality of the measurements.

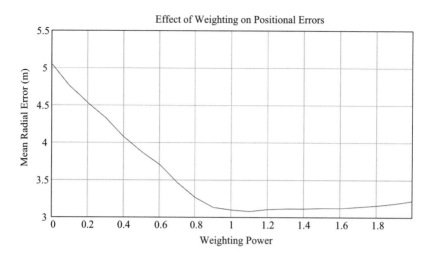

Figure 6.10   Effect of the weighting in the QLS algorithm on positional accuracy near a base station.

To cater for both more than four base stations and weighting of the data, the **B** matrix can be extended with more measurements and base station positions, and then $\mathbf{B}^{-1}$ in (6.72) is replaced with a new 4×4 matrix as follows:

$$\mathbf{B}^{-1} \rightarrow [\mathbf{B}^{\mathrm{T}}\mathbf{W}\mathbf{B}]^{-1}[\mathbf{B}^{\mathrm{T}}\mathbf{W}] \qquad (6.73)$$

The weighting matrix **W** normally reflects the 'quality' of the measurements, but simulations show that the solution becomes sensitive to measurement errors if the mobile node is close to an anchor node/base station. Figure 6.10 shows the results of a Monte Carlo simulation. The simulation is based on four base stations in a square enclosed by a circle of radius 15 m. The position of the mobile device is 1 m from a base station along the radial from the center of the circle to the base station. The simulation uses a measurement STD of 2 m. The diagonal elements of **W** are chosen as $R_n^{-\kappa}$, where κ is the power-law parameter. It is shown that the mean radial error reduces as the power parameter increases from zero, reaching the minimum error when $\kappa \approx 1$. As the power parameter increases beyond $\kappa = 1$, the accuracy decreases slightly. The optimum power-law parameter varies somewhat with the exact details, such as the number of anchor nodes/base stations and their locations, and the position of the mobile device. Overall, the choice of $\kappa = 1$ provides near-optimum performance in almost all situations. As can be observed, the optimum result can be about half the unweighted result ($\kappa = 0$), with the optimum error also being less than that achieved by the classical iterative solution when near a base station. If the measurement pseudo-range data are all of equal quality as assumed in the simulation above, then the optimum weighting for the data is R_n^{-1}. In general, however, other factors such as the SNR need to be considered, so that the weighting factor can be generalized to $R_n^{-\kappa}$, where the power parameter is typically in the range $1 \le \kappa \le 2$.

The requirement for this distance weighting factor means that while no iterations are required for the analytical calculations, the application of the correct weighting requires an initial estimate of position of the mobile device to be made by using equal weighting, thus allowing a reasonable estimate of the weighting factors to be determined for the second iteration. This procedure is only required for the first calculation of position, as the normal situation is for continuous updates of position, so the previous position can be used to estimate the weighting for the next position calculation.

In summary, this analytical solution is useful in seeding other iterative solutions, but as the result is only QLS, other iterative methods usually give superior results. However, when the mobile position is near an anchor node or a base station, this analytical solution gives superior accuracy if the correct weighting is applied. The method also can detect 'bad' data, as the quadratic equation can give a 'no solution' result. This indication is useful when there are large measurement errors, as LS methods can give very poor fits in such cases without any flagging of the poor result. Further, in some cases iterative methods will not converge with poor data, whereas this analytical solution often can give results, albeit with low accuracy. Thus, in some situations, such as emergency applications where a low accuracy result is better than no result, the analytical solution can be useful. In a typical implementation, other methods will normally be used. If they all fail, then the data can be reprocessed using this analytical method.

The accuracy of the QLS algorithm is illustrated as follows. The base station configurations are shown in Figure 6.8 and the distance measurement error is modeled by (6.60). It is necessary that the evaluation of different algorithms is under the same conditions so that their performance can be fairly compared. The offset parameter R_c is set as a random variable uniformly distributed between 0 and 200 m. It is found that the performance of the QLS method is independent of the offset parameter.

Figure 6.11 shows the results of the QLS algorithm when equal weights are employed. Contrary to the SI algorithm, the QLS method performs better when the base stations are located along the boundary of the location area. Also, it seems that the QLS algorithm performs

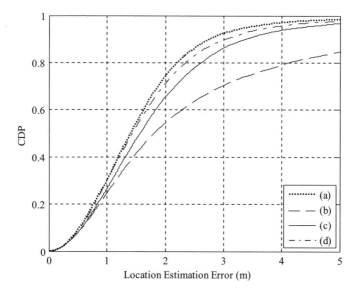

Figure 6.11 Accuracy of the QLS algorithm with four and five base stations under four different configurations. Equal weights are employed. (a)–(d) respectively denote the results under the base station configurations (a)–(d) in Figure 6.8.

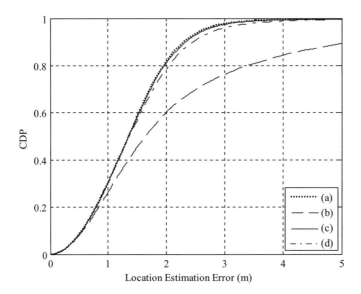

Figure 6.12 Accuracy of the QLS algorithm with four and five base stations under four different configurations. Weights are updated five times with equal weights as starting point. (a)–(d) respectively denote the results under the base station configurations (a)–(d) in Figure 6.8.

particularly well when there are four base stations which are placed at the four corners of the square.

Figure 6.12 shows the accuracy of the QLS method when an iterative procedure is applied. That is, initial location estimates are generated with equal weights. Then, the weights are updated by using the estimated locations and the results are produced through five iterations. The convergence is fast and typically only a few iterations will be enough to approach the steady state. Three configurations generate very similar results, whereas only one configuration performs rather poorly. Therefore, in the presence of a configuration like (b) in Figure 6.8, the QLS method should never be considered. It can also be observed that, for the QLS method, the performance can be improved considerably by updating the weights, compared with using equal weights. Note that the QLS algorithm with weights updating actually becomes an iterative algorithm.

6.5 Linear-Correction Least-Squares Approach

Similar to the SI method described in Section 6.3, the LC-LS method in [6] is also a two-step LS estimation method. First, an initial location estimate is produced based on the linearization-based LS method as described in Section 6.2.1. Then, a correction quantity is produced to compensate for the error incurred in the first-step estimation. The key idea of the method is described as follows.

Rewriting (6.52) as

$$A\theta = \delta + \varepsilon \qquad (6.74)$$

where

$$\mathbf{A} = \begin{bmatrix} x_{2,1} & y_{2,1} & z_{2,1} & \hat{d}_{2,1} \\ x_{3,1} & y_{3,1} & z_{3,1} & \hat{d}_{3,1} \\ \vdots & \vdots & \vdots & \vdots \\ x_{N,1} & y_{N,1} & z_{N,1} & \hat{d}_{N,1} \end{bmatrix} \tag{6.75}$$

$$\boldsymbol{\theta} = \begin{bmatrix} x & y & z & R \end{bmatrix}^{\mathrm{T}}$$

$$\boldsymbol{\delta} = \frac{1}{2} \begin{bmatrix} R_2^2 - \hat{d}_{2,1}^2 & R_3^2 - \hat{d}_{3,1}^2 & \cdots & R_N^2 - \hat{d}_{N,1}^2 \end{bmatrix}^{\mathrm{T}}$$

The LS solution to the location estimation based on (6.74) is produced by minimizing a cost function, which is given by

$$J(\boldsymbol{\theta}) = [\mathbf{A}\boldsymbol{\theta} - \boldsymbol{\delta}]^{\mathrm{T}} [\mathbf{A}\boldsymbol{\theta} - \boldsymbol{\delta}] \tag{6.76}$$

subject to a quadratic constraint:

$$\boldsymbol{\theta}^{\mathrm{T}} \boldsymbol{\Sigma} \boldsymbol{\theta} = 0 \tag{6.77}$$

where

$$\boldsymbol{\Sigma} = \mathrm{diag}\{1, 1, 1, -1\} \tag{6.78}$$

To solve the constrained optimization problem, the Lagrange multiplier λ is introduced and the Lagrangian function is defined as

$$\mathcal{L}(\boldsymbol{\theta}, \lambda) = J(\boldsymbol{\theta}) + \lambda \boldsymbol{\theta}^{\mathrm{T}} \boldsymbol{\Sigma} \boldsymbol{\theta} \tag{6.79}$$

Then, the source location is determined by minimizing the Lagrangian function; that is:

$$\hat{\boldsymbol{\theta}} = \min_{\boldsymbol{\theta}, \lambda} \mathcal{L}(\boldsymbol{\theta}, \lambda) \tag{6.80}$$

Expanding the Lagrangian function produces

$$\mathcal{L}(\boldsymbol{\theta}, \lambda) = \boldsymbol{\theta}^{\mathrm{T}} (\mathbf{A}^{\mathrm{T}} \mathbf{A} + \lambda \boldsymbol{\Sigma}) \boldsymbol{\theta} - 2 \boldsymbol{\delta}^{\mathrm{T}} \mathbf{A} \boldsymbol{\theta} + \boldsymbol{\delta}^{\mathrm{T}} \boldsymbol{\delta} \tag{6.81}$$

Differentiating $\mathcal{L}(\boldsymbol{\theta}, \lambda)$ with respect to $\boldsymbol{\theta}$ and setting the derivative to zero yields

$$(\mathbf{A}^{\mathrm{T}} \mathbf{A} + \lambda \boldsymbol{\Sigma}) \boldsymbol{\theta} - \mathbf{A}^{\mathrm{T}} \boldsymbol{\delta} = 0 \tag{6.82}$$

For a given λ, solving for $\boldsymbol{\theta}$ from (6.82) produces

$$\hat{\boldsymbol{\theta}} = (\mathbf{A}^{\mathrm{T}} \mathbf{A} + \lambda \boldsymbol{\Sigma})^{-1} \mathbf{A}^{\mathrm{T}} \boldsymbol{\delta} \tag{6.83}$$

Let us now determine the Lagrange multiplier λ. Replacing $\boldsymbol{\theta}$ in (6.77) by $\hat{\boldsymbol{\theta}}$ in (6.83), we obtain

$$\boldsymbol{\delta}^{\mathrm{T}} \mathbf{A} (\mathbf{A}^{\mathrm{T}} \mathbf{A} + \lambda \boldsymbol{\Sigma})^{-1} \boldsymbol{\Sigma} (\mathbf{A}^{\mathrm{T}} \mathbf{A} + \lambda \boldsymbol{\Sigma})^{-1} \mathbf{A}^{\mathrm{T}} \boldsymbol{\delta} = 0 \tag{6.84}$$

By making use of eigenvalue decomposition, the matrix $\mathbf{A}^{\mathrm{T}}\mathbf{A}\boldsymbol{\Sigma}$ can be diagonalized as

$$\mathbf{A}^{\mathrm{T}}\mathbf{A}\boldsymbol{\Sigma} = \mathbf{U}\boldsymbol{\Lambda}\mathbf{U}^{-1} \tag{6.85}$$

where

$$\boldsymbol{\Lambda} = \mathrm{diag}\{\kappa_1, \kappa_2, \kappa_3, \kappa_4\} \tag{6.86}$$

where $\{\kappa_i\}$ are the eigenvalues of the matrix $\mathbf{A}^{\mathrm{T}}\mathbf{A}\boldsymbol{\Sigma}$. By substituting (6.85) into (6.84), it can be shown that

$$\mathbf{s}^{\mathrm{T}}(\mathbf{A} + \lambda\mathbf{I})^{-2}\mathbf{q} = 0 \tag{6.87}$$

where

$$\begin{aligned} \mathbf{s} &= \mathbf{U}^{\mathrm{T}}\boldsymbol{\Sigma}\mathbf{A}^{\mathrm{T}}\boldsymbol{\delta} \\ \mathbf{q} &= \mathbf{U}^{-1}\mathbf{A}^{\mathrm{T}}\boldsymbol{\delta} \end{aligned} \tag{6.88}$$

Equation (6.87) can be further written as

$$f(\lambda) = \sum_{i=1}^{4} \frac{s_i q_i}{(\lambda + \kappa_i)^2} = 0 \tag{6.89}$$

It is seen that there is no closed-form solution for λ in general, and a root search is required to find a solution. For instance, the secant method for root searching can be employed. For two given points, $\lambda^{(0)} = 0$ and $\lambda^{(1)}$ which is a small positive number, the root searching is performed iteratively according to

$$\lambda^{(n+1)} = \lambda^{(n)} - \frac{\lambda^{(n)} - \lambda^{(n-1)}}{f(\lambda^{(n)}) - f(\lambda^{(n-1)})} f(\lambda^{(n)}) \tag{6.90}$$

Since λ is a rather small positive number, it may only be necessary to search for the roots close to zero. Once a root is found, the searching can be constrained to the two areas next to the known root if necessary. In the case of a 2D environment, $f(\lambda)$ becomes a fourth-order polynomial given by

$$f(\lambda) = a_1\lambda^4 + a_2\lambda^3 + a_3\lambda^2 + a_4\lambda + a_5 \tag{6.91}$$

where

$$\begin{aligned} a_1 &= s_1 q_1 + s_2 q_2 + s_3 q_3 \\ a_2 &= 2[s_1 q_1(\kappa_2 + \kappa_3) + s_2 q_2(\kappa_1 + \kappa_3) + s_3 q_3(\kappa_1 + \kappa_2)] \\ a_3 &= s_1 q_1[(\kappa_2 + \kappa_3)^2 + 2\kappa_2\kappa_3] + s_2 q_2[(\kappa_1 + \kappa_3)^2 + 2\kappa_1\kappa_3] + s_3 q_3[(\kappa_1 + \kappa_2)^2 + 2\kappa_1\kappa_2] \\ a_4 &= 2[s_1 q_1(\kappa_2 + \kappa_3)\kappa_2\kappa_3 + s_2 q_2(\kappa_1 + \kappa_3)\kappa_1\kappa_3 + s_3 q_3(\kappa_1 + \kappa_2)\kappa_1\kappa_2] \\ a_5 &= s_1 q_1(\kappa_2\kappa_3)^2 + s_2 q_2(\kappa_1\kappa_3)^2 + s_3 q_3(\kappa_1\kappa_2)^2 \end{aligned} \tag{6.92}$$

In this case, we may apply a simple root-finding algorithm to determine the four roots of the polynomial. Substituting the estimated roots into (6.83) produces the corresponding location estimates. The different estimates are tested against the constraints given by (6.77) and the estimate which best matches the constraint is selected as the final source location estimate. In the case of multiple solutions satisfying the constraint, the location dimensional size can be exploited to remove the unsuited solution.

Compared with the SI method, the LC-LS method involves more computation in finding and resolving the multiple roots of the Lagrange multiplier and performing the eigenvalue decomposition. Similar two-step LS algorithms are studied in [7,8].

The evaluation of the location accuracy of the LC-LS algorithm is performed in the same way as evaluating the SI method and the QLS method in the preceding two sections. More specifically, the base station configurations are chosen as in Figure 6.8 and the distance measurement error is modeled by (6.60).

Figure 6.13 shows the location accuracy of the LC-LS algorithm when there are either four or five base stations, corresponding to four different configurations. Compared with the SI algorithm and the QLS method, the LC-LS method is less sensitive to the base station configurations. However, the LC-LS method and the SI algorithm basically share one common property, that the configurations (b) and (c) are more suited than the configurations (a) and (d).

By using these three noniterative algorithms with the given dimensional size of the location area and distance measurement errors, 2 m location accuracy can be achieved with a CDP of 82%, 89%, 96%, and 80% respectively under the four base station configurations (a)–(d). Accordingly, we can compute the average distance between the true position and the estimated position of the mobile node as shown in Table 6.1. Clearly, the average distances are in good accordance with the CDPs. Given the measurement error model, the best accuracy of the three

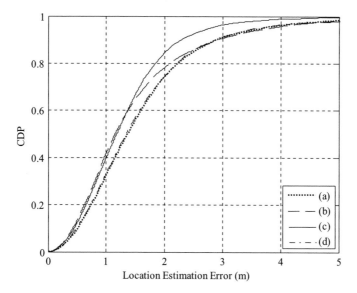

Figure 6.13 Accuracy of the linear-correction algorithm with four and five base stations. (a)–(d) respectively denote the results under the base station configurations (a)–(d) in Figure 6.8.

Table 6.1 Average distances between the true position and the estimated position of the mobile node of the three LS algorithms under the four different base station configurations.

Algorithm	Figure 6.9(a)	Figure 6.9(b)	Figure 6.9(c)	Figure 6.9(d)
SI	4.03 m	1.08 m	0.78 m	1.54 m
LC-LS	1.60 m	1.48 m	1.32 m	1.58 m
QLS-1	1.60 m	3.09 m	1.91 m	1.71 m
QLS-5	1.41 m	2.75 m	1.43 m	1.47 m

LS algorithms is 1.41 m, 1.08 m, 0.78 m and 1.47 m respectively under the four different base station deployments (a)–(d). Apparently, the location accuracy will increase with the TOA or pseudo-range estimation accuracy. In addition, by using the iterative algorithms to be discussed in Chapter 7, the location accuracy can be further enhanced.

References

[1] E. Damosso (ed.), Digital mobile radio towards future generation systems. COST 231 Final Report, 1996.
[2] J.O. Smith and J.S. Abel, 'Closed-form least squares source location estimation from range difference measurements', *IEEE Transactions on Acoustics, Speech and Signal Processing*, **35**(12), 1987, 1661–1669.
[3] C. Gentile and A. Kik, 'An evaluation of ultra wideband technology for indoor ranging', *Proceedings of the IEEE Global Telecommunications Conference (GLOBECOM)*, pp. 1–6, November 2006.
[4] S. Bancroft, 'An algebraic solution of the GPS equations', *IEEE Transactions on Aerospace and Electronic Systems*, **21**(7), 1985, 56–59.
[5] P.J.G. Teunissen and A. Kleusberg (eds), *GPS for Geodesy*, Springer, 1998.
[6] Y. Huang, J. Benesty, G.W. Elko and R.M. Mersereau, 'Real-time passive source localization: a practical linear-correction least-squares approach', *IEEE Transactions on Speech and Audio Processing*, **9**(8), 2001, 943–956.
[7] Y.-T. Chan and K.C. Ho, 'A simple and efficient estimator for hyperbolic location', *IEEE Transactions on Signal Processing*, **42**(8), 1994, 1905–1915.
[8] K.W. Cheung, H.C. So, W.K. Ma and Y.T. Chan, 'Least squares algorithms for time-of-arrival-based mobile location', *IEEE Transactions on Vehicular Technology*, **52**(4), 2004, 1121–1128.

7

Iterative Position Determination

Noniterative position determination techniques and algorithms were presented in Chapter 6. In certain circumstances, noniterative position estimation methods may not produce satisfactory results. Some of the noniterative algorithms may also have the problem of relatively large computational complexity. To generate more accurate position estimates, iterative algorithms may be required. There exist different iterative position determination methods, but the basic ideas are similar: one obtains initial values of the unknown parameters including the position coordinates first. These initial values are usually produced by using the noniterative methods which are studied in Chapter 6. Then the estimates are updated iteratively according to a specific formula until some predefined thresholds are crossed.

In this chapter we first study three typical iterative algorithms in Section 7.1, including the TS-LS method, the iterative optimization method and the ML method. These algorithms can be applied to either static or moving targets. When the target is moving, the motion characteristics such as approximately constant velocity and straight-line track over a short period of time can be exploited to improve positioning accuracy. These filtering-based algorithms are studied in Section 7.2. In general, the raw position estimates vary around the true values, so that the estimated track is rather erratic. To obtain a smooth estimated moving track and improved position accuracy, the raw position data can be smoothed by the techniques described in Section 7.3.

7.1 Iterative Algorithms

In this section, three typical iterative algorithms are presented: the TS-LS method, the iterative optimization method and the ML method. Although the known position points are treated as static, the algorithms can be applied for both static and moving targets. In the case of moving targets, the estimated positions can be further processed by using smoothing techniques, such as those discussed in Section 7.3, provided that the motion characteristics of the target can be approximately predicted.

Ground-Based Wireless Positioning   Kegen Yu, Ian Sharp and Jay Guo
© 2009 John Wiley & Sons, Ltd

7.1.1 Taylor Series Least-Squares Method

The TS-LS method typically requires an initial position estimate close to the true value; otherwise, the convergence cannot be guaranteed [1]. In the TS-LS method, a set of nonlinear measurement equations is linearized by expanding them in a Taylor series at a point, which is an estimate of the true position initially, and keeping only the terms below the second order. The set of linearized equations is solved to produce a new approximate position and the process continues until a prespecified criterion is satisfied.

Consider the TDOA-based distance difference equation as given by $d_{i,1} = c(t_i - t_1) = d_i - d_1$ and define

$$\mathbf{p} = \begin{bmatrix} x & y & z \end{bmatrix}^T$$
$$f_i(\mathbf{p}) = d_{i+1,1} \qquad i = 1, 2, \ldots, N-1 \tag{7.1}$$

Letting the initial position estimate be

$$\mathbf{p}_v = \begin{bmatrix} x_v & y_v & z_v \end{bmatrix}^T \tag{7.2}$$

the position estimation error is then given as

$$\boldsymbol{\delta} = \mathbf{p}_v - \mathbf{p} = \begin{bmatrix} \delta_x & \delta_y & \delta_z \end{bmatrix}^T \tag{7.3}$$

Then, one has

$$f_{i,v} + a_{i,1}\delta_x + a_{i,2}\delta_y + a_{i,3}\delta_z = \hat{d}_{i+1,1} + \varepsilon_i \qquad i = 1, 2, \ldots, N-1 \tag{7.4}$$

where the TDOA measurement errors and the linearization errors are included in ε_i and

$$f_{i,v} = f_i(\mathbf{p}_v)$$

$$a_{i,1} = \left.\frac{\partial f_i(\mathbf{p})}{\partial x}\right|_{\mathbf{p}=\mathbf{p}_v} = \frac{x_1 - x_v}{\hat{d}_1} - \frac{x_{i+1} - x_v}{\hat{d}_{i+1}}$$

$$\hat{d}_i = \sqrt{(x_v - x_i)^2 + (y_v - y_i)^2 + (z_v - z_i)^2} \tag{7.5}$$

$$a_{i,2} = \left.\frac{\partial f_i(\mathbf{p})}{\partial y}\right|_{\mathbf{p}=\mathbf{p}_v} = \frac{y_1 - y_v}{\hat{d}_1} - \frac{y_{i+1} - y_v}{\hat{d}_{i+1}}$$

$$a_{i,3} = \left.\frac{\partial f_i(\mathbf{p})}{\partial z}\right|_{\mathbf{p}=\mathbf{p}_v} = \frac{z_1 - z_v}{\hat{d}_1} - \frac{z_{i+1} - z_v}{\hat{d}_{i+1}}$$

Equation (7.4) can be written in a compact form as

$$\mathbf{A}\boldsymbol{\delta} = \mathbf{g} + \boldsymbol{\varepsilon} \tag{7.6}$$

where

$$[\mathbf{A}]_{i,j} = a_{i,j} \qquad 1 \le i \le N-1, 1 \le j \le 3$$
$$\mathbf{g} = \begin{bmatrix} \hat{d}_{2,1} - f_{1,v} & \hat{d}_{3,1} - f_{2,v} & \cdots & \hat{d}_{N,1} - f_{N-1,v} \end{bmatrix}^T \tag{7.7}$$
$$\boldsymbol{\varepsilon} = \begin{bmatrix} \varepsilon_1 & \varepsilon_2 & \cdots & \varepsilon_{N-1} \end{bmatrix}^T$$

The weighted LS solution of (7.6) is given by

$$\boldsymbol{\delta} = (\mathbf{A}^{\mathrm{T}}\mathbf{W}\mathbf{A})^{-1}\mathbf{A}^{\mathrm{T}}\mathbf{W}\mathbf{g} \tag{7.8}$$

where $\mathbf{W} = \mathrm{cov}^{-1}(\boldsymbol{\varepsilon})$ is the inverse of the covariance matrix of $\boldsymbol{\varepsilon}$. In the case of unknown covariance matrix, $\mathbf{W}$ may be determined based on the SNR, or simply the RSS. Otherwise, $\mathbf{W}$ may be simply taken as

$$\mathbf{W} = \begin{bmatrix} 2 & 1 & 1 & \cdots & 1 & 1 \\ 1 & 2 & 1 & \cdots & 1 & 1 \\ \vdots & \vdots & \vdots & \ddots & \vdots & \vdots \\ 1 & 1 & 1 & \cdots & 1 & 2 \end{bmatrix}^{-1} \tag{7.9}$$

Finally, the parameter vector estimate is updated according to

$$\mathbf{p}_v^{(k+1)} = \mathbf{p}_v^{(k)} + \boldsymbol{\delta} \tag{7.10}$$

where $\mathbf{p}_v^{(k)}$ is the estimate of the position vector $\mathbf{p}$ at the kth iteration. The position coordinate estimates are continually refined until $\boldsymbol{\delta}$ is sufficiently small. To start the iteration, initial position coordinates and initial transmit times are required. The initial values of the position coordinates may be generated by using the noniterative algorithms discussed in Chapter 6.

The advantage of the TS-LS algorithm is that it is rather simple and easy to implement. Provided that an initial position estimate is available, its computational complexity is lower than some of the noniterative algorithms. On the other hand, in some circumstances, particularly when the initial values are far from the true values, the TS-LS algorithm may not converge or may converge to the point that could be dramatically different from the desired one, thus resulting in large position errors. To avoid accepting estimates with abnormal errors it is necessary to examine the reliability of the position estimates. One such method is to calculate the residuals of the estimates. Only when the residuals are below some threshold are the position estimates accepted. Otherwise, the estimates are rejected and a different position determination method may be used, provided that multiple positioning algorithms are available.

7.1.2 Iterative Optimization

Iterative optimization is another iterative approach for position estimation. Typically, the iterative optimization-based algorithms can achieve better location accuracy than noniterative ones, especially when there are relatively large errors in TOA, RSS or AOA measurements. In this method, an objective/cost function is first defined. When pseudo-range (or TDOA) measurements are available, the cost function may be defined as

$$\varepsilon(x, y, z, t_0) = \sum_{i=1}^{N} w_i [c\hat{t}_i - (d_i + ct_0)]^2 \tag{7.11}$$

where $\{w_i\}$ are the weights, c is the speed of propagation and $\hat{t}_i$ is the estimated TOA at the ith base station. The target location estimate

$$\hat{\mathbf{p}} = [\hat{x} \quad \hat{y} \quad \hat{z}]^{\mathrm{T}} \tag{7.12}$$

is produced by minimizing the cost function in (7.11) with respect to the unknown parameter vector

$$\boldsymbol{\theta} = [x \quad y \quad z \quad t_0]^{\mathrm{T}} \tag{7.13}$$

That is:

$$\hat{\mathbf{p}} = \arg \min_{\boldsymbol{\theta}} \varepsilon(x, y, z, t_0) \tag{7.14}$$

There are a range of minimization algorithms that can be used to minimize the cost function. The nonlinear optimization-based location estimation method is particularly suited for scenarios where the measurement error is non-Gaussian and the error is relatively large. The main disadvantage is the relatively large computational complexity. The quasi-Newton method [2] and the Levenberg–Marquardt (LM) method [3] are among the widely exploited minimization algorithms, which are briefly described in the following.

7.1.2.1 Levenberg–Marquardt Method

For notational simplicity, define

$$\begin{aligned} f_i(\boldsymbol{\theta}) &= \sqrt{w_i}\,[c\hat{t}_i - (d_i + ct_0)] \\ \mathbf{f}(\boldsymbol{\theta}) &= [f_1(\boldsymbol{\theta}) \quad f_2(\boldsymbol{\theta}) \quad \cdots \quad f_N(\boldsymbol{\theta})]^{\mathrm{T}} \end{aligned} \tag{7.15}$$

Then the objective function becomes

$$\varepsilon(\boldsymbol{\theta}) = ||\mathbf{f}(\boldsymbol{\theta})||^2 \tag{7.16}$$

Expanding the objective function in the Taylor series at the current parameter estimate $\boldsymbol{\theta}_k$ and retaining the first three terms (that is, up to the second-order term), one has

$$\varepsilon(\boldsymbol{\theta}_k + \mathbf{s}_k) \approx \varepsilon(\boldsymbol{\theta}_k) + \mathbf{g}_k^{\mathrm{T}}\mathbf{s}_k + \frac{1}{2}\mathbf{s}_k^{\mathrm{T}}\mathbf{H}(\boldsymbol{\theta}_k)\mathbf{s}_k \tag{7.17}$$

where $\mathbf{s}_k$ is the directional vector (or location increment vector) to be determined, $\mathbf{g}_k$ is a vector of the first partial derivatives (also called gradient) of $\mathbf{f}(\boldsymbol{\theta})$ at $\boldsymbol{\theta}_k$

$$\begin{aligned} \mathbf{g}_k &= \nabla\varepsilon(\boldsymbol{\theta})|_{\boldsymbol{\theta}=\boldsymbol{\theta}_k} \\ &= \left[\left.\frac{\partial\varepsilon(\boldsymbol{\theta})}{\partial x}\right|_{\boldsymbol{\theta}=\boldsymbol{\theta}_k} \quad \left.\frac{\partial\varepsilon(\boldsymbol{\theta})}{\partial y}\right|_{\boldsymbol{\theta}=\boldsymbol{\theta}_k} \quad \left.\frac{\partial\varepsilon(\boldsymbol{\theta})}{\partial z}\right|_{\boldsymbol{\theta}=\boldsymbol{\theta}_k} \quad \left.\frac{\partial\varepsilon(\boldsymbol{\theta})}{\partial t_0}\right|_{\boldsymbol{\theta}=\boldsymbol{\theta}_k} \right]^{\mathrm{T}} \end{aligned} \tag{7.18}$$

and $\mathbf{H}(\boldsymbol{\theta}_k)$ is the Hessian of the objective function. Minimization of the right-hand side of (7.17) yields

$$\mathbf{H}(\boldsymbol{\theta}_k)\mathbf{s}_k = -\mathbf{g}_k \tag{7.19}$$

The minimization in which $\mathbf{s}_k$ is defined by (7.19) is termed Newton's method. To avoid the calculation of the second-order information in the Hessian, a simplified expression can be obtained from (7.19), resulting in

$$\mathbf{J}_k^{\mathrm{T}}\mathbf{J}_k\mathbf{s}_k = -\mathbf{J}_k^{\mathrm{T}}\mathbf{f}(\boldsymbol{\theta}_k) \tag{7.20}$$

where $\mathbf{J}_k$ is the Jacobian matrix of $\mathbf{f}(\boldsymbol{\theta})$ at $\boldsymbol{\theta}_k$. This is termed the Gauss–Newton method. When $\mathbf{J}_k$ is full rank, which is the usual case of an overdetermined system, we have the linear LS solution

$$\mathbf{s}_k = -(\mathbf{J}_k^{\mathrm{T}}\mathbf{J}_k)^{-1}\mathbf{J}_k^{\mathrm{T}}\mathbf{f}(\boldsymbol{\theta}_k). \tag{7.21}$$

The Gauss–Newton method may get into trouble when the second-order information in the Hessian is not trivial. A method to overcome this problem is the LM method. The LM search direction is defined as the solution of the following equation:

$$(\mathbf{J}_k^{\mathrm{T}}\mathbf{J}_k + \lambda\mathbf{I})\mathbf{s}_k = -\mathbf{J}_k^{\mathrm{T}}\mathbf{f}(\boldsymbol{\theta}_k) \tag{7.22}$$

where the nonnegative damping factor λ is adjusted at each iteration to control both the magnitude and direction of $\mathbf{s}_k$. If the reduction of the cost function is rapid, then a smaller value of the damping factor can be used to increase the magnitude of the increment vector so the search can be speeded up. On the other hand, if an iteration gives insufficient reduction in the cost function, then the damping factor should be increased so that the gradient descent direction can be approached. As a result, the algorithm runs at a speed between that of the Gauss–Newton method and that of the steepest-descent method.

Specifically, the implementation of the LM algorithm can be described as follows. Given a position estimate at iteration $k-1$ as $\boldsymbol{\theta}_{k-1}$ and a positive damping factor as λ_{k-1}, one uses (7.22) to calculate the increment vector $\mathbf{s}_{k-1}$. Then, based on the given position estimate and the increment vector, the predicted cost function is calculated by

$$\varepsilon_{\mathrm{e}}(\boldsymbol{\theta}_k) = \|\mathbf{J}(\boldsymbol{\theta}_{k-1})\mathbf{s}_{k-1} + \mathbf{f}(\boldsymbol{\theta}_{k-1})\|^2 \tag{7.23}$$

Also, the predicted parameter vector is given by

$$\boldsymbol{\theta}_k = \boldsymbol{\theta}_{k-1} + \mathbf{s}_{k-1} \tag{7.24}$$

Then, using (7.16), one obtains the cost function as $\varepsilon(\boldsymbol{\theta}_{k-1})$ and $\varepsilon(\boldsymbol{\theta}_k)$ at $\boldsymbol{\theta}_{k-1}$ and $\boldsymbol{\theta}_k$ respectively. By cubically interpolating the two cost-function points and using gradient and function evaluation, the minimum cost function is produced as $\varepsilon_k(\boldsymbol{\theta}_*)$. Also, a corresponding step length parameter is obtained as α_*. Then, the interpolated minimum cost function is compared with the predicated one as follows. If $\varepsilon_e(\boldsymbol{\theta}_k) > \varepsilon_k(\boldsymbol{\theta}_*)$, the damping factor is reduced by

$$\lambda_k = \frac{\lambda_{k-1}}{1+\alpha_*} \tag{7.25}$$

Alternatively, if $\varepsilon_e(\boldsymbol{\theta}_k) < \varepsilon_k(\boldsymbol{\theta}_*)$, then the damping factor is increased by

$$\lambda_k = \lambda_{k-1} + \frac{\varepsilon_k(\boldsymbol{\theta}_*) - \varepsilon_{\mathrm{e}}(\boldsymbol{\theta}_k)}{1+\alpha_*} \tag{7.26}$$

The updated damping factor is used to update the increment vector by using (7.22). Accordingly, the position estimate is updated according to (7.24). The procedure continues until the position increment is sufficiently small.

7.1.2.2 Quasi-Newton Method

The quasi-Newton method is similar to the Newton's method. The Hessian matrix $\mathbf{H}(\boldsymbol{\theta}_k)$ in (7.19) is now approximated by a symmetric positive definite matrix $\mathbf{B}_k$, which is updated at each iteration. The directional vector, at the kth iteration, is set to be

$$\mathbf{s}_k = -\mathbf{B}_k\mathbf{g}_k \tag{7.27}$$

where $\mathbf{g}_k$ is the gradient of the cost function at $\boldsymbol{\theta}_k$. Then, at iteration $k+1$, the location estimate is updated according to

$$\boldsymbol{\theta}_{k+1} = \boldsymbol{\theta}_k + \alpha_k\mathbf{s}_k \tag{7.28}$$

where α_k is the step size, which is a positive constant and enables

$$\varepsilon(\boldsymbol{\theta}_{k+1}) < \varepsilon(\boldsymbol{\theta}_k) \tag{7.29}$$

The initial matrix $\mathbf{B}_1$ can be any positive definite matrix. It is usually set to be an identity matrix in the absence of any better estimate. There are different ways of updating $\mathbf{B}_k$. One well-known updating formula is the DFP (Davidon–Fletcher–Powell) formula, in which $\mathbf{B}_k$ is updated according to

$$\mathbf{B}_{k+1} = \mathbf{B}_k + \frac{\mathbf{h}_k\mathbf{h}_k^{\mathrm{T}}}{\mathbf{h}_k^{\mathrm{T}}\mathbf{q}_k} - \frac{\mathbf{B}_k\mathbf{q}_k\mathbf{q}_k^{\mathrm{T}}\mathbf{B}_k}{\mathbf{q}_k^{\mathrm{T}}\mathbf{B}_k\mathbf{q}_k} \tag{7.30}$$

where

$$\begin{aligned}\mathbf{h}_k &= \boldsymbol{\theta}_{k+1} - \boldsymbol{\theta}_k \\ \mathbf{q}_k &= \mathbf{g}_{k+1} - \mathbf{g}_k\end{aligned} \tag{7.31}$$

Another important formula is the BFGS (Broyden–Fletcher–Goldfarb–Shanno) formula [4], in which $\mathbf{B}_k$ is updated based on

$$\mathbf{B}_{k+1} = \mathbf{B}_k + \left(1 + \frac{\mathbf{h}_k^{\mathrm{T}}\mathbf{B}_k\mathbf{h}_k}{\mathbf{q}_k^{\mathrm{T}}\mathbf{h}_k}\right)\frac{\mathbf{q}_k\mathbf{q}_k^{\mathrm{T}}}{\mathbf{q}_k^{\mathrm{T}}\mathbf{h}_k} - \frac{\mathbf{q}_k\mathbf{h}_k^{\mathrm{T}}\mathbf{B}_k + \mathbf{B}_k\mathbf{h}_k\mathbf{q}_k^{\mathrm{T}}}{\mathbf{q}_k^{\mathrm{T}}\mathbf{h}_k} \tag{7.32}$$

Clearly, the BFGS formula requires significantly greater computational effort. Compared with the TS-LS methods and the noniterative estimation algorithms, the LM method, the quasi-Newton method and other iterative minimization-based algorithms have higher computational complexity. Certainly, with appropriate parameter selections, it may be feasible to reduce the complexity of the iterative minimization algorithms to some degree through fast convergence when implementing the algorithms in practice.

One of the key issues in implementing the quasi-Newton algorithms is to determine the step size parameter α_k at each iteration. The basic idea of choosing α_k by using the cubic-polynomial-based line search can be described as follows. It is assumed that gradient information of the objective function can be computed. If the gradient is not available, then a different line search method is required. The step size parameter is usually reset to unity after each iteration. When a point along the search direction $\mathbf{s}_k$ as given by (7.27) is found to satisfy (7.29) and $\mathbf{q}_k^T \mathbf{h}_k$ is positive, the position update by (7.28) is accepted. If $\mathbf{q}_k^T \mathbf{h}_k$ is not positive, or equation (7.29) is not satisfied, then further cubic interpolations are performed to satisfy (7.29) and to make $\mathbf{q}_k^T \mathbf{h}_k$ positive. The problem can be dealt with basically under four different cases as follows:

- $\varepsilon(\boldsymbol{\theta}_{k+1}) > \varepsilon(\boldsymbol{\theta}_k)$ and $\mathbf{g}_{k+1}^T \mathbf{s}_k > 0$
  In this case, the step size is reduced according to

$$\alpha_{k+1} = \begin{cases} \alpha_c/2 & \alpha_k < 0.1 \\ \alpha_c & \text{elsewhere} \end{cases} \tag{7.33}$$

  where α_c is the cubically interpolated step size which is calculated at each iterate. In the event that α_c is negative, it is replaced by $2\alpha_k$.
- $\varepsilon(\boldsymbol{\theta}_{k+1}) < \varepsilon(\boldsymbol{\theta}_k)$ and $\mathbf{g}_{k+1}^T \mathbf{s}_k > 0$
  In this case, there are two different sub-cases depending on $\mathbf{q}_k^T \mathbf{h}_k$. When $\mathbf{q}_k^T \mathbf{h}_k \geq 0$ one updates the Hessian, resets the direction vector $\mathbf{s}_k$ and

$$\alpha_{k+1} = \min\{1, \alpha_c\} \tag{7.34}$$

  On the other hand, if $\mathbf{q}_k^T \mathbf{s}_k < 0$, then the step size is reduced to

$$\alpha_{k+1} = 0.9\alpha_c \tag{7.35}$$

- $\varepsilon(\boldsymbol{\theta}_{k+1}) < \varepsilon(\boldsymbol{\theta}_k)$ and $\mathbf{g}_{k+1}^T \mathbf{s}_k < 0$
  This case is also handled based on $\mathbf{q}_k^T \mathbf{h}_k$. If $\mathbf{q}_k^T \mathbf{h}_k \geq 0$, then one updates the Hessian, resets the direction vector and chooses the step size as

$$\alpha_{k+1} = \min\{2, \kappa, \alpha_c\} \tag{7.36}$$

  where

$$\kappa = 1 + \mathbf{q}_k^T \mathbf{h}_k - \mathbf{g}_{k+1}^T \mathbf{s}_k + \min\{0, \alpha_{k+1}\} \tag{7.37}$$

  Alternatively, if $\mathbf{q}_k^T \mathbf{s}_k < 0$, then one temporally changes to the steepest descent method and chooses the step size as

$$\alpha_{k+1} = \min\{2, \max\{1.5, \alpha_k\}, \alpha_c\} \tag{7.38}$$

- $\varepsilon(\boldsymbol{\theta}_{k+1}) > \varepsilon(\boldsymbol{\theta}_k)$ and $\mathbf{g}_{k+1}^T \mathbf{s}_k < 0$
  In this case, the step size is simply reduced by

$$\alpha_{k+1} = \min\{\alpha_k/2, \alpha_c\} \tag{7.39}$$

In the event that the measurement error is a Gaussian random variable with known statistics, the ML estimation can be applied, which is discussed in the next subsection.

7.1.3 Maximum Likelihood Method

Some of the closed-form linear techniques described in Chapter 6 can achieve asymptotic optimum location estimation at high SNR; however, their performance may degrade significantly compared with the optimal estimator at low to medium SNR. In some circumstances the distribution of the measurement errors may be modeled *a priori* based on field measurement data or a database established from trials in the location area. For instance, the TOA measurement error is usually modeled as a Gaussian random variable of zero mean and STD propositional to the true distance in LOS conditions. With the knowledge of the error distribution, the ML estimator can be applied to solve the location estimation problem to achieve the optimal location estimates for any given SNR. Note that in a variety of circumstances, especially in indoor environments, the radio signal propagation is NLOS in general. A comprehensive investigation on how to mitigate the NLOS effect can be found in Chapter 10.

Let us focus on the case where the TDOA or pseudo-range measurements are taken. Assume that the pseudo-range measurements are mutually independent. Note that more information about pseudo-range and other fundamentals of positioning systems can be found in Chapter 5. Then, given the target location vector

$$\mathbf{p} = [x \quad y \quad z]^{\mathrm{T}} \tag{7.40}$$

and that the measurement errors are Gaussian random variables with zero mean, the conditional PDF of the TDOA-based range difference measurement vector

$$\hat{\mathbf{r}}_d = [\hat{d}_{2,1} \quad \hat{d}_{3,1} \quad \cdots \quad \hat{d}_{N,1}]^{\mathrm{T}} \tag{7.41}$$

is then given by

$$p(\hat{\mathbf{r}}_d|\mathbf{p}) = (2\pi)^{-(N-1)/2} [\det(\mathbf{Q}_d)]^{-1/2} \exp\left[-\frac{1}{2}(\hat{\mathbf{r}}_d - \mathbf{r}_d)^{\mathrm{T}} \mathbf{Q}_d^{-1} (\hat{\mathbf{r}}_d - \mathbf{r}_d)\right] \tag{7.42}$$

where $\mathbf{r}_d$ is the true range difference vector defined as

$$\mathbf{r}_d = [d_{2,1} \quad d_{3,1} \quad \cdots \quad d_{N,1}]^{\mathrm{T}} \tag{7.43}$$

and $\mathbf{Q}_d$ is the covariance matrix of the measurement vector $\hat{\mathbf{r}}_d$. The corresponding log-likelihood function, after ignoring the irrelevant constants, is

$$\Lambda(\mathbf{p}) = -\frac{1}{2}(\hat{\mathbf{r}}_d - \mathbf{r}_d)^{\mathrm{T}} \mathbf{Q}_d^{-1} (\hat{\mathbf{r}}_d - \mathbf{r}_d). \tag{7.44}$$

The ML estimate of the unknown location is yielded by maximizing the above log-likelihood function. That is:

$$\hat{\mathbf{p}} = \arg \max_{\mathbf{p}} \Lambda(\mathbf{p}) \tag{7.45}$$

Since it is a nonlinear function, there is no closed-form solution, and only a numerical solution can be obtained. The quasi-Newton method and the LM algorithm described earlier can be employed to carry out the minimization. In the following subsection, a simple iterative algorithm is employed to realize the ML estimation.

7.1.4 Approximate Maximum Likelihood Method

In the presence of nonlinear observation equations, the ML estimator involves maximization of a log-likelihood function in general. The operation of iterative nonlinear maximization or minimization is usually time consuming. Therefore, it is desirable to develop a simple iterative algorithm to achieve location accuracy close to that of the ML method. One such algorithm is proposed in [5], which is described in this subsection. Compared with the iterative nonlinear minimization, this iterative algorithm has lower computational complexity and can be readily implemented. The algorithm can also be employed without knowledge of the error covariance matrix, but with some performance degradation.

Let us still focus on positioning based on pseudo-range or TDOA measurements. Differentiating the log-likelihood function in (7.44) with respect to $\mathbf{p}$ and setting the results to zero produces

$$\mathbf{U}\boldsymbol{\alpha} = 0 \tag{7.46}$$

where

$$\mathbf{U} = \left[\frac{\partial \mathbf{r}_d(\mathbf{p})}{\partial \mathbf{p}} \right]^{\mathrm{T}} \mathbf{Q}_d^{-1} \tag{7.47}$$

$$\boldsymbol{\alpha} = [d_{2,1} - \hat{d}_{2,1} \quad d_{3,1} - \hat{d}_{3,1} \quad \cdots \quad d_{N,1} - \hat{d}_{N,1}]^{\mathrm{T}}$$

Making use of

$$d_{i,1} - \hat{d}_{i,1} = \frac{d_i^2 - (d_1 + \hat{d}_{i,1})^2}{d_i + d_1 + \hat{d}_{i,1}} \tag{7.48}$$

equation (7.46) can be written as

$$\boldsymbol{\Phi}\boldsymbol{\beta} = 0 \tag{7.49}$$

where

$$\begin{aligned} \boldsymbol{\Phi} &= \mathbf{U}\boldsymbol{\Omega} \\ \boldsymbol{\Omega} &= \mathrm{diag}\left\{ \frac{1}{d_2 + d_1 + \hat{d}_{2,1}} \quad \frac{1}{d_3 + d_1 + \hat{d}_{3,1}} \quad \cdots \quad \frac{1}{d_N + d_1 + \hat{d}_{N,1}} \right\} \\ \boldsymbol{\beta} &= [d_2^2 - (d_1 + \hat{d}_{2,1})^2 \quad d_3^2 - (d_1 + \hat{d}_{3,1})^2 \quad \cdots \quad d_N^2 - (d_1 + \hat{d}_{N,1})^2]^{\mathrm{T}} \end{aligned} \tag{7.50}$$

The elements of β can be expanded as

$$d_i^2 - (d_1 + \hat{d}_{i,1})^2 = 2(x_1 - x_i)x + 2(y_1 - y_i)y + 2(z_1 - z_i)z + R_i^2 - R_1^2 - 2d_1\hat{d}_{i,1} - \hat{d}_{i,1}^2 \tag{7.51}$$

which are then substituted into (7.46), producing

$$2\mathbf{\Phi}\mathbf{D}\mathbf{p} = \mathbf{\Phi}\mathbf{v} \tag{7.52}$$

where

$$\mathbf{D} = \begin{bmatrix} x_1 - x_2 & y_1 - y_2 & z_1 - z_2 \\ \vdots & \vdots & \vdots \\ x_1 - x_N & y_1 - y_N & z_1 - z_N \end{bmatrix} \tag{7.53}$$

$$\mathbf{v} = [\hat{d}_{2,1}^2 + 2d_1\hat{d}_{2,1} + R_1^2 - R_2^2 \quad \cdots \quad \hat{d}_{N,1}^2 + 2d_1\hat{d}_{N,1} + R_1^2 - R_N^2]^T$$

Decomposing $\mathbf{v}$ into

$$\mathbf{v} = \mathbf{v}_1 + 2d_1\hat{\mathbf{r}}_d \tag{7.54}$$

where

$$\mathbf{v}_1 = [\hat{d}_{2,1}^2 + R_1^2 - R_2^2 \quad \cdots \quad \hat{d}_{N,1}^2 + R_1^2 - R_N^2]^T \tag{7.55}$$

and letting $\mathbf{\Phi}$ be an identity matrix, Equation (7.52) can be simplified as

$$2\mathbf{D}\mathbf{p} = \mathbf{v}_1 + 2d_1\hat{\mathbf{r}}_d \tag{7.56}$$

The LS solution for (7.56) (that is, the estimate of the position vector) is given by

$$\hat{\mathbf{p}} = \frac{1}{2}(\mathbf{D}^T\mathbf{Q}_d^{-1}\mathbf{D})^{-1}\mathbf{Q}_d^{-1}\mathbf{D}(\mathbf{v}_1 + 2d_1\hat{\mathbf{r}}_d) \tag{7.57}$$

where, on the right-hand side, d_1 is the only unknown parameter. Clearly, each of the three coordinate estimates is a linear function of d_1. Then, the three unknown position coordinates (x, y and z) in

$$d_1 = \sqrt{(x - x_1)^2 + (y - y_1)^2 + (z - z_1)^2} \tag{7.58}$$

are replaced by the three coordinate estimates obtained by (7.57), producing a quadratic equation in d_1, resulting in two roots. The best root may be selected as follows. If only one root is positive, then the positive root is selected. If both roots are positive, then the root that gives larger $\Lambda(\mathbf{p})$ (see (7.44)) is selected. In the case of two negative or complex roots, the absolute values of the roots are taken for root selection. Giving the selected root, $\hat{\mathbf{p}}$ can be determined by (7.57) and the results are then used to calculate $\mathbf{\Phi}$ by (7.50). From (7.52), $\hat{\mathbf{p}}$ is recomputed by

$$\hat{\mathbf{p}} = \frac{1}{2}(\mathbf{\Phi}\mathbf{D})^{-1}\mathbf{\Phi}(\mathbf{v}_1 + 2d_1\hat{\mathbf{r}}_d) \tag{7.59}$$

which is given in d_1 again. In the same procedure, a quadratic equation is generated and then a root is selected. Repeating (7.59) with new values of $\hat{\mathbf{p}}$ for updating $\mathbf{\Phi}$ and d_1 for a number

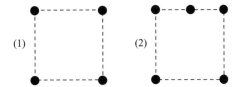

Figure 7.1 Two different base station configurations.

(say κ) of times produces κ values of $\hat{\mathbf{p}}$ and $\Lambda(\hat{\mathbf{p}})$ as well. It is not necessary to choose the latest update as the final location estimate. Instead, the location estimate with the largest $\Lambda(\hat{\mathbf{p}})$ is selected as the final location estimate.

An illustrative example is provided below to show the performance of the iterative algorithms described above. The algorithms are evaluated under the same simulation setup used in Section 6.3 (see Chapter 6, but with only two base station configurations as shown in Figure 7.1). The results of the three iterative algorithms, namely the TS-LS method, the LM-based nonlinear minimization method and the ML method are presented. The initial location estimate for the LM-based method and the TS-LS method is produced by using the QLS algorithm described in Section 6.4. The results of the ML method are produced by using the approximate ML algorithm with knowledge of the error statistics, which are the mean and variance.

Figures 7.2 and 7.3 show the CDPs of the three iterative algorithms with four base stations deployed as in diagrams (1) and (2) respectively of Figure 7.1. For comparison, the results of the

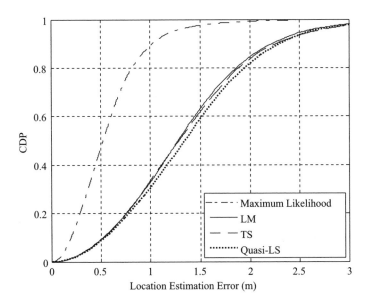

Figure 7.2 CDP of three iterative algorithms with four base stations deployed according to diagram (1) in Figure 7.1. 'LM' denotes the LM-based nonlinear minimization algorithm, 'TS' denotes the Taylor-series-based LS algorithm, 'Maximum Likelihood' denotes the iterative approximate ML algorithm with full knowledge of the error statistics.

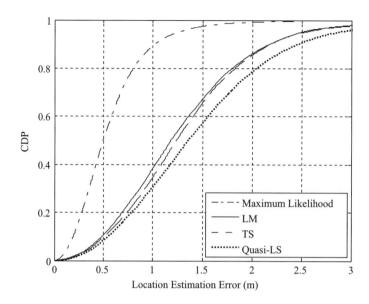

Figure 7.3 CDP of the iterative algorithms with five base stations deployed according to diagram (2) in Figure 7.1. Legend as in Figure 7.2.

QLS method are also plotted. It can be seen that the TS-LS algorithm and the LM nonlinear minimization method produce virtually the same performance. In this case the TS-LS method is preferred owing to its lower computational complexity. The TS-LS algorithm and the LM method outperform the QLS method considerably only in Figure 7.3. Therefore, using iterative algorithms of even high complexity does not mean a significant accuracy gain. In the presence of complexity constraints, noniterative algorithms or low-complexity iterative algorithms should first be considered. When knowledge of the error statistics is used, the performance gain is huge. Therefore, it is desirable to develop techniques to estimate the error statistics and then use them to achieve accurate position estimation. The error statistics can also be used to mitigate the NLOS errors directly, which will be addressed in Chapter 10.

7.2 Filtering-based Methods

In a variety of circumstances, the mobile node is moving at certain speed along a specific route, such as a vehicle travelling on highway or a person walking along a corridor in an office building. Then the motion characteristics can be employed to assist the localization process. To this end, Kalman filtering and particle filtering techniques for positioning are presented in this section.

7.2.1 Extended Kalman Filtering

Kalman filtering has been widely used in modern control systems, tracking and navigation of vehicles. There are a large number of references about using Kalman filters for tracking moving targets or smoothing position/velocity estimates of moving objects, such as [6–8]. Kalman filtering is studied in several chapters throughout the book for performing different tasks. Here, it is used to locate the mobile nodes, which are either static or moving, by using distance measurements. In the case of TDOA and AOA measurements, the Kalman filter can be used in a similar manner. It is assumed that the signal propagation between the anchor nodes and the mobile node is in LOS conditions. The issue of multipath mitigation will be addressed in Chapter 10 and the NLOS identification will be discussed in Chapter 13.

Suppose that the distance measurement equations are given as

$$\hat{d}_i^{(k)} = d_i^{(k)} + \varepsilon_i^{(k)} \tag{7.60}$$

where i indexes the anchor nodes or base stations, k indexes the measurement time instants and

$$d_i^{(k)} = \sqrt{(x^{(k)} - x_i)^2 + (y^{(k)} - y_i)^2 + (z^{(k)} - z_i)^2} \tag{7.61}$$

In general, the mobile location $(x^{(k)}, y^{(k)}, z^{(k)})$ changes with time. On the other hand, the base station locations (x_i, y_i, z_i) are usually fixed in cellular networks, whereas the anchor nodes in wireless WSNs can vary with time. Define the state vector at time instant k as

$$\boldsymbol{\theta}^{(k)} = \begin{bmatrix} x^{(k)} & y^{(k)} & z^{(k)} & v_x^{(k)} & v_y^{(k)} & v_z^{(k)} \end{bmatrix}^{\mathrm{T}} \tag{7.62}$$

where $v_x^{(k)}$, $v_y^{(k)}$ and $v_z^{(k)}$ are the velocity components along the x, y, and z coordinates respectively. Then the system dynamic model, called the process equation, becomes

$$\boldsymbol{\theta}^{(k)} = \boldsymbol{\Phi}^{(k-1)} \boldsymbol{\theta}^{(k-1)} + \mathbf{B}^{(k-1)} \boldsymbol{\varepsilon}^{(k-1)} \tag{7.63}$$

where the matrices $\boldsymbol{\Phi}^{(k-1)}$ and $\mathbf{B}^{(k-1)}$ are given respectively by

$$\boldsymbol{\Phi}^{(k-1)} = \begin{bmatrix} 1 & 0 & 0 & \delta t & 0 & 0 \\ 0 & 1 & 0 & 0 & \delta t & 0 \\ 0 & 0 & 1 & 0 & 0 & \delta t \\ 0 & 0 & 0 & 1 & 0 & 0 \\ 0 & 0 & 0 & 0 & 1 & 0 \\ 0 & 0 & 0 & 0 & 0 & 1 \end{bmatrix}$$

$$\mathbf{B}^{(k-1)} = \begin{bmatrix} 0.5\delta t^2 & 0 & 0 \\ 0 & 0.5\delta t^2 & 0 \\ 0 & 0 & 0.5\delta t^2 \\ \delta t & 0 & 0 \\ 0 & \delta t & 0 \\ 0 & 0 & \delta t \end{bmatrix} \tag{7.64}$$

Here, δt is the sampling time increment, $\boldsymbol{\varepsilon}^{(k-1)}$ is the acceleration noise modeled as a white Gaussian random vector of zero mean and the covariance matrix is denoted by $\mathbf{Q}^{(k-1)}$. Also, let us define the observation vector at time instant k as

$$\mathbf{p}^{(k)} = \begin{bmatrix} x^{(k)} & y^{(k)} & z^{(k)} \end{bmatrix}^{\mathrm{T}}. \tag{7.65}$$

Then, the measurement model becomes

$$\mathbf{p}^{(k)} = \mathbf{g}^{(k)}(\boldsymbol{\theta}^{(k)}) + \mathbf{n}^{(k)} \qquad (7.66)$$

where $\mathbf{n}^{(k)}$ is the observation noise vector, also modeled as Gaussian with zero mean, the covariance matrix is denoted by $\mathbf{R}^{(k-1)}$ and

$$\mathbf{g}^{(k)}(\boldsymbol{\theta}^{(k)}) = \begin{bmatrix} d_1^{(k)} & d_2^{(k)} & \cdots & d_N^{(k)} \end{bmatrix}^{\mathrm{T}} \qquad (7.67)$$

For notational simplicity, we drop the superscripts of $\boldsymbol{\Phi}^{(k-1)}$, $\mathbf{B}^{(k-1)}$, $\mathbf{Q}^{(k-1)}$ and $\mathbf{R}^{(k-1)}$ from now on, since they are constant for the problem being studied. Unlike the system dynamic equation given by (7.63), the observation equation given by (7.66) is nonlinear with respect to the state vector. As a result, the linear Kalman filter (LKF) cannot be applied directly. To cope with the nonlinear relation, linearization can be employed. Specifically, (7.66) can be approximated as

$$\mathbf{p}^{(k)} \approx \mathbf{G}^{(k)}\boldsymbol{\theta}^{(k)} + \mathbf{n}^{(k)} + \left[\mathbf{g}^{(k)}(\hat{\boldsymbol{\theta}}^{(k|k-1)}) - \mathbf{G}^{(k)}\hat{\boldsymbol{\theta}}^{(k|k-1)} \right] \qquad (7.68)$$

where $\hat{\boldsymbol{\theta}}^{(k|k-1)}$ is the estimate of the state vector at time instant k based on the previous estimates and observations including that at time instant $k-1$, and

$$\mathbf{G}^{(k)} = \left. \frac{\partial \mathbf{g}^{(k)}(\boldsymbol{\theta})}{\partial \boldsymbol{\theta}} \right|_{\boldsymbol{\theta} = \hat{\boldsymbol{\theta}}^{(k|k-1)}} \qquad (7.69)$$

The linearization and the subsequent application of the LKF result in the extended Kalman filter (EKF). Under the approximate observation model, the implementation of the Kalman filter can be summarized as follows. The initial estimate $\hat{\boldsymbol{\theta}}^{(0|0)}$ and its error covariance matrix $\mathbf{C}^{(0|0)}$ are first given. For example, the initial velocities can be simply set at zero, the initial position estimate can be obtained by using the noniterative methods described in Chapter 6, and the error covariance matrix can be set at an identity matrix multiplied by a relatively large positive number. The measurement error covariance matrix $\mathbf{R}$ can be estimated based on the accuracy of the distance measurements, while the system noise covariance matrix $\mathbf{Q}$ should be chosen much smaller than $\mathbf{R}$. Then, the prediction, Kalman gain computation and estimate updating/correction stages are performed sequentially. Specifically, the state estimate is predicted by

$$\hat{\boldsymbol{\theta}}^{(k|k-1)} = \boldsymbol{\Phi}\hat{\boldsymbol{\theta}}^{(k-1|k-1)} \qquad k \geq 1 \qquad (7.70)$$

The minimum prediction of the error covariance matrix is computed according to

$$\mathbf{C}^{(k|k-1)} = \boldsymbol{\Phi}\mathbf{C}^{(k-1|k-1)}\boldsymbol{\Phi}^{\mathrm{T}} + \mathbf{BQB}^{\mathrm{T}} \qquad (7.71)$$

Then, the Kalman gain matrix can be determined by

$$\mathbf{K}^{(k)} = \mathbf{C}^{(k|k-1)}(\mathbf{G}^{(k)})^{\mathrm{T}}(\mathbf{G}^{(k)}\mathbf{C}^{(k|k-1)}(\mathbf{G}^{(k)})^{\mathrm{T}} + \mathbf{R})^{-1} \qquad (7.72)$$

where $\mathbf{G}^{(k)}$ is calculated by using (7.69). The state estimate is updated according to

$$\hat{\boldsymbol{\theta}}^{(k|k)} = \hat{\boldsymbol{\theta}}^{(k|k-1)} + \mathbf{K}^{(k)}(\mathbf{p}^{(k)} - \mathbf{G}^{(k)}\hat{\boldsymbol{\theta}}^{(k|k-1)}) \tag{7.73}$$

and the error covariance matrix is updated by

$$\mathbf{C}^{(k|k)} = (\mathbf{I} - \mathbf{K}^{(k)}\mathbf{G}^{(k)})\mathbf{C}^{(k|k-1)} \tag{7.74}$$

It is seen that the EKF is updated recursively from equations (7.70) to (7.74) with $\mathbf{G}^{(k)}$ updated by (7.69). It is worth mentioning that the Kalman filter is an optimal estimator in the sense of minimizing the mean square error under linear system and observation models. However, the EKF does not have optimality properties and its performance will depend on the accuracy of linearization. In some cases, the higher order terms in the Taylor series expansion may become significant, so that the linearization can be rather inaccurate, thus leading to filter instability. For this reason, the unscented Kalman filter (UKF) and the particle filter (PF) are widely considered for target tracking, and will be studied in the following sections.

7.2.2 Unscented Kalman Filtering

The performance of the EKF may be particularly poor in the presence of highly nonlinear functions in both the system and observation equations. One of the techniques developed to compensate for the drawback of the EKF is the UKF [9]. The UKF has been widely considered a better choice than the EKF for localization and target tracking. In the UKF, a set of points, referred to as *sigma points*, is constructed as a given measurement or state estimate, which is deterministically constrained to have the same known statistics, such as the first and second moments. A nonlinear transformation can be applied to each sigma point and the unscented estimate can be obtained by computing the statistics of the transformed set. For example, the mean and covariance of the transformed set approximate the nonlinear transformation of the original mean and covariance estimate. The following presents the fundamental principles of the UKF.

Without loss of generality, let us consider a system with both nonlinear discrete-time process and observation equations which are described by

$$\begin{aligned} \boldsymbol{\theta}^{(k)} &= \mathbf{f}(\boldsymbol{\theta}^{(k-1)}) + \boldsymbol{\varepsilon}^{(k-1)} \\ \mathbf{p}^{(k)} &= \mathbf{g}(\boldsymbol{\theta}^{(k)}) + \mathbf{n}^{(k)} \end{aligned} \tag{7.75}$$

Let L be the length of the state vector $\boldsymbol{\theta}$ and let $\mathbf{Q}$ and $\mathbf{R}$ be the covariance matrixes of $\boldsymbol{\varepsilon}$ and $\mathbf{n}$ respectively. The implementation of the UKF may be realized in three steps as follows.

1. *Predicting the vector state and covariance matrix* The state vector is augmented by including the mean of the process noise as

$$\boldsymbol{\Theta}^{(k-1|k-1)} = \begin{bmatrix} \boldsymbol{\theta}^{(k-1|k-1)} \\ E[\boldsymbol{\varepsilon}^{(k)}] \end{bmatrix} \tag{7.76}$$

and the covariance matrix of the state vector estimation error is augmented by including the covariance of the process noise as

$$\Xi^{(k-1|k-1)} = \begin{bmatrix} \mathbf{C}^{(k-1|k-1)} & 0 \\ 0 & \mathbf{Q}^{(k)} \end{bmatrix} \tag{7.77}$$

A set of $2(L+1)$ sigma points is generated according to

$$\begin{aligned} \boldsymbol{\chi}_0^{(k-1|k-1)} &= \boldsymbol{\Theta}^{(k-1|k-1)} \\ \boldsymbol{\chi}_i^{(k-1|k-1)} &= \boldsymbol{\chi}_0^{(k-1|k-1)} + (\sqrt{(L+\lambda)\Xi^{(k-1)}})_i & i = 1, 2, \ldots, L \\ \boldsymbol{\chi}_i^{(k-1|k-1)} &= \boldsymbol{\chi}_0^{(k-1|k-1)} - (\sqrt{(L+\lambda)\Xi^{(k-1)}})_{i-L} & i = L+1, L+2, \ldots, 2L \end{aligned} \tag{7.78}$$

where $\lambda = \alpha^2(L+\kappa) - L$, α determines the spread of the sigma points and satisfies $10^{-4} \leq \alpha \leq 1$, κ is a scaling parameter, which is usually set at zero, and $(\sqrt{(L+\kappa)\Xi^{(k-1)}})_i$ is the ith row or column of the matrix square root of $(L+\lambda)\Xi^{(k-1)}$. The sigma points are then propagated through the process model or the state transit function as

$$\boldsymbol{\chi}_i^{(k|k-1)} = \mathbf{f}(\boldsymbol{\chi}_i^{(k-1|k-1)}) \qquad i = 0, 1, \ldots, 2L \tag{7.79}$$

By using the propagated sigma points, the state vector and the covariance matrix are predicted according to

$$\begin{aligned} \hat{\boldsymbol{\theta}}^{(k|k-1)} &= \sum_{i=0}^{2L} w_i^{(s)} \boldsymbol{\chi}_i^{(k|k-1)} \\ \mathbf{C}^{(k|k-1)} &= \sum_{i=0}^{2L} w_i^{(c)} (\boldsymbol{\chi}_i^{(k|k-1)} - \hat{\boldsymbol{\theta}}^{(k|k-1)})(\boldsymbol{\chi}_i^{(k|k-1)} - \hat{\boldsymbol{\theta}}^{(k|k-1)})^{\mathrm{T}} \end{aligned} \tag{7.80}$$

where $\{w_i^{\{s\}}\}$ and $\{w_i^{\{c\}}\}$ are the weights for the state and the covariance respectively, given as

$$\begin{aligned} w_0^{(s)} &= \frac{\lambda}{L+\lambda} \\ w_0^{(c)} &= \frac{\lambda}{L+\lambda} + (1 - \alpha^2 + \beta) \\ w_i^{(s)} &= w_i^{(s)} = \frac{1}{2(L+\lambda)} \qquad i = 1, 2, \ldots, 2L \end{aligned} \tag{7.81}$$

where β is related to the statistics of the process noise and usually set at 2.

2. *Updating measurements* Similar to equations (7.76) and (7.77), the predicted state vector and covariance matrix can be augmented by including the mean and covariance of the observation noise as

$$\boldsymbol{\Theta}^{(k|k-1)} = \begin{bmatrix} \hat{\boldsymbol{\theta}}^{(k|k-1)} \\ E[\mathbf{n}^{(k)}] \end{bmatrix} \qquad \Xi^{(k|k-1)} = \begin{bmatrix} \mathbf{C}^{(k|k-1)} & 0 \\ 0 & \mathbf{R}^{(k)} \end{bmatrix} \tag{7.82}$$

Based on the augmented state and covariance in (7.82), the sigma points are recomputed as

$$
\begin{aligned}
\boldsymbol{\chi}_0^{(k|k-1)} &= \boldsymbol{\Theta}^{(k|k-1)} \\
\boldsymbol{\chi}_i^{(k|k-1)} &= \boldsymbol{\chi}_0^{(k|k-1)} + (\sqrt{(L+\lambda)\Xi^{(k|k-1)}})_i && i = 1,2,\dots,L \\
\boldsymbol{\chi}_i^{(k|k-1)} &= \boldsymbol{\chi}_0^{(k|k-1)} - (\sqrt{(L+\lambda)\Xi^{(k|k-1)}})_{i-L} && i = L+1, L+2, \dots, 2L
\end{aligned}
\tag{7.83}
$$

The sigma points are then propagated through the observation function as

$$
\boldsymbol{\eta}_i^{(k)} = \mathbf{g}(\boldsymbol{\chi}_i^{(k|k-1)}) \qquad i = 0, 1, \dots, 2L
\tag{7.84}
$$

Accordingly, the measurements are predicted as

$$
\hat{\mathbf{p}}^{(k)} = \sum_{i=0}^{2L} w_i^{(s)} \boldsymbol{\eta}_i^{(k)}
\tag{7.85}
$$

3. *Computing Kalman gain and correcting state vector and covariance matrix*   From equations (7.84) and (7.85), the predicted covariance of the measurements can be computed according to

$$
\mathbf{C}_{\hat{\mathbf{p}}\hat{\mathbf{p}}}^{(k|k)} = \sum_{i=0}^{2L} w_i^{(c)} (\boldsymbol{\eta}_i^{(k)} - \hat{\mathbf{p}}^{(k)})(\boldsymbol{\eta}_i^{(k)} - \hat{\mathbf{p}}^{(k)})^{\mathrm{T}}
\tag{7.86}
$$

The state-observation cross-correlation matrix is calculated by

$$
\mathbf{C}_{\hat{\boldsymbol{\theta}}\hat{\mathbf{p}}}^{(k|k)} = \sum_{i=0}^{2L} w_i^{(c)} (\boldsymbol{\chi}_i^{(k|k-1)} - \hat{\boldsymbol{\theta}}^{(k|k-1)})(\boldsymbol{\eta}_i^{(k)} - \hat{\mathbf{p}}^{(k)})^{\mathrm{T}}
\tag{7.87}
$$

Then, the UKF Kalman gain is determined by

$$
\mathbf{K}^{(k)} = \mathbf{C}_{\hat{\boldsymbol{\theta}}\hat{\mathbf{p}}}^{(k|k)} (\mathbf{C}_{\hat{\mathbf{p}}\hat{\mathbf{p}}}^{(k|k)})^{-1}
\tag{7.88}
$$

Consequently, the state vector and the covariance matrix are corrected according to

$$
\begin{aligned}
\hat{\boldsymbol{\theta}}^{(k|k)} &= \hat{\boldsymbol{\theta}}^{(k|k-1)} + \mathbf{K}^{(k)}(\mathbf{p}^{(k)} - \hat{\mathbf{p}}^{(k)}) \\
\mathbf{C}^{(k|k)} &= \mathbf{C}^{(k|k-1)} + \mathbf{K}^{(k)} \mathbf{C}_{\hat{\mathbf{p}}\hat{\mathbf{p}}}^{(k|k)} (\mathbf{K}^{(k)})^{\mathrm{T}}
\end{aligned}
\tag{7.89}
$$

To start the recursive process of the UKF, initialization is required. For localization, the initial values can be chosen in the same way as for the EKF.

7.2.3 Particle Filtering

Another alternative to the EKF are the PFs, which are also known as the sequential Monte Carlo methods. Particle filtering is a technique for implementing a recursive Bayesian filter through

Monte Carlo simulations. The key idea is to represent the required posterior density function by a set of random samples with associated weights and to estimate the parameters based on these samples and weights. As the number of samples becomes very large, the functional description of the posterior PDF would be equivalently represented by the Monte Carlo characterization so that the sequential Monte Carlo methods approach the optimal Bayesian estimator [10].[1] Note that when the observation function and the state function in (7.75) are linear and both ε and $\mathbf{n}$ are Gaussian, the Kalman filter achieves the exact Bayesian filtering, i.e. the optimum filtering. However, when one or both the functions are nonlinear, the EKF makes use of the first-order approximation, which may result in significant performance degradation and the problem of divergence. There are different types of PFs and the sampling importance resampling method is presented in the following.

Filtering is the problem of sequentially estimating the parameters of a system based on a sequence of observations. Let $p(\boldsymbol{\theta}_{0:k}|\mathbf{p}_{1:k})$ denote the posterior PDF of the system, where $\mathbf{p}_{1:k} = \{\mathbf{p}_j, j = 1, 2, \ldots, k\}$ is the set of all observations up to time k and $\boldsymbol{\theta}_{1:k} = \{\boldsymbol{\theta}_j, j = 1, 2, \ldots, k\}$ is the set of all states up to time k. Given the PDF, the moments of different orders, including the mean and variance of the system states, can be determined. In reality, it may be difficult to handle the true posterior PDF directly. Instead, a known *proposal distribution*, denoted by $q(\boldsymbol{\theta}_{0:k}|\mathbf{p}_{1:k})$, which is also called the *importance distribution*, is employed to substitute the true posterior PDF. Let $\{\boldsymbol{\theta}_{0:k}^{(i)}, w_k^{(i)}\}_{i=1}^{N_s}$ denote a weighted set of particles (samples), which characterizes the posterior PDF $p(\boldsymbol{\theta}_{0:k}|\mathbf{p}_{1:k})$. By drawing samples from the proposal distribution, the importance weights can be estimated recursively by

$$w_k^{(i)} = w_{k-1}^{(i)} \frac{p(\mathbf{p}_k|\boldsymbol{\theta}_k^{(i)})p(\boldsymbol{\theta}_k^{(i)}|\boldsymbol{\theta}_{k-1}^{(i)})}{q(\boldsymbol{\theta}_k^{(i)}|\boldsymbol{\theta}_{0:k-1}^{(i)}, \mathbf{p}_{1:k})} \tag{7.90}$$

where it is assumed that the states are a first-order Markov process. The importance weights $\{w_k^{(i)}\}$ are approximations to the relative posterior PDF of the particles and they are normalized, thus resulting in $\sum_{i=1}^{N_s} w_k^{(i)} = 1$. Consequently, one has

$$\tilde{w}_k^{(i)} = \frac{w_{k-1}^{(i)}}{\sum_{j=1}^{N_s} w_k^{(j)}} \tag{7.91}$$

Then, the posterior PDF $p(\boldsymbol{\theta}_k|\mathbf{p}_{1:k})$ can be approximated as

$$p(\boldsymbol{\theta}_k|\mathbf{p}_{1:k}) \approx \sum_{i=1}^{N_s} \tilde{w}_k^{(i)} \delta(\boldsymbol{\theta}_k - \boldsymbol{\theta}_k^{(i)}) \tag{7.92}$$

where $\delta(.)$ is the Dirac delta function. A common problem of computing the importance weights by using (7.91) is the degeneracy phenomenon. This means that, after a few iterations, all but one particle will have negligible weight, thus resulting in a large computational effort

---

[1] One of the drawbacks of a PF is that in some circumstances a large number of particles is required in order to approximate the posterior PDF accurately, resulting in large computational complexity.

being devoted to updating particles which virtually contribute nothing. A suitable measure of degeneracy is the effective sample size, which can be approximated as

$$N_{\text{eff}} \approx \frac{N_s}{1 + \sum_{i=1}^{N_s} (\tilde{w}_k^{(i)})^2} \tag{7.93}$$

Clearly, $N_{\text{eff}} \leq N_s$, and small N_{eff} indicates severe degeneracy. When the computed effective sample size is smaller than some predefined threshold, resampling is performed. The aim of resampling is to eliminate particles that have rather small weights. A number of resampling algorithms can be found in [11,12]. Another important issue is the selection of the important distribution. Although the optimal importance density can be found, it is usually difficult to make use of it. For convenience, the importance density is often chosen to be the prior density as

$$q(\boldsymbol{\theta}_k | \boldsymbol{\theta}_{k-1}^{(i)}, \mathbf{p}_k) = p(\boldsymbol{\theta}_k | \boldsymbol{\theta}_{k-1}^{(i)}) \tag{7.94}$$

When $\boldsymbol{\theta}_k$ belongs to a finite set of states such as in a Jump–Markov linear system, the discrete modal state can be tracked by using a PF, and the Kalman filter may be employed to track the continuous base state [13]. The performance of the PF depends largely on the selection of the proposal/importance distribution. The EKF has been considered by a number of researchers for generating the importance distribution [14]. The EKF-based PF outperforms the standard PF where the importance density is chosen according to (7.94) in some circumstances. For highly nonlinear problems, however, the EKF tends to be inaccurate, thus resulting in poor performance and even divergence. To overcome the EKF-related problems, the UKF may be used for the generation of the importance density, which leads to the unscented PF (UPF). The UPF combines the merits of unscented transformation and particle filtering [15].

7.2.4 Discussion

When the motion statistics can be accurately predicted, the mobile track is smooth without abrupt turns and the measurement noise is Gaussian, Kalman filtering is preferred. On the other hand, when the mobile track involves frequent turns or the moving speed varies frequently, the filtering-based methods may not produce satisfactory results. In this case, the methods studied in Section 7.1 may be more suited. Therefore, it would be wise to do performance comparison by using field measurement data to select an appropriate method for real-time target tracking.

7.3 Data Smoothing

In many circumstances, the targets are in sporadic or continuous motion, such as roaming animals and moving vehicles. To achieve accurate object tracking and produce accurate node position information for efficient network operation and management, node/object position information needs to be updated regularly when the target is moving. At different time instants, a number of signal measurements are collected from a specific set of nearby nodes with known locations. The selected nodes can be different from time to time as the target moves. Usually, the nearby nodes closest to the target are selected to provide accurate parameter measurements, since, in general, shorter distance results in higher received signal power and so better performance can be obtained.

It is worth noting that updating the position information too often might waste useful resources, such as power, which is so important for battery-powered ordinary sensor nodes. On the other hand, not updating the position estimation often enough would lead to not being able to track the nodes to provide accurate position information quickly. The updating frequency should be chosen based on the mobility/speed of the nodes of interest. Suppose that the maximum speed of the desired nodes in a sensor network is 18 km/h (that is, 5 m/s). If the position error is required to be below 1 m, then the position information needs to be updated at least five times per second. To achieve higher position accuracy will require a higher updating frequency. Different updating frequencies may be applied for different nodes if the maximum velocities of the nodes are different and they can be predicted *a priori*. This strategy would improve the network efficiency.

In the preceding section, the EKF, the UKF and the PF for tracking moving targets are presented. This section focuses on improving location accuracy by smoothing the initial position estimation results produced by using either the noniterative algorithms presented in Chapter 6 or the iterative or recursive methods in the preceding two sections. There are a range of smoothing techniques and methods. In this section, three algorithms, namely the LKF, the linear LS smoothing algorithm and the sinc function are employed to smooth the sequence of location estimates. Figure 7.4 shows the simplified block diagram of a generic positioning and tracking system.

7.3.1 Kalman Filtering for Position Smoothing

In the presence of a nonlinear system model and/or observation model, the LKF cannot be directly applied. Instead, the EKF, the UKF or the PF can be considered as described in the preceding sections. However, when smoothing a sequence of position estimates, the LKF can be directly exploited under the assumption of a linear motion model. Implementing the

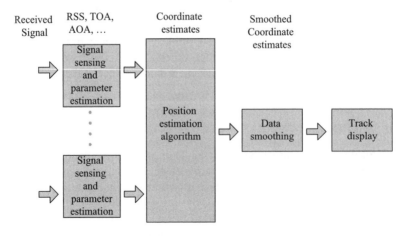

Figure 7.4 Simplified functional block diagram of radio positioning and tracking.

LKF for smoothing follows the same procedure as implementing the EKF (see equations (7.70)–(7.74)), except that the observation equation now becomes

$$\mathbf{p}^{(k)} = \tilde{\mathbf{F}}^{(k)}\boldsymbol{\theta}^{(k)} + \mathbf{n}^{(k)} \tag{7.95}$$

where $\tilde{\mathbf{F}}^{(k)}$ is the observation matrix given by

$$\tilde{\mathbf{F}}^{(k)} = \begin{bmatrix} 1 & 0 & 0 & 0 & 0 & 0 \\ 0 & 1 & 0 & 0 & 0 & 0 \\ 0 & 0 & 1 & 0 & 0 & 0 \end{bmatrix} \tag{7.96}$$

Also, $\mathbf{R}$ now denotes the covariance matrix of the coordinate estimation errors instead of the distance measurement errors. The system noise covariance matrix $\mathbf{Q}$ may be chosen as different from that in the EKF. In practice, these parameters would be tuned in advance based on field measurements.

Alternatively, the estimates of each coordinate can be smoothed by an individual LKF. Then, the process equation and the observation equation for one of the coordinates, say the x coordinate, can be written as

$$\begin{bmatrix} x^{(k)} \\ v_x^{(k)} \end{bmatrix} = \begin{bmatrix} 1 & \delta t \\ 0 & 1 \end{bmatrix} \begin{bmatrix} x^{(k-1)} \\ v_x^{(k-1)} \end{bmatrix} + \begin{bmatrix} 0.5\delta t^2 \\ \delta t \end{bmatrix} w^{(k)}$$

$$p_x^{(k)} = \begin{bmatrix} 1 & 0 \end{bmatrix} \begin{bmatrix} x^{(k)} \\ v_x^{(k)} \end{bmatrix} + n_x^{(k)} \tag{7.97}$$

In this case, the system noise, the observation noise and the observation are all scalar. The accuracy of the two different implementations would be exactly the same. However, using three individual filters may have the advantage of reduced complexity.

7.3.2 Least-Squares Approach for Position Smoothing

The linear LS approach has been used in [16] to obtain smoothed estimates of the position and the speed of the moving target simultaneously. Assume that the target is in a linear motion with a constant velocity along each coordinate. This assumption would be reasonable when a short distance is considered, although the whole track may not be linear and the velocity can be time varying. Then, the true target position at time t_k can be described by

$$\mathbf{p}_k = \mathbf{p}_0 + \mathbf{v}t_k \qquad k \geq 1 \tag{7.98}$$

where $\mathbf{p}_0$ is the position at time t_0. The estimated target position at time t_k is modeled as

$$\hat{\mathbf{p}}_k = \mathbf{p}_k + \boldsymbol{\varepsilon}_k \tag{7.99}$$

where $\boldsymbol{\varepsilon}_k$ is the estimation error vector. Let us make use of a sequence of position estimates $\hat{\mathbf{p}}_k$, $1 \leq k \leq K$, for position smoothing. The LS estimator is found by minimizing

$$\sum_{k=1}^{K} ||\hat{\mathbf{p}}_k - (\mathbf{p}_0 + \mathbf{v}t_k)||^2 \tag{7.100}$$

This minimization can be expanded into

$$\min_{p_{x,0},v_x} \sum_{k=1}^{K}[\hat{p}_{x,k}-(p_{x,0}+v_xt_k)]^2 + \min_{p_{y,0},v_y} \sum_{k=1}^{K}[\hat{p}_{y,k}-(p_{y,0}+v_yt_k)]^2 + \min_{p_{z,0},v_z} \sum_{k=1}^{K}[\hat{p}_{z,k}-(p_{z,0}+v_zt_k)]^2$$

(7.101)

The first term in (7.101) can be written as

$$\min_{p_{x,0},v_x} (\mathbf{p}_x - \mathbf{A}\boldsymbol{\theta}_x)^{\mathrm{T}}(\mathbf{p}_x - \mathbf{A}\boldsymbol{\theta}_x)$$

(7.102)

where

$$\mathbf{p}_x = \begin{bmatrix} p_{x,1} & p_{x,2} & \cdots & p_{x,K} \end{bmatrix}^{\mathrm{T}}$$
$$\boldsymbol{\theta}_x = \begin{bmatrix} p_{x,0} & v_x \end{bmatrix}^{\mathrm{T}}$$
$$\mathbf{A} = \begin{bmatrix} 1 & 1 & \cdots & 1 \\ t_1 & t_2 & \cdots & t_K \end{bmatrix}^{\mathrm{T}}$$

(7.103)

The minimization in (7.102) results in

$$\hat{\boldsymbol{\theta}}_x = (\mathbf{A}^{\mathrm{T}}\mathbf{A})^{-1}\mathbf{A}^{\mathrm{T}}\mathbf{p}_x$$

(7.104)

After some mathematical manipulations, (7.104) can be written as

$$\hat{\boldsymbol{\theta}}_x = \frac{1}{K\sum_{k=1}^{K}t_k^2 - \left(\sum_{k=1}^{K}t_k\right)^2} \begin{bmatrix} \sum_{k=1}^{K}t_k^2\sum_{k=1}^{K}p_{x,k} - \sum_{k=1}^{K}t_k\sum_{k=1}^{K}t_kp_{x,k} \\ K\sum_{k=1}^{K}t_kp_{x,k} - \sum_{k=1}^{K}t_k\sum_{k=1}^{K}p_{x,k} \end{bmatrix}$$

(7.105)

So one has

$$\hat{p}_{x,0} = \frac{\sum_{k=1}^{K}t_k^2\sum_{k=1}^{K}p_{x,k} - \sum_{k=1}^{K}t_k\sum_{k=1}^{K}t_kp_{x,k}}{K\sum_{k=1}^{K}t_k^2 - \left(\sum_{k=1}^{K}t_k\right)^2}$$

(7.106)

and

$$\hat{v}_x = \frac{K\sum_{k=1}^{K}t_kp_{x,k} - \sum_{k=1}^{K}t_k\sum_{k=1}^{K}p_{x,k}}{K\sum_{k=1}^{K}t_k^2 - \left(\sum_{k=1}^{K}t_k\right)^2}$$

(7.107)

Then, we have the smoothed x-coordinate position estimate at time t_k as

$$\hat{p}_{x,k} = \hat{p}_{x,0} + \hat{v}_x t_k \tag{7.108}$$

Similarly, one can obtain $\hat{p}_{y,0}$ and $\hat{p}_{z,0}$ in the form of (7.106), $\hat{v}_y$ and $\hat{v}_z$ in the form of (7.107) and $\hat{p}_{y,k}$ and $\hat{p}_{x,k}$ in the form of (7.108).

7.3.3 Sinc Smoothing

When the track is rather irregular or nonlinear and/or the velocity of the moving nodes is time varying, the LS smoothing approach may not perform well. Kalman filtering requires that the variances of the system noise and observation noise are known *a priori*. However, in practice, the position estimation noise variances are usually unknown and the implementation of the Kalman filter at low-complexity nodes may be not desirable due to its relatively high computational requirement. In this case, the sinc function may be employed to smooth the mobile tracks and improve the accuracy of the position estimates. Note that sinc function has been exploited to interpolate pilot symbol-aided channel estimates in [17].

The principle of sinc smoothing is rather simple. The smoothing function is given by

$$h(j) = \frac{\sin(2j\pi f_M T)}{2j\pi f_M T} \qquad -\infty < j < +\infty \tag{7.109}$$

where f_M is the maximum frequency of the mobile (function of time) and T is the position updating period. In practice, truncation is used and the length of the window is chosen to be equal to K so that

$$-J \leq j \leq J \qquad J = \frac{K-1}{2} \tag{7.110}$$

When K is not large, such as from 10 to 20, windowing techniques such as the Hanning window and Hamming window may be employed to reduce the effect of the abrupt truncation. Let $\tilde{x}_i$ be the sequence of the estimated coordinates (any of the three coordinates) and $\hat{x}_i$ be the corresponding results after smoothing. Then:

$$\hat{x}_i = \sum_{j=-J}^{J} h(j)\tilde{x}_{i+j} \qquad i \geq -J \tag{7.111}$$

It has been shown in [17] that sinc smoothing is optimal in the case where the power spectrum of the signal is bandlimited and the signal has a flat spectrum up to f_M. Usually, f_M is unknown, but it can be chosen empirically. The advantage of sinc smoothing is its simplicity, whereas the drawback is that the current smoothed estimate is not the current position of the target. As a result, it may not be suited for fast tracking.

Let us consider tracking people in an indoor environment. The monitored area is in a square shape with the edge length equal to 30 m. Four base stations are placed at the four corners of the area. It is assumed that the person is walking in the corridor at a speed of 1.5 m/s. The sampling rate of the distance measurements is five times per second. The range measurement error model is the same as (6.92) in Section 6.3. The QLS algorithm is first used to produce the raw position estimates at the first phase and an LKF is employed to smooth the raw data at the

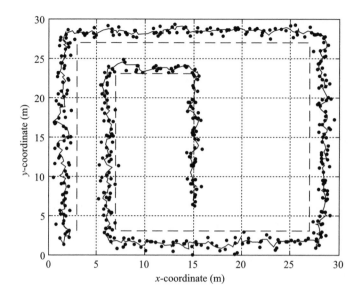

Figure 7.5   Target tracking with QLS algorithm producing the raw location estimates and a Kalman filter smoothing the raw data. The dashed line represents the true track, the dotted line denotes the raw location estimates and the solid line denotes the smoothed estimates.

second phase. The second phase is performed immediately when a raw location estimate is available.

Figure 7.5 shows the true track that the assumed person walks on, denoted by the dashed line, the estimated raw locations denoted by the dots and the smoothed location estimates denoted by the solid line. The starting point is set at (3, 3) and the stopping point is at (15, 7). In this case the accuracy measure is defined as the average distance between the true location and the estimated one. It is found that the accuracy of the raw data and the estimated ones is 1.41 m and 1.44 m respectively. In this case the smoothing does not improve accuracy but only smooth the estimated track. This may be because the estimation error bias cannot be effectively suppressed through smoothing. To reduce the NLOS effect effectively, NLOS mitigation techniques must be considered, which will be studied in Chapter 10. In Figure 7.5 we can also see that the accuracy of the location estimates, especially the raw estimates, can be greatly enhanced if the building map or road map is used. For instance, a person can only walk in corridors, instead of crossing walls. The width of the corridor may be just 1.5 m, so the location estimates can be constrained accordingly.

References

[1] W.H. Foy, 'Position-location solutions by Taylor-series estimation', *IEEE Transactions on Aerospace and Electronic Systems*, **12**(2), 1976, 187–194.
[2] R. Fletcher and M.J.D. Powell, 'A rapidly convergent descent method for minimization', *Computer Journal*, **6**, 1963, 163–168.
[3] D. Marquardt, 'Algorithm for least-squares estimation of nonlinear parameters', *SIAM Journal on Applied Mathematics*, **11**, 1963, 431–441.

[4]   R. Fletcher, *Practical Methods of Optimization*, John Wiley & Sons, Ltd., Chichester, 1987.

[5]   Y.-T. Chan, H.Y.C. Hang and P.-C. Ching, 'Exact and approximate maximum likelihood localization algorithms', *IEEE Transactions on Vehicular Technology*, **55**(1), 2006, 10–16.

[6]   R. Doraiswami, 'A novel Kalman filter-based navigation using beacons', *IEEE Transactions on Aerospace and Electronic Systems*, **32**(2), 1996, 830–840.

[7]   M. Hellebrandt and R. Mathar, 'Location tracking of mobiles in cellular radio networks', *IEEE Transactions on Vehicular Technology*, **48**(5), 1999, 1558–1562.

[8]   M. McGuire and K.N. Plataniotis, 'Dynamic model-based filtering for mobile terminal location estimation', *IEEE Transactions on Vehicular Technology*, **52**(4), 2003, 1012–1031.

[9]   S.J. Julier and J.K. Uhlmann, 'Unscented filtering and nonlinear estimation', *Proceedings of the IEEE*, **92**(3), 2004, 401–422.

[10] M.S. Arulampalam, S. Maskell, N. Gordon and T. Clapp, 'A tutorial on particle filters for online nonlinear/non-Gaussian Bayesian tracking', *IEEE Transactions on Signal Processing*, **50**(2), 2002, 174–188.

[11] G. Kitagawa, 'Monte Carlo filter and smoother for non-Gaussian and nonlinear state space models', *Journal of Computational and Graphical Statistics*, **5**(1), 1996, 1–25.

[12] J.S. Liu and R. Chen, 'Sequential Monte Carlo methods for dynamic systems', *Journal of the American Statistical Association*, **93**(443), 1998, 1032–1044.

[13] A. Doucet, N.Gordon and V. Krishnamurthy, 'Particle filters for state estimation of jump Markov linear systems', *IEEE Transactions on Signal Processing*, **49**(3), 2001, 613–624.

[14] M.K. Pitt and N. Shephard, 'Filtering via simulation: auxiliary particle filters', *Journal of the American Statistical Association*, **94**(446), 1999, 590–599.

[15] R. van der Merwe, A. Doucet, N. de Freitas and E.A. Wan, 'The unscented particle filter', Technical Report, Department of Engineering, University of Cambridge, UK, 2000.

[16] K. Yu, J.P. Montillet, A. Rabbachin, P. Cheong, and I. Oppermann, 'UWB location and tracking for wireless embedded networks', *Signal Processing*, **86**(9), 2006, 2153–2171.

[17] K. Yu, J.S. Evans and I.B. Collings, 'Performance analysis of LMMSE receiver for M-ary QAM in Rayleigh faded CDMA channels', *IEEE Transactions on Vehicular Technology*, **52**(5), 2003, 1242–1253.

8

Positioning Accuracy Evaluation

The accuracy of a particular position fix depends on a number of factors, including the radio ranging measurement accuracy, the algorithm used to process the measurements and the geometry of the nodes in the system. To measure the accuracy of various algorithms and positioning systems, a number of accuracy measures can be employed. The Cramer–Rao lower bound (CRLB) benchmarks the location accuracy of any unbiased estimators. The GDOP describes the geometric impact of node configuration on positioning accuracy. Furthermore, given a specific location algorithm, its accuracy can be measured by the RMSE of the location estimates or the CDP of the location errors. In this chapter, some of the main positioning accuracy measures are studied. After a brief introduction in Section 8.1, the CRLB is presented for LOS scenarios in Section 8.2. Then, the CRLB analysis is performed for NLOS scenarios in Section 8.3. The performance of the LS estimator and the iterative optimization-based location algorithm is evaluated against the CRLB. Also, the impact of the anchor location errors is investigated. GDOP is considered in detail in Chapter 9.

8.1 Accuracy Measures

Before proceeding to the accuracy analysis for particular positioning methods, let us first briefly examine the key accuracy measures which can be employed to evaluate positioning performance for either navigation or tracking algorithms and systems.

8.1.1 Cramer–Rao Lower Bound

The CRLB has been widely used in signal processing and parameter estimation to set a lower bound on any unbiased estimators. The CRLB is also a key performance measure in radio positioning [1–3]. Assume that the measurement equation is given by

$$\mathbf{r} = \mathbf{f}(\boldsymbol{\theta}) + \mathbf{n} \tag{8.1}$$

where $\mathbf{r}$ is the observation vector, $\boldsymbol{\theta}$ is the unknown parameter vector, $\mathbf{f}(\boldsymbol{\theta})$ is the vector-form function of the parameter vector and $\mathbf{n}$ is the observation noise vector. The CRLB for the kth parameter of the position estimation vector $\hat{\boldsymbol{\theta}}$, that is $\hat{\theta}_k$, is defined by

$$\mathrm{CRLB}(\hat{\theta}_k) = [\mathbf{F}^{-1}(\boldsymbol{\theta})]_{k,k} \tag{8.2}$$

$\mathbf{F}(\boldsymbol{\theta})$ is the Fisher information matrix (FIM), whose elements are defined by

$$[\mathbf{F}(\boldsymbol{\theta})]_{k,k} = -E\left[\frac{\partial^2 \ln p(\mathbf{r}|\boldsymbol{\theta})}{\partial \theta_k \theta_\ell}\right] \quad k, \ell = 1, 2, \ldots, K \tag{8.3}$$

where K is the length of the parameter vector and $p(\mathbf{r}|\boldsymbol{\theta})$ is the PDF of the observation vector for a given parameter vector. The CRLB can be used as a performance reference, but it cannot be directly used to represent the performance of a specific estimation algorithm. Only when the accuracy of the estimation results from an algorithm is already calculated based on an accuracy measure such as the RMSE or the STD can one see the gap between the CRLB and the accuracy of the algorithm. A large gap indicates that a better estimation method may be pursued to achieve better estimation results. The CRLB will be studied later in this chapter and in Chapters 10–12 under different scenarios.

8.1.2 Geometric Dilution of Precision

The GDOP concept was originally developed in the 1940s to describe the performance of terrestrial navigation systems [4]. Later, it became associated with the geometric effect of a satellite configuration on GPS accuracy. When the angular positions of satellites are close together in the sky, the GDOP value is high, resulting in poor positioning accuracy. To achieve good positioning accuracy with GPS, the satellites should be spread out around the mobile terminal. GDOP can also be used as an accuracy measure for terrestrial positioning systems [5,6]. The concept of GDOP will be studied in detail in Chapter 9 with a focus on GDOP analysis for the design and testing of a practical location system.

8.1.3 Root-Mean-Square Error

The RMSE is often employed to measure the accuracy of the location estimates of specific positioning algorithms and systems [7–9]. Assume that there are L locations examined with the true coordinates denoted by (x_ℓ, y_ℓ, z_ℓ), $1 \le \ell \le L$, and that the corresponding coordinate estimates denoted by $(\hat{x}_\ell, \hat{y}_\ell, \hat{z}_\ell)$, then the RMSE of the location estimation can be defined as

$$\varepsilon = \sqrt{\frac{1}{3L}\sum_{\ell=1}^{L}\left[(\hat{x}_\ell - x_\ell)^2 + (\hat{y}_\ell - y_\ell)^2 + (\hat{z}_\ell - z_\ell)^2\right]} \tag{8.4}$$

Caution is needed when using RMSE for accuracy comparison. For example, given a set of very accurate estimation results except for only a few abnormal points with extremely large errors, the RMSE can still be very large. To provide a fair comparison among different algorithms, therefore, it is often necessary to exclude a few abnormal location estimates from the data set. Similar to RMSE, the mean and variance of the location errors can also be used as an accuracy measure.

8.1.4 Cumulative Distribution Probability

Another accuracy measure is the CDP of the location errors [10,11]. Here, the location error is typically defined as the distance between the true and the estimated positions. From the CDP curve, one can easily obtain the percentage of test points within a radius of a circle centered at the true location of the target. This percentage may be the location accuracy requirement for the design and testing of some realistic positioning systems. For instance, the enhanced 911 services may be required to provide location information to 911 operators when an emergency call is made. FCC rules may require some specific systems to identify a caller's location within 120 m 95% of the time.

A similar accuracy measure is the circular error probability (CEP), which is defined as the radius of a circle that has its center at the mean of the estimates and contains half of the location estimates [5]. If the location estimator is unbiased, then the CEP is a measure of the estimator uncertainty relative to the true target location.

8.2 Cramer–Rao Lower Bound in Line-of-Sight Conditions

In this section the CRLB is derived for scenarios where both the distance and 2D angular measurements are provided. It is assumed that the anchor locations are known.

8.2.1 Signal Model

For convenience, nodes with known positions are called anchor nodes, whereas nodes with unknown positions are called ordinary nodes. It is also assumed that the range measurements are obtained, for example, by estimating the RTT or by measuring the received signal strength. To obtain AOA information, an antenna array is required, which may be implemented at the anchor nodes. The anchor nodes' orientation is required to make use of the AOA information from different anchors which can be obtained by implementing a digital compass at each anchor. For instance, the HMR3000 digital compass has a static accuracy of $\pm 0.5°$ and a resolution of $\pm 0.1°$. It is seen that the AOA errors come from two sources: the angular measurement error and orientation measurement error. For analytical and notational simplicity, we model the two errors as one random variable throughout this chapter. It is assumed that the AOA measurements are accurate enough so that they can be used to enhance location accuracy. As shown in Figure 8.1, (x, y, z) are the unknown coordinates of the ordinary node of interest, while (x_i, y_i, z_i) are the known coordinates of the ith anchor. The azimuth and elevation angles of the signal impinging on the receiver of the ith anchor are denoted by ϕ_i and α_i respectively, and the distance between the ordinary node and the ith anchor is denoted d_i. Then, it can be readily shown that

$$d_i = \sqrt{(x - x_i)^2 + (y - y_i)^2 + (z - z_i)^2}$$

$$\phi_i = \tan^{-1} \frac{y - y_i}{x - x_i}$$

$$\alpha_i = \cos^{-1} \frac{z - z_i}{d_i}$$

(8.5)

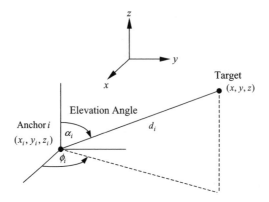

Figure 8.1   Illustration of the angle and distance in LOS condition.

In the presence of measurement errors, the measurement equations are given as

$$\hat{d}_i = d_i + n_{\hat{d}_i}$$
$$\hat{\phi}_i = \phi_i + n_{\hat{\phi}_i} \qquad (8.6)$$
$$\hat{\alpha}_i = \alpha_i + n_{\hat{\alpha}_i}$$

where $n_{\hat{d}_i}$, $n_{\hat{\phi}_i}$ and $n_{\hat{\alpha}_i}$ are the measurement errors of the distance, azimuth angle and elevation angle respectively.

8.2.2 *Effect of Distance and Angle-of-Arrival Errors*

From (8.3), it can be seen that the CRLB is related to the distribution of the parameter estimation errors. To have a clear understanding on how the different errors affect the CRLB, two different scenarios are considered. The first one is that there are distance and AOA measurement errors, whereas the anchor locations are error free. The second one is that the anchor location error is the only error source. In this subsection, we consider the first case.

In the presence of measurement noise, (8.6) can be rewritten as

$$r_j = f_j(x,y,z) + n_j \qquad j = 1, 2, \ldots, 3N \qquad (8.7)$$

where

$$f_{3(i-1)+1}(x,y,z) = \sqrt{(x - x_i)^2 + (y - y_i)^2 + (z - z_i)^2} \qquad i = 1, 2, \ldots, N$$
$$f_{3(i-1)+2}(x,y,z) = \tan^{-1}\frac{y - y_i}{x - x_i} \qquad (8.8)$$
$$f_{3i}(x,y,z) = \cos^{-1}\frac{z - z_i}{d_i}$$

and $\{n_j\}$ are the mutually independent measurement noise variables of mean zero and variances given by[1]

$$
\begin{aligned}
\sigma^2_{n_{3(i-1)+1}} &= \sigma^2_{\hat{d}_i} \qquad i = 1, 2, \ldots, N \\
\sigma^2_{n_{3(i-1)+2}} &= \sigma^2_{\hat{\phi}_i} \\
\sigma^2_{n_{3i}} &= \sigma^2_{\hat{\alpha}_i}
\end{aligned}
\tag{8.9}
$$

Define

$$
\begin{aligned}
\mathbf{p} &= \begin{bmatrix} x & y & z \end{bmatrix}^{\mathrm{T}} \in \mathbb{R}^{3\times 1} \\
\mathbf{f}(\mathbf{p}) &= \begin{bmatrix} f_1(\mathbf{p}) & f_2(\mathbf{p}) & \cdots & f_{3N}(\mathbf{p}) \end{bmatrix}^{\mathrm{T}} \in \mathbb{R}^{3N\times 1} \\
\mathbf{r} &= \begin{bmatrix} r_1 & r_2 & \cdots & r_{3N} \end{bmatrix}^{\mathrm{T}} \in \mathbb{R}^{3N\times 1} \\
\mathbf{n} &= \begin{bmatrix} n_1 & n_2 & \cdots & n_{3N} \end{bmatrix}^{\mathrm{T}} \in \mathbb{R}^{3N\times 1}
\end{aligned}
\tag{8.10}
$$

Then, (8.8) can be written in compact form as

$$
\mathbf{r} = \mathbf{f}(\mathbf{p}) + \mathbf{n}
\tag{8.11}
$$

When the errors are mutually independent Gaussian random variables, the conditional PDF $p(\mathbf{r}|\mathbf{p})$ is given by

$$
p(\mathbf{r}|\mathbf{p}) = \frac{1}{\sqrt{(2\pi)^{3N}} \prod_{i=1}^{3N} \sigma_{n_i}} \exp\left\{ -\sum_{i=1}^{3N} \frac{1}{2\sigma^2_{n_i}} [r_i - f_i(\mathbf{p})]^2 \right\}
\tag{8.12}
$$

With some mathematical manipulations, it can be shown that the FIM is given as

$$
\begin{aligned}
[\mathbf{F}(\mathbf{p})]_{1,1} &= \sum_{i=1}^{N} \left[\frac{(x-x_i)^2}{\sigma^2_{\hat{d}_i} d_i^2} + \frac{(y-y_i)^2}{\sigma^2_{\hat{\phi}_i} d_{2,i}^4} + \frac{(x-x_i)^2(z-z_i)^2}{\sigma^2_{\hat{\alpha}_i} d_i^2 d_{2,i}^2} \right] \\
[\mathbf{F}(\mathbf{p})]_{1,2} = [\mathbf{F}(\mathbf{p})]_{2,1} &= \sum_{i=1}^{N} (x-x_i)(y-y_i) \left[\frac{1}{\sigma^2_{\hat{d}_i} d_i^2} - \frac{(y-y_i)^2}{\sigma^2_{\hat{\phi}_i} d_{2,i}^4} + \frac{(z-z_i)^2}{\sigma^2_{\hat{\alpha}_i} d_i^4 d_{2,i}^2} \right] \\
[\mathbf{F}(\mathbf{p})]_{1,3} = [\mathbf{F}(\mathbf{p})]_{3,1} &= \sum_{i=1}^{N} \frac{(x-x_i)(z-z_i)}{d_i^2} \left(\frac{1}{\sigma^2_{\hat{d}_i}} - \frac{1}{\sigma^2_{\hat{\alpha}_i} d_i^2} \right) \\
[\mathbf{F}(\mathbf{p})]_{2,2} &= \sum_{i=1}^{N} \left[\frac{(y-y_i)^2}{\sigma^2_{\hat{d}_i} d_i^2} + \frac{(x-x_i)^2}{\sigma^2_{\hat{\phi}_i} d_{2,i}^4} + \frac{(y-y_i)^2(z-z_i)^2}{\sigma^2_{\hat{\alpha}_i} d_i^4 d_{2,i}^2} \right] \\
[\mathbf{F}(\mathbf{p})]_{2,3} = [\mathbf{F}(\mathbf{p})]_{3,2} &= \sum_{i=1}^{N} \frac{(y-y_i)(z-z_i)}{d_i^2} \left(\frac{1}{\sigma^2_{\hat{d}_i}} - \frac{1}{\sigma^2_{\hat{\alpha}_i} d_i^2} \right) \\
[\mathbf{F}(\mathbf{p})]_{3,3} &= \sum_{i=1}^{N} \frac{1}{d_i^2} \left(\frac{(z-z_i)^2}{\sigma^2_{\hat{d}_i}} + \frac{d_{2,i}^2}{\sigma^2_{\hat{\alpha}_i} d_i^2} \right)
\end{aligned}
\tag{8.13}
$$

---

[1] Note that range measurements based on RTT may be biased from the actual distance such that this leads to nonzero-mean noise measurements. Also, in the event of NLOS signal propagation, the TOA (and hence distance) and the AOA errors could be correlated.

where

$$d_{2,i} = \sqrt{(x-x_i)^2 + (y-y_i)^2} \tag{8.14}$$

From the above equations, one obtains the CRLB for the three position coordinate estimates $(\hat{x}, \hat{y}, \hat{z})$ as

$$\text{CRLB}(\hat{x}) = \frac{1}{\det(\mathbf{F}(\mathbf{p}))}\left\{[\mathbf{F}(\mathbf{p})]_{2,2}[\mathbf{F}(\mathbf{p})]_{3,3} - ([\mathbf{F}(\mathbf{p})]_{2,3})^2\right\}$$

$$\text{CRLB}(\hat{y}) = \frac{1}{\det(\mathbf{F}(\mathbf{p}))}\left\{[\mathbf{F}(\mathbf{p})]_{1,1}[\mathbf{F}(\mathbf{p})]_{3,3} - ([\mathbf{F}(\mathbf{p})]_{1,3})^2\right\} \tag{8.15}$$

$$\text{CRLB}(\hat{z}) = \frac{1}{\det(\mathbf{F}(\mathbf{p}))}\left\{[\mathbf{F}(\mathbf{p})]_{1,1}[\mathbf{F}(\mathbf{p})]_{2,2} - ([\mathbf{F}(\mathbf{p})]_{1,2})^2\right\}$$

where the following characteristic of the FIM has been exploited:

$$[\mathbf{F}(\mathbf{p})]_{i,j} = [\mathbf{F}(\mathbf{p})]_{j,i} \tag{8.16}$$

Clearly, the CRLB is dependent on the geometry of the anchors and the target and the accuracy of the range and angle estimation. Note that dropping the angle-related terms in the above expressions of the CRLB results in the CRLB in the absence of angle measurements.

8.2.3 Effect of Anchor Location Errors

In ad hoc networks and WSNs, the anchor locations may not be permanently fixed and they may change position. Typically, the anchor locations would be estimated by GPS and there would be errors in the estimates, especially for low-cost commercial GPS devices. Therefore, it would be useful to derive the CRLB for the node location estimation in the presence of anchor location error. To show the effect of the anchor location errors clearly, it is assumed that anchor position errors are the only source of the error in the following analysis. In reality, the sensor node positioning error will come from both the parameter (TOA, AOA or RSS) estimation errors and the anchor location errors.

8.2.3.1 Approximate Error Model

When the distance and angular measurements are assumed error free and errors exist in the anchor location estimates, the equivalent observation equations can be written as

$$\hat{d}_i = \sqrt{(x-\hat{x}_i)^2 + (y-\hat{y}_i)^2 + (z-\hat{z}_i)^2} = d_i + n_{\hat{d}_i} \qquad i = 1, 2, \ldots, N$$

$$\hat{\phi}_i = \tan^{-1}\frac{y-\hat{y}_i}{x-\hat{x}_i} = \phi_i + n_{\hat{\phi}_i} \tag{8.17}$$

$$\hat{\alpha}_i = \cos^{-1}\frac{z-\hat{z}_i}{d_i} = \alpha_i + n_{\hat{\alpha}_i}$$

Denoting the coordinate errors of the ith anchor as

$$\begin{aligned}
\delta x_i &= x_i - \hat{x}_i \\
\delta y_i &= y_i - \hat{y}_i \\
\delta z_i &= z_i - \hat{z}_i
\end{aligned} \tag{8.18}$$

which are assumed to be Gaussian random variables of mean zero and variances $\sigma_{\delta x_i}^2$, $\sigma_{\delta y_i}^2$ and $\sigma_{\delta z_i}^2$ respectively. Using the approximations

$$\begin{aligned}
\tan(\phi_i + n_{\hat{\phi}_i}) &\approx \frac{\sin \phi_i + n_{\hat{\phi}_i} \cos \phi_i}{\cos \phi_i - n_{\hat{\phi}_i} \sin \phi_i} \\
\cos(\alpha_i + n_{\hat{\alpha}_i}) &\approx \cos \alpha_i - n_{\hat{\alpha}_i} \sin \alpha_i
\end{aligned} \tag{8.19}$$

and ignoring the square error terms, one obtains

$$\begin{aligned}
n_{\hat{d}_i} &\approx \frac{(x - x_i)\delta x_i + (y - y_i)\delta y_i + (z - z_i)\delta z_i}{d_i} \\
n_{\hat{\phi}_i} &\approx \frac{\delta y_i \cos \phi_i - \delta x_i \sin \phi_i}{(x - x_i) \cos \phi_i + (y - y_i)\sin \phi_i} \\
n_{\hat{\alpha}_i} &\approx -\frac{\delta z_i}{d_i \sin \alpha_i}
\end{aligned} \tag{8.20}$$

The above variables can be grouped together to form a vector as follows:

$$\varepsilon_i = [n_{\hat{d}_i}, n_{\hat{\phi}_i}, n_{\hat{\alpha}_i}]^{\mathrm{T}} \tag{8.21}$$

Given the positions of all the anchors and the target ordinary node, the three error variables $n_{\hat{d}_i}$, $n_{\hat{\phi}_i}$ and $n_{\hat{\alpha}_i}$ are Gaussian random variables with zero mean and covariance matrix defined as

$$\sum{}_i = E[\varepsilon_i \varepsilon_i^{\mathrm{T}}] \tag{8.22}$$

whose elements are given by

$$\begin{aligned}
[\textstyle\sum_i]_{1,1} &= \frac{(x - x_i)^2 \sigma_{\delta x_i}^2 + (y - y_i)^2 \sigma_{\delta y_i}^2 + (z - z_i)^2 \sigma_{\delta z_i}^2}{d_i^2} \\
[\textstyle\sum_i]_{1,2} = [\textstyle\sum_i]_{2,1} &= \frac{-(x - x_i) \sin \phi_i \sigma_{\delta x_i}^2 + (y - y_i)\cos \phi_i \sigma_{\delta y_i}^2}{[(x - x_i)\cos \phi_i + (y - y_i)\sin \phi_i]d_i} \\
[\textstyle\sum_i]_{1,3} = [\textstyle\sum_i]_{3,1} &= -\frac{(z - z_i)\sigma_{\delta z_i}^2}{d_i^2 \sin \alpha_i} \\
[\textstyle\sum_i]_{2,2} &= \frac{\sin^2 \phi_i \, \sigma_{\delta x_i}^2 + \cos^2 \phi_i \, \sigma_{\delta y_i}^2}{[(x - x_i)\cos \phi_i + (y - y_i)\sin \phi_i]^2} \\
[\textstyle\sum_i]_{2,3} = [\textstyle\sum_i]_{3,2} &= 0 \\
[\textstyle\sum_i]_{3,3} &= \frac{\sigma_{\delta z_i}^2}{d_i^2 \sin^2 \alpha_i}
\end{aligned} \tag{8.23}$$

8.2.3.2 Cramer–Rao Lower Bound Derivation

Recall that the effect of the distance and angle errors on the CRLB is studied in Section 8.2.2. Here, let us study how the anchor location errors affect the CRLB. Let

$$\mathbf{r}_i = \begin{bmatrix} \hat{d}_i & \hat{\phi}_i & \hat{\alpha}_i \end{bmatrix}^{\mathrm{T}} \qquad i = 1, 2, \ldots, N \tag{8.24}$$

One has

$$\mathbf{r}_i = \mathbf{f}_i(\mathbf{p}) + \varepsilon_i \tag{8.25}$$

where

$$\mathbf{f}_i(\mathbf{p}) = \begin{bmatrix} f_{i,1}(\mathbf{p}) & f_{i,2}(\mathbf{p}) & f_{i,3}(\mathbf{p}) \end{bmatrix}^{\mathrm{T}} \tag{8.26}$$

whose elements are given by

$$f_{i,j}(\mathbf{p}) = \begin{cases} \sqrt{(x - x_i)^2 + (y - y_i)^2 + (z - z_i)^2} & j = 1 \\ \tan^{-1} \dfrac{y - y_i}{x - x_i} & j = 2 \\ \cos^{-1} \dfrac{z - z_i}{d_i} & j = 3 \end{cases} \tag{8.27}$$

and ε_i is the measurement error vector of mean zero and covariance matrix given by (8.22). The conditional PDF of the observation vector $\mathbf{r}_i$ at the ith anchor becomes

$$p(\mathbf{r}_i | \mathbf{p}) = \frac{1}{\sqrt{(2\pi)^3 |\Sigma_i|}} \exp\left[-\frac{1}{2} (\mathbf{r}_i - \mathbf{f}_i(\mathbf{p}))^{\mathrm{T}} \Sigma_i^{-1} (\mathbf{r}_i - \mathbf{f}_i(\mathbf{p})) \right] \tag{8.28}$$

where $|\Sigma_i|$ is the determinant of the covariance matrix Σ_i. Since the observations at one anchor are independent from those at other anchors, the conditional PDF of the observation vector,

$$\mathbf{r} = \begin{bmatrix} \mathbf{r}_1^{\mathrm{T}}, & \mathbf{r}_2^{\mathrm{T}} & \ldots, & \mathbf{r}_N^{\mathrm{T}} \end{bmatrix}^{\mathrm{T}} \in \mathbb{R}^{3N \times 1} \tag{8.29}$$

is given by

$$p(\mathbf{r} | \mathbf{p}) = \prod_{i=1}^{N} p(\mathbf{r}_i | \mathbf{p}) = \frac{1}{\sqrt{(2\pi)^{3N} \displaystyle\prod_{i=1}^{N} |\Sigma_i|}} \exp\left[-\frac{1}{2} \sum_{i=1}^{N} (\mathbf{r}_i - \mathbf{f}_i(\mathbf{p}))^{\mathrm{T}} \Sigma_i^{-1} (\mathbf{r}_i - \mathbf{f}_i(\mathbf{p})) \right]$$

$$\tag{8.30}$$

The entries of the corresponding FIM can be determined by

$$[\mathbf{F}(\mathbf{p})]_{1,1} = \sum_{i=1}^{N} \frac{\partial \mathbf{f}_i^T(\mathbf{p})}{\partial x} \Sigma_i^{-1} \frac{\partial \mathbf{f}_i(\mathbf{p})}{\partial x}$$

$$[\mathbf{F}(\mathbf{p})]_{1,2} = [\mathbf{F}(\mathbf{p})]_{2,1} = \sum_{i=1}^{N} \frac{\partial \mathbf{f}_i^T(\mathbf{p})}{\partial y} \Sigma_i^{-1} \frac{\partial \mathbf{f}_i(\mathbf{p})}{\partial x}$$

$$[\mathbf{F}(\mathbf{p})]_{2,2} = \sum_{i=1}^{N} \frac{\partial \mathbf{f}_i^T(\mathbf{p})}{\partial y} \Sigma_i^{-1} \frac{\partial \mathbf{f}_i(\mathbf{p})}{\partial y} \qquad (8.31)$$

$$[\mathbf{F}(\mathbf{p})]_{2,3} = [\mathbf{F}(\mathbf{p})]_{3,2} = \sum_{i=1}^{N} \frac{\partial \mathbf{f}_i^T(\mathbf{p})}{\partial z} \Sigma_i^{-1} \frac{\partial \mathbf{f}_i(\mathbf{p})}{\partial y}$$

$$[\mathbf{F}(\mathbf{p})]_{3,3} = \sum_{i=1}^{N} \frac{\partial \mathbf{f}_i^T(\mathbf{p})}{\partial z} \Sigma_i^{-1} \frac{\partial \mathbf{f}_i(\mathbf{p})}{\partial z}$$

where

$$\frac{\partial \mathbf{f}_i(\mathbf{p})}{\partial x} = \left[\frac{x - x_i}{d_i} \quad -\frac{y - y_i}{d_{2,i}^2} \quad \frac{(x - x_i)(z - z_i)}{d_i^2 d_{2,i}} \right]^T$$

$$\frac{\partial \mathbf{f}_i(\mathbf{p})}{\partial y} = \left[\frac{y - y_i}{d_i} \quad \frac{x - x_i}{d_{2,i}^2} \quad \frac{(y - y_i)(z - z_i)}{d_i^2 d_{2,i}} \right]^T \qquad (8.32)$$

$$\frac{\partial \mathbf{f}_i(\mathbf{p})}{\partial z} = \left[\frac{z - z_i}{d_i} \quad 0 \quad -\frac{d_{2,i}}{d_i^2} \right]^T$$

Substituting the above entries into (8.15) produces the CRLB for the respective three position coordinate estimates.

8.3 Derivation of Cramer–Rao Lower Bound in Non-Line-of-Sight Conditions

In this section we derive the CRLB in NLOS conditions in the presence of both distance and angular measurements assuming the anchor locations are error free. The derivation of CRLB in the presence of anchor location errors can be performed in the same way as in Section 8.2.3.

8.3.1 Signal Model

As shown in Figure 8.2, the radio propagation between each of the anchors and the target node is now in a NLOS condition. There is a single scatterer between each anchor and the target so that the signal is scattered only once before arriving at the receiver. For convenience, it is assumed that the signal received by or transmitted from the ith anchor is scattered by the ith scatterer. By letting

- $d_{S_i,a}$ be the distance between the ith scatterer and ith anchor
- $d_{S_i,t}$ be the distance from the ith scatterer to the target

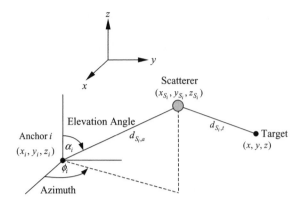

Figure 8.2 Illustration of the angle and distance in NLOS condition.

- d_i be the total distance from the target, through the ith scatterer, and to the ith anchor
- ϕ_i be the azimuth angle of the signal waveform impinging on the receiver at the ith anchor, which is transmitted from the target and then reflected by the ith scatterer
- α_i be the elevation angle of the signal waveform impinging on the receiver at the ith anchor, which is transmitted from the target and then reflected by the ith scatterer
- $(x_{S_i}, y_{S_i}, z_{S_i})$ be the coordinates of the ith scatterer

one can establish the relationships among the parameters as

$$
\begin{aligned}
d_i &= d_{S_i,a} + d_{S_i,t} \\
d_{S_i,a} &= \sqrt{(x_{S_i} - x_i)^2 + (y_{S_i} - y_i)^2 + (z_{S_i} - z_i)^2} \\
d_{S_i,t} &= \sqrt{(x_{S_i} - x)^2 + (y_{S_i} - y)^2 + (z_{S_i} - z)^2} \\
\phi_i &= \tan^{-1} \frac{y_{S_i} - y_i}{x_{S_i} - x_i} \\
\alpha_i &= \cos^{-1} \frac{z_{S_i} - z_i}{d_{S_i,a}}
\end{aligned}
\tag{8.33}
$$

The measurement equations have the same form as in (8.6), but the parameters have different definitions, as can be seen from (8.5) and (8.33).

8.3.2 Cramer–Rao Lower Bound Derivation

In Section 8.2 we studied how the parameter errors affect the CRLB in LOS conditions. In this subsection we study the CRLB in NLOS conditions. For clarity, we redefine the observation vector as

$$
\mathbf{r} = \begin{bmatrix} \hat{\mathbf{d}} & \hat{\boldsymbol{\varphi}} & \hat{\boldsymbol{\alpha}} \end{bmatrix}^{\mathrm{T}}
\tag{8.34}
$$

where

$$\hat{\mathbf{d}} = \begin{bmatrix} \hat{d}_1 & \hat{d}_2 & \cdots & \hat{d}_N \end{bmatrix}^{\mathrm{T}}$$
$$\hat{\boldsymbol{\varphi}} = \begin{bmatrix} \hat{\phi}_1 & \hat{\phi}_2 & \cdots & \hat{\phi}_N \end{bmatrix}^{\mathrm{T}} \qquad (8.35)$$
$$\hat{\boldsymbol{\alpha}} = \begin{bmatrix} \hat{\alpha}_1 & \hat{\alpha}_2 & \cdots & \hat{\alpha}_N \end{bmatrix}^{\mathrm{T}}$$

and the parameter vector as

$$\vartheta = \begin{bmatrix} x & y & z & \mathbf{x}_s^{\mathrm{T}} & \mathbf{y}_s^{\mathrm{T}} & \mathbf{z}_s^{\mathrm{T}} \end{bmatrix}^{\mathrm{T}} \qquad (8.36)$$

where

$$\mathbf{x}_s = \begin{bmatrix} x_{S_1} & x_{S_2} & \cdots & x_{S_N} \end{bmatrix}^{\mathrm{T}}$$
$$\mathbf{y}_s = \begin{bmatrix} y_{S_1} & y_{S_2} & \cdots & y_{S_N} \end{bmatrix}^{\mathrm{T}} \qquad (8.37)$$
$$\mathbf{z}_s = \begin{bmatrix} z_{S_1} & z_{S_2} & \cdots & z_{S_N} \end{bmatrix}^{\mathrm{T}}$$

Assume that the distance estimation errors and the angular estimation errors are all Gaussian distributed with zero mean and variances $\sigma_{\hat{d}_i}^2$, $\sigma_{\hat{\phi}_i}^2$ and $\sigma_{\hat{\alpha}_i}^2$ respectively. This assumption is purely for analytical simplicity. In reality, both the distance and angular measurements would be biased to a certain degree, depending on the environmental conditions, the measuring equipment and the signal detection and processing techniques. Then, given the positions of all the anchors and scatterers, we obtain the joint conditional probability density function of the $3N$ observation variables as

$$p(\mathbf{r}|\vartheta) = \frac{1}{\sqrt{(2\pi)^{3N}} \displaystyle\prod_{i=1}^{N} \sigma_{\hat{d}_i} \sigma_{\hat{\phi}_i} \sigma_{\hat{\alpha}_i}} \exp\left\{ -\frac{1}{2} \sum_{i=1}^{N} \left[\frac{1}{\sigma_{\hat{d}_i}^2}(\hat{d}_i - d_i)^2 + \frac{1}{\sigma_{\hat{\phi}_i}^2}(\hat{\phi}_i - \phi_i)^2 + \frac{1}{\sigma_{\hat{\alpha}_i}^2}(\hat{\alpha}_i - \alpha_i)^2 \right] \right\}$$

$$(8.38)$$

Taking natural logarithms on both sides of (8.38) and ignoring the irrelevant constants, we obtain the log-likelihood function as

$$\ln p(\mathbf{r}|\vartheta) = -\frac{1}{2} \sum_{i=1}^{N} \left[\frac{1}{\sigma_{\hat{d}_i}^2}(\hat{d}_i - d_i)^2 + \frac{1}{\sigma_{\hat{\phi}_i}^2}(\hat{\phi}_i - \phi_i)^2 + \frac{1}{\sigma_{\hat{\alpha}_i}^2}(\hat{\alpha}_i - \alpha_i)^2 \right] \qquad (8.39)$$

Based on the definition of the FIM, it can be shown that the FIM of the parameter vector ϑ is given by

$$\mathbf{F}(\vartheta) = \begin{bmatrix} \mathbf{F}_1 & \mathbf{F}_2 \\ \mathbf{F}_3 & \mathbf{F}_4 \end{bmatrix} \in \mathbb{R}^{3(N+1) \times 1} \qquad (8.40)$$

where $\{\mathbf{F}_i\}$ are defined in Annex 8.A. Making use of the lemma of matrix inversion in block form, we obtain

$$\mathbf{F}^{-1}(\mathbf{p}) = [\mathbf{F}_1 - \mathbf{F}_2 \mathbf{F}_4^{-1} \mathbf{F}_3]^{-1} \qquad (8.41)$$

As a result, the CRLB for each coordinate estimate is given by

$$
\begin{aligned}
\text{CRLB}(\hat{x}) &= [\mathbf{F}^{-1}(\mathbf{p})]_{1,1} \\
\text{CRLB}(\hat{y}) &= [\mathbf{F}^{-1}(\mathbf{p})]_{2,2} \\
\text{CRLB}(\hat{z}) &= [\mathbf{F}^{-1}(\mathbf{p})]_{3,3}
\end{aligned}
\tag{8.42}
$$

8.4 Approximate Variance of Linear Least-Squares Algorithm

In Chapter 6, a number of noniterative LS-based position estimation algorithms are described. Those simple noniterative algorithms are often employed for position estimation owing to their simplicity of implementation. This section is focused on the theoretical analysis of the accuracy of one of the noniterative LS position estimators in terms of the variance of the estimation errors. Defining

$$
\begin{aligned}
x_{i,1} &= x_i - x_1 \\
y_{i,1} &= y_i - y_1 \\
z_{i,1} &= z_i - z_1 \\
g_{i,1} &= 0.5\{x_i^2 + y_i^2 + z_i^2 - (x_1^2 + y_1^2 + z_1^2) + \hat{d}_1^2 - \hat{d}_i^2\}
\end{aligned}
\tag{8.43}
$$

the distance measurement equations become

$$
x_{i,1}x + y_{i,1}y + z_{i,1}z \approx g_{i,1} \qquad i = 2, 3, \ldots, N
\tag{8.44}
$$

which can be written in a compact form as

$$
\mathbf{Ap} \approx \mathbf{g}
\tag{8.45}
$$

where

$$
\mathbf{A} = \begin{bmatrix} x_{2,1} & y_{2,1} & z_{2,1} \\ \vdots & \vdots & \vdots \\ x_{N,1} & y_{N,1} & z_{N,1} \end{bmatrix} \qquad \mathbf{p} = \begin{bmatrix} x \\ y \\ z \end{bmatrix} \qquad \mathbf{g} = \begin{bmatrix} g_{2,1} \\ \vdots \\ g_{N,1} \end{bmatrix}
\tag{8.46}
$$

The linear LS estimator considered[2] produces the estimate of the position coordinates by minimizing the sum of the squares of the difference between the two sides of (8.44). That is:

$$
\varepsilon = \sum_{i=2}^{N} w_i (x_{i,1}x + y_{i,1}y + z_{i,1}z - g_{i,1})^2
\tag{8.47}
$$

where $\{w_i\}$ are the weights which are introduced to reflect the reliability of the measurements. For instance, anchors with a higher received SNR (or simply a higher received signal power) should have larger weights placed on their parameter measurements. In the absence of *a priori*

---

[2] In this case, the linear LS estimator is not optimal due to the fact that the linearization results in a loss of information.

confidence on the distance estimation, the weights may be simply chosen to be unity, which corresponds to the ordinary LS estimator. The weighted LS solution to (8.45) is given by

$$\hat{\mathbf{p}} = (\mathbf{A}^T \mathbf{W} \mathbf{A})^{-1} \mathbf{A}^T \mathbf{W} \mathbf{g} \tag{8.48}$$

where $\mathbf{A}$ is required to have a full rank to enable the matrix inversion and

$$\mathbf{W} = \text{diag}\{w_2, w_3, \ldots, w_N\} \tag{8.49}$$

is the diagonal weighting matrix.

In the event that an antenna array is implemented at anchors, the AOA (both azimuth and elevation) measurements can be obtained. When both the distance and the AOA measurements are available, linear observation equations can be obtained as

$$\begin{aligned}
x &\approx x_i + \hat{d}_i \sin \hat{\alpha}_i \cos \hat{\phi}_i \qquad i = 1, 2, \ldots, N \\
y &\approx y_i + \hat{d}_i \sin \hat{\alpha}_i \sin \hat{\phi}_i \\
z &\approx z_i + \hat{d}_i \cos \hat{\alpha}_i
\end{aligned} \tag{8.50}$$

where the approximations are obtained by dropping both the distance and angle estimation errors. As expected, a position estimate can be produced by using even one anchor or base station when both the distance and angle measurements are available. When there is more than one anchor, the location estimates are produced by applying the LS estimator

$$\hat{\mathbf{p}} = (\mathbf{B}^T \mathbf{V} \mathbf{B})^{-1} \mathbf{B}^T \mathbf{V} \mathbf{h} \tag{8.51}$$

where

$$\begin{aligned}
\mathbf{B} &= \text{diag}\{\mathbf{e}_N, \mathbf{e}_N, \mathbf{e}_N\} \in \mathbb{R}^{3N \times 3} \\
\mathbf{V} &= \text{diag}\{v_1, v_2, \ldots, v_{3N}\} \in \mathbb{R}^{3N \times 3N} \\
\mathbf{h} &= [x_1 + \hat{d}_1 \sin \hat{\alpha}_1 \cos \hat{\phi}_1, \ldots, x_N + \hat{d}_N \sin \hat{\alpha}_N \cos \hat{\phi}_N, y_1 + \hat{d}_1 \sin \hat{\alpha}_1 \sin \hat{\phi}_1, \ldots, y_N \\
&\quad + \hat{d}_N \sin \hat{\alpha}_N \sin \hat{\phi}_N, z_1 + \hat{d}_1 \cos \hat{\alpha}_1, \ldots, z_N + \hat{d}_N \cos \hat{\alpha}_N]^T \in \mathbb{R}^{3N \times 1}
\end{aligned} \tag{8.52}$$

in which $\mathbf{e}_N$ is a column vector of N ones. Note that $\mathbf{V}$ is a diagonal weighting matrix which can be determined based on the reliability of the distance and angle measurements at each anchor node.

8.4.1 Effect of Distance and Angle-of-Arrival Errors

The CRLB provides an accuracy benchmark that no unbiased estimators can surpass; however, it may not tell how much the variance of an estimator is. The variance of an estimator, similar to the RMSE, is one of the accuracy measures in evaluating the performance of an estimation algorithm. Note that when the estimation is unbiased, the STD of the error is equivalent to the RMSE. For this reason, the analytical expressions to approximate the variances of the LS estimator described above are derived first.

Let us deal with the distance-based LS algorithm first. Expanding the estimated position vector in (8.48) into Taylor series at the true target location and retaining the first two terms, one obtains the position error vector

$$\hat{\mathbf{p}} - \mathbf{p} \approx \sum_{i=1}^{N} \frac{\partial \mathbf{p}}{\partial d_i} (\hat{d}_i - d_i) \tag{8.53}$$

where

$$\frac{\partial \mathbf{p}}{\partial u} = (\mathbf{A}^{\mathrm{T}} \mathbf{W} \mathbf{A})^{-1} \mathbf{A}^{\mathrm{T}} \mathbf{W} \frac{\partial \mathbf{g}}{\partial u} \tag{8.54}$$

Here, the partial derivatives are given by

$$\begin{aligned}
\left[\frac{\partial \mathbf{g}}{\partial d_1} \right]_j &= d_1 \qquad j = 1, 2, \ldots, N \\
\left[\frac{\partial \mathbf{g}}{\partial d_i} \right]_j &= \begin{cases} -d_i & j = i - 1, i > 1 \\ 0 & j \neq i - 1, i > 1 \end{cases}
\end{aligned} \tag{8.55}$$

Then, from (8.53), the approximate variances of the estimated coordinates can be obtained as

$$\begin{aligned}
\mathrm{var}(\hat{x}) &\approx \sum_{i=1}^{N} \left(\left[\frac{\partial \mathbf{p}}{\partial d_i} \right]_1 \right)^2 \sigma_{\hat{d}_i}^2 \\
\mathrm{var}(\hat{y}) &\approx \sum_{i=1}^{N} \left(\left[\frac{\partial \mathbf{p}}{\partial d_i} \right]_2 \right)^2 \sigma_{\hat{d}_i}^2 \\
\mathrm{var}(\hat{z}) &\approx \sum_{i=1}^{N} \left(\left[\frac{\partial \mathbf{p}}{\partial d_i} \right]_3 \right)^2 \sigma_{\hat{d}_i}^2
\end{aligned} \tag{8.56}$$

Similarly, we can determine the approximate variances of the LS estimation that is based on both distance and angle measurements as

$$\begin{aligned}
\mathrm{var}(\hat{x}) &\approx \sum_{i=1}^{N} \left\{ \left[\left(\frac{\partial \mathbf{p}}{\partial d_i} \right)_1 \right]^2 \sigma_{\hat{d}_i}^2 + \left[\left(\frac{\partial \mathbf{p}}{\partial \alpha_i} \right)_1 \right]^2 \sigma_{\hat{\alpha}_i}^2 + \left[\left(\frac{\partial \mathbf{p}}{\partial \phi_i} \right)_1 \right]^2 \sigma_{\hat{\phi}_i}^2 \right\} \\
\mathrm{var}(\hat{y}) &\approx \sum_{i=1}^{N} \left\{ \left[\left(\frac{\partial \mathbf{p}}{\partial d_i} \right)_2 \right]^2 \sigma_{\hat{d}_i}^2 + \left[\left(\frac{\partial \mathbf{p}}{\partial \alpha_i} \right)_2 \right]^2 \sigma_{\hat{\alpha}_i}^2 + \left[\left(\frac{\partial \mathbf{p}}{\partial \phi_i} \right)_2 \right]^2 \sigma_{\hat{\phi}_i}^2 \right\} \\
\mathrm{var}(\hat{z}) &\approx \sum_{i=1}^{N} \left\{ \left[\left(\frac{\partial \mathbf{p}}{\partial d_i} \right)_3 \right]^2 \sigma_{\hat{d}_i}^2 + \left[\left(\frac{\partial \mathbf{p}}{\partial \alpha_i} \right)_3 \right]^2 \sigma_{\hat{\alpha}_i}^2 + \left[\left(\frac{\partial \mathbf{p}}{\partial \phi_i} \right)_3 \right]^2 \sigma_{\hat{\phi}_i}^2 \right\}
\end{aligned} \tag{8.57}$$

where the distance and AOA estimation errors are assumed mutually independent and the partial derivatives are given by

$$\frac{\partial \mathbf{p}}{\partial u} = (\mathbf{B}^{\mathrm{T}} \mathbf{V} \mathbf{B})^{-1} \mathbf{B}^{\mathrm{T}} \mathbf{V} \frac{\partial \mathbf{h}}{\partial u} \tag{8.58}$$

In this case the partial derivatives in (8.58) are given by

$$\left[\frac{\partial \mathbf{h}}{\partial d_i}\right]_j = \begin{cases} \sin \alpha_i \cos \phi_i & j = i \\ \sin \alpha_i \sin \phi_i & j = N+i \\ \cos \alpha_i & j = 2N+i \\ 0 & \text{elsewhere} \end{cases}$$

$$\left[\frac{\partial \mathbf{h}}{\partial \alpha_i}\right]_j = \begin{cases} d_i \cos \alpha_i \cos \phi_i & j = i \\ d_i \cos \alpha_i \sin \phi_i & j = N+i \\ -d_i \sin \alpha_i & j = 2N+i \\ 0 & \text{elsewhere} \end{cases} \tag{8.59}$$

$$\left[\frac{\partial \mathbf{h}}{\partial \phi_i}\right]_j = \begin{cases} -d_i \sin \alpha_i \sin \phi_i & j = i \\ d_i \sin \alpha_i \cos \phi_i & j = N+i \\ 0 & \text{elsewhere} \end{cases}$$

Note that the distance and AOA estimates are replaced by their true values in computing the partial derivatives. The results presented in this subsection are under the assumption that the anchor position information is perfectly known. The effect of the anchor position error is addressed in the following subsection.

8.4.2 Effect of Anchor Location Errors

When evaluating the accuracy of a positioning algorithm, it is usually assumed that anchor/base station position information is error free. In practice, this assumption may be invalid. For instance, the anchor position information may come from GPS measurements perhaps with an accuracy of several meters. Another example is in distributed networks where a group of nodes may be selected as anchors. The positions of the selected anchors may be first estimated and their position estimates are then used to locate other nodes. For high-accuracy positioning, small anchor-location errors might have a nontrivial impact on the accuracy of ordinary node positioning. In this subsection, the impact of anchor location errors when using the simple LS estimation method is analyzed theoretically.

Using the error model described in Section 8.2.3.1, the variance of the LS estimator with anchor location errors can be derived in the same way as in Section 8.4.1. When the estimation is based on the distance measurements, the approximate variances of the coordinate estimation errors have the same expressions as (8.56), but with

$$\sigma_{\hat{d}_i}^2 = \frac{1}{d_i^2}\left[(x-x_i)^2\sigma_{\delta x_i}^2 + (y-y_i)^2\sigma_{\delta y_i}^2 + (z-z_i)^2\sigma_{\delta z_i}^2\right] \tag{8.60}$$

When the estimation is based on the distance and angular measurements, the variances can be approximated as

$$\text{var}(\hat{x}) \approx \left(\sum_{i=1}^{N} w_i \right)^{-2} \sum_{i=1}^{N} w_i^2 \mathbf{q}_i^{\mathrm{T}} \Sigma_i \mathbf{q}_i$$

$$\text{var}(\hat{y}) \approx \left(\sum_{i=1}^{N} w_{N+i} \right)^{-2} \sum_{i=1}^{N} w_{N+i}^2 \mathbf{s}_i^{\mathrm{T}} \Sigma_i \mathbf{s}_i \qquad (8.61)$$

$$\text{var}(\hat{z}) \approx \left(\sum_{i=1}^{N} w_{2N+i} \right)^{-2} \sum_{i=1}^{N} w_{2N+i}^2 \mathbf{u}_i^{\mathrm{T}} \Sigma_i \mathbf{u}_i$$

where the entries of Σ_i are given by (8.23) and

$$\begin{aligned}
\mathbf{q}_i &= \left[\sin \alpha_i \cos \phi_i \quad -d_i \sin \alpha_i \sin \phi_i \quad d_i \cos \alpha_i \cos \phi_i \right]^{\mathrm{T}} \\
\mathbf{s}_i &= \left[\sin \alpha_i \sin \phi_i \quad d_i \sin \alpha_i \cos \phi_i \quad d_i \cos \alpha_i \sin \phi_i \right]^{\mathrm{T}} \qquad (8.62) \\
\mathbf{u}_i &= \left[\cos \alpha_i \quad 0 \quad -d_i \sin \alpha_i \right]^{\mathrm{T}}
\end{aligned}$$

Note that the approximate variance can also be derived in another way. In the case of distance-based LS estimation, (8.53) can be written in another form as

$$\hat{\mathbf{p}} - \mathbf{p} \approx \sum_{i=1}^{N} \left[\frac{\partial \mathbf{p}}{\partial x_i} (\hat{x}_i - x_i) + \frac{\partial \mathbf{p}}{\partial y_i} (\hat{y}_i - y_i) + \frac{\partial \mathbf{p}}{\partial z_i} (\hat{z}_i - z_i) \right] \qquad (8.63)$$

Therefore, the variances of the location estimates can be approximated as

$$\text{var}(\hat{x}) \approx \sum_{i=1}^{N} \left\{ \left[\left(\frac{\partial \mathbf{p}}{\partial x_i} \right)_1 \right]^2 \sigma_{\hat{x}_i}^2 + \left[\left(\frac{\partial \mathbf{p}}{\partial y_i} \right)_1 \right]^2 \sigma_{\hat{y}_i}^2 + \left[\left(\frac{\partial \mathbf{p}}{\partial z_i} \right)_1 \right]^2 \sigma_{\hat{z}_i}^2 \right\}$$

$$\text{var}(\hat{y}) \approx \sum_{i=1}^{N} \left\{ \left[\left(\frac{\partial \mathbf{p}}{\partial x_i} \right)_2 \right]^2 \sigma_{\hat{x}_i}^2 + \left[\left(\frac{\partial \mathbf{p}}{\partial y_i} \right)_2 \right]^2 \sigma_{\hat{y}_i}^2 + \left[\left(\frac{\partial \mathbf{p}}{\partial z_i} \right)_2 \right]^2 \sigma_{\hat{z}_i}^2 \right\} \qquad (8.64)$$

$$\text{var}(\hat{z}) \approx \sum_{i=1}^{N} \left\{ \left[\left(\frac{\partial \mathbf{p}}{\partial x_i} \right)_3 \right]^2 \sigma_{\hat{x}_i}^2 + \left[\left(\frac{\partial \mathbf{p}}{\partial y_i} \right)_3 \right]^2 \sigma_{\hat{y}_i}^2 + \left(\left[\frac{\partial \mathbf{p}}{\partial z_i} \right]_3 \right)^2 \sigma_{\hat{z}_i}^2 \right\}$$

Letting $\theta_i \in \{x_i, y_i, z_i\}$, one obtains

$$\begin{aligned}
\frac{\partial \mathbf{p}}{\partial \theta_i} &= (\mathbf{A}^{\mathrm{T}} \mathbf{W} \mathbf{A})^{-1} \left(\frac{\partial \mathbf{A}^{\mathrm{T}}}{\partial \theta_i} \mathbf{W} \mathbf{A} + \mathbf{A}^{\mathrm{T}} \mathbf{W} \frac{\partial \mathbf{A}}{\partial \theta_i} \right) (\mathbf{A}^{\mathrm{T}} \mathbf{W} \mathbf{A})^{-1} \mathbf{A}^{\mathrm{T}} \mathbf{W} \mathbf{g} + (\mathbf{A}^{\mathrm{T}} \mathbf{W} \mathbf{A})^{-1} \\
&\quad \times \left(\frac{\partial \mathbf{A}^{\mathrm{T}}}{\partial \theta_i} \mathbf{W} \mathbf{g} + \mathbf{A}^{\mathrm{T}} \mathbf{W} \frac{\partial \mathbf{g}}{\partial \theta_i} \right) \qquad (8.65) \\
&= (\mathbf{A}^{\mathrm{T}} \mathbf{W} \mathbf{A})^{-1} \left[\mathbf{A}^{\mathrm{T}} \mathbf{W} \left(\frac{\partial \mathbf{A}}{\partial \theta_i} \mathbf{p} + \frac{\partial \mathbf{g}}{\partial \theta_i} \right) + 2 \frac{\partial \mathbf{A}^{\mathrm{T}}}{\partial \theta_i} \mathbf{W} \mathbf{g} \right]
\end{aligned}$$

where we used the formula

$$\frac{\partial \mathbf{Q}^{-1}}{\partial \theta_i} = \mathbf{Q}^{-1} \frac{\partial \mathbf{Q}}{\partial \theta_i} \mathbf{Q}^{-1} \tag{8.66}$$

Note that (8.66) is valid for any invertible parametric matrix $\mathbf{Q}$. Also, the anchor error effect on the distance and AOA-based LS estimation can be analyzed in the same way. That is, the variances have the same expressions as given by (8.64) but with

$$\frac{\partial \mathbf{p}}{\partial x_i} = \beta \frac{\partial \mathbf{h}}{\partial x_i} \qquad \frac{\partial \mathbf{p}}{\partial y_i} = \beta \frac{\partial \mathbf{h}}{\partial y_i} \qquad \frac{\partial \mathbf{p}}{\partial z_i} = \beta \frac{\partial \mathbf{h}}{\partial z_i} \tag{8.67}$$

where

$$\beta = (\mathbf{B}^T \mathbf{V} \mathbf{B})^{-1} \mathbf{B}^T \mathbf{V}$$

$$\left[\frac{\partial \mathbf{h}}{\partial \theta_i} \right]_j = \begin{cases} 1 & j = i \\ 0 & \text{elsewhere} \end{cases} \qquad \theta_i \in \{x_i, y_i, z_i\} \tag{8.68}$$

In the design of a positioning system, it is necessary to have some knowledge about the anchor location accuracy and its effect on the location accuracy of ordinary nodes in advance. Only when both anchor location errors and the radio parameter estimation errors are considered can the sensor node localization performance be reliably predicted.

8.5 Accuracy Comparison

In the following, we use the theory presented in the preceding sections to examine a number of accuracy measures, including the CRLB, RMSE and STD. Simulation results on GDOP will be presented in Chapter 9. Two estimation algorithms are chosen: the LS algorithm and the iterative optimization method. A cubic region of dimension $100\,\text{m} \times 100\,\text{m} \times 100\,\text{m}$ is considered,[3] where both anchors and ordinary nodes are randomly deployed. For each simulation run, 1000 node configurations are examined and one single ordinary node is tested for each realization and the performance is then averaged. The theoretical performance considered is the CRLB that benchmarks all the algorithms and the STD of the LS algorithm. The RMSE of the two algorithms is computed based on the simulation results. The CRLB is calculated according to

$$\sqrt{\frac{1}{3L} \sum_{\ell=1}^{L} \sum_{k=1}^{3} [\mathbf{F}^{-1}(\mathbf{p})]_{k,k}^{(\ell)}} \tag{8.69}$$

where ℓ indexes the node configurations and $L = 1000$.

Figure 8.3 shows the simulated and approximated location accuracy versus the number of anchors. The STDs of the distance and AOA measurements are 2 m and 2° respectively. Note that the 2° error STD affects both the azimuth and the elevation angular measurements. The

---

[3] In practice, the actual dimensions of the monitored area should be used. This is just an example to show how the accuracy measures vary with the accuracy of the radio parameter measurements and the anchor location estimates.

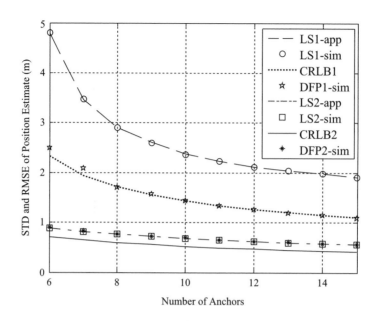

Figure 8.3   Approximated and simulated positioning accuracy versus the number of anchors with error-free anchor position information. The STD of the distance error is set at 2 m and the STD of both the azimuth and elevation angle errors is set at 2°.

centroid of the anchors is employed as the starting point for the DFP optimization algorithm. More details about the DFP algorithm can be found in Section 7.1.2.2. It is found that the DFP algorithm is not sensitive to the initial position guesses in this case. The DFP optimization solutions are found to be very similar when using either the centroid of the anchors or the LS solutions as the initial position estimates. In the legends

- 'LS1-app' denotes the analytical STD of the LS algorithm based on distance measurements
- 'LS1-sim' denotes the simulated RMSE of the distance-based LS estimation
- 'LS2-app' denotes the analytical STD of the LS estimation based on both distance and AOA
- 'LS2-sim' denotes the simulated RMSE of the LS estimation based on both distance and AOA
- 'DFP1-sim' denotes the simulated RMSE of the DFP algorithm by using the cost function defined by $\sum_{i=1}^{N} w_i (\hat{d}_i - d_i)^2$, where $\{w_i\}$ are simply set at unity
- 'DFP2-sim' denotes the simulated RMSE of the DFP algorithm by using the cost function defined by $\sum_{i=1}^{N} \{w_i (\hat{d}_i - d_i)^2 + w_{N+i} [x \tan \hat{\phi}_i - y - (x_i \tan \hat{\phi}_i - y_i)]^2 + w_{N+i} [z - (z_i + d_i \cos \hat{\alpha}_i)]^2\}$, where $\{w_i\}$ are also set at unity
- 'CRLB2' denotes the CRLB computed based on both distance and AOA
- 'CRLB1' denotes the CRLB computed based only on distance.

From Figure 8.3, a couple of observations can be made. When only the distance measurements are employed, the optimization algorithm outperforms the linear LS method considerably. This may be due to the fact that the squaring operations in the LS algorithm would enlarge the measurement error, as mentioned earlier. When both distance and angular measurements

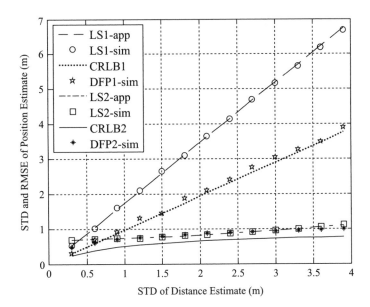

Figure 8.4 Approximated and simulated positioning accuracy versus the STD of the distance measurements with error-free anchor position information. There are seven anchors and the STD of both the azimuth and elevation angle measurement error is set at 2°.

are used, the LS algorithm and the optimization method have very similar accuracy and both approach the CRLB. This may be due to the fact that linear measurement equations can be directly produced without any squaring operation. Even with a rough starting point, the optimization-based method closely approaches the CRLB so that the CRLB can be used to predict the performance of the optimization-based approaches. The performance improves as the number of anchors increases, although the gain gradually gets smaller.

Figure 8.4 shows the accuracy versus the STD of the distance measurements when there are seven anchors and the STD of the AOA measurements is set at 2°. The accuracy of the distance-based LS algorithm degrades rather sharply (approximately linearly) as the distance error increases, whereas the curve of the DFP algorithm and the distance-and-angle-based LS algorithm remain quite flat. The location accuracy of the DFP algorithm and that of the distance-and-angle-based LS algorithm also approach the CRLB. The results may tell us that the optimization-based method is particularly suited for positioning in the presence of only distance measurements. With both distance and angle information, it is sufficient to use the LS algorithm, instead of the iterative optimization algorithm. In the absence of angular measurements, the linear LS-based approach is simple and has a low complexity, so it would be more suited for situations in which low complexity is an important issue and the resultant accuracy is acceptable. The accuracy of the DFP algorithm can be much higher, but it also has a much higher computational complexity, so the algorithm may be employed when the required computational capacity is available. Note that there are a range of other position determination algorithms, as presented in Chapters 6 and 7. The final choice of one or two positioning algorithms in practice will be dependent on the specific application, the environmental conditions and even the favorite of the system designer. Certainly, it is important that the positioning algorithms should be

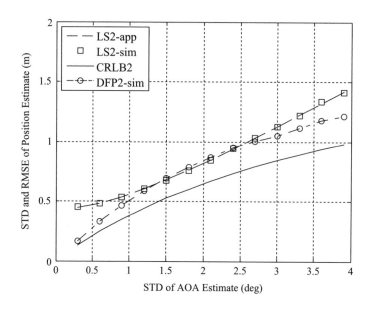

Figure 8.5 Approximated and simulated positioning accuracy versus the STD of the AOA measurements with error-free anchor position information. There are seven anchors and the STD of the distance errors is set at 2 m.

examined based on measurement data or directly tested in real-time operations. It is always advisable to avoid blindly or randomly selecting a positioning algorithm.

Figure 8.5 shows the location accuracy versus the STD of the AOA measurements when there are also seven anchors and the STD of the distance measurement error is 2 m. Both the STDs of the angular measurement errors are up to 4°, and the accuracy of both the LS and the DFP algorithms is better than 1.5 m. It is seen that both algorithms approach the CRLB.

Let us examine how anchor location errors affect sensor node location accuracy. Figure 8.6 shows the position accuracy versus the number of anchors. The errors of the x, y and z coordinates of each anchor are modeled as independent Gaussian random variables of zero mean and STDs of 1.5 m.[4] The results presented in this plot include the simulated RMSE of the distance-based LS algorithm and the two DFP algorithms. Also shown are the theoretical STD of the distance-based algorithm and the CRLBs. The simulated and analytical results of the distance-and-angle-based LS algorithm approach the CRLB, so they are not presented. Apparently, the impact of the anchor coordinate errors is particularly large on the distance-based LS algorithm. For the given parameters, the minimum position error is about 0.5 m. When sub-meter location accuracy is required, this anchor-caused error would pose a major challenge.

Figure 8.7 shows the location accuracy versus the STD of the anchor coordinate errors. The STDs of the three coordinate estimations of each anchor are equal, and there are seven anchors. It can be seen that the impact of the anchor position errors is similar to that of the distance errors. The two location errors resulting from the measurement errors and the anchor position errors

---

[4] When the anchors are set up manually beforehand and a local coordinate system is used, the anchor position error could be as small as centimeters. On the other hand, when a global system is employed and the anchor position information is provided by GPS, the error could be as large as 10 m.

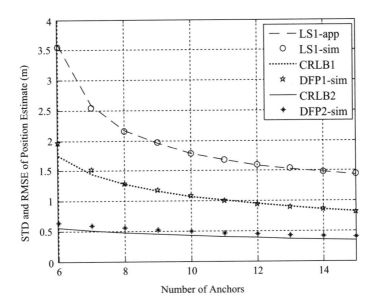

Figure 8.6   Approximated and simulated location accuracy versus the number of anchors with imperfect anchor position information. The STDs of the coordinate errors of the anchor locations are set at 1.5 m.

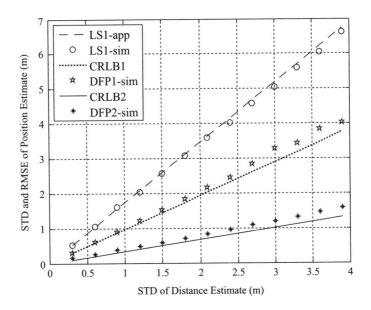

Figure 8.7   Approximated and simulated location accuracy versus the STD of the anchor coordinate estimates when there are seven anchors. The STDs of the three coordinate estimation errors are the same.

will be accumulated to form the final position estimation error. Therefore, it is necessary to make the anchor position errors as small as possible.

Annex 8.A: Components of the Fisher Information Matrix

In Section 8.3 we studied the CRLB in NLOS conditions. For clarity, the detailed formulas of the CRLB are given in this annex. The components of the FIM in (8.41) in Section 8.3 are defined as

$$
\begin{aligned}
\mathbf{F}_1 &= \begin{bmatrix} f_{xx} & f_{xy} & f_{xz} \\ f_{yx} & f_{yy} & f_{yz} \\ f_{zx} & f_{zy} & f_{zz} \end{bmatrix} \\
\mathbf{F}_2 = \mathbf{F}_3^T &= \begin{bmatrix} \mathbf{F}_{xx_s} & \mathbf{F}_{xy_s} & \mathbf{F}_{xz_s} \\ \mathbf{F}_{yx_s} & \mathbf{F}_{yy_s} & \mathbf{F}_{yz_s} \\ \mathbf{F}_{zx_s} & \mathbf{F}_{zy_s} & \mathbf{F}_{zz_s} \end{bmatrix} \in \mathbb{R}^{3 \times 3N} \\
\mathbf{F}_4 &= \begin{bmatrix} \mathbf{F}_{x_s x_s} & \mathbf{F}_{x_s y_s} & \mathbf{F}_{x_s z_s} \\ \mathbf{F}_{y_s x_s} & \mathbf{F}_{y_s y_s} & \mathbf{F}_{y_s z_s} \\ \mathbf{F}_{z_s x_s} & \mathbf{F}_{z_s y_s} & \mathbf{F}_{z_s z_s} \end{bmatrix} \in \mathbb{R}^{3N \times 3N}
\end{aligned}
\tag{8.70}
$$

The components of $\mathbf{F}_1$ are given as

$$
\begin{aligned}
f_{xx} &= \sum_{i=1}^{N} \left[\frac{(x_{S_i} - x)^2}{\sigma_{\hat{d}_i}^2 d_{S_i,t}^2} + \frac{(y_{S_i} - y)^2}{\sigma_{\hat{\phi}_i}^2 d_{2S_i,t}^4} + \frac{(x_{S_i} - x)^2 (z_{S_i} - z)^2}{\sigma_{\hat{\alpha}_i}^2 d_{S_i,t}^4 d_{2S_i,t}^2} \right] \\
f_{xy} &= f_{yx} = \sum_{i=1}^{N} (x_{S_i} - x)(y_{S_i} - y) \left[\frac{1}{\sigma_{\hat{d}_i}^2 d_{S_i,t}^2} - \frac{(y_{S_i} - y)^2}{\sigma_{\hat{\phi}_i}^2 d_{2S_i,t}^4} + \frac{(z_{S_i} - z)^2}{\sigma_{\hat{\alpha}_i}^2 d_{S_i,t}^4 d_{2S_i,t}^2} \right] \\
f_{xz} &= f_{zx} = \sum_{i=1}^{N} \frac{(x_{S_i} - x)(z_{S_i} - z)}{d_{S_i,t}^2} \left(\frac{1}{\sigma_{\hat{d}_i}^2} - \frac{1}{\sigma_{\hat{\alpha}_i}^2 d_{S_i,t}^2} \right) \\
f_{yy} &= \sum_{i=1}^{N} \left[\frac{(y_{S_i} - y)^2}{\sigma_{\hat{d}_i}^2 d_{S_i,t}^2} + \frac{(x_{S_i} - x)^2}{\sigma_{\hat{\phi}_i}^2 d_{2S_i,t}^4} + \frac{(y_{S_i} - y)^2 (z_{S_i} - z)^2}{\sigma_{\hat{\alpha}_i}^2 d_{S_i,t}^4 d_{2S_i,t}^2} \right] \\
f_{yz} &= f_{zy} = \sum_{i=1}^{N} \frac{(y_{S_i} - y)(z_{S_i} - z)}{d_{S_i,t}^2} \left(\frac{1}{\sigma_{\hat{d}_i}^2} - \frac{1}{\sigma_{\hat{\alpha}_i}^2 d_{S_i,t}^2} \right) \\
f_{zz} &= \sum_{i=1}^{N} \frac{1}{d_i^2} \left(\frac{(z_{S_i} - z)^2}{\sigma_{\hat{d}_i}^2} + \frac{d_{2,i}^2}{\sigma_{\hat{\alpha}_i}^2 d_{S_i,t}^2} \right) \\
f_{yx} &= f_{xy}, f_{yz} = f_{zy}, f_{zx} = f_{xz} \\
d_{S_i,t} &= \sqrt{(x_{S_i} - x)^2 + (y_{S_i} - y)^2 + (z_{S_i} - z)^2} \\
d_{2S_i,t} &= \sqrt{(x_{S_i} - x)^2 + (y_{S_i} - y)^2}
\end{aligned}
\tag{8.71}
$$

The components of $\mathbf{F}_2$ are given by

$$[\mathbf{F}_2]_{1,i} = \frac{x - x_{S_i}}{\sigma_{\hat{d}_i}^2 \, d_{S_i,t}} \left(\frac{x_{S_i} - x}{d_{S_i,t}} + \frac{x_{S_i} - x_i}{d_{S_i,a}} \right) \qquad i = 1, 2, \ldots, N$$

$$[\mathbf{F}_2]_{1,N+i} = \frac{x - x_{S_i}}{\sigma_{\hat{d}_i}^2 \, d_{S_i,t}} \left(\frac{y_{S_i} - y}{d_{S_i,t}} + \frac{y_{S_i} - y_i}{d_{S_i,a}} \right) \qquad [\mathbf{F}_2]_{1,2N+i} = \frac{x - x_{S_i}}{\sigma_{\hat{d}_i}^2 \, d_{S_i,t}} \left(\frac{z_{S_i} - z}{d_{S_i,t}} + \frac{z_{S_i} - z_i}{d_{S_i,a}} \right)$$

$$[\mathbf{F}_2]_{2,i} = \frac{y - y_{S_i}}{\sigma_{\hat{d}_i}^2 \, d_{S_i,t}} \left(\frac{x_{S_i} - x}{d_{S_i,t}} + \frac{x_{S_i} - x_i}{d_{S_i,a}} \right) \qquad [\mathbf{F}_2]_{2,N+i} = \frac{y - y_{S_i}}{\sigma_{\hat{d}_i}^2 \, d_{S_i,t}} \left(\frac{y_{S_i} - y}{d_{S_i,t}} + \frac{y_{S_i} - y_i}{d_{S_i,a}} \right)$$

$$[\mathbf{F}_2]_{2,2N+i} = \frac{y - y_{S_i}}{\sigma_{\hat{d}_i}^2 \, d_{S_i,t}} \left(\frac{z_{S_i} - z}{d_{S_i,t}} + \frac{z_{S_i} - z_i}{d_{S_i,a}} \right) \qquad [\mathbf{F}_2]_{3,i} = \frac{z - z_{S_i}}{\sigma_{\hat{d}_i}^2 \, d_{S_i,t}} \left(\frac{x_{S_i} - x}{d_{S_i,t}} + \frac{x_{S_i} - x_i}{d_{S_i,a}} \right)$$

$$[\mathbf{F}_2]_{3,N+i} = \frac{z - z_{S_i}}{\sigma_{\hat{d}_i}^2 \, d_{S_i,t}} \left(\frac{y_{S_i} - y}{d_{S_i,t}} + \frac{y_{S_i} - y_i}{d_{S_i,a}} \right) \qquad [\mathbf{F}_2]_{3,2N+i} = \frac{z - z_{S_i}}{\sigma_{\hat{d}_i}^2 \, d_{S_i,t}} \left(\frac{z_{S_i} - z}{d_{S_i,t}} + \frac{z_{S_i} - z_i}{d_{S_i,a}} \right).$$

$$(8.72)$$

Also, the components of $\mathbf{F}_4$ are given by

$$[\mathbf{F}_{x_s x_s}]_{i,i} = \frac{1}{\sigma_{\hat{d}_i}^2} \left(\frac{x_{S_i} - x}{d_{S_i,t}} + \frac{x_{S_i} - x_i}{d_{S_i,a}} \right)^2 + \frac{(y_{S_i} - y_i)^2}{\sigma_{\hat{\phi}_i}^2 \, d_{2S_i,a}^4} + \frac{[(x_{S_i} - x_i)(z_{S_i} - z_i)]^2}{\sigma_{\hat{\alpha}_i}^2 \, d_{2S_i,a}^2 d_{S_i,a}^4}$$

$$i = 1, \ldots, N$$

$$[\mathbf{F}_{y_s y_s}]_{i,i} = \frac{1}{\sigma_{\hat{d}_i}^2} \left(\frac{y_{S_i} - y}{d_{S_i,t}} + \frac{y_{S_i} - y_i}{d_{S_i,a}} \right)^2 + \frac{(x_{S_i} - x_i)^2}{\sigma_{\hat{\phi}_i}^2 \, d_{2S_i,a}^4} + \frac{[(y_{S_i} - y_i)(z_{S_i} - z_i)]^2}{\sigma_{\hat{\alpha}_i}^2 \, d_{2S_i,a}^2 d_{S_i,a}^4}$$

$$[\mathbf{F}_{z_s z_s}]_{i,i} = \frac{1}{\sigma_{\hat{d}_i}^2} \left(\frac{z_{S_i} - z}{d_{S_i,t}} + \frac{z_{S_i} - z_i}{d_{S_i,a}} \right)^2 + \frac{d_{2S_i,a}^2}{\sigma_{\hat{\alpha}_i}^2 \, d_{S_i,a}^4}$$

$$[\mathbf{F}_{x_s y_s}]_{i,i} = \frac{1}{\sigma_{\hat{d}_i}^2} \left(\frac{x_{S_i} - x}{d_{S_i,t}} + \frac{x_{S_i} - x_i}{d_{S_i,a}} \right) \left(\frac{y_{S_i} - y}{d_{S_i,t}} + \frac{y_{S_i} - y_i}{d_{S_i,a}} \right) - \frac{(x_{S_i} - x_i)(y_{S_i} - y_i)}{d_{2S_i,a}^2}$$

$$\times \left[\frac{1}{\sigma_{\hat{\phi}_i}^2 \, d_{2S_i,a}^2} - \frac{(z_{S_i} - z_i)^2}{\sigma_{\hat{\alpha}_i}^2 \, d_{S_i,a}^4} \right]$$

$$[\mathbf{F}_{x_s z_s}]_{i,i} = \frac{1}{\sigma_{\hat{d}_i}^2} \left(\frac{x_{S_i} - x}{d_{S_i,t}} + \frac{x_{S_i} - x_i}{d_{S_i,a}} \right) \left(\frac{z_{S_i} - z}{d_{S_i,t}} + \frac{z_{S_i} - z_i}{d_{S_i,a}} \right) - \frac{(x_{S_i} - x_i)(z_{S_i} - z_i)}{\sigma_{\hat{\alpha}_i}^2 \, d_{S_i,a}^4}$$

$$[\mathbf{F}_{y_s z_s}]_{i,i} = \frac{1}{\sigma_{\hat{d}_i}^2} \left(\frac{y_{S_i} - y}{d_{S_i,t}} + \frac{y_{S_i} - y_i}{d_{S_i,a}} \right) \left(\frac{z_{S_i} - z}{d_{S_i,t}} + \frac{z_{S_i} - z_i}{d_{S_i,a}} \right) - \frac{(y_{S_i} - y_i)(z_{S_i} - z_i)}{\sigma_{\hat{\alpha}_i}^2 \, d_{S_i,a}^4}$$

$$\mathbf{F}_{y_s x_s} = \mathbf{F}_{x_s y_s}^{\mathrm{T}} \qquad \mathbf{F}_{z_s x_s} = \mathbf{F}_{x_s z_s}^{\mathrm{T}} \qquad \mathbf{F}_{z_s y_s} = \mathbf{F}_{y_s z_s}^{\mathrm{T}}$$

Note that the nondiagonal elements of all the nine sub-matrices of $\mathbf{F}_4$ are equal to zero.

References

[1] N. Patwari, A.O. Hero III, M. Perkins, N.S. Correal and R.J. O'Dea, 'Relative location estimation in wireless sensor networks', *IEEE Transactions on Signal Processing*, **51**(8), 2003, 2137–2148.

[2] E.G. Larsson, 'Cramer–Rao bound analysis of distributed positioning in sensor networks', *IEEE Signal Processing Letters*, **11**(3), 2004, 334–337.

[3] H. Miao, K. Yu and M. Juntti, 'Positioning for NLOS propagation: algorithm derivations and Cramer–Rao bounds', *IEEE Transactions on Vehicular Technology*, **56**(5), 2007, 2568–2580.

[4] J.A. Pierce, A.A. McKenzie and R.H. Woodward, *Loran*, MIT Radiation Laboratory Series, volume 4, McGraw-Hill, New York, 1948.

[5] D.J. Torieri, 'Statistical theory of passive location systems', *IEEE Transactions on Aerospace and Electronic Systems*, **20**(2), 1984, 183–198.

[6] M.A. Spirito, 'On the accuracy of cellular mobile station location estimation', *IEEE Transactions on Vehicular Technology*, **50**(3), 2001, 674–685.

[7] H. Saarnisaari, 'TLS-ESPRIT in a time delay estimation', *Proceedings of 47th IEEE Vehicular Technology Conference (VTC)*, pp. 1619–1623, May 1997.

[8] A. Savvides, W.L. Garber, R.L. Moses and M.B. Srivastava, 'An analysis of error inducing parameters in multihop sensor node localization', *IEEE Transactions on Mobile Computing*, **4**(6), 2005, 567–577.

[9] L. Xiao, L.J. Greenstein and N.B. Mandayam, 'Sensor-assisted localization in cellular systems', *IEEE Transactions on Wireless Communications*, **6**(12), 2007, 4244–4248.

[10] A.Ward, A. Jones and A. Hopper, 'A new location technique for the active office,' *IEEE Personal Communications*, **4**(5), 1997, 42–47.

[11] S. Tekinay, E. Chao and R. Richton, 'Performance benchmarking for wireless location systems', *IEEE Communications Magazine*, **36**(4), 1998, 72–76.

9

Geometric Dilution of Precision Analysis

The accuracy of a position fix is dependent on many factors, the most important of which are the accuracy of the ranging measurements and geometric factors relating the relative positions of the base stations or anchor nodes and the mobile device whose position is to be determined. The accuracy of measurements has been addressed elsewhere (Chapters 2 and 5); this chapter analyzes the characteristics of the GDOP for a location system. GDOP defines how the positioning accuracy is reduced due to geometric factors, so that the positional accuracy can be much worse than the ranging measurement accuracy. These GDOP effects can be the dominant factor in the accuracy of a system if the location and number of base stations in the coverage area are not carefully planned. The detailed calculations described in the chapter allow the GDOP to be computed for an arbitrary arrangement of base stations and mobile devices. The analysis is essentially the classical theory of hyperbolic navigation positional accuracy, but with specific reference to the medium-range outdoor and indoor locating system. The class of applications of specific interest are related to sporting applications, both outdoors (horse racing, car racing such as NASCAR, football and track athletics) and indoors (basketball, velodrome and swimming). In these applications the base stations must be deployed around the sporting arena, which is typically rectangular or oval in shape.

9.1 Geometric Error Analysis

The geometry of the mobile unit and base stations is shown in Figure 9.1. The true position of the mobile unit is located at the origin of the x–y axes. The true range between the mobile unit and the base stations is given by $d_i = \sqrt{(x - x_i)^2 + (y - y_i)^2}, i = 1, 2, \ldots, N$. The base stations do not directly measure these ranges, but determine an estimate of the pseudo-range from the epoch of a pn-code in the receiver at each base station. As discussed in Chapter 5, the pseudo-range is defined as the sum of the propagation delay from the mobile unit to the base station and the unknown phase (time offset) of the pn-code in the mobile unit relative to the pn-code in each

Ground-Based Wireless Positioning Kegen Yu, Ian Sharp and Jay Guo
© 2009 John Wiley & Sons, Ltd

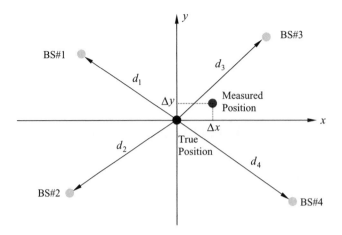

Figure 9.1   Geometry of the mobile unit in relationship to the base stations.

base station. It is assumed that all the base stations are time synchronized. Then, the pseudo-range is given by

$$r_i = d_i + c\phi_0 \tag{9.1}$$

where c is the speed of propagation and ϕ_0 is the phase of the pn-code in the mobile unit relative to the base station.

In practice, owing to receiver noise in each base station and errors associated with multipath and signal shadowing effects, there will be errors in the measurement of both the ranges and the mobile unit code phase, and the pseudo-range measurement errors will result in errors in the position estimation. If the position coordinate estimation errors Δx and Δy are small, and the mobile unit clock phase estimate is in error by $\Delta\phi_0$, then the corresponding error in pseudo-range to the ith base station can be approximated as

$$\Delta r_i \approx \frac{\partial r_i}{\partial x}\Delta x + \frac{\partial r_i}{\partial y}\Delta y + \frac{\partial r_i}{\partial \Delta\phi_0}\Delta\phi_0$$
$$= \alpha_i\Delta x + \beta_i\Delta y + c\Delta\phi_0 \qquad i = 1, 2, \ldots, N \tag{9.2}$$

where

$$\alpha_i = \frac{x - x_i}{d_i} = \cos\theta_i$$
$$\beta_i = \frac{y - y_i}{d_i} = \sin\theta_i \tag{9.3}$$

Equation (9.3) gives the error in the pseudo-range as a function of the positional and code phase errors; however, what is required is the positional error in terms of the pseudo-range errors. This can be achieved by performing an LS fit on the linear equations represented by (9.2), so that the difference between the measured pseudo-range errors and that given by (9.2) is minimized. For analytical convenience, (9.2) is written in a compact form as

$$\mathbf{A\delta} \approx \mathbf{h} \tag{9.4}$$

where

$$
\mathbf{A} = \begin{bmatrix} \alpha_1 & \beta_1 & c \\ \vdots & \vdots & \vdots \\ \alpha_N & \beta_N & c \end{bmatrix} \qquad \boldsymbol{\delta} = \begin{bmatrix} \Delta x \\ \Delta y \\ \Delta \phi_0 \end{bmatrix} \qquad \mathbf{h} = \begin{bmatrix} \Delta r_1 \\ \vdots \\ \Delta r_N \end{bmatrix} \tag{9.5}
$$

Note that (9.4) defines a set of linear equations; furthermore, if the number of base stations N is greater than three, then the LS solution for $\boldsymbol{\delta}$ can be obtained by solving

$$
\tilde{\mathbf{A}}\boldsymbol{\delta} = \tilde{\mathbf{h}} \tag{9.6}
$$

where

$$
\tilde{\mathbf{A}} = \mathbf{A}^T\mathbf{A} = \begin{bmatrix} \sum_{i=1}^{N}\alpha_i^2 & \sum_{i=1}^{N}\alpha_i\beta_i & c\sum_{i=1}^{N}\alpha_i \\ \sum_{i=1}^{N}\alpha_i\beta_i & \sum_{i=1}^{N}\beta_i^2 & c\sum_{i=1}^{N}\beta_i \\ c\sum_{i=1}^{N}\alpha_i & c\sum_{i=1}^{N}\beta_i & Nc^2 \end{bmatrix} \qquad \tilde{\mathbf{h}} = \mathbf{A}^T\mathbf{h} = \begin{bmatrix} \sum_{i=1}^{N}\alpha_i\Delta r_i \\ \sum_{i=1}^{N}\beta_i\Delta r_i \\ c\sum_{i=1}^{N}\Delta r_i \end{bmatrix} \tag{9.7}
$$

The solution to these linear equations can be conveniently expressed in terms of matrix determinants. That is:

$$
\Delta x = \frac{\det(\mathbf{B}_x)}{\det(\tilde{\mathbf{A}})}
$$
$$
\Delta y = \frac{\det(\mathbf{B}_y)}{\det(\tilde{\mathbf{A}})} \tag{9.8}
$$

where $\det(\mathbf{B})$ denotes the determinant of matrix $\mathbf{B}$ and

$$
\mathbf{B}_x = \begin{bmatrix} \sum_{i=1}^{N}\alpha_i\Delta r_i & \sum_{i=1}^{N}\alpha_i\beta_i & c\sum_{i=1}^{N}\alpha_i \\ \sum_{i=1}^{N}\beta_i\Delta r_i & \sum_{i=1}^{N}\beta_i^2 & c\sum_{i=1}^{N}\beta_i \\ c\sum_{i=1}^{N}\Delta r_i & c\sum_{i=1}^{N}\beta_i & Nc^2 \end{bmatrix}
$$

$$
\mathbf{B}_y = \begin{bmatrix} \sum_{i=1}^{N}\alpha_i^2 & \sum_{i=1}^{N}\alpha_i\Delta r_i & c\sum_{i=1}^{N}\alpha_i \\ \sum_{i=1}^{N}\alpha_i\beta_i & \sum_{i=1}^{N}\beta_i\Delta r_i & c\sum_{i=1}^{N}\beta_i \\ c\sum_{i=1}^{N}\alpha_i & c\sum_{i=1}^{N}\Delta r_i & Nc^2 \end{bmatrix} \tag{9.9}
$$

Note that in calculating the three determinants $\det(\tilde{\mathbf{A}})$, $\det(\mathbf{B}_x)$ and $\det(\mathbf{B}_y)$, the factor c in the last row and column can be taken outside the determinates, so the expressions for Δx and Δy become independent of c as follows:

$$\Delta x = \frac{\det(\mathbf{B}'_x)}{\det(\tilde{\mathbf{A}}')}$$

$$\Delta y = \frac{\det(\mathbf{B}'_y)}{\det(\tilde{\mathbf{A}}')}$$

(9.10)

where

$$\tilde{\mathbf{A}}' = \begin{bmatrix} \sum_{i=1}^{N}\alpha_i^2 & \sum_{i=1}^{N}\alpha_i\beta_i & \sum_{i=1}^{N}\alpha_i \\ \sum_{i=1}^{N}\alpha_i\beta_i & \sum_{i=1}^{N}\beta_i^2 & \sum_{i=1}^{N}\beta_i \\ \sum_{i=1}^{N}\alpha_i & \sum_{i=1}^{N}\beta_i & N \end{bmatrix} \quad \mathbf{B}'_x = \begin{bmatrix} \sum_{i=1}^{N}\alpha_i\Delta r_i & \sum_{i=1}^{N}\alpha_i\beta_i & \sum_{i=1}^{N}\alpha_i \\ \sum_{i=1}^{N}\beta_i\Delta r_i & \sum_{i=1}^{N}\beta_i^2 & \sum_{i=1}^{N}\beta_i \\ \sum_{i=1}^{N}\Delta r_i & \sum_{i=1}^{N}\beta_i & N \end{bmatrix}$$

$$\mathbf{B}'_y = \begin{bmatrix} \sum_{i=1}^{N}\alpha_i^2 & \sum_{i=1}^{N}\alpha_i\Delta r_i & \sum_{i=1}^{N}\alpha_i \\ \sum_{i=1}^{N}\alpha_i\beta_i & \sum_{i=1}^{N}\beta_i\Delta r_i & \sum_{i=1}^{N}\beta_i \\ \sum_{i=1}^{N}\alpha_i & \sum_{i=1}^{N}\Delta r_i & N \end{bmatrix}$$

(9.11)

It is seen that the positional error can be expressed solely in geometric terms and the accuracy of the pseudo-range measurements. As can be observed, the constant c does not contribute to the calculation of the positional accuracy; thus, it is convenient to set it to unity in subsequent calculations.

9.2 Statistical Error Analysis

Section 9.1 provides a deterministic estimate of the positional errors as a function of geometry and pseudo-range measurement errors. However, the pseudo-range errors $\{\Delta r_i\}$ are random variables, so the positional errors are also random variables. The following statistical properties are assumed for the pseudo-range errors:

$$E[\Delta r_i] = 0 \quad i = 1, 2, \ldots, N$$
$$E[\Delta r_i \Delta r_j] = 0$$
$$E[\Delta r_i^2] = \sigma_r^2$$

(9.12)

The first equation states that the mean of the pseudo-range errors is zero. This assumption is valid if there are LOS propagation conditions with random errors due to multipath and receiver noise. The first equation is also valid for NLOS conditions with constant bias errors, because the constant bias in the measurements can be incorporated into the clock phase term, which is determined as part of the processing. If the bias errors in the pseudo-ranges are not constant, then the data can be organized such that the mean of the bias errors is considered as the constant bias and the remaining component of the bias is considered another random variable with zero mean; this random variable will be uncorrelated with the random error component (described in the LOS case), and these two components can be combined into a new random variable with a zero mean. Thus, in all cases, with the proper interpretation, the assumption of a zero mean random error in the pseudo-range is valid. The second equation in (9.12) states that the ranging errors are statistically independent. Since the errors associated with separate paths from the mobile device to the base stations are physically well separated in the coverage area, it is reasonable to expect these errors to be independent. Finally, the third equation states that the variances of the pseudo-range errors for each base station are all equal. The basis of this hypothesis is that the propagation environment is quasi-homogeneous, so the statistical variance in the measurements is essentially independent of position. This conclusion is only partially true; the actual error distributions will be somewhat different from this simplified model. However, this assumption is required for the simple geometric interpretation of GDOP developed below.

Let us now consider the statistical properties of the positional error Δx. From (9.6), it can be seen that

$$\Delta x = [\tilde{\mathbf{A}}^{-1}]_{1,1} \sum_{i=1}^{N} \alpha_i \Delta r_i + [\tilde{\mathbf{A}}^{-1}]_{1,2} \sum_{i=1}^{N} \beta_i \Delta r_i + [\tilde{\mathbf{A}}^{-1}]_{1,3} \sum_{i=1}^{N} \Delta r_i$$

$$= \sum_{i=1}^{N} \rho_i \Delta r_i \tag{9.13}$$

where

$$\rho_i = [\tilde{\mathbf{A}}^{-1}]_{1,1} \alpha_i + [\tilde{\mathbf{A}}^{-1}]_{1,2} \beta_i + [\tilde{\mathbf{A}}^{-1}]_{1,3} \tag{9.14}$$

The mean and variance of Δx are thus given by

$$E[\Delta x] = E\left[\sum_{i=1}^{N} \rho_i \Delta r_i\right] = \sum_{i=1}^{N} E[\rho_i \Delta r_i] = \sum_{i=1}^{N} \rho_i E[\Delta r_i] = 0$$

$$\sigma_x^2 = E[\Delta x^2] = E\left[\left(\sum_{i=1}^{N} \rho_i \Delta r_i\right)^2\right] = E\left[\sum_{i=1}^{N} (\rho_i \Delta r_i)^2\right] = \sigma_r^2 \sum_{i=1}^{N} \rho_i^2 \tag{9.15}$$

It is observed that the mean of the x positional error is zero, and the variance is related to the pseudo-ranging variance by a constant which is only a function of the geometry. In deriving the expression for the variance, use has been made of the statistical independence of the pseudo-ranges measured by each base station and that the mean of the positional errors is zero. Similar expressions apply for the statistics of the y-coordinate positional error as

$$E[\Delta y] = 0$$

$$\sigma_y^2 = E[\Delta y^2] = \sigma_r^2 \sum_{i=1}^{N} \kappa_i^2 \tag{9.16}$$

where

$$\kappa_i = [\tilde{\mathbf{A}}^{-1}]_{2,1}\,\alpha_i + [\tilde{\mathbf{A}}^{-1}]_{2,2}\,\beta_i + [\tilde{\mathbf{A}}^{-1}]_{2,3} \tag{9.17}$$

The expression for the statistical covariance can also be determined as

$$\sigma_{xy} = E[\Delta x \Delta y] = \sigma_{\mathrm{r}}^2 \sum_{i=1}^{N} \rho_i \kappa_i \tag{9.18}$$

Thus, in summary, we have the following relationships:

$$\frac{\sigma_x^2}{\sigma_{\mathrm{r}}^2} = \sum_{i=1}^{N} \rho_i^2$$

$$\frac{\sigma_y^2}{\sigma_{\mathrm{r}}^2} = \sum_{i=1}^{N} \kappa_i^2 \tag{9.19}$$

$$\frac{\sigma_{xy}}{\sigma_{\mathrm{r}}^2} = \sum_{i=1}^{N} \rho_i \kappa_i$$

The measurement errors will have both Gaussian and systematic (multipath, shadowing) components. However, as can be observed from (9.4), the positional error is the sum of random measurement errors; thus, the distribution will be approximately Gaussian due to the CLT. Also note that, in general, Δx and Δy will not be statistically independent, as the vectors $\begin{bmatrix} \rho_1 & \rho_2 & \cdots & \rho_N \end{bmatrix}^{\mathrm{T}}$ and $\begin{bmatrix} \kappa_1 & \kappa_2 & \cdots & \kappa_N \end{bmatrix}^{\mathrm{T}}$ are linearly independent, so their product is usually nonzero.

9.3 Calculation of Geometric Dilution of Precision

Because the errors in the pseudo-range translate into errors in Δx and Δy, the position estimates will have a 2D statistical distribution. However, in specifying the accuracy of the position estimation, it is convenient to define a single figure for the accuracy; GDOP is such a measure of accuracy.

Before considering the calculation of the GDOP, let us first consider the 2D error distribution in more detail. As noted above, Δx and Δy are not statistically independent in general, but, by use of suitable rotation of the axes, the rotated orthogonal errors $\Delta x'$ and $\Delta y'$ can be made statistically independent. Specifically, rotating the axes by an angle φ produces

$$x' = x \cos \varphi + y \sin \varphi$$
$$y' = y \cos \varphi - x \sin \varphi \tag{9.20}$$

Accordingly, we have

$$\Delta x' = \Delta x \cos \varphi + \Delta y \sin \varphi$$
$$\Delta y' = \Delta y \cos \varphi - \Delta x \sin \varphi \tag{9.21}$$

Then, the covariance of $\Delta x'$ and $\Delta y'$ becomes

$$
\begin{aligned}
\sigma'^2_{xy} &= E[\Delta x' \Delta y'] - E[\Delta x']E[\Delta y'] \\
&= E[\Delta x' \Delta y'] \\
&= E[(\Delta x \cos \varphi + \Delta y \sin \varphi)(\Delta y \cos \varphi - \Delta x \sin \varphi)] \\
&= E[\Delta x \Delta y (\cos^2 \varphi - \sin^2 \varphi) + (\Delta y^2 - \Delta x^2)\sin \varphi \cos \varphi] \\
&= \sigma_{xy} \cos 2\varphi + \left(\frac{\sigma_y^2 - \sigma_x^2}{2}\right) \sin 2\varphi
\end{aligned}
\tag{9.22}
$$

By setting the rotation angle at

$$
\varphi_r = \frac{1}{2} \tan^{-1} \left(\frac{2\sigma_{xy}}{\sigma_x^2 - \sigma_y^2}\right)
\tag{9.23}
$$

the covariance of $\Delta x'$ and $\Delta y'$ becomes zero; that is, $\Delta x'$ and $\Delta y'$ are uncorrelated. Because Δx and Δy have an (approximate) Gaussian distribution with zero mean, the distribution of $\Delta x'$ and $\Delta y'$ will also be Gaussian with zero mean. As a result, $\Delta x'$ and $\Delta y'$ are independent. The variances in the rotated domain are given by

$$
\begin{aligned}
\sigma'^2_x &= \cos^2 \varphi_r \sigma_x^2 + \sin^2 \varphi_r \sigma_y^2 + \sin 2\varphi_r \sigma_{xy} \\
\sigma'^2_y &= \cos^2 \varphi_r \sigma_y^2 + \sin^2 \varphi_r \sigma_x^2 - \sin 2\varphi_r \sigma_{xy}
\end{aligned}
\tag{9.24}
$$

Thus the joint PDF of the position estimation errors relative to the rotated axes becomes

$$
p(\Delta x', \Delta y') = \frac{1}{2\pi \sigma'_x \sigma'_y} \exp\left\{-\frac{1}{2}\left[\left(\frac{\Delta x'}{\sigma'_x}\right)^2 + \left(\frac{\Delta y'}{\sigma'_y}\right)^2\right]\right\}
\tag{9.25}
$$

From (9.25) it can be observed that the contours of equal joint PDF are given by the ellipse as

$$
\xi(m) = \left(\frac{\Delta x'}{\sigma'_x}\right)^2 + \left(\frac{\Delta y'}{\sigma'_y}\right) = m^2
\tag{9.26}
$$

where m defines the size of the ellipse.

The GDOP is now defined by

$$
\begin{aligned}
GDOP &= \sqrt{\frac{\sigma'^2_x + \sigma'^2_y}{\sigma_r^2}} \\
&= \sqrt{\frac{\sigma_x^2 + \sigma_y^2}{\sigma_r^2}}
\end{aligned}
\tag{9.27}
$$

where the second equation can be derived from (9.24). Therefore, the GDOP can be defined as the STD of the error radius normalized by the STD in the measurements of the pseudo-range.

Note that the GDOP concept was originally developed to describe the performance of terrestrial navigation systems [1], but later it is used to evaluate the geometric effect of satellite configuration on GPS accuracy. When the angular positions of satellites are close together in the sky, the GDOP value is high, thus resulting in poor positioning accuracy. For good positional accuracy with GPS, the angular positions of the satellites should be such that the GPS receiver is 'surrounded' by the satellites. GDOP has been investigated by some researchers for GPS [2–4] and for terrestrial positioning systems [5–7].

9.4 Accuracy Probabilities

The accuracy of the position estimation can be specified in terms of the probability that the position estimate lies within a region **R**. This probability is computed by integrating the joint PDF (9.25) over the region **R**; that is:

$$P_f(\Delta x', \Delta y' \subset \mathbf{R}) = \iint_{\mathbf{R}} p(\Delta x', \Delta y') d\Delta x' \, d\Delta y' \tag{9.28}$$

The most obvious region to evaluate the integral over is the ellipse defined by (9.26). By changing to polar coordinates, the ellipse equations become

$$\begin{aligned} \Delta x' &= \sigma'_x \lambda \cos \phi \\ \Delta y' &= \sigma'_y \lambda \sin \phi \end{aligned} \tag{9.29}$$

where $0 \leq \lambda \leq m$. Then, (9.28) becomes

$$\begin{aligned} P_f(m) &= \sigma'_x \sigma'_y \int_0^m \int_0^{2\pi} p(\sigma'_x \lambda \cos \phi, \sigma'_y \lambda \sin \phi) d\phi \, d\lambda \\ &= \int_0^m \lambda \exp\left(-\frac{\lambda^2}{2}\right) d\lambda \\ &= 1 - \exp\left(-\frac{m^2}{2}\right) \end{aligned} \tag{9.30}$$

Clearly, the probability is independent of the ellipse shape but completely determined by the size parameter m.

Generally speaking, the shape (eccentricity) of the ellipse is unknown, and it is more convenient to express the probability that the position fix lies within the region **R** defined as a multiple n times the GDOP radius. In particular, the circular error probability CEP is defined as the probability that the position fix lies within the radius defined by GDOP. Thus, from (9.27), the CEP radius is given by

$$R_{CEP} = (\text{GDOP})\sigma_r \tag{9.31}$$

No closed-form expression is available for an arbitrary eccentricity of the ellipse, but upper and lower estimates can be computed for eccentricity $e = 0$ and $e = 1$. As the eccentricity is not usually known, the CEP upper and lower estimates are useful for a general specification of positional fixing accuracy.

For the case when $e = 0$ (circular error distribution, $\sigma'_x = \sigma'_y$), by changing to polar coordinates with $\Delta x' = \sqrt{2}\sigma'_x \rho \cos \theta$ and $\Delta y' = \sqrt{2}\sigma'_x \rho \sin \theta$, the probability that the position is within n CEP radii is given by the integral

$$P_f(n) = \sigma'^2_x \int_0^n \int_0^{2\pi} p(\sqrt{2}\sigma'_x \rho \cos \theta, \sqrt{2}\sigma'_x \rho \sin \theta) d\theta \, d\rho = \int_0^n 2\rho e^{-\rho^2} d\rho = 1 - e^{-n^2} \tag{9.32}$$

For the case when the eccentricity is equal to one, the ellipse is flattened out (say along the x-axis). As the STD along the y-axis approaches zero, the Gaussian distribution in the y-direction approaches a delta function $\delta(y)$. Thus, from (9.25), the PDF becomes

$$p(\Delta x', \Delta y') = \frac{1}{\sqrt{2\pi}\sigma_x'} \exp\left[-\frac{1}{2}\left(\frac{\Delta x'}{\sigma_x'}\right)^2\right] \delta(\Delta y') \qquad (9.33)$$

Thus, the PDF is zero except along the x-axis, so that (9.28) reduces to a one-dimensional (1D) integral as

$$P_f(m) = \frac{1}{\sqrt{2\pi}} \int_{-m}^{m} e^{-x'^2/2}\, dx' \qquad (9.34)$$

$$= 1 - 2Q(m)$$

where $Q(x)$ is the standard Q function, defined as

$$Q(x) = \frac{1}{\sqrt{2\pi}} \int_{x}^{\infty} e^{-t^2/2}\, dt \qquad (9.35)$$

The probability that the position estimation error is outside the specified region is plotted in Figure 9.2. A log scale for the y-axis is used so that the small values for large m can be observed.

A simulated example of the distribution of errors is shown in Figure 9.3. Also shown are the CEP circle and the rotation of the elliptical error region.

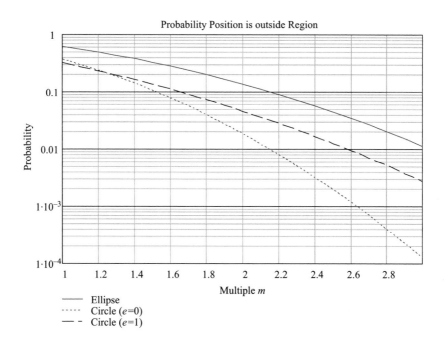

Figure 9.2 Probability that the position error is outside a region.

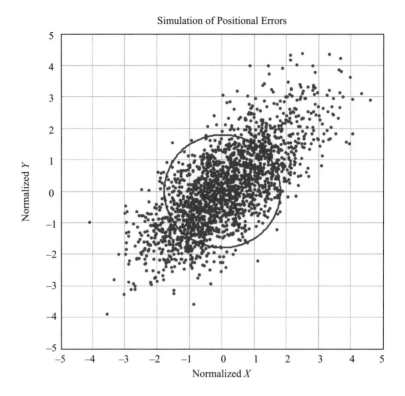

Figure 9.3 Simulation of position errors and CEP. The GDOP is 1.8 and the rotation angle is 45°. The scales are normalized to the measurement error STD.

9.5 Special Cases: Analytical Solutions to Geometric Dilution of Precision

The theory in the previous section can be applied to particular geometric cases of interest. Before considering particular configurations, such as sporting applications, it is informative to analyze simple geometric cases for which analytical solutions are possible. In particular, two cases will be analyzed; in both cases the base stations are organized into a circle with equal angular spacing. This arrangement of base stations is an approximation to the configuration applicable to sport arenas, where the arena is approximately circular or oval in shape. The solution can also apply approximately to a more cellular arrangement of base stations, particularly for indoor applications. Because of the limited range from the mobile device to the fixed base stations, the base stations in range tend to surround the mobile device in a quasi-circular fashion. Although the base stations in range will change as the mobile device moves through the coverage area, this quasi-circular arrangement will always apply. As in the cellular case, there may be additional base stations at longer ranges, but the simplification of assuming a circular pattern of base stations will usually overestimate the actual GDOP, so assuming the simple circular pattern will give conservative estimates of GDOP.

Because of the complexity of the matrix algebra that defines the GDOP solution, the analytical solutions will be limited to simple geometric cases, where the symmetry of the geometry results in considerable simplification in the analytical solutions. In particular, these solutions will be limited to the points at the center of the circle, points on the circle defined by the base stations and those on a few special radials where symmetry is present.

Two important assumptions are made in the following analyses: the mobile device signal can be received at all base stations and the statistics of the errors are the same for each base station. If the size (diameter) of arena is less than the range of the mobile device, then it is expected that all base stations will receive the signal. Further, it is anticipated that the ranging errors will be dominated by signal shadowing effects rather than the SNR, so it is reasonable to assume that the ranging accuracy is similar to all base stations.

9.5.1 Mobile Device at Center of Circle

The first case considered is with the mobile device at the center of the circle of base stations evenly spaced around the circle; in this location the GDOP is the minimum and, thus, the GDOP for all other locations must be greater. Further, the computed GDOP will be a function of the number of base stations N, so the effect of the number of base stations on position accuracy can be determined.

For a mobile device at the center of a circle, the error analysis is greatly simplified. Figure 9.4 shows an example of four base stations on a circle and the mobile device located at the center. For convenience, the origin of the coordinate system is set at the center of the circle and the BS#1 is on the positive x-axis.

Based on the GDOP parameters introduced previously, we have for this case

$$\alpha_i = \cos\left[(i-1)\frac{2\pi}{N}\right]$$

$$\beta_i = \sin\left[(i-1)\frac{2\pi}{N}\right]$$

(9.36)

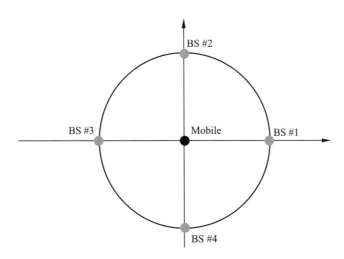

Figure 9.4   The mobile device is set at the center of the circle and four base stations are equally spaced around the circle. The high degree of symmetry in this case makes the analytical solution particularly simple.

By making use of the formulas (1.342.1), (1.351.1) and (1.351.2) in [8], we obtain

$$\sum_{i=1}^{N} \alpha_i \beta_i = 0$$

$$\sum_{i=1}^{N} {\alpha_i}^2 = \sum_{i=1}^{N} {\beta_i}^2 = \frac{N}{2}$$

(9.37)

Accordingly, the matrix in (9.7) and its inverse simply become

$$\tilde{\mathbf{A}} = \begin{bmatrix} N/2 & 0 & 0 \\ 0 & N/2 & 0 \\ 0 & 0 & N \end{bmatrix} \qquad \tilde{\mathbf{A}}^{-1} = \begin{bmatrix} 2/N & 0 & 0 \\ 0 & 2/N & 0 \\ 0 & 0 & 1/N \end{bmatrix}$$

(9.38)

Applying (9.19), the normalized variances of the position estimates are given by

$$\frac{\sigma_x^2}{\sigma_r^2} = \sum_{i=1}^{N} \rho_i^2 = \left(\frac{2}{N}\right)^2 \sum_{i=1}^{N} \alpha_i^2 = \frac{2}{N}$$

$$\frac{\sigma_y^2}{\sigma_r^2} = \sum_{i=1}^{N} \kappa_i^2 = \left(\frac{2}{N}\right)^2 \sum_{i=1}^{N} \beta_i^2 = \frac{2}{N}$$

(9.39)

$$\frac{\sigma_{xy}}{\sigma_r^2} = \sum_{i=1}^{N} \rho_i \kappa_i = \left(\frac{2}{N}\right)^2 \sum_{i=1}^{N} \alpha_i \beta_i = 0$$

Thus, from (9.27), the GDOP at the center of the circle is given by

$$\text{GDOP}_{\text{center}} = \sqrt{\frac{\sigma_x^2 + \sigma_y^2}{\sigma_r^2}}$$

$$= \frac{2}{\sqrt{N}}$$

(9.40)

It is seen that the GDOP decreases as the inverse square root of the number of base stations, so the estimated positional accuracy only improves slowly with the number of base stations. With the minimum practical number of base stations being four, the maximum GDOP is unity. If the number of base stations is increased to nine, then the GDOP only reduces marginally to 0.66. Note also that the covariance is zero and the x- and y-coordinate estimation variances are equal, so the error distribution will be circular.

9.5.2 Mobile Device on the Circle

The second case analyzed has the base stations equally spaced on a circle, but with the mobile device located anywhere on the circumference of the same circle. It is shown in the following that the GDOP is a constant for this particular geometry regardless of the exact position of the mobile device. The actual GDOP within the arena will thus be between the

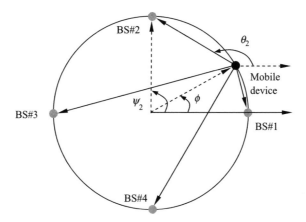

Figure 9.5   An example of a mobile unit and base stations on a circle.

minimum value at the center and the maximum value on the periphery of the arena (on circle).

Figure 9.5 shows an example of mobile device on a circle with four equally spaced base stations. However, the solution is applicable to more than four base stations. For convenience, BS#1 is placed on the positive x-axis. By simple geometric calculations, it can be shown that angular direction θ_i of the mobile device relative to the ith base station anywhere on the circle is related to the angular position of the base station ψ_i and the angular position of the mobile device ϕ by the simple relationship

$$\theta_i = \pm\frac{\pi}{2} + \frac{\psi_i + \phi}{2} \tag{9.41}$$

where the plus/minus sign depends on the particular position on the circle. Later calculations will show that the GDOP is actually independent of the plus/minus sign. Using (9.41), the GDOP parameters α and β are given by

$$\alpha_i = \cos\theta_i = \pm\sin\left(\frac{\psi_i + \phi}{2}\right)$$
$$\beta_i = \sin\theta_i = \pm\cos\left(\frac{\psi_i + \phi}{2}\right) \tag{9.42}$$

Using these definitions for the GDOP parameters, it can be shown that (9.37) still holds. In addition, let

$$a = \sum_{i=1}^{N}\alpha_i$$
$$b = \sum_{i=1}^{N}\beta_i \tag{9.43}$$

Using the formulas (1.342.1) and (1.342.2) in [8], we obtain

$$\left(\sum_{i=1}^{N}\alpha_i\right)^2 + \left(\sum_{i=1}^{N}\beta_i\right)^2 = a^2 + b^2 = 1 + S_N^2 \tag{9.44}$$

where

$$S_N = \sum_{i=1}^{N}\sin\left[\frac{\pi}{N}(i-1)\right]$$
$$= \cot\left(\frac{\pi}{2N}\right) \tag{9.45}$$

For a large N we have

$$S_N \approx \frac{2N}{\pi} \tag{9.46}$$

Thus, for a mobile device on the circle, the matrix $\tilde{\mathbf{A}}$ in (9.7) and its inverse $\tilde{\mathbf{A}}^{-1}$ are given by

$$\tilde{\mathbf{A}} = \begin{bmatrix} N/2 & 0 & a \\ 0 & N/2 & b \\ a & b & N \end{bmatrix} \qquad \tilde{\mathbf{A}}^{-1} = \frac{1}{\eta}\begin{bmatrix} 2N^2 - 4b^2 & 4ab & -2aN \\ 4ab & 2N^2 - 4a^2 & -2bN \\ -2aN & -2bN & N^2 \end{bmatrix} \tag{9.47}$$

where

$$\eta = 4\det(\tilde{\mathbf{A}}) = N[N^2 - 2(1 + S_N^2)]$$
$$\approx N^3\left(1 - \frac{8}{\pi^2}\right) \qquad \text{for large } N \tag{9.48}$$

Then, from (9.15) and (9.16), one obtains

$$\frac{\sigma_x^2}{\sigma_r^2} = \sum_{i=1}^{N}\rho_i^2$$
$$= \eta^{-1}\sum_{i=1}^{N}[(2N^2 - 4b^2)\alpha_i + 4ab\beta_i - 2aN]^2 \tag{9.49}$$
$$= \eta^{-1}\left\{2N[(N^2 - 2b^2)^2 + 4a^2b^2 + 2a^2N^2] - 8\sum_{i=1}^{N}[aN(N^2 - 2b^2)\alpha_i + 2a^2bN\beta_i]\right\}$$

and

$$\frac{\sigma_y^2}{\sigma_r^2} = \sum_{i=1}^{N}\kappa_i^2$$
$$= \eta^{-1}\sum_{i=1}^{N}[(2N^2 - 4a^2)\beta_i + 4ab\alpha_i - 2bN]^2 \tag{9.50}$$
$$= \eta^{-1}\left\{2N[(N^2 - 2a^2)^2 + 4a^2b^2 + 2b^2N^2] - 8\sum_{i=1}^{N}[bN(N^2 - 2a^2)\beta_i + 2ab^2N\alpha_i]\right\}$$

Combining the results in (9.49) and (9.50) produces

$$\frac{\sigma_x^2 + \sigma_y^2}{\sigma_r^2} = \frac{4N}{\eta} \left[N^4 - 3N^2(a^2 + b^2) + 2(a^2 + b^2)^2 \right] \tag{9.51}$$

From (9.27), we can obtain the GDOP as

$$\text{GDOP}(N) = \frac{2}{N^2 - 2C_N} \sqrt{\frac{N^4 - 3C_N N^2 + 2C_N^2}{N}} \tag{9.52}$$

where

$$C_N = 1 + S_N^2 = \text{cosec}^2\left(\frac{\pi}{2N}\right) \tag{9.53}$$

Note that the expression for GDOP on the circle is solely a function of the number of base stations and is independent of the location on the circle. As the number of base stations becomes large, the approximation for S_N (see (9.46)) can be used, resulting in

$$\text{GDOP}(N) \approx \frac{2\sqrt{\pi^4 - 12\pi^2 + 32}}{\pi^2 - 8} \frac{1}{\sqrt{N}}$$
$$= \frac{3.544}{\sqrt{N}} \tag{9.54}$$

The GDOP based on the exact expression in (9.52) and that based on the approximate expression in (9.54) are shown in Figure 9.6. Note that the asymptotic solution has a GDOP

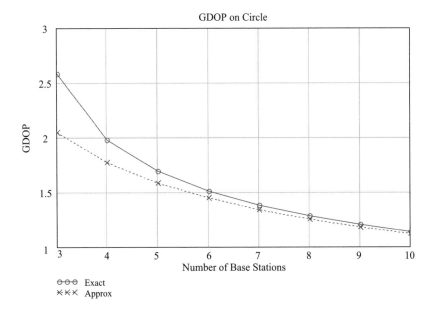

Figure 9.6 Exact and approximate expressions of GDOP when the mobile unit is on the circumference of a circle, with N base stations evenly spaced around the circle.

which is the inverse square root of the number of base stations, as was GDOP when the mobile device is at the center of the circle. Thus, it is probable that this inverse square-root behavior is approximately true throughout the GDOP map.

9.5.3 Mobile Device on Radials

The third analytical calculation of GDOP is along radials, both within the circle and beyond the circle. These formulae allow the determination of GDOP within the coverage area inside the circle as well as outside the coverage area outside the circle. As will be shown, outside the circular coverage area the GDOP increases rapidly; thus, this area is unsuitable for accurate position determination. This principle applies, in general, to shapes other than a circular arrangement of base stations; thus, for practical applications, the mobile must be 'surrounded' by base stations, which is intuitively correct.

As with the previous analytical calculations, the complexity of the calculations means that the analysis is restricted to symmetrical geometries. In particular, these symmetry conditions imply that, in terms of the GDOP conditions in (9.7), $\sum_i \beta_i = 0$ and $\sum_i \alpha_i \beta_i = 0$. Further, in the case of radials, the analytical calculations will be restricted to the special but practically important case of four base stations. The calculations are for the radial along the x-axis and at an angle midway between the base stations (45°). However, as the geometry has four-way symmetry, the calculations are applicable to eight radials spaced by 45°. Thus, the combined effect is a good radial coverage of GDOP. The analysis can be extended to more base stations, but the analytical expressions become increasingly complicated.

9.5.4 Radial along the x-Axis

The geometry for the calculation of GDOP along a radial defined by the x-axis is shown in Figure 9.7. In this case, the position can be specified by one angular parameter θ, from which

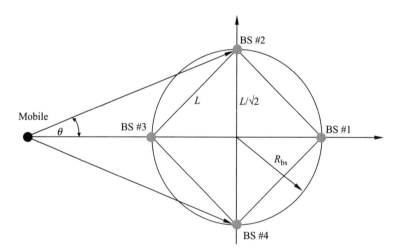

Figure 9.7   Geometry of the radial line passing through the location of a base station. The mobile point in this case is assumed to be outside the circle enclosing the base stations. The base station separation is L.

$c = \cos(\theta)$ and $s = \sin(\theta)$. Accordingly, the matrix in (9.7) and its inverse become

$$
\tilde{A} = \begin{bmatrix} 2(1+c^2) & 0 & 2(1+c) \\ 0 & 2s^2 & 0 \\ 2(1+c) & 0 & 4 \end{bmatrix} \qquad \tilde{A}^{-1} = \frac{1}{(c-1)^2} \begin{bmatrix} 1 & 0 & -\left(\dfrac{c+1}{2}\right) \\ 0 & \dfrac{(c-1)^2}{2(1-c^2)} & 0 \\ -\left(\dfrac{c+1}{2}\right) & 0 & \dfrac{c^2+1}{2} \end{bmatrix}
$$

$$(9.55)$$

Applying these matrices to (9.14) and (9.17) gives

$$
\rho_i = \frac{1}{(c-1)^2}\alpha_i - \frac{c+1}{2}
$$

$$
\kappa_i = \frac{1}{2(1-c^2)}\beta_i
$$

$$(9.56)$$

Then, from (9.15) and (9.16) we obtain

$$
\frac{\sigma_x^2}{\sigma_r^2} = \sum_{i=1}^{N} \rho_i^2 = \frac{1}{(c-1)^2}
$$

$$(9.57)$$

and

$$
\frac{\sigma_y^2}{\sigma_r^2} = \sum_{i=1}^{N} \kappa_i^2 = \frac{1}{2s^2}
$$

$$(9.58)$$

Combining the results in (9.57) and (9.58) produces

$$
\text{GDOP} = \frac{\sqrt{2(1+c)(3+c)}}{2s^2}
$$

$$(9.59)$$

On the outside edge of the circle ($\theta = \pi/4$), GDOP is calculated to be 3.558. A similar calculation when the point (mobile) is inside the circle gives

$$
\text{GDOP} = \sqrt{\frac{3+s^2}{2s^2(3-s^2)}}
$$

$$(9.60)$$

At the center ($\theta = \pi/2$), GDOP is unity, in agreement with (9.40). On the inside edge of the circle ($\theta = \pi/4$), GDOP is calculated to be $\sqrt{7/5} = 1.183$; thus, there is a sharp transition to the value 1.987 on the circle, as derived from (9.52). These discontinuities at the base station are discussed in more detail in Section 9.5.7.

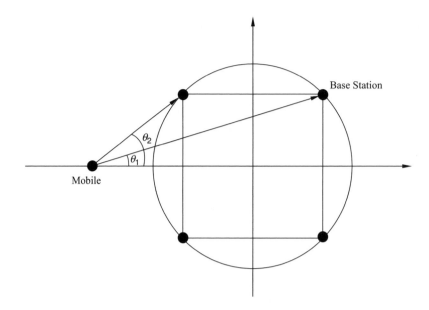

Figure 9.8 Geometry of the radial line defined along the x-axis rotated 45° relative to that defined in Figure 9.7.

9.5.5 The 45° Radial

The second radial on which GDOP is calculated is at an angle of 45° relative to the x-axis in Figure 9.7; this angle is thus midway between the radial passing through the base station on the x-axis and one on the y-axis. To preserve the symmetry, however, the following analysis uses a new reference axis system rotated by 45°, so that the square defined by the base stations is aligned with the new axes, as shown in Figure 9.8. In this case, the position of the base stations relative to the mobile device can be specified by two angular parameters θ_1 and θ_2, from which $c_1 = \cos(\theta_1), c_2 = \cos(\theta_2)$ and $s_1 = \sin(\theta_1), s_2 = \sin(\theta_2)$. Accordingly, the matrix in (9.7) and its inverse become

$$\tilde{A} = \begin{bmatrix} 2(c_1^2 + c_2^2) & 0 & 2(c_1 + c_2) \\ 0 & 2(s_1^2 + s_2^2) & 0 \\ 2(c_1 + c_2) & 0 & 4 \end{bmatrix}$$

$$\tilde{A}^{-1} = \frac{1}{(c_1 - c_2)^2} \begin{bmatrix} 1 & 0 & -\left(\dfrac{c_1 + c_2}{2}\right) \\ 0 & \dfrac{(c_1 - c_2)^2}{2(s_1^2 + s_2^2)} & 0 \\ -\left(\dfrac{c_1 + c_2}{2}\right) & 0 & \dfrac{c_1^2 + c_2^2}{2} \end{bmatrix} \qquad (9.61)$$

Applying these matrices to (9.14) and (9.17) gives

$$\rho_i = \frac{1}{(c_1 - c_2)^2}\left(\alpha_i - \frac{c_1 + c_2}{2}\right)$$

$$\kappa_i = \frac{1}{2(s_1^2 + s_2^2)}\beta_i$$

(9.62)

Then, from (9.15) and (9.16) we obtain

$$\frac{\sigma_x^2}{\sigma_r^2} = \sum_{i=1}^{N} \rho_i^2 = \frac{1}{2(c_1 - c_2)^2}$$

(9.63)

and

$$\frac{\sigma_y^2}{\sigma_r^2} = \sum_{i=1}^{N} \kappa_i^2 = \frac{1}{2(s_1^2 + s_2^2)}$$

(9.64)

Combining the results in (9.57) and (9.58) produces

$$\text{GDOP} = \sqrt{\frac{4 - (c_1 + c_2)^2}{2(s_1^2 + s_2^2)(c_1 - c_2)^2}}$$

(9.65)

9.5.6 Geometric Dilution of Precision Radial Characteristics

The GDOP characteristics defined by (9.59), (9.60) and (9.65) are plotted in Figure 9.9, where the radial distance is normalized by the base station radius parameter R_{bs}. In addition, the GDOP given by (9.52) at the point of intersection of the circle and the 45° radial is also plotted. As can be observed, the GDOP for these two radials is somewhat different, but the general characteristic is that GDOP increases with radial distance. However, the GDOP at the radial distance at a base station exhibits an interesting behavior, with three different values of GDOP. Just inside the circle the GDOP is 1.183, while just on the outside of the circle the GDOP has a step increase to 3.588. Furthermore, approaching the position of the base station along the circle results in a value of 1.987. Thus, GDOP has the curious feature of being triple valued near a base station location. The reason for this characteristic, and the consequences for positional accuracy near a base station, is discussed in Section 9.5.7.

At large radial distances r from the origin ($s \to 0, c \to 1$), (9.59) shows that on the 0° radial the GDOP approaches the function

$$\text{GDOP}_0 \approx \frac{2}{s^2} = 2\left(\frac{r}{R_{bs}}\right)^2$$

(9.66)

Thus, at large distances from the origin (outside the enclosing circle), GDOP increases as the square of the radial distance measured in units of the square size. Similarly, from (9.65) at large

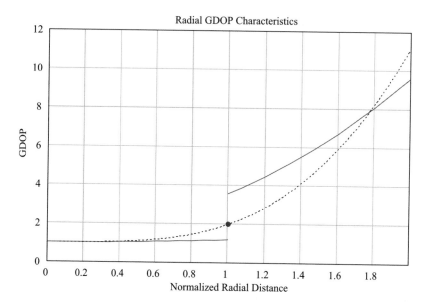

Figure 9.9 GDOP calculated along the radial at $0°$ (solid) and $45°$ (dotted). The dot is the GDOP on the circle as calculated by (9.52). The radial distance is normalized to the base station radius parameter R_{bs}. Thus, at a base station, the GDOP has discontinuities, but the GDOP radial function is continuous at other angles.

radial distances at $45°$ orientation GDOP approaches the function

$$\text{GDOP}_{45} \approx \frac{1}{c_1 - c_2} = \sqrt{2}\left(\frac{r}{R_{bs}}\right)^3 \qquad (9.67)$$

Thus, the GDOP on the two radials has different asymptotic functions: one (radial through the base station) being proportional to the square of the normalized radial distance and the other ($45°$ radial) being proportional to the cube of the radial distance. For example, at twice the base station radius parameter the respective exact/approximate GDOP values are (8/9.6) and (11.16/11.31). As these numbers are large, the effective coverage area typically will be limited to inside the circle.

9.5.7 Geometric Dilution of Precision near a Base Station

The above analysis shows that near a base station the GDOP function is discontinuous and multiple valued, depending on how the limit is approached. To help understand this concept without mathematical complexity, a simple 1D case with two base stations is considered. In this example, two base stations are separated by a distance D and the mobile position is either between the two base stations or on the extension of this line beyond the base stations. The GDOP is sought for these two cases.

In this first case, with the mobile between the base stations, the pseudo-range equations are

$$P_1 = d + c + \varepsilon_1$$
$$P_2 = D - d + c + \varepsilon_2 \tag{9.68}$$

where d is the distance to BS#1, c is the pseudo-range constant and ε_1 and ε_2 are the random measurement errors. The above equations can be solved for the distance d, resulting in the expression

$$d = \frac{1}{2}(P_1 - P_2 + D - \varepsilon_1 + \varepsilon_2) \tag{9.69}$$

Thus, an estimate of the position of the mobile device can be found. Assuming the measurement errors have zero mean with the same variance and are statistically independent, the GDOP for this case is $1/\sqrt{2}$.

Now consider the case where the mobile is on the extension of the line joining the base stations. In this case the pseudo-range equations are

$$P_1 = d + c + \varepsilon_1 = c' + \varepsilon_1$$
$$P_2 = D + d + c + \varepsilon_2 = D + c' + \varepsilon_2 \tag{9.70}$$

where a new pseudo-range constant c' is defined to incorporate the distance d. As these modified equations no longer contain any information about the position d, the conclusion is that GDOP $= \infty$. Thus, as the mobile position passes from just inside the base station line to just outside, GDOP has an infinite step change. This principle is also true if there are more base stations (as analyzed in previous sections), but the step size is moderated by the presence of other base stations. As small measurement errors can cause this transition, the accuracy of the position fix when the mobile is near a base station is severely degraded. However, if the mobile is, say, two STDs or further from the base station on the inside line, then the accuracy will be defined from (9.69), as even with the measurement errors the calculated position will remain on the inside.

In systems such as GPS or other wide-area systems, this position accuracy degradation is of little or no importance, as the mobile device cannot be close to a base station (or satellite for GPS). However, for short-range systems, particularly those associated with WSNs and indoor positioning systems, the mobile can often be quite close to a base station. The definition of 'near' is related to the accuracy of the system ranging measurement, but as a rough guide the mobile can be considered 'near' a base station when the radial distance to that base station is less than (say) twice the ranging accuracy. As the accuracy indoors can be of the order of, say, 2 m, a mobile within 4 m of the base station can expect to have reduced positional accuracy when compared with the linearized GDOP theory. Thus, in general, the proportion of positions in the coverage area which can be considered 'near' a base station is approximately given by

$$p_{\text{near}} = 2N_{\text{bs}}\left(\frac{\sigma_{\text{d}}}{R_{\text{bs}}}\right)^2 \tag{9.71}$$

where σ_{d} is the measurement STD of the pseudo-range. In the practical case where the base stations are not on a circle, the minimum size of circle enclosing the base stations can be used for the estimate.

9.6 Geometric Dilution of Precision Performance

In the following, the GDOP performance from simulations and geometry found in real tracking applications is investigated. The simulations are limited to the classical iterative solution (see Chapter 7) and the results compared with the theoretical estimates developed in the previous sections.

9.6.1 Comparison with Simulation Results

The theoretical analysis of GDOP is based on the linearization of the error equations, so the results are strictly valid only if the measurement errors are small relative to the geometry of the base stations; that is, the radius of the circle enclosing the base stations.[1] This condition certainly holds in cases such as GPS and other wide-area systems, but the measurement errors can be a significant fraction of the size of the coverage range in the case of short-range systems for indoor and WSN applications. In particular, from the theoretical analysis, the GDOP 'near' a base station is somewhat indeterminate. To investigate these effects, a simulation was performed, based on the classical iterative solution closely similar to the methods used for the theoretical GDOP calculations. The parameters for the simulation are chosen to be representative of indoor tracking applications, with the radius of the circle being 15 m and the number of base stations limited to four. This arrangement means that the maximum range is 30 m, typical of indoor NLOS applications. The measurement accuracy was set at 2 m,[2] which can be achieved by indoor systems with bandwidths of the order of 100 MHz.

To improve the performance of the simulation, the LS-fit positioning algorithm has checks on the 'quality' of the position fix. For example, the solution requires the inversion of a matrix (see (9.64)), which for 'bad' data becomes indeterminate or the iterative solution to (9.64) fails to converge. In both cases the position fix attempt is flagged as 'bad', although the simulations show that convergence failure is the most common reason. Further, as the GDOP increases rapidly if the radial distance is much greater than the base station circle radius, the algorithm places an upper limit (2 in this simulation) on radial distance to base station circle radius ratio and a position fix is flagged as 'bad' if the upper limit is exceeded. A combination of these techniques in a practical implementation means that the simulation results will differ from the theoretical estimates, as some of the position-fix attempts are rejected. As a consequence of flagging 'bad' data, the simulation probability (see Figure 9.11) of obtaining a position fix reduces considerably at radial distances beyond the circle radius, so that the mean GDOP of the simulation beyond the circle will be less due to the rejection of some of the 'bad' position fixes.

Comparison of an LS-fit iterative algorithm and the theoretical GDOP along a radial at $0°$ is shown in Figure 9.10. The LS-fit data are obtained from a Monte Carlo simulation, with only the 'good' position data included in calculating the mean performance. For normalized radial distances up to about 0.7 (when the distance to the base station is about two STDs of the measurement errors) the theoretical and LS-fit simulation agree well, as the assumptions that

---

[1] From equation (9.71), the proportion of the nominal coverage area where the theoretical GDOP does not apply in this example is about 14%.

[2] There is a tradeoff between the range and the accuracy. For example, UWB systems of at least 500 MHz bandwidth can achieve a measurement accuracy of the order of 30 cm, but the range is typically less than 10 m.

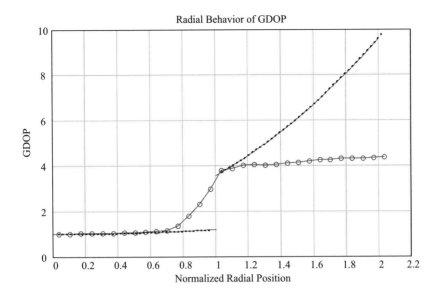

Figure 9.10 Comparison of LS-fit algorithm (circles) and the theoretical GDOP along a radial at $0°$. Also shown (dotted) is the simulation result with small statistical range errors. The radial distances are normalized by the radius of the base station circle.

the ranging errors are small relative to the range are approximately true. However, at greater radial distances there is a divergence between the results, with the simulation results considerably worse than the theoretical results. As the position approaches the location of a base station, there is an increasing probability that the computed position fix will be beyond the base station where the GDOP is considerably greater. Thus, the simulation GDOP is the weighted sum of the theoretical GDOP inside the circle and outside the circle. From the probability results in Figure 9.11, it can be observed that up to a normalized radial distance of 0.7 there is a 100% probability of obtaining a position fix. Beyond this radial distance the probability reduces to a minimum of 40% at the radial distance of the base station, and it then increases slightly at greater radial distances to about 50%. This 50% limit is a consequence of choosing a radial distance to circle radius ratio of 2:1, so that at normalized radial distances greater than 2 the position fix is flagged as invalid; thus, only about 50% of the position fixes are considered valid at the normalized radial distance of 2. For radial distances greater than the circle radius the GDOP in the simulation is approximately constant. This result is a consequence of the 'bad' position fixes being eliminated in the simulation, but not in the analytical analysis. Note, however, that as the STD in the ranging error approaches zero (implicitly assumed in the analytical derivation), the simulation results closely approach the analytical estimates, as would be expected. This is shown as the dotted line in Figure 9.10.

The results of this simulation show that the theoretical estimation of GDOP is reliable if the mobile position is well within an area surrounded by base stations, but provided the position is not too close to a base station. When the position is beyond the area enclosed by the base stations, other factors influence the positional accuracy (GDOP), so that the theoretical estimates typically overestimate the GDOP outside the coverage area, because practical position-determination algorithms try to eliminate poor position fixes.

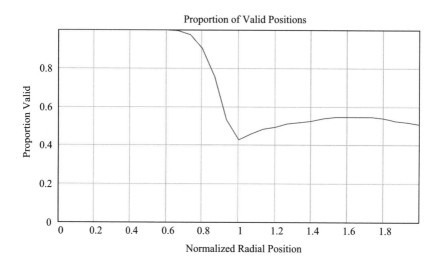

Figure 9.11   Probability of obtaining a 'good' position fix for the example in Figure 9.10.

9.6.2 Oval Track Example

Let us now look at an example of a GDOP map based on the location of base stations around a sporting arena. The particular example is the simulation of the geometry of the typical race track (Harold Park trotting/pacing track in Sydney, Australia). Arenas associated with outdoor sports are similar in size, while indoor sports are much smaller. However, GDOP is a function of the *shape* of the arena rather than its size, so the results can apply to all arenas of similar (oval) shape and base station deployment positions.

The location of the base stations is a compromise between accuracy and complexity. Increasing the number of base stations improves accuracy roughly as the square root of the number of base stations, but the improvement beyond about six base stations will be rather minor compared with the cost of base stations and the time required for temporary installation.

Another consideration is the location of the mobile antenna. If the base stations are on the inside of the track and the mobile antenna points sideways, then the antenna will always point at the base stations approximately. In contrast, if the base stations are on the outside of the track to minimize GDOP, then the antenna will point at the base station on one side of the track, but point away from the base station on the other side of the oval track. In this latter case, the radiation must diffract around the body/vehicles, resulting in positional errors. Therefore, there is a strong incentive to install the base stations on the inside of the track. However, practical limitations on where the base stations can be placed usually restrict the locations to the outside of the playing area of the arena.

Based on the above discussions, the following arrangement of base stations was made:

1. One base station will be located at the end of each straight of the track, or a total of four base stations.
2. One base station will be placed at the center of the track. This base station is optimally located to receive signals from all positions on the track.

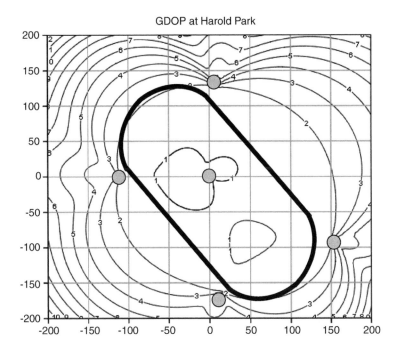

Figure 9.12 GDOP for Harold Park. The map scale is in meters. Note that the GDOP in the center is about 1 and is about 2 at the edge of the region enclosed by the base stations. Beyond this area the GDOP increases rapidly, and it is particularly poor immediately behind the base stations.

The proposed layout of base stations for Harold Park is shown in Figure 9.12. In this case the layout includes a base station at the center of the track. The GDOP for this layout of base stations is shown in Figure 9.12. As can be observed, the GDOP is near unity at the center of the arena, but is near to 2 on the track; this result is in agreement with the expectations for four base stations from the above theory.

References

[1] J.A. Pierce, A.A. McKenzie and R.H. Woodward. *Loran*, MIT Radiation Laboratory Series, volume 4, McGraw-Hill, New York, 1948.

[2] J. Zhu, 'Calculation of geometric dilution of precision', *IEEE Transactions on Aerospace and Electronic Systems*, **28**(3), 1992, 893–895.

[3] R. Yarlagadda, I. Ali and J. Hershey, 'GPS GDOP metric', *IEE Proceedings: Radar, Sonar and Navigation*, **147**(5), 2000, 259–264.

[4] M.S. Phatak, 'Recursive method for optimum GPS satellite selection', *IEEE Transactions on Aerospace and Electronic Systems*, **37**(2), 2001, 751–754.

[5] D.J. Torieri, 'Statistical theory of passive location systems', *IEEE Transactions on Aerospace and Electronic Systems*, **20**(2), 1984, 183–198.

[6] H.B. Lee, 'A novel procedure for assessing the accuracy of hyperbolic multilateration systems', *IEEE Transactions on Aerospace and Electronic Systems*, **11**(1), 1975, 2–15.

[7] N. Levanon, 'Lowest GDOP in 2-D scenarios', *IEE Proceedings: Radar, Sonar and Navigation*, **147**(3), 2000, 149–155.

[8] I.S. Gradshteyn and I.M. Ryzhik, *Table of Integrals, Series, and Products*, 7th edition, Academic Press, 2007.

10

Multipath Mitigation

One of the major error sources in wireless positioning is the multipath radio propagation. In the presence of LOS propagation, the multipath effect can be significantly reduced by signal processing techniques at the receiver. For instance, using suitable leading-edge detection techniques, the multipath effect of the TOA detection can be substantially reduced, as shown in Chapter 4. However, it poses a greater challenge to mitigate the NLOS impact through signal processing at the receiver. When TOA or range measurements are used, the NLOS propagation results in excessive traveling time/distance relative to the LOS condition. When AOA measurements are employed, the NLOS propagation produces an additional angular bias compared with the LOS propagation. When using RSS measurements, the NLOS propagation results in an extra power loss due to scattering and diffraction, and longer traveling distance. In a variety of scenarios, such as in indoor environments and urban areas, NLOS propagation is virtually inevitable. Substantial location performance degradation would be incurred if NLOS-corrupted measurements are directly employed for position determination without any NLOS error mitigation. To cope with the problem of NLOS propagation, various NLOS mitigation techniques and methods have been proposed. In this chapter, a range of such mitigation methods are presented.

The remainder of this chapter is organized as follows. Section 10.1 studies the use of position estimation residuals for weighting the combination-based intermediate position estimates to generate final position estimates. Section 10.2 presents a number of filtering-based methods to smooth out or exclude the NLOS biases. Section 10.3 makes use of constrained minimization techniques to reduce the NLOS effect. Section 10.4 gives a number of methods which exploit scatterer and angle information. Section 10.5 employs error statistics to perform ML estimation under a number of scenarios. Section 10.6 discusses how to use propagation models for NLOS mitigation. Section 10.7 describes three database-based pattern-matching position-determination methods. Finally, analytical performance is derived in Section 10.8.

10.1 Residual-Weighting-based Method

Consider scenarios where the positioning system is overdetermined. That is, the number of anchor nodes participating in positioning is greater than the required minimum number of anchor nodes. When the number of NLOS-corrupted measurements is small compared with the number of NLOS-free measurements, the measurements can be divided into a number of groups/combinations. For instance, in the case of five measurements, if each group has at least three measurements, then there will be 16 different combinations. Then, one may use each combination to produce a location estimate by using an algorithm described in Chapter 6. Assume that there are K location estimates from K combinations of measurements, denoted by $\hat{\mathbf{p}}_k, k = 1, 2, \ldots, K$.

The normalized residual of location estimate from the kth combination is defined as

$$s_k = \frac{1}{J^{(k)}} \sum_{j=1}^{J^{(k)}} (r_j^{(k)} - \hat{d}_j^{(k)})^2 \tag{10.1}$$

where $J^{(k)}$ is the number of anchor nodes of the kth combination, $r_j^{(k)}$ is the distance measurement and $\hat{d}_j^{(k)}$ is the distance between the jth anchor node of the kth combination and the location estimate $\hat{\mathbf{p}}_k$. In the case of five anchor nodes, if only one anchor node is in NLOS condition, then eight out of the 16 combinations does not have NLOS-corrupted measurements. Consequently, they produce more accurate position estimates and smaller residuals in general. Therefore, the residuals can be used to weight the intermediate position estimates. Specifically, the final target location estimate is determined as a weighted sum of all the combination-based location estimates [1]; that is:

$$\hat{\mathbf{p}} = \frac{\sum_{k=1}^{K} (s_k)^{-1} \hat{\mathbf{p}}_k}{\sum_{k=1}^{K} (s_k)^{-1}} \tag{10.2}$$

Clearly, the estimation result in (10.2) emphasizes the contributions of the more reliable combination-based location estimates which are evaluated based on their residuals defined in (10.1). The method does not require *a prior* knowledge of the measurement noise or NLOS error statistics, so that it can be readily implemented. It is efficient when the number of NLOS-corrupted measurements is relatively small. In the case that there are more NLOS-corrupted measurements than NLOS-free measurements, other NLOS-mitigation algorithms should be considered.

10.2 Filtering-based Method

In a variety of scenarios, such as transportation, car racing, skiing, or just walking and running, the target is in a motion state. Therefore, the TOA or range measurements at different time instants correspond to different mobile locations. Based on a sequence of TOA or distance measurements, NLOS identification and mitigation can be enhanced by using various filtering techniques.

10.2.1 General Smoothing

One of the basic ways to reduce the errors caused by multipath including NLOS propagation is to smooth the location estimates by making use of motion characteristics of the mobile node. For instance, the mobile node may move along a smooth path and at a certain constant speed during any short period of time. For outdoor vehicle tracking and indoor people tracking, a map can be exploited to further correct the location errors. The basic principle is that vehicle runs on a highway whose characteristics are known and a person can only walk along a corridor instead of going through a wall. Smoothing can be employed to mitigate the impact of the occasional abnormal estimation errors due to NLOS propagation and other factors. The general smoothing techniques include different filters, such as the LKF studied in Chapter 7, to which the interested reader is referred for more details.

10.2.2 Mean Excess Delay

In wideband systems, channel impulse measurements can be employed for reducing errors caused by NLOS propagation. For instance, from the correlator output of the receiver, one may be able to estimate the root-mean-squared (RMS) delay spread τ_{rms} of the received wideband signal. The mean excess delay τ_m is then determined by making use of the empirical relationship between τ_m and τ_{rms}. Once the mean excess delay is estimated, it is removed from the raw measurements so that the corrected data approximately has a zero mean [2].

10.2.3 Tracking Bias

Another basic method to mitigate the NLOS effect is to filter out the biases from the TOA or distance measurements which are corrupted by NLOS errors. Once the data has been filtered, it can be employed for position calculation. If we suppose that the target is in continuous linear movement at a certain speed, then one models the distance measurement at time instant k at an anchor node as

$$r_k = d_k + b_k + n_k \tag{10.3}$$

where d_k is the distance between the anchor node and the target, b_k is the NLOS bias modeled as a random variable and n_k is the measurement noise. Then, the discrete-time state equation may be modeled as

$$\begin{bmatrix} d_{k+1} \\ v_{k+1} \\ b_{k+1} \end{bmatrix} = \begin{bmatrix} 1 & \delta t & 0 \\ 0 & 1 & 0 \\ 0 & 0 & 1 \end{bmatrix} \begin{bmatrix} d_k \\ v_k \\ b_k \end{bmatrix} + \begin{bmatrix} 0.5\,\delta t^2 \\ \delta t \\ 0 \end{bmatrix} \varepsilon_k \tag{10.4}$$

where δt is the sampling time increment, v_k is the speed of the target and ε is the acceleration noise. From (10.3), the observation equation can be readily written as

$$r_k = \begin{bmatrix} 1 & 0 & 1 \end{bmatrix} \begin{bmatrix} d_k \\ v_k \\ b_k \end{bmatrix} + n_k \tag{10.5}$$

Accordingly, the discrete-time Kalman filter can be implemented to track the bias and produce the cleaned distance measurements. Alternatively, a continuous-time Kalman filter can be considered. In this case, the continuous-time state equation may be defined as

$$
\begin{bmatrix} \dfrac{\partial d}{\partial t} \\[2mm] \dfrac{\partial v}{\partial t} \\[2mm] \dfrac{\partial b}{\partial t} \end{bmatrix} = \begin{bmatrix} 0 & 1 & 0 \\ 0 & 0 & 0 \\ 0 & 0 & 0 \end{bmatrix} \begin{bmatrix} d \\ v \\ b \end{bmatrix} + \begin{bmatrix} 0 \\ 1 \\ 0 \end{bmatrix} \varepsilon \tag{10.6}
$$

The observation equation can be obtained in the same form as (10.5), but with the subscripts dropped. Tracking and removing the bias (or the mean of the bias) can also be performed in other ways. For instance, the process can be performed in two steps. The Kalman filter is first used to get an initial position estimate. Then, the NLOS bias is estimated from the residuals of the initial estimates. In the event of using both TOA and AOA measurements, the AOA measurements should not be corrupted by NLOS errors. Otherwise, large location estimation errors would be produced.

10.2.4 Extended Kalman Filtering

Recall that, in earlier chapters, we studied how to apply the EKF for position estimation and tracking in LOS conditions. By including the NLOS bias variables, the state vector of the EKF is enlarged, so that the dimensions of the state transition matrix and the observation vector are increased accordingly. Without loss of generality, suppose that the TOA-based distance measurements are available, the process equation and the measurement equation of the Kalman filter would then become

$$
\begin{aligned}
\mathbf{\Theta}^{(k)} &= \tilde{\mathbf{\Phi}}^{(k-1)} \mathbf{\Theta}^{(k-1)} + \tilde{\mathbf{B}}^{(k-1)} \boldsymbol{\varepsilon}^{(k-1)} \\
\mathbf{p}^{(k)} &= \tilde{\mathbf{g}}^{(k)}(\mathbf{\Theta}^{(k)}) + \mathbf{n}^{(k)}
\end{aligned} \tag{10.7}
$$

where, in the case of 2D positioning:

$$
\begin{aligned}
\tilde{\mathbf{g}}^{(k)}(\mathbf{\Theta}^{(k)}) &= \mathbf{g}^{(k)}(\boldsymbol{\theta}^{(k)}) + \mathbf{b}^{(k)} \\
\boldsymbol{\theta}^{(k)} &= \begin{bmatrix} x & y & v_x & v_y \end{bmatrix}^{\mathrm{T}} \\
\mathbf{b}^{(k)} &= \begin{bmatrix} b_1^{(k)} & b_2^{(k)} & \cdots & b_N^{(k)} \end{bmatrix}^{\mathrm{T}} \\
\mathbf{\Theta}^{(k)} &= \begin{bmatrix} x^{(k)} & y^{(k)} & v_x^{(k)} & v_y^{(k)} & \mathbf{b}^{(k)} \end{bmatrix}^{\mathrm{T}} \\
\tilde{\mathbf{\Phi}}^{(k)} &= \begin{bmatrix} \mathbf{\Phi}^{(k)} & \mathbf{0} \\ \mathbf{0} & \mathbf{I}_N \end{bmatrix} \\
\tilde{\mathbf{B}}^{(k)} &= \begin{bmatrix} \mathbf{B}^{(k)} & \mathbf{0} \\ \mathbf{0} & \mathbf{I}_N \end{bmatrix}
\end{aligned} \tag{10.8}
$$

I_N is the identity matrix of dimensions $N \times N$ and

$$\Phi^{(k)} = \begin{bmatrix} 1 & 0 & \delta t & 0 \\ 0 & 1 & 0 & \delta t \\ 0 & 0 & 1 & 0 \\ 0 & 0 & 0 & 1 \end{bmatrix} \qquad B^{(k)} = \begin{bmatrix} 0.5\,\delta t^2 & 0 \\ 0 & 0.5\,\delta t^2 \\ \delta t & 0 \\ 0 & \delta t \end{bmatrix} \qquad (10.9)$$

Note that δt is the time interval between samples and the covariance matrix of ε may be modeled as

$$\mathrm{cov}(\varepsilon^{(k)}) = \begin{bmatrix} \mathbf{Q}_a^{(k)} & \mathbf{0} \\ \mathbf{0} & \mathbf{Q}_b^{(k)} \end{bmatrix} \qquad (10.10)$$

where $\mathbf{Q}_a^{(k)}$ is the covariance matrix of the acceleration noise and $\mathbf{Q}_b^{(k)}$ is the covariance matrix of the noise related to the NLOS biases. After linearization, the observation equation can be expressed as

$$\mathbf{p}^{(k)} \approx \tilde{\mathbf{G}}^{(k)} \boldsymbol{\Theta}^{(k)} + \mathbf{n}^{(k)} \qquad (10.11)$$

where

$$\tilde{\mathbf{G}}^{(k)} = \begin{bmatrix} \mathbf{G}^{(k)} & \mathbf{0} \\ \mathbf{0} & \mathbf{I}_N \end{bmatrix}$$

$$\mathbf{G}^{(k)} = \left. \frac{\partial \mathbf{g}^{(k)}(\boldsymbol{\theta})}{\partial \boldsymbol{\theta}} \right|_{\boldsymbol{\theta} = \hat{\boldsymbol{\theta}}^{(k|k-1)}} \qquad (10.12)$$

Given the state equation and the observation equations, the EKF can be implemented according to the standard procedure.

As discussed in Chapter 7, there are other filtering-based positioning methods. including the UKF, the PF, and the UPF, which may be more suited for nonlinear systems than the EKF. Compared with the implementation of these filters in LOS conditions, implementing these filters in NLOS conditions requires modifying the state and observation equations by including the NLOS bias variables. Then, these filters can be implemented in the same way as for the LOS conditions.

10.3 Constrained Optimization

Another method to mitigate the NLOS effect is to exploit constrained optimization, which is particularly well suited for producing desirable parameter estimates when there is no or limited information about measurement errors. Optimization or minimization is performed over an objective/cost function, which is usually defined as the weighted sum of the squares of the difference between the measured distance and the true distance; that is:

$$\varepsilon(x, y) = \sum_{i=1}^{N} w_i (r_i - d_i)^2 \qquad (10.13)$$

where N is the number of anchor nodes, $\{w_i\}$ are the weights to emphasize the importance of more reliable measurements, d_i is the distance between the target mobile node and the ith anchor node and r_i is the corresponding distance measurement. In the case of lack of confidence in the reliability of the measurements, the weights may be simply set at unity. The location estimate is then determined by minimizing the cost function

$$\hat{\mathbf{p}} = \arg\min_{x,y} \varepsilon(x,y) \tag{10.14}$$

Constraints can be applied to the minimization to improve performance. One constraint comes from the fact that NLOS estimation errors are always positive when time and/or distance measurements are used. Another constraint is based on the knowledge of the dimensions of the monitored area and the established coordinate system. This constraint may be particularly useful for indoor positioning when the location area is relatively small. In the following, a number of different constrained optimization algorithms are studied.

10.3.1 Linear Quadratic Programming

In the presence of measurement noise and NLOS errors, the distance measurements can be modeled as

$$r_i = d_i(\mathbf{p}) + b_i + n_i \qquad i = 1, 2, \ldots, N \tag{10.15}$$

where $\mathbf{p}$ is the unknown position vector and

$$d_i(\mathbf{p}) = \sqrt{(x - x_i)^2 + (y - y_i)^2} \tag{10.16}$$

b_i is the extra distance (positive bias) relative to the straight-line distance, so that $b_i + d_i(\mathbf{p})$ is the actual distance length that the signal travels from the ith anchor to the mobile node or vice versa, and n_i is the measurement noise. In the presence of a straight-line path between the ith anchor and the mobile node, $b_i = 0$. The NLOS distance bias b_i is often modeled as a random variable that is exponentially distributed, whereas the measurement noise is generally modeled as a Gaussian random variable. Equation (10.15) can be written in a compact form as

$$\mathbf{r} = \mathbf{d}(\mathbf{p}) + \mathbf{b} + \mathbf{n} \tag{10.17}$$

where

$$
\begin{aligned}
\mathbf{r} &= \begin{bmatrix} r_1 & r_2 & \cdots & r_N \end{bmatrix}^{\mathrm{T}} \\
\mathbf{d}(\mathbf{p}) &= \begin{bmatrix} d_1(\mathbf{p}) & d_2(\mathbf{p}) & \cdots & d_N(\mathbf{p}) \end{bmatrix}^{\mathrm{T}} \\
\mathbf{b} &= \begin{bmatrix} b_1 & b_2 & \cdots & b_N \end{bmatrix}^{\mathrm{T}} \\
\mathbf{n} &= \begin{bmatrix} n_1 & n_2 & \cdots & n_N \end{bmatrix}^{\mathrm{T}}
\end{aligned}
\tag{10.18}
$$

As in the approximate ML method described in Chapter 7, by introducing a new variable $R = \sqrt{x^2 + y^2}$, the measurement equations can be linearized. That is, (10.17) can be expressed as

$$\mathbf{G}\boldsymbol{\theta} = \mathbf{h} + \boldsymbol{\varepsilon} \tag{10.19}$$

where

$$
\mathbf{G} = \begin{bmatrix} -2x_1 & -2y_1 & 1 \\ -2x_2 & -2y_2 & 1 \\ \vdots & \vdots & \vdots \\ -2x_N & -2y_N & 1 \end{bmatrix} \qquad \boldsymbol{\theta} = \begin{bmatrix} x \\ y \\ R^2 \end{bmatrix} \qquad \mathbf{h} = \begin{bmatrix} r_1^2 - (x_1^2 + y_1^2) \\ r_2^2 - (x_2^2 + y_2^2) \\ \vdots \\ r_N^2 - (x_N^2 + y_N^2) \end{bmatrix} \tag{10.20}
$$

and $\boldsymbol{\varepsilon}$ is the error vector whose components are given as $\varepsilon_i = -(b_i + n_i)^2 - 2d_i(b_i + n_i)$. The LS solution for (10.19) is given by

$$
\hat{\boldsymbol{\theta}} = (\mathbf{G}^T \boldsymbol{\Psi}^{-1} \mathbf{G})^{-1} \mathbf{G}^T \boldsymbol{\Psi}^{-1} \mathbf{h} \tag{10.21}
$$

where

$$
\begin{aligned}
\boldsymbol{\Psi} &= E[\boldsymbol{\varepsilon}\boldsymbol{\varepsilon}^T] \\
&\approx 4c^2 \mathbf{B} \mathbf{Q}_n \mathbf{B}
\end{aligned} \tag{10.22}
$$

where c is the speed of propagation, $\mathbf{Q}_n$ is the covariance matrix of $\mathbf{n}$ and $\mathbf{B}$ is a diagonal true-distance matrix defined as

$$
\mathbf{B} = \mathrm{diag}\{ d_1 \quad d_2 \quad \cdots \quad d_N \} \tag{10.23}
$$

Note that the approximation in (10.22) results from ignoring the positive biases and the squared error terms. Initially, the location estimate can be obtained from (10.21) by setting $\mathbf{B}$ as an identity matrix. Then, $\mathbf{B}$ can be estimated by using the estimated locations. The process may be iterated to generate an improved location estimate. In the absence of NLOS propagation or when the NLOS conditions are not severe, this method can produce very accurate location estimates. However, as the NLOS propagation becomes severe, significant performance loss may occur. To enhance the location accuracy further, the linear quadratic programming (LQP) optimization [3] may be applied. Then, the location estimate is produced by

$$
\hat{\boldsymbol{\theta}} = \arg \min_{\boldsymbol{\theta}} \{ (\mathbf{h} - \mathbf{G}\boldsymbol{\theta})^T \boldsymbol{\Psi}^{-1} (\mathbf{h} - \mathbf{G}\boldsymbol{\theta}) \} \tag{10.24}
$$

subject to $\mathbf{h} \geq \mathbf{G}\boldsymbol{\theta}$, where it is assumed that the sum of the measurement noise and the bias is nonnegative. Compared with nonlinear optimization algorithms, such as those to be discussed in the following subsections, the LQP-based method has the advantage of low complexity.

10.3.2 Three-Step Approach

An alternative way to deal with the NLOS problem is to determine how much the NLOS error is and then compensate for it. One such method is the three-step optimization-based method, which operates as follows [4]. First, the nonlinear measurement equations are linearized by using a Taylor series expansion and an initial mobile node location estimate is produced by employing the linear LS estimator. The NLOS error (bias) is then estimated based on an iterative optimization method under the constraint of positive biases. The initial mobile node location estimate is then corrected using the estimated bias. More specifically, let us consider the compact measurement equation in (10.17). Since $\mathbf{d}(\mathbf{p})$ is nonlinear with respect to $\mathbf{p}$, we may expand it at a reference point into a Taylor series and retain the terms below the second order, resulting in

$$\mathbf{d}(\mathbf{p}) \approx \mathbf{d}(\mathbf{p}_0) + \mathbf{G}_0(\mathbf{p} - \mathbf{p}_0) \tag{10.25}$$

where $\mathbf{G}_0$ is the Jacobian matrix of $\mathbf{d}(\mathbf{p})$ at $\mathbf{p}_0$, which is defined as

$$\mathbf{G}_0 = \begin{bmatrix} \dfrac{\partial d_1(\mathbf{p})}{\partial x} & \dfrac{\partial d_1(\mathbf{p})}{\partial y} \\ \vdots & \vdots \\ \dfrac{\partial d_N(\mathbf{p})}{\partial x} & \dfrac{\partial d_N(\mathbf{p})}{\partial y} \end{bmatrix}_{\mathbf{p}=\mathbf{p}_0} \tag{10.26}$$

Note that in the case of 3D positioning, the dimensions of $\mathbf{G}_0$ would be $N \times 3$. The initial reference point may be obtained by using a noniterative method such as the simple linear LS algorithm described in earlier chapters. Substituting (10.25) into (10.17) produces

$$\boldsymbol{\eta} \approx \mathbf{G}_0 \mathbf{p} + \mathbf{b} + \mathbf{n} \tag{10.27}$$

where

$$\boldsymbol{\eta} \approx \mathbf{r} - (d(\mathbf{p}_0) - \mathbf{G}_0 \mathbf{p}_0) \tag{10.28}$$

Then, the cost function becomes

$$J(\mathbf{p}) \approx (\boldsymbol{\eta} - \mathbf{G}_0 \mathbf{p} + \mathbf{b})^{\mathrm{T}} \mathbf{Q}_n^{-1} (\boldsymbol{\eta} - \mathbf{G}_0 \mathbf{p} + \mathbf{b}) \tag{10.29}$$

where $\mathbf{Q}_n = \mathrm{cov}(\mathbf{n})$ is the covariance matrix of noise vector $\mathbf{n}$. For a given bias vector $\mathbf{b}$, the minimization of the cost function in (10.29) results in

$$\hat{\mathbf{p}} = \tilde{\mathbf{p}} + \mathbf{U}\mathbf{b} \tag{10.30}$$

where

$$\mathbf{U} = (\mathbf{G}_0^{\mathrm{T}} \mathbf{Q}_n^{-1} \mathbf{G}_0^{\mathrm{T}})^{-1} \mathbf{G}_0^{\mathrm{T}} \mathbf{Q}_n^{-1} \tag{10.31}$$

and

$$\tilde{\mathbf{p}} = -\mathbf{U}\mathbf{r} \tag{10.32}$$

is the corresponding bias-free location estimate. Since the bias vector $\mathbf{b}$ is unknown, it has to be estimated. Define

$$\boldsymbol{\xi} = \boldsymbol{\eta} - \mathbf{G}_0 \tilde{\mathbf{p}} \tag{10.33}$$

Then, substituting (10.27) and (10.30) into (10.33), we obtain

$$\boldsymbol{\xi} = \mathbf{S}\mathbf{b} + \boldsymbol{\varepsilon} \tag{10.34}$$

where

$$S = I + G_0 U$$
$$\varepsilon = G_0(p - \hat{p}) + n \tag{10.35}$$

It can be seen that the covariance matrix of ε is given by

$$\text{cov}(\varepsilon) = G_0 \text{cov}(\hat{p}) G_0^T + Q_n \tag{10.36}$$

where

$$\text{cov}(\hat{p}) = (G_0^T Q_n^{-1} G_0)^{-1} \tag{10.37}$$

From (10.34), we can define the cost function for estimating b as

$$J(b) \approx (\xi - Sb)^T \text{cov}^{-1}(\varepsilon)(\xi - Sb) \tag{10.38}$$

and the estimate of b is produced by minimizing the above cost function; that is:

$$\hat{b} = \arg\min J(b) \tag{10.39}$$

subject to the constraint given by

$$b^L \leq b \leq b^U \tag{10.40}$$

The lower bound b^L may be simply set at

$$b^L = 0 \tag{10.41}$$

whereas the upper bound b^U may be chosen based on the geometry of the nodes layout and the range measurements; that is:

$$b_i^U = \min\{r_i + r_j - d_{i,j}, j = 1, 2, \ldots, N, j \neq i\} \qquad i = 1, 2, \ldots, N \tag{10.42}$$

Based on the cost function given by (10.38) and the constraint given by (10.40), a range of constrained optimization algorithms can be applied to produce an estimate of the bias vector. The estimated bias vector is then substituted into (10.30) to produce the corrected location estimate. The process can be iterated to achieve an improved location estimate.

A different way to estimate the NLOS error is to weight the distance measurements by scale factors so that the scaled measurement is equal to the true distance [5]. That is:

$$d_i = \kappa_i \hat{d}_i \tag{10.43}$$

where $\{\kappa_i\}$ are the scale factors to be determined. The cost function is defined as the sum of the square of ranges between the mobile node location and a number of intersection points of the range circles, which are closest to the mobile node location. The mobile node coordinates are approximated to be the linear LS solution and they are a function of the scale factors. Constrained optimization of the cost function produces the estimates of the scale factors. Consequently, the mobile node location can be determined by using the estimated scale factors. This scale-factor-based method may be efficient when the number of mobile nodes is small, such as three for 2D scenarios; however, it may be tedious to determine the inner intersection points of a relatively large number of range circles.

10.3.3 Sequential Quadratic Programming-based Method

Employing optimization techniques to deal with the NLOS problem can also be realized by jointly estimating the unknown coordinates and NLOS biases. That is, the optimization is carried out with respect to both the unknown coordinates and the NLOS biases. Intuitively, the joint optimization scheme would produce better results than the other schemes that either only estimate the coordinates or estimate the coordinates and biases separately.

The SQP-based method makes use of the SQP algorithm which is a nonlinear iterative minimization algorithm. This method is intended to minimize a cost function with respect to both the position coordinates and the measurement biases. Specifically, the cost function is defined as

$$S(\boldsymbol{\theta}) = \sum_{i=1}^{N} w_i [d_i(\mathbf{p}) + b_i - r_i]^2 \tag{10.44}$$

where

$$\boldsymbol{\theta} = \begin{bmatrix} \mathbf{p}^{\mathrm{T}} & \mathbf{b}^{\mathrm{T}} \end{bmatrix}^{\mathrm{T}} \tag{10.45}$$

and $\{w_i\}$ are the weights for emphasizing the more reliable range measurements. By using the knowledge of the dimensions of the location area and the established coordinate system, we can readily obtain a constraint as

$$(x^{\mathrm{L}}, y^{\mathrm{L}}) \leq (x, y) \leq (x^{\mathrm{U}}, y^{\mathrm{U}}) \tag{10.46}$$

where $(x^{\mathrm{L}}, y^{\mathrm{L}})$ and $(x^{\mathrm{U}}, y^{\mathrm{U}})$ are the lower and upper bounds of the unknown coordinates respectively. Since the NLOS error is always positive when using TOA or distance measurements, we have another constraint as

$$r_i \geq d_i(\mathbf{p}) + n_i^{\mathrm{L}} \tag{10.47}$$

where n_i^{L} is the possible lower bound of the measurement noise which may be determined in advance based on experimental results. Additionally, constraints can be applied to the biases as given by (10.40). Therefore, the coordinate estimates are produced by

$$\hat{\boldsymbol{\theta}} = \arg \min_{\boldsymbol{\theta}} S(\boldsymbol{\theta}) \tag{10.48}$$

subject to the constraints given by (10.46), (10.47) and (10.40). The minimization is carried out by using the SQP algorithm. A brief description of the SQP algorithm is given in Annex 10.A. To begin the process, initial values of the unknown coordinates and biases are required. Given that the NLOS bias is zero under an LOS condition, the bias initial values may be simply set at zero as mentioned previously. The initial coordinate estimates can be determined by using the simple noniterative position determination methods as described in Chapter 6. The main shortcoming of the SQP algorithm is the high computational complexity.

10.3.4 Taylor Series Expansion-based Linear Quadratic Programming Algorithm

The SQP-based algorithm is more complex than the LQP method described in Section 10.3.1. Reduction of computational complexity is generally desirable, especially for devices with limited computational power. To this end, a Taylor series expansion-based LQP (TS-LQP) algorithm can be employed to reduce the computational complexity significantly. Expanding $d_i(\mathbf{p})$ in (10.16) in a Taylor series at the initial position estimate $\mathbf{p}_0 = \begin{bmatrix} x_0 & y_0 \end{bmatrix}^T$ and retaining the first two terms, we obtain

$$d_i(\mathbf{p}) = d_i(\mathbf{p}_0) + f_x^{(i)}(x - x_0) + f_y^{(i)}(y - y_0) + \varepsilon_i \tag{10.49}$$

where ε_i is the error resulting from the linearization and

$$
\begin{aligned}
f_x^{(i)} &= \left.\frac{\partial d_i(\mathbf{p})}{\partial x}\right|_{\mathbf{p}=\mathbf{p}_0} = \frac{x_0 - x_i}{d_i(\mathbf{p}_0)} \\
f_y^{(i)} &= \left.\frac{\partial d_i(\mathbf{p})}{\partial y}\right|_{\mathbf{p}=\mathbf{p}_0} = \frac{y_0 - y_i}{d_i(\mathbf{p}_0)}
\end{aligned}
\tag{10.50}
$$

Substituting (10.49) into (10.15) produces

$$\tilde{r}_i = f_x^{(i)} x + f_y^{(i)} y + b_i + \tilde{n}_i \tag{10.51}$$

where

$$
\begin{aligned}
\tilde{r}_i &= r_i + f_x^{(i)} x_0 + f_y^{(i)} y_0 - d_i(\mathbf{p}_0) \\
\tilde{n}_i &= n_i + \varepsilon_i
\end{aligned}
\tag{10.52}
$$

Instead of considering N unknown NLOS parameters $\{b_i\}_{i=1}^{N}$, one can just use one NLOS variable (b_s). This option would be equivalent to using one parameter to represent all the N parameters. The variable b_s plays the role as a 'balancing' parameter to partially reduce the effect of the biases. Relatively speaking, this one NLOS variable-based scheme would be most efficient when all the bias values are similar. When the biases are significantly different from each other, the 'balancing' variable would approach the mean of the bias values so that the effect of the biases can also be reduced. Using one NLOS variable instead of N NLOS variables would result in some accuracy degradation. Nevertheless, the problem is greatly simplified since there are only three unknowns, instead of $N+2$ unknowns. Rewrite (10.51) as

$$\tilde{\mathbf{r}} = \mathbf{A}\boldsymbol{\theta} + \mathbf{v} \tag{10.53}$$

where $\mathbf{v}$ is the error vector and

$$
\tilde{\mathbf{r}} = \begin{bmatrix} \tilde{r}_1 \\ \tilde{r}_2 \\ \vdots \\ \tilde{r}_N \end{bmatrix} \qquad
\boldsymbol{\theta} = \begin{bmatrix} x \\ y \\ b_s \end{bmatrix} \qquad
\mathbf{A} = \begin{bmatrix} f_x^{(1)} & f_y^{(1)} & 1 \\ f_x^{(2)} & f_y^{(2)} & 1 \\ \vdots & \vdots & \vdots \\ f_x^{(N)} & f_y^{(N)} & 1 \end{bmatrix}
\tag{10.54}
$$

Since (10.53) is linear with respect to the unknown parameter vector, we can readily apply the LQP algorithm to solve the location problem. The position estimate is produced by

$$\hat{\boldsymbol{\theta}} = \arg\min_{\boldsymbol{\theta}}\{(\tilde{\mathbf{r}} - \mathbf{A}\boldsymbol{\theta})^{\mathrm{T}}\boldsymbol{\Phi}^{-1}(\tilde{\mathbf{r}} - \mathbf{A}\boldsymbol{\theta})\} \tag{10.55}$$

where $\boldsymbol{\Phi} = \mathrm{cov}(\mathbf{v})$ is the covariance matrix of the error vector. Simulation results show that the computational time of the TS-LQP algorithm is greatly reduced compared with the SQP-based algorithm.

10.3.5 Closest Intersection Points-based Algorithm

Another joint estimation method makes use of the closest intersection points of the range circles [6]. Compared with other optimization-based algorithms, this method is simple since it does not require any iterative minimization method and a few simple iterates would be sufficient in general. On the other hand, it may be tedious to determine and compute the intersection points which are closest to the mobile node when there is a relatively large number of range circles. In the 3D scenario, it would be more difficult to use this method, since the intersection planes instead of points will be involved. This method is briefly described in the following way. Equation (10.49) can be written in a compact form as

$$\mathbf{d}(\mathbf{p}) = \mathbf{d}(\mathbf{p}_0) + \mathbf{G}_0(\mathbf{p} - \mathbf{p}_0) + \boldsymbol{\varepsilon} \tag{10.56}$$

where $\boldsymbol{\varepsilon}$ is the error vector, resulting from the linearization, and

$$\mathbf{G}_0 = \begin{bmatrix} \dfrac{\partial d_1(\mathbf{p})}{\partial x} & \dfrac{\partial d_1(\mathbf{p})}{\partial y} \\ \vdots & \vdots \\ \dfrac{\partial d_N(\mathbf{p})}{\partial x} & \dfrac{\partial d_N(\mathbf{p})}{\partial y} \end{bmatrix}_{\mathbf{p}=\mathbf{p}_0}. \tag{10.57}$$

Replacing $\mathbf{d}(\mathbf{p})$ in (10.56) by the measurement vector produces

$$\mathbf{r} = \mathbf{d}(\mathbf{p}_0) + \mathbf{G}(\mathbf{p}_0)(\mathbf{p} - \mathbf{p}_0) + \mathbf{v} \tag{10.58}$$

where $\mathbf{v}$ is the combination of $\boldsymbol{\varepsilon}$ and the difference between the true distance vector $\mathbf{d}(\mathbf{p})$ and the measurement vector $\mathbf{r}$. Equation (10.58) can be rewritten as

$$\mathbf{G}_a\boldsymbol{\theta} = \mathbf{h} \tag{10.59}$$

where

$$\begin{aligned} \mathbf{G}_a &= \begin{bmatrix} \mathbf{G}_0 & \mathbf{I}_N \end{bmatrix} \\ \boldsymbol{\theta} &= \begin{bmatrix} \mathbf{p}^{\mathrm{T}} & \mathbf{v}^{\mathrm{T}} \end{bmatrix}^{\mathrm{T}} \\ \mathbf{h} &= \mathbf{r} - \mathbf{d}(\mathbf{p}_0) + \mathbf{G}_0\mathbf{p}_0 \end{aligned} \tag{10.60}$$

where $\mathbf{I}_N$ is the identity matrix of dimensions $N \times N$. As mentioned previously, owing to the fact that the true position of the target is usually located in the overlapped area of the distance circles, the cost function may be defined as the sum of the square of ranges between the mobile node location and the intersection points of the range circles; that is:

$$S(\boldsymbol{\theta}) = \sum_{k=1}^{K} ||\boldsymbol{\theta} - \mathbf{u}_i||^2 \tag{10.61}$$

where

$$\mathbf{u}_i = \begin{bmatrix} x_i^{(c)} & y_i^{(c)} & 0 & \cdots & 0 \end{bmatrix}^{\mathrm{T}} \in \mathbb{R}^{(N+2)\times 1} \tag{10.62}$$

where $(x_i^{(c)}, y_i^{(c)})$ are the coordinates of the intersection points. The location estimate is then produced by minimizing the cost function subject to the constraint given by (10.59); that is:

$$\hat{\mathbf{p}} = \arg \min_{\substack{\boldsymbol{\theta} \\ \mathbf{G}_a\boldsymbol{\theta}=\mathbf{h}}} S(\boldsymbol{\theta}). \tag{10.63}$$

It can be shown that an explicit expression for the location estimate based on (10.63) is given by

$$\hat{\boldsymbol{\theta}} = \frac{1}{K} \sum_{k=1}^{K} \mathbf{u}_k + \frac{1}{K} \mathbf{G}_a^{\mathrm{T}} (\mathbf{G}_a \mathbf{G}_a^{\mathrm{T}})^{-1} \left(K\mathbf{h} - \mathbf{G}_a \sum_{k=1}^{K} \mathbf{u}_k \right) \tag{10.64}$$

Clearly, the first term is the centroid of the intersection points, whereas the second term can be treated as the adjustment. Since $\mathbf{u}_k$ is dependent on the true locations of the target, a numerical method is required to produce a numerical solution for $\boldsymbol{\theta}$ from (10.64). For instance, the initial estimates of the intersection points can be obtained by using the distance measurements to replace the true distances so that the initial location estimate can be obtained. The estimated locations are then used to produce the updated estimates of the intersection points. The process continues until the variations of the location estimates or the intersection points are below some predefined threshold.

10.3.6 Accuracy Comparison

In this subsection a few iterative optimization-based algorithms are selected for performance evaluation by carrying out the following simulation. The location area is a square with dimensions of $40\,\mathrm{m} \times 40\,\mathrm{m}$. The dimensions correspond to an indoor environment. The results in outdoor conditions with cellular deployment will be presented later. Four fixed anchor nodes are assumed to be located at $(0, 0)$, $(40, 0)$, $(0, 40)$, and $(40, 40)$ (units in meters). This is based on the well-established fact that placing the anchors along the boundary of the location area produces a better performance. The target location is randomly selected in the area and 3000 different locations are examined for each simulation run. The measurement noise is Gaussian with zero mean and an STD being proportional to the distance; that is, 2 % of the true distance. The NLOS error is exponentially distributed with a mean equal to 8 % of the true distance. The selection of these values is in accordance with the range estimation accuracy when using TOA

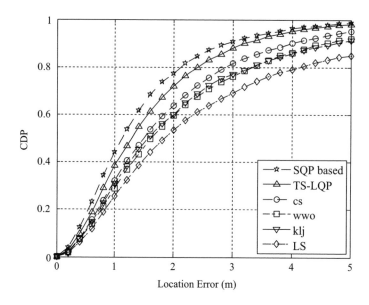

Figure 10.1 Error CDP of six different location algorithms in the case of two NLOS anchors among four anchors.

measurements, such as in [7,8]. The NLOS anchors are randomly selected among all the anchors and the weighting matrices are simply chosen to be identity matrices. In practice, information such as SNR can be used to determine the weighting matrix. The location estimates from the linear LS algorithm are used as the initial position values for all the optimization/ minimization algorithms studied. Two different performance measures are considered: variance of the coordinate estimation errors and the CDP of the location error, which is defined as $\sqrt{(\hat{x} - x)^2 + (\hat{y} - y)^2}$ for each location (x, y).

Figure 10.1 shows the CDPs of six different algorithms when there are two NLOS fixed nodes. The six algorithms are: the method in [9], but with the minimization performed based on the SQP algorithm, and the results denoted by 'cs'; the method in [3] with results denoted by 'wwo'; the method in [4] with results denoted by 'klj'; the joint location and bias estimation method with results denoted by 'SQP based'; the TS-LQP algorithm with results denoted by 'TS-LQP'; and the linear LS algorithm with results denoted by 'LS'. It is shown that the SQP-based joint location and bias estimation method outperforms the other methods. The TS-LQP algorithm also performs well and the accuracy loss compared with the SQP-based method is not substantial in general. On the other hand, as mentioned earlier, the computational complexity of the TS-LQP algorithm is much lower than the SQP-based method.

Figure 10.2 illustrates the effect of the number of NLOS anchors (from one to four) when the location error is up to 3 m. Basically, the accuracy degrades approximately linearly as the number of NLOS anchors increases. The SQP-based joint estimation and the TS-LQP methods achieve an accuracy of 3 m at a probability greater than 78 % in all the cases. Table 10.1 shows the STD of the x-coordinate estimation error. Very similar results are obtained for the y-coordinate estimation error. It is seen that the STD increases as the number of NLOS

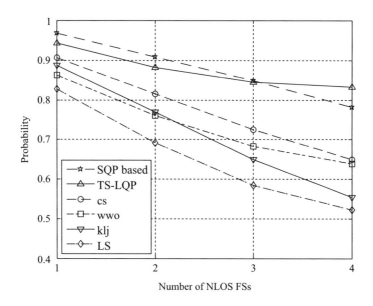

Figure 10.2   Impact of number of NLOS anchors (from one to four) on the accuracy of the six algorithms when location error is up to 3 m. The legend items have the same definitions as in Figure 10.1.

anchors increases, and the SQP-based joint estimation method produces the smallest STD on average.

The above results are produced under indoor environments. In this following we examine the accuracy of the same algorithms in outdoor conditions for cellular deployment. When positioning under outdoor cellular deployment, one of the main challenges is the near–far (hearability) issue. This may be caused by one or several factors. For instance, when the mobile node is very close to one of the fixed nodes and far from the others, the mobile node may be able to detect the signal only from the closest fixed node, either because the signals from other fixed nodes are too weak to be detectable or because the other signals are overwhelmed by the strongest signal. Another example is that the signals from one or more fixed nodes that are within radio range are blocked by obstacles so that the mobile node cannot receive signals from more than two fixed nodes. The hearability problem means that the mobile may have communication links with less than three fixed nodes so that the mobile node location cannot be determined.

Table 10.1   STD of the x-coordinate estimation errors of the six different algorithms under different number of NLOS anchors N_n.

N_n	SQP based	TS-LQP	cs	wwo	klj	LS
1	0.93 m	1.11 m	1.35 m	1.62 m	1.43 m	2.12 m
2	1.31 m	1.45 m	1.74 m	2.02 m	2.05 m	2.86 m
3	1.58 m	1.62 m	2.06 m	2.36 m	2.58 m	3.37 m
4	1.79 m	1.70 m	2.35 m	2.46 m	3.02 m	3.61 m

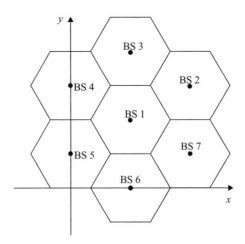

Figure 10.3 Illustration of hexagonal cellular deployment.

A number of techniques have been proposed to overcome the hearability problem. In the idle period down link scheme [10], idle periods are created during which the serving fixed node/ anchor node turns off transmissions so that the hearability of the signals from other distant fixed nodes can be improved. In the cumulative virtual blanking scheme [11,12], the strongest signals are removed successively in a software process which is performed in the serving mobile location center. Here, we assume that the near–far/hearability problem has been solved, so the mobile node is able to receive signals from at least three fixed nodes.

Let us consider a hexagonal cellular deployment as shown in Figure 10.3. Base station (BS) 1 is surround by six other base stations. The distance between each pair of adjacent base stations is 1 km. The coordinates (units in meters) of the seven base stations are (866, 1000), (1732, 1500), (866, 2000), (0, 1500), (0, 500), (866, 0), (1732, 500). The radio range of the mobile node is assumed to be 1 km after the hearability issue is resolved. Without loss of generality, it is assumed that the mobile node is randomly located in the center cell with BS 1 as its home base station so that it can be within radio range of either three or four base stations including BS 1. The mean of the exponential NLOS distance biases is set at 10% of the corresponding distance and the STD of the Gaussian distance measurement errors is set at 1.5% of the corresponding distance.

Figure 10.4 shows the CDPs of the six algorithms when there is one NLOS base station among either three or four base stations. The six algorithms are the same algorithms that are already examined in indoor conditions. The SQP-based algorithm achieves accuracy of 80 m with a probability of about 90 %. The other five algorithms achieve the same accuracy with a probability of around 85 %. Figure 10.5 shows the CDPs of the same six algorithms in the presence of two base stations among the three or four base stations. Although some performance degradation occurs compared with the case of one base station, both the SQP-based and the TS-LQP algorithms approach the accuracy of 100 m with a probability of about 88 %, while the other algorithms about 80 %.

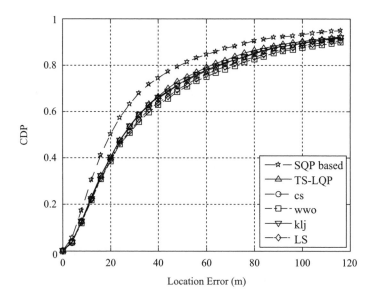

Figure 10.4   Location error CDPs of six different location algorithms in the case of one NLOS anchor node among either three or four anchor nodes with cellular deployment. The legend for Figure 10.1 also applies here.

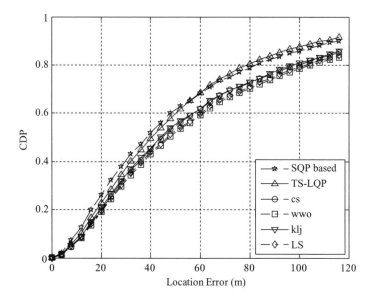

Figure 10.5   Location error CDPs of six different location algorithms in the case of two base stations among either three or four base stations with cellular deployment. The legend for Figure 10.1 also applies here.

10.4 Scatterer-based Method

In the case of a wireless link, the received signal is usually diffracted by scatterers in the case of NLOS propagation. The spatial locations of scatterers and the AOA of the incoming signal can be employed to enhance position determination. This section presents several approaches which make use of spatial location and angle information of scatterers.

10.4.1 Angle Measurement at Mobile Device

When a mobile device is equipped with a directional antenna or an antenna array, the AOA can also be measured at the mobile device. Although the assumption of AOA measurements taken at the mobile may not be realistic for many applications, it is still feasible if the carrier frequency is high enough. In particular, it is interesting to evaluate the location accuracy when the AOA information is available at both the transmitter and the receiver.

Figure 10.6 shows a simple example of two NLOS paths between an anchor node located at point $\mathbf{k}$ and the target at point $\mathbf{t}$ in the absence of a LOS path. Each of the two components is diffracted by only one scatterer. The transmitted signal is diffracted at points $\mathbf{a}$ by the first scatterer and $\mathbf{b}$ by the second scatterer, both of which have unknown coordinates. The total distance traversing path $\mathbf{kbt}$ or path $\mathbf{kat}$ can be estimated by performing RTT measurements under the assumption that accurate calibration has been carried out so that the internal radio delays have been compensated.

When both the anchor node and the target share the same orientation, such as by using a compass, the angles $\alpha_{1,1}$ and $\beta_{1,1}$ are measured under the same orientation. Giving the five unknown parameters – that is, the four coordinates of points $\mathbf{a}$ and $\mathbf{t}$ and the distance $\mathbf{ka}$ (or $\mathbf{ta}$) – four independent linear equations can be produced by applying the basic trigonometric theorems; that is:

$$
\begin{aligned}
x_a - x_k &\approx d_{ka} \sin \hat{\beta}_{1,1} \\
y_a - y_k &\approx d_{ka} \cos \hat{\beta}_{1,1} \\
x_a - x_t &\approx (\hat{d}_{kat} - d_{ka}) \sin \hat{\alpha}_{1,1} \\
y_a - y_t &\approx (\hat{d}_{kat} - d_{ka}) \cos \hat{\alpha}_{1,1}
\end{aligned}
\tag{10.65}
$$

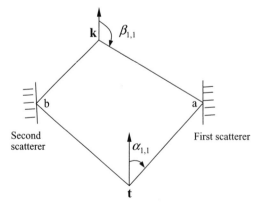

Figure 10.6   An example of two-path propagation in absence of an LOS path between an anchor node at point $\mathbf{k}$ and the target at point $\mathbf{t}$.

where (x_a, y_a) are the unknown coordinates of the scatterer at point $\mathbf{a}$, (x_t, y_t) are the unknown coordinates of the target, (x_k, y_k) are the known coordinates of the anchor node, $\hat{d}_{kat}$ is the distance estimate and d_{ka} is the unknown distance between points $\mathbf{k}$ and $\mathbf{a}$. As a result, the target location is constrained on a line segment, given by

$$\hat{y}_t = k_{1,1}\hat{x}_t + b_{1,1} \tag{10.66}$$

where

$$k_{1,1} = \frac{\sin \hat{\beta}_{1,1} + \sin \hat{\alpha}_{1,1}}{\cos \hat{\beta}_{1,1} + \cos \hat{\alpha}_{1,1}} \tag{10.67}$$
$$b_{1,1} = y_k - k_{1,1}x_k + \hat{d}_{kat}(k_{1,1}\sin \hat{\alpha}_{1,1} - \cos \hat{\alpha}_{1,1})$$

Using the measurements related to the second scatterer produces another four equations and a second line segment. Therefore, we can obtain a unique solution for the target location as the intersection point of the two line segments under the condition that the anchor node and the two scatterers do not lie in a straight line. Alternately, in the case that only one path measurement is available, another anchor node is required to obtain a unique solution. When extra path measurements are available, and/or in the event that measurements are made at multiple base stations, accuracy enhancement through diversity gain would be expected. To make efficient use of extra measurements, an LS estimator can be employed to produce a final target location estimate by using measurements from all paths and all anchor nodes. Note that when the statistics of measurement errors are known, the ML estimator may be applied to produce the optimal target location estimate [13].

10.4.2 Doppler Frequency of a Moving Target

In many circumstances, the target may not be equipped with a directional antenna or an antenna array but the target is moving at a certain speed, as shown in Figure 10.7. When the anchor node is able to estimate the Doppler frequency, the angle information related to the target can still be

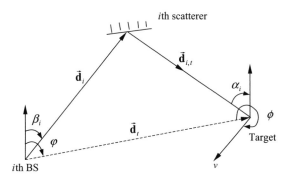

Figure 10.7   Single reflection scatterer geometry with a moving target.

exploited [14]. Specifically, from Figure 10.7, and based on the geometric relations, we can obtain an angle relation as

$$\alpha_i = \pi - \left[\varphi + \phi \pm \cos^{-1} \left(\frac{f_{D,i}\lambda}{v} \right) \right] \tag{10.68}$$

where $f_{D,i}$ is the measured Doppler frequency of the ith diffracted signal, λ is the known carrier wavelength and v is the unknown speed of the target. Note that both angles φ and ϕ in (10.68) are unknown parameters which are independent of the multipath and the plus/minus sign is used to include the two situations where the target is moving either toward or from the anchor node.

Suppose that, at the anchor node, the AOA β_i and the TOA τ_i of each diffracted signal are measured. Let $\tau_{1,0}$ denote the TDOA between the first path and the LOS path and let $\tau_{i,1}$ denote the TDOA between the ith path and the first path. Clearly, $\tau_{1,0}$ is unknown, whereas $\tau_{i,1}$ can be estimated. It can be seen that

$$d_i + d_{i,t} - d_t = c\tau_{i,1} + c\tau_{1,0} \tag{10.69}$$

where c is the speed of propagation, d_i is the magnitude of the distance vector from the anchor node to the scatterer, $d_{i,t}$ is the magnitude of the distance vector from the scatterer to the target and d_t is the magnitude of the distance vector from the anchor node to the target. Using the Pythagoras's theorem produces

$$d_{i,t} = \sqrt{d_i^2 + d_t^2 - 2d_i d_t \cos(\varphi - \beta_i)} \tag{10.70}$$

Substituting (10.70) into (10.69) yields

$$d_i = \frac{(c\tau_{i,1} + c\tau_{1,0})^2 + 2(c\tau_{i,1} + c\tau_{1,0})d_t}{2[d_t + c\tau_{i,1} + c\tau_{1,0} - d_t \cos(\varphi - \beta_i)]} \tag{10.71}$$

It is known that the summation of the three distance vectors is equal to zero; that is:

$$\vec{\mathbf{d}}_i + \vec{\mathbf{d}}_{i,t} - \vec{\mathbf{d}}_t = \vec{\mathbf{0}} \tag{10.72}$$

By decomposing each of the distance vectors into two components with one on the vertical axis and the other on the horizontal axis, (10.72) becomes

$$E_i = d_i \cos\beta_i - d_{i,t} \cos\alpha_i - d_t \cos\varphi + j(d_i \sin\beta_i + d_{i,t} \sin\alpha_i - d_t \sin\varphi) = 0 \tag{10.73}$$

where complex notation is used to denote the two components in the vertical and horizontal directions. Note that the unknown parameters of interest are $\{d_t, \phi, \varphi, v, \tau_{1,0}\}$ and other unknown parameters are functions of them. In the presence of measurement errors, the second equality in (10.73) does not hold in general. Clearly, it is a nonlinear equation and an iterative minimization algorithm is needed to solve the problem. The cost function may be simply defined as

$$S(d_t, \phi, \varphi, v, \tau_{1,0}) = \sum_{i=1}^{L} w_i |E_i|^2 \tag{10.74}$$

where it is assumed that there are L scatterers producing L diffracted signals which are received by the anchor node and $\{w_i\}$ are the weights for emphasizing the more reliable path parameters' estimates. To obtain a unique solution for the unknown parameters, L must be at least equal to six, since there is uncertainty about the moving direction of the target. The advantage of the method is that only one anchor node is needed to produce a mobile location estimate and there is no requirement of a directional antenna or an antenna array at the mobile. The drawback of the method is that the mobile must be in motion and angle estimation is required at the anchor node.

10.4.3 Environmental Structure

Another single anchor-node-based positioning technique is to make use of the information about the environment in the neighborhood of the anchor node [15]. In this method, a wideband signaling scheme is required. Also, an antenna array and the associated parameter estimation algorithms are required at the anchor node. It requires knowledge of the AOAs and the relevant absolute propagation delays for a given number of multipath components impinging on the anchor node. Accordingly, it requires that the mobile terminal and the anchor node are synchronized or the RTT measurement is performed. A sentinel function (SF) is introduced and defined as the Euclidian distance between the anchor node and the nearest obstacle that lies along a certain azimuth direction. The SF might be sampled, then tabulated and eventually stored at the anchor node in advance.

In real time, the absolute distance traveled by the first path of the signal received at the anchor node is calculated by

$$\hat{d}_1 = \sqrt{(c\hat{\tau}_1)^2 + (h_{\mathrm{BS}} - h_{\mathrm{MS}})^2} \qquad (10.75)$$

where $\hat{\tau}_1$ is the estimate of the propagation delay of the first-path signal and h_{BS} and h_{MS} are the heights of the anchor node and the mobile node respectively. Then, the SF, $f(\alpha)$ and α_1 (the AOA of the first impinging path) are used to decide whether the mobile node is in a LOS condition. If $f(\alpha_1) > \hat{d}_1$, then the LOS condition is assumed and the mobile node location can be readily determined. On the other hand, if $f(\alpha_1) < \hat{d}_1$, then the NLOS condition is assumed. In this case, the method requires knowledge of the coordinates of the scatterers found along the AOA for the first L multipath components. Then, the distances from the mobile node to the scatterers can be calculated by

$$\hat{d}_i^{\mathrm{m\text{-}s}} = c\hat{\tau}_i - f(\alpha_i). \qquad (10.76)$$

The cost function is defined as

$$S(x, y) = \sum_{i=1}^{L} w_i [\hat{d}_i^{\mathrm{m\text{-}s}} - \sqrt{(x - x_i^{\mathrm{s}})^2 + (y - y_i^{\mathrm{s}})^2}]^2 \qquad (10.77)$$

where $\{(x_i^{\mathrm{s}}, y_i^{\mathrm{s}})\}$ are the coordinates of the scatterers and (x, y) are the coordinates of the mobile node. Finally, the mobile node location estimate is produced by

$$(\hat{x}, \hat{y}) = \arg\min_{x,y} S(x, y) \qquad (10.78)$$

subject to $\sqrt{(x-x_i^s)^2+(y-y_i^s)^2} \leq c\hat{\tau}_1$. This technique is suited for operating in a micro-cellular environment. The main disadvantage of the method is that it is difficult or even impossible to obtain accurate locations of all the relevant scatterers, so that only the main scatterers would be considered in practice.

10.5 Error Statistics

In some cases, the type of distribution and parameters of NLOS distance/TOA errors may be determined in advance from field measurements. In that case, the well-known statistical detection theory can be applied to solve the location estimation problem. The NLOS error is typically an exponentially distributed random variable, while the measurement noise usually behaves as a Gaussian random variable. Depending on how much information of the NLOS distribution is available, different joint conditional density functions can be obtained. In the ideal case where the NLOS conditions of each node and the distribution parameters are perfectly known, an optimal location estimate of the target can be produced by maximizing the likelihood function. In the presence of uncertainties in the NLOS condition and/or distribution parameters, a suboptimal location estimate can be obtained.

Let us focus on scenarios where the distance measurement noise n_i at the ith anchor node is Gaussian with zero mean and variance σ_i^2 and the NLOS error b_i is exponentially distributed with the mean λ_i and variance λ_i^2.

Assuming that the measurement noise and the NLOS distance bias are independent random variables, the joint distribution of the sum of the two random variables

$$\varepsilon_i = b_i + n_i \tag{10.79}$$

can be derived as

$$p_{\varepsilon_i} = \frac{1}{2\lambda_i} \exp\left[-\frac{1}{\lambda_i}\left(\varepsilon_i - \frac{\sigma_i^2}{2\lambda_i}\right)\right] \mathrm{erfc}\left(\frac{\sigma_i^2 - \lambda_i \varepsilon_i}{\sqrt{2}\lambda_i \sigma_i}\right) \tag{10.80}$$

where $\mathrm{erfc}(.)$ is the complementary error function. For ease of optimization and theoretical derivation, the sum of a Gaussian and an exponentially distributed random variable may be approximated as a Gaussian random variable with mean equal to λ_i and variance equal to $\sigma_i^2 + \lambda_i^2$ provided that λ_i is not much greater than σ_i. The accuracy of such an approximation decreases as λ_i increases. Note that simulation results show that direct use of (10.80) without the approximation does not produce better results, perhaps due to the complexity and the high nonlinearity of the function.

10.5.1 Idealized Case: Known Non-Line-of-Sight Conditions and Distribution Parameters

We first consider the ideal case in which NLOS anchor nodes and distribution parameters are known perfectly. Without loss of generality, it is supposed that among the N anchors, N_{los} anchors have LOS links with the target node and N_{nlos} anchors have NLOS links. Then, when all the measurements are mutually independent, we have the joint conditional density function (likelihood)

$$p(\mathbf{r}|\mathbf{p}) = \prod_{i=1}^{N} \frac{1}{\sqrt{2\pi}\tilde{\sigma}_i} \exp\left(-\frac{(\tilde{r} - d_i)^2}{2\tilde{\sigma}_i^2}\right) \tag{10.81}$$

where $\mathbf{p} = \begin{bmatrix} x & y \end{bmatrix}^\mathrm{T}$ is the unknown position vector, $\mathbf{r} = \begin{bmatrix} r_1 & r_2 & \cdots & r_N \end{bmatrix}^\mathrm{T}$ is the distance measurement vector and

$$\tilde{\sigma}_i = \begin{cases} \sigma_i & 1 \le i \le N_{\mathrm{los}} \\ \sqrt{\sigma_i^2 + \lambda_i^2} & \text{elsewhere} \end{cases}$$

$$\tilde{r}_i = \begin{cases} r_i & 1 \le i \le N_{\mathrm{los}} \\ r_i - \lambda_i & \text{elsewhere} \end{cases} \tag{10.82}$$

and $d_i = \sqrt{(x_i - x)^2 + (y_i - y)^2}$ is the straight-line distance between anchor i and the target. Taking logarithms on both sides of (10.81) and ignoring the irrelevant constants, we obtain the following log likelihood:

$$\Lambda(\mathbf{r}|\mathbf{p}) = -\sum_{i=1}^{N} \frac{(\tilde{r} - d_i)^2}{2\tilde{\sigma}_i^2} \tag{10.83}$$

The ML location estimate of the target is produced by maximizing the log likelihood in (10.83). Similarly, the ML location estimate is achieved by

$$\hat{\mathbf{p}} = \arg\min_{\mathbf{p}}\{-\Lambda(\mathbf{r}|\mathbf{p})\} \tag{10.84}$$

subject to the constraints of the dimensions of the location area and that the distance NLOS bias is always positive. In this case, a variety of minimization algorithms may be employed. It is found that both the unconstrained optimization and constrained optimization yield very similar results for this specific case.

10.5.2 Known Non-Line-of-Sight Conditions but Unknown Non-Line-of-Sight Statistics

In practice, some of the parameters are known while others are not known. One such case is that we know the noise variance and which anchors are in NLOS conditions, but there are uncertainties in the distribution parameters $\{\lambda_i\}$, which are bounded by

$$\lambda_i^{\mathrm{L}} \le \lambda_i \le \lambda_i^{\mathrm{U}} \qquad i = N_{\mathrm{los}} + 1, N_{\mathrm{los}} + 2, \ldots, N \tag{10.85}$$

Let

$$\boldsymbol{\lambda}_{\mathrm{b}} = \begin{bmatrix} \lambda_{N_{\mathrm{los}}+1} & \lambda_{N_{\mathrm{los}}+2} & \cdots & \lambda_N \end{bmatrix}^\mathrm{T} \tag{10.86}$$

Then, the location estimate can be produced by maximizing the log likelihood given by

$$\Lambda(\mathbf{r}|\mathbf{p},\lambda_{\mathrm{b}}) = -\left\{\sum_{i=1}^{N_{\mathrm{los}}} \frac{(r_i - d_i)^2}{2\sigma_i^2} + \sum_{i=N_{\mathrm{los}}+1}^{N} \frac{[r_i - (d_i + \lambda_i)]^2}{2(\sigma_i^2 + \lambda_i^2)}\right\}$$ (10.87)

with respect to $\mathbf{p}$ and λ_{b}. That is, the target location estimate is produced by

$$\hat{\mathbf{p}} = \arg\min_{\mathbf{p},\lambda_{\mathrm{b}}}\{-\Lambda(\mathbf{r}|\mathbf{p},\lambda_{\mathrm{b}})\}$$ (10.88)

subject to the constraints on the coordinates and the NLOS biases.

10.5.3 Known Non-Line-of-Sight Probability and Distribution Parameters

An alternative scenario is that we do not know which anchors are in NLOS conditions, but we know the NLOS probability and the distribution parameters at each anchor. Then, the PDF of the measurements is given by

$$\tilde{p}_i = P_i^{(\mathrm{los})} p_{n_i} + (1 - P_i^{(\mathrm{los})}) p_{\varepsilon_i}$$ (10.89)

where $P_i^{(\mathrm{los})}$ is the probability that the ith anchor has an LOS link with the target, p_{n_i} is a Gaussian density function with zero mean and variance σ_i^2 and p_{ε_i} is also a Gaussian density function with mean λ_i and variance $\sigma_i^2 + \lambda_i^2$. Similarly, we can have the joint conditional PDF and log likelihood of the measurements as

$$p(\mathbf{r}|\mathbf{p}) = \prod_{i=1}^{N} \tilde{p}_i$$
$$\Lambda(\mathbf{r}|\mathbf{p}) = \sum_{i=1}^{N} \ln \tilde{p}_i$$ (10.90)

Therefore, the location estimate can be produced by maximizing the log likelihood under the constraints of the dimensions of the location area and the fact of positive biases.

10.5.4 Known Non-Line-of-Sight Probability but Unknown λ_i

A more likely scenario is that the NLOS probability of each anchor and the variance of the measurement noise are known, but we do not know which anchors are in NLOS conditions and there are also uncertainties in the distribution parameters. Assume we keep the distribution parameter bounds given by (10.85). The location estimate can be produced by maximizing the log likelihood in (10.90) with respect to both the unknown coordinates and the NLOS biases as follows:

$$\hat{\mathbf{p}} = \arg\max_{\mathbf{p},\lambda_{\mathrm{b}}}\{\Lambda(\mathbf{r}|\mathbf{p},\lambda_{\mathrm{b}})\}$$
$$= \arg\min_{\mathbf{p},\lambda_{\mathrm{b}}}\left\{-\sum_{i=1}^{N}\ln\tilde{p}_i\right\}$$ (10.91)

Note that another way to handle the problem of unknown NLOS conditions is to identify the NLOS anchors first. For example, the LOS/NLOS identification methods presented in Chapter 14 may be used to solve the identification problem. Then, (10.84) or (10.88) can be applied to produce the ML location estimate. This method largely depends on the probability of successfully identifying the NLOS anchors. Also, it is worth noting that the worst case is that there are uncertainties in all the parameters: the NLOS probabilities and the distribution parameters in both LOS and NLOS conditions. In this case, the error-statistics-based algorithms described above may not be suited. Instead, the constrained-optimization-based methods, such as the SQP-based joint location and bias estimation method, the TS-LQP algorithm and the LQP algorithm would be more appropriate.

10.5.5 Accuracy Evaluation

This subsection presents accuracy evaluation of the ML estimator algorithms described in the preceding subsections. The simulation setup is the same as in Section 10.3.6. Figure 10.8 shows the error CDPs of the ML estimator algorithms in the presence of two NLOS anchors. Also shown are the results of the SQP-based joint estimation method for comparison. The case of distribution parameter mismatch is also examined and the results are denoted by 'with mismatch', and it is assumed that the parameter error is uniformly distributed between the negative half and the positive half of the true value. It is seen that the ML estimator can effectively handle the uncertainties in the distribution parameters as long as the NLOS status is

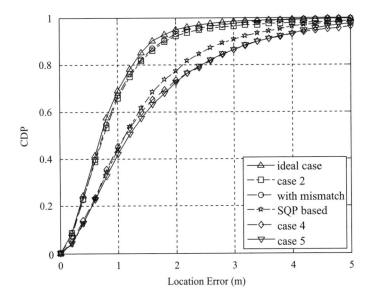

Figure 10.8 CDP of ML estimation algorithms when there are two NLOS anchors: 'ideal case', results with known LOS/NLOS conditions and distribution parameters; 'case 2', results with uncertainties only in distribution parameters; 'with mismatch', results with error in distribution parameters; 'case 4', results with known NLOS probability and distribution parameters; 'case 5', results with known NLOS probability and noise variance.

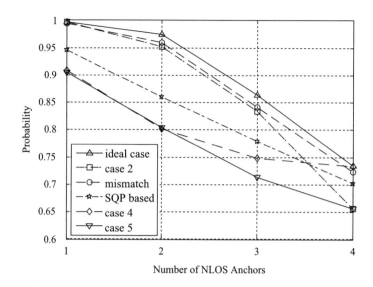

Figure 10.9 Probability versus number of NLOS anchors of the ML estimation algorithms when the location error is up to 2.5 m. The legend items have the same definitions as in Figure 10.8.

given. The SQP-based method yields better accuracy than that of the ML estimator when the knowledge of NLOS conditions is not available. It can be concluded that knowledge of the NLOS conditions is crucial for the ML estimator to achieve superior performance.

Figure 10.9 shows the cumulative probability of the location error versus the number of NLOS anchors of the ML estimator algorithms when the location error is up to 2.5 m. As expected, the probability goes down as the number of NLOS anchors increases. The accuracy drops only about 3 % under the given distribution parameter error. Note that, when all anchors are in NLOS conditions, case 4 approaches the ideal case, while case 5 becomes case 2.

Figure 10.10 shows the effect of parameter mismatch under different levels of errors when the error is uniformly distributed between $-u \times$ (true value) and $u \times$ (true value) with u ranging from 0.5 to 0.9. Apparently, the accuracy loss is rather minor for $u < 0.5$. This may indicate that, in practice, whenever an approximate model can be obtained, the ML estimator may be applied to produce an accurate location estimate.

10.6 Propagation-Model-based Method

A variety of statistical models exist to describe radio signal propagation in different scenarios. Some of the models may also be employed for location estimation. The purpose of using propagation models for positioning is mainly to estimate the LOS distance between the transmitter and receiver under NLOS conditions. Once the LOS distance estimates are available, various position-estimation algorithms can be employed to determine the unknown coordinates. The use of propagation models reduces the dependence on the empirical data, so it is simple compared with the database-related methods, which will be discussed in Section 10.7.

One such model is based on signal power loss when a signal travels through different media. In an indoor environment, the *wall attenuation factor* model [16] and the log-distance

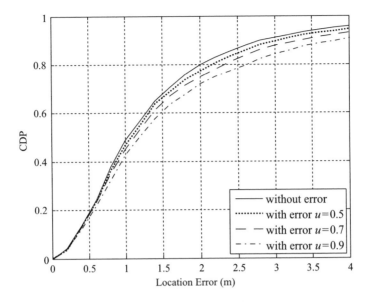

Figure 10.10   Impact of distribution parameter error on the ML estimation when there are three NLOS anchors among four anchors.

(log-loss) model [17] may be adopted to characterize the received power. In an outdoor environment, the Walfisch–Ikegami path loss model [18] is suitable for radio propagation in suburban areas and metropolitan centers. In each of the models, two or more model parameters need to be determined *a priori*. These parameters are typically dependent on the specific environment, so field-specific experiments are required to make measurements, which are processed to tune the relevant parameters.

Another type of model in describing radio propagation relies on the topology of scatterers. In macrocellular environments, for instance, the established propagation models include the ring of scatterers (ROS), the disk of scatterers and the clipped Gaussian model. Given a scattering model, the density function of the propagation distance can be determined and the variance of the distance can be derived, which is a function of the unknown location. Also, different moments, such as the mean and variance of TOA measurements, can be determined. The unknown location can then be estimated by equating the measured and derived TOA variances. As an illustration, let us make use of the ROS model. As shown in Figure 10.11, the scatterers are located on a ring centered at the mobile node with a radius r_c.

The angle between the direct path and the scattered path, denoted by ϕ in Figure 10.11, is typically modeled as a random variable uniformly distributed in $[0, 2\pi]$ and the scatterers are uniformly distributed around the ring whose radius is assumed fixed but unknown. From the geometry it can be shown that the multipath (NLOS) distance, denoted by η_i, from the anchor node to the mobile node is given by

$$\eta_i = r_c + \sqrt{r_c^2 + d_i^2 - 2r_c d_i \cos \phi}. \tag{10.92}$$

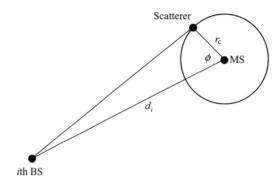

Figure 10.11   Geometry of the anchor node, mobile node and scatterer for the ROS model.

It can be shown that the PDF of η_i is given by

$$p(\eta_i) = \frac{\eta_i - r_c}{\pi \sqrt{1 - \left(\frac{d_i^2 + 2r_c \eta_i - \eta_i^2}{2 d_i r_c}\right)^2}} \qquad d_i \leq \eta_i \leq d_i + 2r_c \qquad (10.93)$$

To obtain an expression for the variance of η_i in terms of d_i and r_c, $p(\eta_i)$ may be expanded in a Taylor series and a few terms are retained. Then, the predicted variance is equated to the variance of the TOA-based distance measurements to form a variance equation denoted by $\hat{\sigma}_i^2$; that is:

$$\int_{d_i}^{d_i + 2r_c} \eta_i^2 p(\eta_i) \, d\eta_i - \left(\int_{d_i}^{d_i + 2r_c} \eta_i p(\eta_i) \, d\eta_i\right)^2 = \hat{\sigma}_i^2. \qquad (10.94)$$

To produce a unique solution for the mobile node location, at least three anchor nodes are required to generate three independent variance equations. Since (10.94) is nonlinear in terms of the coordinates of the mobile node location, an iterative method, such as the Newton–Raphson method, is required to solve it.

Alternatively, when the density function of the distances/TOAs is derived from the model, the ML and Bayesian estimators may be applied to estimate the LOS distance based on the multipath TOA measurements [19]. The scattering model may also be used to derive the mean of the NLOS error, which is then subtracted from the corresponding measurement.

The propagation-model-based approaches described above are largely dependent on the accuracy of the models employed. When the model accurately represents the actual channel/environment and the propagation parameters are well tuned on site, satisfactory location accuracy would be expected. In the absence of reliable models, other approaches, such as the following pattern matching approach, may be considered.

10.7 Pattern Matching

In some applications, an accurate propagation model may not exist, or it may be difficult to predict. An alternative way to efficiently reduce the impact of the NLOS effect is to use a database. In this approach, an extensive survey must be made *a priori* to create a signature (or fingerprint) database. For instance, the area concerned can be divided into nonoverlapping zones or grids. A mobile node is carried through each grid and sends a signal to a group of receivers with known locations. The received signals are then processed to produce a unique signature/pattern for each grid, which is recorded as a 'fingerprint' in a database. The fingerprint can be a single parameter or a combination of several signal characteristics, such as TOA, RSS, SNR, AOA or channel impulse response. In many cases, such as tracking people in buildings, the orientation of the object should also be taken into account. For instance, at the same location, the RSS observed at one receiver can be significantly different when the person with a tag/transmitter in their pocket is facing different directions. The grid resolution of the fingerprints should be comparable to the accuracy that can be achieved with the method. Once the database is established, one can perform the positioning and tracking tasks by sending the real-time measurements to the location server, in which the target location is estimated by comparing the measured data with the recorded fingerprints. The basic idea of pattern mapping was investigated more than three decades ago [20]. Since then, various approaches have been proposed to match the measured data with the database fingerprints. Three approaches are briefly described in Sections 10.7.1–10.7.3.

10.7.1 Nearest-Neighbor-in-Signal-Space Method

The first approach, perhaps the simplest one, is the nearest-neighbor-in-signal-space (NNSS) method originally proposed in [16]. In this method, each measurement vector is represented as a point in the signal vector space. At the real-time phase, the Euclidean distance measure is used to search for the location. To be specific, suppose that the RSS measurements are employed, the monitored area is divided into L grids and at each grid four different orientations of the object are examined. Consequently, there are $4L$ fingerprints related to each anchor, denoted by

$$s_i^{(\ell,k)} \qquad i = 1,\ldots,N, \ell = 1,\ldots,L, k = 1,\ldots,4$$

where i indexes the anchor node, ℓ indexes the grids/locations and k indexes the orientation. The fingerprints are stored in a database, which is done in the first survey phase. In the second phase, real-time RSS measurements are observed at all anchor nodes, which are denoted as r_i, $i = 1,\ldots,N$. Then, the Euclidean distance is computed as

$$\rho^{(\ell,k)} = \sqrt{\sum_{i=1}^{N}(r_i - s_i^{(\ell,k)})^2} \qquad \ell = 1,\ldots,L, k = 1,\ldots,4 \qquad (10.95)$$

The basic NNSS method is that one picks the location/grid whose recorded data is closest to the observed data in terms of Euclidean distance defined in (10.95); that is:

$$\hat{\mathbf{p}} = \arg\min_{\ell,k} \rho^{(\ell,k)} \qquad (10.96)$$

Alternatively, in the case that there are multiple Euclidean distances that are very similar, multiple nearest neighbors may be considered. One way to make use of the multiple nearest neighbors is to average the corresponding location estimates. It is nontrivial to point out that the number of nearest neighbors should be set small to avoid including the locations that are far away from the desired one. That is, a small group of locations that produce the smallest Euclidean distances and are close to each other should be selected and averaged to produce the location estimate. Note that, for positioning in cellular networks, the database search can be limited to a relatively small area by using location area code and identification.

In wideband systems, such as 3G and WLANs, the channel power delay profile (PDP) can be employed [21]. The measured PDP of the signal transmitted from the mobile node is correlated with the PDPs stored in the database. The database point with the highest correlation output corresponds to the target location estimate. Note that channel impulse response (CIR) can be employed in a similar way. It is obvious that the database based on PDP and CIR would be much more complex than those using TOA and/or RSS as fingerprints.

10.7.2 Kernel-Function-based Method

Another approach is the nonparametric estimation method which uses kernel functions [22]. The basic idea is to assign an approximate PDF to the surveyed data at each survey point. The PDF can be constructed with respect to the RSS, TOA, TDOA or a combination of two or more of them. The approximate PDF is a sum of kernel functions. Each kernel function value decreases monotonically as the distance from the survey point increases, so that points close to survey points have higher PDF values than points far away from survey points. With a good selection of kernel functions, good approximations of the PDF can be generated. In general, the kernel function for a survey point is a function of the location of the survey point and the measurement at that point. Therefore, it would be convenient to model the joint PDF of measurements and locations as a sum of the product of the kernel functions for each survey measurement; that is:

$$\hat{p}(\boldsymbol{\theta}, \mathbf{r}) = \frac{1}{L} \sum_{i=1}^{L} (h_r)^{-m_r} (h_\theta)^{-m_p} K_\mathbf{r} \left(\frac{\mathbf{r} - \mathbf{r}_i}{h_r} \right) K_\mathbf{p} \left(\frac{\boldsymbol{\theta} - \boldsymbol{\theta}_i}{h_\theta} \right) \tag{10.97}$$

where L is the number of survey points, $K_\mathbf{r}(\mathbf{r})$ and $K_\boldsymbol{\theta}(\boldsymbol{\theta})$ are the kernel functions for the measurements and the locations respectively, h_r and h_θ are the smoothing constants that determine the width of each of the kernel functions and m_r and m_θ are the lengths of $\mathbf{r}$ and $\boldsymbol{\theta}$ respectively. Let us consider the minimum-mean-square-error (MMSE) estimator, which produces the location estimate given by

$$\hat{\boldsymbol{\theta}}_{\text{MMSE}} = E[\boldsymbol{\theta}|\mathbf{r}] = \int_S \boldsymbol{\theta} \, p(\boldsymbol{\theta}|\mathbf{r}) \mathrm{d}\boldsymbol{\theta} \tag{10.98}$$

where $p(\boldsymbol{\theta}|\mathbf{r})$ is the posterior conditional PDF of the location vector given the observation vector and S is the location region. Equation (10.98) can be written as

$$\hat{\boldsymbol{\theta}}_{\text{MMSE}} = \frac{\int_S \boldsymbol{\theta} \, p(\boldsymbol{\theta}, \mathbf{r}) \mathrm{d}\boldsymbol{\theta}}{\int_S p(\boldsymbol{\theta}, \mathbf{r}) \mathrm{d}\boldsymbol{\theta}} \tag{10.99}$$

Table 10.2 Kernel functions.

Kernel name	Kernel function $K_{\mathbf{r}}(\mathbf{r})$
Parzen Gaussian	$\left(\frac{1}{\sqrt{2\pi}}\right)^{m_r} \mathrm{cov}^{-1/2}(\mathbf{r})\exp\left(-\frac{\mathbf{r}^T\mathrm{cov}^{-1}(\mathbf{r})\mathbf{r}}{2}\right)$
Parzen Laplace	$\frac{1}{2}\exp(-\|\mathbf{r}\|)$
Distance based	$\prod_{i=1}^{m_r}\frac{1}{\pi}\frac{1}{1+(\mathbf{r}_i)^2}$

Substituting (10.97) into (10.98) produces

$$\hat{\boldsymbol{\theta}}_{\mathrm{MMSE}} = \sum_{i=1}^{L} w_i(\mathbf{r})\boldsymbol{\theta}_i \tag{10.100}$$

where

$$w_i(\mathbf{r}) = \frac{K_{\mathbf{r}}\left(\frac{\mathbf{r}-\mathbf{r}_i}{h_r}\right)}{\sum_{j=1}^{L} K_{\mathbf{r}}\left(\frac{\mathbf{r}-\mathbf{r}_j}{h_r}\right)} \tag{10.101}$$

The performance of the kernel-function-based method depends mainly on the selection of the kernel functions and the number and distribution of survey points. Three kernel functions are listed in Table 10.2.

10.7.3 Histogram-based Method

The third database-based approach is to use a histogram. The histogram-based method uses a different way of estimating density functions, which is closely related to the conversion of continuous values to discrete ones. In the survey phase, data are observed over a period of time under different conditions for each individual anchor node and every given survey point. Then, we can define a set of nonoverlapping intervals (bins) that cover a range given by the minimum and maximum observed values. The density value (or mass probability) of each bin is determined by the number of samples that are inside the bin. To handle the case where a bin has zero probability, a constant that is smaller than the minimum nonzero mass probability is assigned to all bins. As a result, for each survey point and an anchor node, there is a probability mass function (PMF). Under the assumption of independence among the anchor nodes, an approximate joint PMF can be determined for each survey point. At the real-time phase, the joint PMF is computed based on the observed vector and the point with the maximum joint PMF is chosen as the location estimate. In [23], the three methods (nearest neighbor, kernel function and histogram) are employed for a case study in a typical one-floor office. The results demonstrated that the histogram method outperforms the others, while the nearest-neighbor-based method has the poorest performance.

The advantage of the database method is that accuracy improvement is expected as the number of survey points increases. It is found that the database-based approach may greatly outperform the traditional trilateration method and Kalman filter-based tracking techniques in practice [24]. The disadvantage is that the creation and maintenance of such a database is time consuming. Therefore, it may not be suited for situations where either the number of points to be surveyed is too large to be feasible or the structure of the area varies too often.

10.8 Performance Analysis

This section provides some theoretical tools for evaluating the accuracy of positioning algorithms in NLOS scenarios. The CRLB is first derived and then the approximate mean and variance of the ML position determination are presented for both error-free modeling and with model mismatch.

10.8.1 Cramer–Rao Lower Bound

As mentioned in Chapter 8, the CRLB sets a lower bound on any unbiased estimators. The CRLB for the kth parameter of $\mathbf{p}$ is defined by

$$CRLB(\hat{\mathbf{p}}_k) = [\mathbf{F}^{-1}(\mathbf{p})]_{k,k} \qquad (10.102)$$

where $\mathbf{F}(\mathbf{p})$ is the FIM whose elements are defined as

$$[\mathbf{F}(\mathbf{p})]_{k,\ell} = -E\left[\frac{\partial^2 \ln p(\mathbf{r}|\mathbf{p})}{\partial \mathbf{p}_k \partial \mathbf{p}_\ell}\right] \qquad (10.103)$$

where the likelihood function $p(\mathbf{r}|\mathbf{p})$ is given by (10.81). It can be readily shown that

$$[\mathbf{F}(\mathbf{p})]_{1,1} = \sum_{i=1}^{N} \frac{(x - x_i)^2}{\tilde{\sigma}_i^2 d_i^2}$$

$$[\mathbf{F}(\mathbf{p})]_{2,2} = \sum_{i=1}^{N} \frac{(y - y_i)^2}{\tilde{\sigma}_i^2 d_i^2} \qquad (10.104)$$

$$[\mathbf{F}(\mathbf{p})]_{1,2} = [\mathbf{F}(\mathbf{p})]_{2,1} = \sum_{i=1}^{N} \frac{(x - x_i)(y - y_i)}{\tilde{\sigma}_i^2 d_i^2}$$

The variances of the coordinate estimates satisfy

$$\mathrm{var}(\hat{x}) \geq CRLB(\hat{x}) = \frac{[\mathbf{F}(\mathbf{p})]_{2,2}}{[\mathbf{F}(\mathbf{p})]_{1,1}[\mathbf{F}(\mathbf{p})]_{2,2} - [\mathbf{F}(\mathbf{p})]_{1,2}^2}$$

$$\mathrm{var}(\hat{y}) \geq CRLB(\hat{y}) = \frac{[\mathbf{F}(\mathbf{p})]_{1,1}}{[\mathbf{F}(\mathbf{p})]_{1,1}[\mathbf{F}(\mathbf{p})]_{2,2} - [\mathbf{F}(\mathbf{p})]_{1,2}^2} \qquad (10.105)$$

10.8.2 Approximate Mean and Variance of Maximum Likelihood Estimator

It is known that the ML estimation of an unknown parameter vector is Gaussian distributed asymptotically with mean equal to the true value and covariance equal to the CRLB, provided that the received data satisfies certain regularity conditions [25]. The ML estimator is optimum when the observation model is linear and the noise is Gaussian. The CRLB given above may not provide an accurate measure for the ML location estimation algorithms described in Section 10.5 due to the nonlinearity. In this section, we derive alternative theoretical expressions to approximate the mean and variance of the ML location estimation described in Section 10.5.1 for the cases with and without model parameter errors.

10.8.2.1 With Error-free Model Parameters

The ML estimation produces a location estimate by maximizing the log likelihood given by (10.83). Differentiating $\Lambda(\mathbf{p})$ with respect to x and to y and setting them equal to zero yields

$$
\begin{aligned}
g_x(\mathbf{p}) &= -\sum_{i=1}^{N} \frac{(d_i - \tilde{r}_i)(x - x_i)}{\tilde{\sigma}_i^2 d_i} = 0 \\
g_y(\mathbf{p}) &= -\sum_{i=1}^{N} \frac{(d_i - \tilde{r}_i)(y - y_i)}{\tilde{\sigma}_i^2 d_i} = 0
\end{aligned}
\tag{10.106}
$$

The solution of x and y in (10.106) gives the ML position estimate. Since these are nonlinear equations, we cannot obtain an exact closed-form expression for the solution. To overcome the problem, the Taylor expansion can be used to linearize the equations. Expanding $g_x(\mathbf{p})$ and $g_y(\mathbf{p})$ in a Taylor series at the true location, which is denoted by $\mathbf{p}^o = \begin{bmatrix} x^o & y^o \end{bmatrix}^{\mathrm{T}}$, and retaining the terms below the second order produces

$$
\begin{aligned}
a_{11}\delta x + a_{12}\delta y &\approx a_{13} \\
a_{21}\delta x + a_{22}\delta y &\approx a_{23}
\end{aligned}
\tag{10.107}
$$

where

$$
a_{11} = \left.\frac{\partial g_x(\mathbf{p})}{\partial x}\right|_{\mathbf{p}=\mathbf{p}^o} \qquad a_{12} = \left.\frac{\partial g_x(\mathbf{p})}{\partial y}\right|_{\mathbf{p}=\mathbf{p}^o} \qquad a_{13} = -g_x(\mathbf{p}^o)
$$

$$
a_{21} = \left.\frac{\partial g_y(\mathbf{p})}{\partial x}\right|_{\mathbf{p}=\mathbf{p}^o} \qquad a_{22} = \left.\frac{\partial g_y(\mathbf{p})}{\partial y}\right|_{\mathbf{p}=\mathbf{p}^o} \qquad a_{23} = -g_y(\mathbf{p}^o) \tag{10.108}
$$

$$
\delta x = x - x^o \qquad \delta y = y - y^o
$$

Taking expectation over $\tilde{r}_i$ on both sides of (10.107) yields two equations with two unknowns: $E[\delta x]$ and $E[\delta y]$. Since the means of a_{13} and a_{23} are both zero, it can be readily seen that

$$
E[\delta x] = 0 \qquad E[\delta y] = 0 \tag{10.109}
$$

That is, the means of the ML coordinate estimates are equal to the true values of the coordinates as

$$
\hat{x}_{\mathrm{ML}} = x^o \qquad \hat{y}_{\mathrm{ML}} = y^o \tag{10.110}
$$

This is in accordance with the asymptotic properties of the ML estimation. Under the assumption of uncorrelated δx and δy, and with some mathematical manipulations of (10.107), we obtain

$$
\begin{aligned}
f_{11}\text{var}(\delta x) + f_{12}\text{var}(\delta y) &\approx f_{13} \\
f_{21}\text{var}(\delta x) + f_{22}\text{var}(\delta y) &\approx f_{23}
\end{aligned}
\tag{10.111}
$$

where $\{f_{ij}\}$ are given in Annex 10.B. Consequently, we obtain the approximate variances of the ML coordinate estimates as

$$
\begin{aligned}
\text{var}(\delta x) &\approx \frac{f_{13}f_{22} - f_{12}f_{23}}{f_{11}f_{22} - f_{12}^2} \\
\text{var}(\delta y) &\approx \frac{f_{11}f_{23} - f_{13}f_{21}}{f_{11}f_{22} - f_{12}^2}
\end{aligned}
\tag{10.112}
$$

10.8.2.2 With Model Parameters Mismatch

When an empirical model is used to characterize the NLOS bias or the measurement noise, a certain mismatch exists between the actual data and the model. When the environment structure or the system parameters are changed, additional modeling error would be incurred. In practice, one may be interested in how much accuracy degradation would be incurred due to the model mismatch. The error in the distribution parameter estimation is not the only source that results in the difference between the model and the actual data, since the actual data may not perfectly fit to a specific distribution due to environmental complexity and variety. However, for simplicity, let us assume that the mismatch is only caused by the distribution parameter error. That is, the estimated parameters are

$$
\begin{aligned}
\hat{\sigma}_i &= \sigma_i^o + \delta\sigma_i \\
\hat{\lambda}_i &= \lambda_i^o + \delta\lambda_i
\end{aligned}
\tag{10.113}
$$

where $\{\sigma_i^o\}$ and $\{\lambda_i^o\}$ are the true parameters and $\{\delta\sigma_i\}$ and $\{\delta\lambda_i\}$ are the corresponding parameter estimation/matching errors. It is interesting to obtain the variances of the ML estimation under this circumstance of model parameter mismatch by following the same procedure in the preceding subsection. In the presence of the distribution parameter errors, expanding $g_x(\mathbf{p})$ and $g_y(\mathbf{p})$ in (10.106) in a Taylor series at the true position $\mathbf{p}^o$ and the true distribution parameters $\{\sigma_i^o\}$ and $\{\lambda_i^o\}$, one obtains

$$
\begin{aligned}
a_{11}\delta x + a_{12}\delta y &\approx a_{13} + \sum_{i=1}^{N}\kappa_{1i}\delta\sigma_i + \sum_{i=N_{\text{los}}+1}^{N}\rho_{1i}\delta\lambda_i \\
a_{21}\delta x + a_{22}\delta y &\approx a_{23} + \sum_{i=1}^{N}\kappa_{2i}\delta\sigma_i + \sum_{i=N_{\text{los}}+1}^{N}\rho_{2i}\delta\lambda_i
\end{aligned}
\tag{10.114}
$$

where $\{a_{ij}\}$ have the same definitions as in (10.107) and

$$
\kappa_{1i} = \left.\frac{\partial g_x(\boldsymbol{\theta})}{\partial \sigma_i}\right|_{\boldsymbol{\theta}=\boldsymbol{\theta}^{\mathrm{o}}} \qquad \rho_{1i} = \left.\frac{\partial g_x(\boldsymbol{\theta})}{\partial \lambda_i}\right|_{\boldsymbol{\theta}=\boldsymbol{\theta}^{\mathrm{o}}}
$$

$$
\kappa_{2i} = \left.\frac{\partial g_y(\boldsymbol{\theta})}{\partial \sigma_i}\right|_{\boldsymbol{\theta}=\boldsymbol{\theta}^{\mathrm{o}}} \qquad \rho_{2i} = \left.\frac{\partial g_y(\boldsymbol{\theta})}{\partial \lambda_i}\right|_{\boldsymbol{\theta}=\boldsymbol{\theta}^{\mathrm{o}}} \tag{10.115}
$$

$$
\boldsymbol{\theta} = \begin{bmatrix} x & y & \sigma_1 & \cdots & \sigma_N & \lambda_{N_{\mathrm{los}}+1} & \cdots & \lambda_N \end{bmatrix}^{\mathrm{T}}
$$

$$
\boldsymbol{\theta}^{\mathrm{o}} = \begin{bmatrix} x^{\mathrm{o}} & y^{\mathrm{o}} & \sigma_1^{\mathrm{o}} & \cdots & \sigma_N^{\mathrm{o}} & \lambda_{N_{\mathrm{los}}+1}^{\mathrm{o}} & \cdots & \lambda_N^{\mathrm{o}} \end{bmatrix}^{\mathrm{T}}
$$

Under the assumption that δx, δy, $\{\delta \sigma_i\}$ and $\{\delta \lambda_i\}$ are uncorrelated, and that $\{\delta \lambda_i\}$ have a zero mean, it can be readily seen that the mean of the ML estimation still equals zero. Similar to (10.111), we have

$$
\begin{aligned}
f_{11}\mathrm{var}(\delta x) + f_{12}\mathrm{var}(\delta y) &\approx f_{13} + g_{13} \\
f_{21}\mathrm{var}(\delta x) + f_{22}\mathrm{var}(\delta y) &\approx f_{23} + g_{23}
\end{aligned} \tag{10.116}
$$

where $\{f_{ij}\}$ have the same definitions as in (10.111) and

$$
\begin{aligned}
g_{13} &= \sum_{i=1}^{N_{\mathrm{los}}} (x^{\mathrm{o}} - x_i)^2 E_{\mathrm{s}} + \sum_{i=N_{\mathrm{los}}+1}^{N_{\mathrm{los}}} (x^{\mathrm{o}} - x_i)^2 E_{\mathrm{c}} \\
g_{23} &= \sum_{i=1}^{N_{\mathrm{los}}} (y^{\mathrm{o}} - y_i)^2 E_{\mathrm{s}} + \sum_{i=N_{\mathrm{los}}+1}^{N_{\mathrm{los}}} (y^{\mathrm{o}} - y_i)^2 E_{\mathrm{c}}
\end{aligned} \tag{10.117}
$$

where

$$
E_{\mathrm{s}} = \frac{4E[\delta \sigma_i^2]}{(d_i^{\mathrm{o}})^2 (\sigma_i^{\mathrm{o}})^4}
$$

$$
E_{\mathrm{c}} = \frac{4\{4(\sigma_i^{\mathrm{o}})^2 E[\delta \sigma_i^2] + [(\tilde{\sigma}_i^{\mathrm{o}})^4 + 4(\lambda_i^{\mathrm{o}} \sigma_i^{\mathrm{o}})^2]E[\delta \lambda_i^2]\}}{(d_i^{\mathrm{o}})^2 (\tilde{\sigma}_i^{\mathrm{o}})^8} \tag{10.118}
$$

$$
\tilde{\sigma}_i^{\mathrm{o}} = \sqrt{(\sigma_i^{\mathrm{o}})^2 + (\lambda_i^{\mathrm{o}})^2}
$$

The only difference between (10.111) and (10.116) is that an extra term appears on the right-hand side of each of the two equations in (10.116). As a result, the variances of the coordinate estimation in the presence of a distribution parameter mismatch are given by (10.112) with f_{i3}, $i = 1, 2$, replaced by $f_{i3} + g_{i3}$.

Table 10.3 shows the theoretical and simulated STDs of the ML estimation algorithm with and without parameter mismatch. Also listed are the CRLBs for comparison. A main observation is that, on average, the derived theoretical variances may provide a better measure for the accuracy of the ML estimation than the CRLB.

Table 10.3 Simulated and theoretical STDs of the *x*-coordinate estimation errors of the ML estimation algorithms (the idealized case and the case with parameter mismatch) with different number of NLOS anchors N_n.

N_n	Without mismatch		With mismatch		
	Simulated	Analytical	Simulated	Analytical	CRLB
1	0.54 m	0.43 m	0.58 m	0.50 m	0.2 m
2	0.83 m	0.59 m	0.90 m	0.64 m	0.29 m
3	1.31 m	0.88 m	1.44 m	0.93 m	0.47 m
4	1.77 m	1.46 m	1.90 m	1.72 m	0.65 m

Annex 10.A: Sequential Quadratic Programming Algorithm

The SQP algorithm for the optimization-based location estimation is described as follows. First, a Lagrangian function is defined as

$$\mathcal{L}(\boldsymbol{\theta}_k, \boldsymbol{\lambda}) = S(\boldsymbol{\theta}_k) + \sum_{i=1}^{3N+4} \lambda_i g_i(\boldsymbol{\theta}_k) \tag{10.119}$$

where $\boldsymbol{\theta}_k$ is the estimate of $\boldsymbol{\theta}$ at the kth iteration, $\{g_i\}$ ($g_i \le 0$) are defined as

$$g_i = \begin{cases} \sqrt{(x-x_i)^2 + (y-y_i)^2} - (r_i + n_i^{\mathrm{L}}) & 1 \le i \le N \\ x - x^{\mathrm{U}} & i = N+1 \\ y - y^{\mathrm{U}} & i = N+2 \\ x^{\mathrm{L}} - x & i = N+3 \\ y^{\mathrm{L}} - y & i = N+4 \\ b_i - b_i^{\mathrm{U}} & N+5 \le i \le 2N+4 \\ b_i^{\mathrm{L}} - b_i & 2N+5 \le i \le 3N+4 \end{cases} \tag{10.120}$$

and $\{\{\boldsymbol{\lambda}\}_i = \lambda_i\}$ are the Lagrange multipliers that are estimated based on the Kuhn–Tucker equations, that is,

$$\begin{aligned} \nabla S(\boldsymbol{\theta}_k) + \sum_{i=1}^{3N+4} \lambda_i g_i(\boldsymbol{\theta}_k) &= 0 \\ \lambda_i g_i(\boldsymbol{\theta}_k) &= 0 \\ \lambda_i \ge 0 \quad &1 \le i \le 3N+4 \end{aligned} \tag{10.121}$$

where ∇ is the gradient (differentiation) operator. Then, the quasi-Newton method is used to update/approximate the Hessian of the Lagrangian function by using the BFGS formula as follows:

$$\mathbf{H}_{k+1} = \mathbf{H}_k + \frac{\mathbf{q}_k \mathbf{q}_k^{\mathrm{T}}}{\mathbf{q}_k^{\mathrm{T}} \mathbf{s}_k} - \frac{\mathbf{H}_k^{\mathrm{T}} \mathbf{H}_k}{\mathbf{s}_k^{\mathrm{T}} \mathbf{H}_k \mathbf{s}_k} \tag{10.122}$$

where

$$\mathbf{s}_k = \boldsymbol{\theta}_{k+1} - \boldsymbol{\theta}_k$$
$$\mathbf{q}_k = \nabla S(\boldsymbol{\theta}_{k+1}) + \sum_{i=1}^{3N+4} \lambda_i \nabla g_i(\boldsymbol{\theta}_{k+1}) - \left(\nabla S(\boldsymbol{\theta}_k) + \sum_{i=1}^{3N+4} \lambda_i \nabla g_i(\boldsymbol{\theta}_k) \right) \tag{10.123}$$

and $\mathbf{H}_k$ is initialized with a positive definite matrix. Next, the updated Hessian $\mathbf{H}_{k+1}$ is used to produce a quadratic programming solution to the search direction, $\boldsymbol{v}_k \in \mathbb{R}^{(N+2) \times 1}$, according to

$$\min_{\boldsymbol{v}_k}(\boldsymbol{v}_k) = \frac{1}{2} \boldsymbol{v}_k^{\mathrm{T}} \mathbf{H}_{k+1} \boldsymbol{v}_k + \nabla S^{\mathrm{T}}(\boldsymbol{\theta}_k) \boldsymbol{v}_k \tag{10.124}$$

subject to

$$\nabla g_i^{\mathrm{T}}(\boldsymbol{\theta}_k) \boldsymbol{v}_k + g_i(\boldsymbol{\theta}_k) \leq 0 \tag{10.125}$$

The search direction is then used to update the position estimate based on

$$\boldsymbol{\theta}_{k+1} = \boldsymbol{\theta}_k + \delta_k \boldsymbol{v}_k \tag{10.126}$$

where δ_k is the step length parameter that is determined to produce a sufficient decrease in a merit/penalty function defined as

$$F(\boldsymbol{\theta}, \boldsymbol{\rho}) = S(\boldsymbol{\theta}) + \sum_{i=1}^{3N+4} \rho_i \max(g_i(\boldsymbol{\theta}), 0) \tag{10.127}$$

where $\{\boldsymbol{\rho}\}_i = \rho_i > 0$ are the penalty parameters initialized to $||\nabla S(\boldsymbol{\theta})|| / ||g_i(\boldsymbol{\theta}_k)||$ (where $||.||$ is the Euclidean norm) and updated according to $\rho_i^{(k+1)} = \max\{\lambda_i, (\rho_i^{(k)} + \lambda_i)/2\}$. The merit function is used as a criterion for determining whether a point is better than another in terms of decreasing $S(\boldsymbol{\theta})$ and being closer to satisfying the constraints. The process repeats until a predesigned criterion, such as the number of iterations or the increment of $\boldsymbol{v}_k$, is satisfied.

Annex 10.B: Equation Coefficients

The $\{f_{ij}\}$ in (10.111) are given as follows:

$$f_{11} = \sum_{i=1}^{N} \frac{1}{\tilde{\sigma}_i^4} \left[1 - \frac{2(y-y_i)^2}{d_i^2} + \frac{(\tilde{\sigma}_i^2 + d_i^2)(y-y_i)^4}{d_i^6} \right] + 2\sum_{i=1}^{N-1} \sum_{j=i+1}^{N} \frac{(x-x_i)^2(x-x_j)^2}{\tilde{\sigma}_i^2 \tilde{\sigma}_j^2 d_i^2 d_j^2}$$

$$f_{22} = \sum_{i=1}^{N} \frac{1}{\tilde{\sigma}_i^4} \left[1 - \frac{2(x-x_i)^2}{d_i^2} + \frac{(\tilde{\sigma}_i^2 + d_i^2)(x-x_i)^4}{d_i^6} \right] + 2\sum_{i=1}^{N-1} \sum_{j=i+1}^{N} \frac{(y-y_i)^2(y-y_j)^2}{\tilde{\sigma}_i^2 \tilde{\sigma}_j^2 d_i^2 d_j^2}$$

$$f_{12} = f_{21} = \sum_{i=1}^{N} \frac{(\tilde{\sigma}_i^2 + d_i^2)(x-x_i)^2(y-y_i)^2}{\tilde{\sigma}_i^4 d_i^6} + 2\sum_{i=1}^{N-1} \sum_{j=i+1}^{N} \frac{(x-x_i)(x-x_j)(y-y_i)(y-y_j)}{\tilde{\sigma}_i^2 \tilde{\sigma}_j^2 d_i^2 d_j^2}$$

$$f_{13} = \sum_{i=1}^{N} \frac{(x-x_i)^2}{\tilde{\sigma}_i^2 d_i^2}$$

$$f_{23} = \sum_{i=1}^{N} \frac{(y-y_i)^2}{\tilde{\sigma}_i^2 d_i^2} \tag{10.128}$$

where

$$\begin{aligned} x &= x^o \\ y &= y^o \\ d_i &= \sqrt{(x^o - x_i)^2 + (y^o - y_i)^2} \end{aligned} \tag{10.129}$$

References

[1] P.-C. Chen, 'A non-line-of-sight error mitigation algorithm in location estimation', in *Proceedings of IEEE Wireless Communications and Networking Conference (WCNC)*, pp. 316–320, September 1999.

[2] Y. Jeong, H. You and C. Lee, 'Calibration of NLOS error for positioning systems', in *Proceedings of IEEE Vehicular Technology Conference (VTC)*, pp. 2605–2608, May 2001.

[3] W. Wang, Z. Wang and B. O'Dea, 'A TOA-based location algorithm reducing the errors due to nonline-of-sight (NLOS) propagation', *IEEE Transactions on Vehicular Technology*, **52**(1), 2003, 112–116.

[4] W. Kim, J.G. Lee and G.-I. Jee, 'The interior-point method for an optimal treatment of bias in trilateration location', *IEEE Transactions on Vehicular Technology*, **55**(4), 2006, 1291–1301.

[5] S. Venkatraman, J. CafferyJr. and H.-R. You, 'Location using LOS range estimation in NLOS environments', in *Proceedings of IEEE Vehicular Technology Conference (VTC)*, pp. 856–860, May 2002.

[6] W. Wang, J.-Y. Xiong and Z.-L. Zhu, 'A new NLOS error mitigation algorithm in location estimation', *IEEE Transactions on Vehicular Technology*, **54**(6), 2005, 2048–2053.

[7] B. Alavi and K. Pahlavan, 'Modeling of the distance error for indoor geolocation', in *Proceedings of IEEE Wireless Communications and Networking*, pp. 668–672, March 2003.

[8] X. Li and K. Pahlavan, 'Super-resolution TOA estimation with diversity for indoor geolocation', *IEEE Transactions on Wireless Communications*, **3**(1), 2004, 224–234.

[9] J.J. CafferyJr. and G.L. Stuber, 'Subscriber location in CDMA cellular systems', *IEEE Transactions on Vehicular Technology*, **47**(2), 1998, 406–416.

[10] J.H. Yap S. Ghaheri-Niri and R. Tafazolli, 'Accuracy hearability of mobile positioning in GSM and CDMA networks', in *Proc. 3G Mobile Communication Technologies*, pp. 350–354, May 2002.

[11] P. J. Duffett-Smith and M. D. Macnaughtan, 'Precise UE positioning in UMTS using cumulative virtual blanking', in *Proceedings of 3G Mobile Communication Technologies*, pp. 355–359, May 2002.

[12] Y. Jay Guo, *Advances in Mobile Radio Access Networks*, Artech House, 2003.

[13] H. Miao, K. Yu and M. Juntti, 'Positioning for NLOS propagation: algorithm derivations and Cramer–Rao bounds', *IEEE Transactions on Vehicular Technology*, **56**(5), 2007, 2568–2580.

[14] N.J. Thomas, D.G.M. Cruickshank and D.I. Laurenson, 'Calculation of mobile location using scatterer information', *Electronics Letters*, **37**(19), 2001, 1193–1194.

[15] M. Porretta, P. Nepa, G. Manara, F. Giannetti, M. Dohler, B. Allen and A.H. Aghvami, 'User positioning technique for microcellular wireless networks', *Electronics Letters*, **39**(9), 2003, 745–747.

[16] P. Bahl and V. Padmanabhan, 'RADAR: an in-building RF-based user location and tracking system', in *Proceedings of IEEE Conference on Computer Communications (INFOCOM)*, pp. 775–784, 2000.

[17] T.S. Rappaport, *Wireless Communications – Principles and Practice*, Prentice Hall, 2001.

[18] E. Damosso (ed.), Digital mobile radio towards future generation systems. COST 231 Final Report, 1996.

[19] S. Al-Jazzar, J. Caffery Jr and H.R. You, 'A scattering model based approach to NLOS mitigation in TOA location systems', in *Proceedings of IEEE Vehicular Technology Conference (VTC)*, pp. 861–865, May 2002.

[20] W.G. Figel, N.H. Shepherd and W.F. Trammell, 'Personal location services emerge', *IEEE Transactions on Vehicular Technology*, **18**(3), 1969, 105–109.

[21] S. Juurakko and W. Backman, 'Database correlation method with error correlation for emergency location', *Wireless Personal Communications*, **30**(2–4), 2004, 183–194.

[22] M. McGuire, K. Plataniotis and A. Venetsanopoulos, 'Location of mobile terminals using time measurements and survey points', *IEEE Transactions on Vehicular Technology*, **52**(4), 2003, 999–1011.

[23] T. Roos, P. Myllymaki, H. Tirri, P. Misikangas and J. Sievanen, 'A probabilistic approach to WLAN user location estimation', *International Journal of Wireless Information Networks*, **9**(1), 2002, 155–164.

[24] K. Pahlavan, X. Li and J.-P. Makela, 'Indoor geolocation science and technology', *IEEE Communications Magazine*, **40**(2), 2002, 112–118.

[25] S.M. Kay, *Fundamentals of Statistical Signal Processing: Estimation Theory*, Prentice Hall, Upper Saddle River, NJ, 1993

11

Anchor-based Localization for Wireless Sensor Networks

For WSNs, localization has direct relevance in such applications as personal and assets tracking, emergence rescue operations and healthcare. In many circumstances, the position information is a key component in interpreting the sensor data. Furthermore, the location information can be used to improve networking performance.

Generally speaking, there are anchor nodes in most medium to large-scale WSNs, which serve as access points, gateways or base stations. The locations of the anchor nodes are known *a priori*, either from another location system such as GPS or being determined manually by some surveying or mapping technique. The objective of localization is to estimate the positions of the 'ordinary' sensor nodes by using the known locations of the anchors.

In previous chapters, position determination is accomplished based mainly on TOA techniques. While these methods can achieve good accuracy, both indoors and outdoors, the signal processing load can be very high and, thus, inappropriate for the simple low-power nodes employed in WSNs. For outdoor applications, small low-cost GPS receivers can be used, albeit with extra cost, power consumption and difficulties associated with having adequate GPS antenna performance in a small package. In many cases, particularly indoors, GPS reception is simply not possible, so that an alternative localization method based solely on the simple radios used for data communications is highly desirable. Such methods will necessarily be of lower positional accuracy; but, as will be shown, accuracy in the order of 5 m can still be achieved. Such positional accuracy is comparable with GPS, but is considerably worse than the best TOA-based systems described in previous chapters.

The chapter is organized as follows. The general characteristics of WSNs are described in Section 11.1. In Section 11.2, some basic position determination methods are discussed, which at the simplest only require the identity of nearby anchor nodes. Section 11.3 presents a number of typical multihop localization algorithms. Finally, TOA-based localization is studied in detail in Section 11.4. A number of practical parameters are considered, including the clock frequency offsets, system internal delays and clock time offsets.

Ground-Based Wireless Positioning   Kegen Yu, Ian Sharp and Jay Guo
© 2009 John Wiley & Sons, Ltd

11.1 Characteristics of Wireless Sensor Networks

A WSN is comprised of spatially distributed autonomous nodes. Each node consists of a two-way radio with one or more sensors which perform sensing of the environmental parameters, such as temperature, vibration, moisture, pressure, motion, chemicals or pollutants. One common characteristic of typical sensor data is the low rate of change, so that the update rate can be quite low; hence, the data bandwidth required in the network can be modest. The sensed data are typically cooperatively forwarded to a fixed node for storage and processing. The development of WSNs was originally motivated by military applications, such as battlefield surveillance. However, WSNs are now used in many civilian applications, including environment and habitat monitoring, healthcare, home automation and traffic control [1].

A WSN may possess one or more of the following characteristics:

- *Mobility*   Sensor nodes may change their location after initial deployment. Mobility can result from environmental influences, such as wind or the flow of water, and the fact that sensor nodes may be attached to or be carried by mobile entities. In other words, mobility may be either an incidental side effect or intentional, such as nodes moving to different physical locations. In the latter case, mobility may be either active (sensor-directed motion) or passive (attached to a moving object not under the control of the sensor node).
- *Limited battery power*   Sensor nodes are typically powered by batteries. Owing to the constraints on the physical size of the devices, the battery-stored energy will limit the operational life of the node. Typical batteries can only store about 1 J/mm^3, so that the total available energy may be as low as 1 kJ. Therefore, power consumption is a crucial issue in WSN.
- *Low computational power and limited memory space*   Sensor nodes are usually built at low cost and with a small dimensional size. As a result, running complicated algorithms involving intensive computation at ordinary sensor nodes is usually not feasible. Further, because of the limited memory including both volatile (RAM) and nonvolatile (flash memory), intensive storage and holding data for a long period of time should be avoided.
- *Low communications bandwidth*   The communications bandwidth is usually low (say a few kilobits a second), and often must be shared with adjacent nodes for mesh communications. Further, it is important to avoid too much overhead traffic in the network protocol.
- *Limited lifetime*   The lifetime of a sensor node is often determined by the battery if no other source of power is available. As the lifetime is defined by the ratio of the battery capacity to the average power consumption, low data sampling rates, low communications rates and long hibernation times are desirable. The powering of WSNs is currently one of the limiting factors in their development. It is often desired that the sensor nodes are deployed in harsh environments and on a large scale. Therefore, manually recharging or changing the battery is not desirable due to the logistic difficulties, cost and potential risks.
- *Large-scale deployment*   Sensor nodes can be deployed on a large scale. For environmental monitoring, such as soil and water quality monitoring, hundreds or even thousands of sensors can be scattered over a large area. The deployment can be either random, by dropping the sensor nodes from an aircraft for instance, or done according to some plan.
- *Heterogeneity of nodes*   Early visions of sensor networks were that they would typically consist of homogeneous devices that were mostly identical from the hardware and software

points of view. However, in many prototypical systems available today, sensor networks consist of a variety of different devices.

- *Harsh environmental conditions*   Sensor nodes may be deployed in harsh environmental conditions, such as high temperature, high voltage and high pressure.
- *Unattended operation*   Once the sensor nodes are deployed, they are usually unattended.

Localization algorithm development and system implementation of sensor networks need to take into account of the above characteristics. The most important limitations are the processing power, the time to process the data (limiting battery power consumption) and the limited capability of signal processing hardware due to cost.

11.2 Coarse Localization Methods

Owing to their simplicity and low demand on the sensor nodes, coarse localization methods are especially import for WSNs. In some cases, the coarse position information can be employed as the initial values for more complex algorithms to produce more accurate position estimates. Three such basic distance or radio-range-based localization methods are described in this section.

11.2.1 Centroid Method

The centroid-based approach [2] is a simple connectivity metric method for coarse localization in a heterogeneous network containing powerful data-processing anchor nodes with established location information. The method assumes perfect spherical radio propagation and identical transmission range of all anchors, even though this is known to be inaccurate. In this method, anchors transmit their positions to neighboring anchors and ordinary sensor nodes keep an account of all received transmissions. Using the proximity information, sensor nodes assume their location is at the centroid of the anchors from which they receive transmissions. Note that this method does not use explicitly any ranging information, but it does require a number of anchor nodes to 'surround' the ordinary node. The number of such anchor nodes will be limited by the range of the transmission system and, thus, the anchor nodes must be of sufficient spatial density for the method to work.

Consider a typical situation, as shown in Figure 11.1. The sensor node denoted by a square can receive transmissions from anchors 2 and 4, but not from anchors 1 and 3. Clearly, the ambiguity region is the shaded area, and the sensor node simply computes its own position according to

$$\hat{x} = \frac{1}{2}(x_2 + x_4)$$
$$\hat{y} = \frac{1}{2}(y_2 + y_4)$$

(11.1)

where (x_i, y_i) are the coordinates of anchor i. Note that the actual position may well be outside the shaded region due to different radio propagation characteristics from the assumed equal-range circles, but the most probable position is the centroid position. This method can be generalized to the case where there are more than two anchor nodes in range.

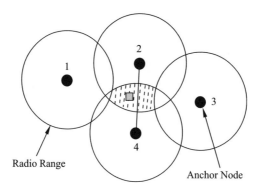

Figure 11.1 Illustration of centroid based localization. The WSN node denoted by square can receive from nodes 2 and 4 but not nodes 1 and 3. If all nodes have the same radio range, then the WSN node must be in the shaded area.

11.2.2 Anchor-Combination-based Point-in-Triangulation Test Method

The anchor-combination-based point-in-triangulation (APIT) test [3] is an area-based approach to perform location estimation by using triangular regions defined by three anchor nodes. This method requires a heterogeneous network where a certain percentage of the sensing nodes are anchor nodes equipped with high-power transmitters and with receivers that can measure the signal strength. The percentages can vary depending on network and node density. Note that, because of fading effects in a multipath environment, some local averaging of the RSS is required, as the method is based on the large-scale variation of signal strength with distance. The method assumes that, on average, the radio propagation environment is quasi-homogeneous. Unlike the centroid method above, this method requires the measurement of the RSS and, hence, requires a more complex radio receiver. The position estimation then requires the distribution of data to the neighboring nodes.

The APIT method operates as follows. Each ordinary node keeps a table of anchor nodes in the propagation range, which includes the anchor identification, anchor location and the RSS. After receiving the signal from an anchor node, the ordinary node updates its table. If there are at least three such anchor entries in the table, then the node transmits a message to exchange the anchor data with its neighbors. The table is managed at every node to maintain a neighborhood state, which is then used to determine whether there is a neighboring node that has consistently larger/smaller signal strengths associated with three anchors.

The position estimation method using the neighborhood table is as follows. If the signal strengths measured by one or more neighbors of the particular node are all greater than the corresponding ones measured by the node, then it is assumed that the node is outside the triangle region formed by the three anchors. If no such neighbor is found, then the node is assumed to be inside this triangle region. The test is repeated for varying combinations of three anchors. Next, aggregation is used to determine the area with the maximum overlap of all the triangular regions associated with the node. The center of gravity of this area is the location estimate. Figure 11.2 shows a simple example to illustrate the APIT approach. The shaded area is the overlapping region of the three triangles (123, 124 and 135) within which the target node is assumed to be. Clearly, the smaller the overlapping region is, the more accurate the

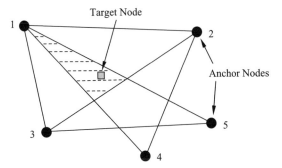

Figure 11.2 Illustration of APIT approach. The (square) node can receive from nodes 1–5, and in each case it is assumed the node must be inside the triangle associated with the three particular nodes (in this case 123, 124 and 135). The estimated position is thus assumed to be within the shaded area, with the most probable position assumed to be the center of gravity of the area.

location estimates are. The position is then assumed to be located at the centroid of the overlapping region.

11.2.3 Bounding Box Method

If a distance estimate is available, say using the signal strength method, then the position can be estimated using a bounding box method. In this method, the range circle is replaced by the corresponding enclosing square box. This approach greatly simplifies the required calculations for position determination.

Consider the case shown in Figure 11.3 with three nodes numbered 1, 2 and 3 as anchors. After obtaining the distance measurements between the target and the anchors, the coordinates of the target node are bounded by

$$x_i - \hat{d}_i \leq x \leq x_i + \hat{d}_i$$
$$y_i - \hat{d}_i \leq y \leq y_i + \hat{d}_i$$

(11.2)

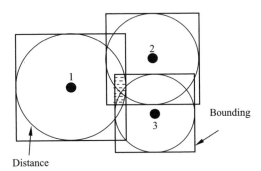

Figure 11.3 Bounding-box-based localization method. The distance from the node to the base stations is measured by the node, so that three bounding boxes can be defined. The node position is assumed to be within the area associated with the overlapping of the three boxes.

where (x_i, y_i) are the coordinates of anchor i and $\hat{d}_i$ is the measured distance between the target and anchor i. That is, for each distance measurement, the target node is constrained to the inside of the square, called the bounding box, which is tangential to the distance circle and aligned with the Cartesian coordinate system. The maximum overlapping area (the shaded area in the figure) of the bounding boxes is the area in which the target is assumed to be. The position estimate of the node is taken to be the centroid of the area.

Because this method would typically be used when the range data are of low quality, the corresponding boxes may not overlap, or may overlap in an incorrect fashion. In the first case, no position would be determined by this method, so that one of the simpler methods described previously would be used. In the latter case, a position will be obtained, but the accuracy may be poor.

11.3 Global Localization Methods

The three basic localization methods described in Section 11.2 provide rather rough position information. To localize a large number of sensor nodes, more accurate and efficient methodologies are required. In this section, a number of global localization methods are analyzed.

11.3.1 Incremental Scheme

In a network where there is a relatively low density of anchor nodes, position determination to ordinary (mobile) nodes may be difficult due to range limitations of the transmissions. However, if one particular ordinary node can determine its position using anchors, then it can act as a quasi-anchor node. By repeating this process the positions of all nodes in a network can potentially be determined. This technique is referred to as the incremental scheme [4]. A summary of the scheme protocol is as follows. Ordinary nodes attempt to determine their positions using the anchor nodes. Any suitable position determination technique described previously can be used. In the case of a 2D environment, at least three anchor nodes must be in range, while at least four are required for a 3D environment. Once a node has determined its location, it broadcasts its location information to other nearby nodes, potentially enabling them to estimate their positions when they are not able to communicate with three or more anchors. This process repeats for all nodes, each of which would be able to communicate with at least three (2D) or four (3D) anchors and/or localized ordinary nodes. The drawbacks of the approach are:

- in the case of relatively low anchor density, the method may get stuck in regions of the network that are relatively sparse;
- error propagation becomes an issue in large networks, a consequence of using localized nodes as anchors.

To improve performance, error propagation can be reduced by using weighted multilateration. In this scheme, the locations of anchors are weighted more than those of the localized ordinary nodes, as anchor location information is more accurate. For the localized ordinary nodes, an estimate of the positional accuracy can normally be obtained by the position determination process, and this can be used to determine the weighting factors for other position determination using this node. Error propagation can also be mitigated by obtaining an overdetermined set of equations, usually based on using more than the minimum number of anchors or localized ordinary nodes for the position determination process.

In the event that a node cannot communicate with sufficient neighboring anchors or localized ordinary nodes, it will not be able to estimate its position by using the basic multilateration method. When this occurs, a node may attempt to estimate its position by making use of the location information of anchors that are two hops away. For example, consider the case when inter-nodal distances can be measured (rather than the more usual pseudo-ranges). In this situation, two or more ordinary nodes can jointly utilize the anchor location information, the intermediate distance measurements between ordinary nodes and those between an ordinary node and the anchors to estimate their locations. Typically, when communications of a node are limited to only two other nodes (2D case) whose location is known, the estimated position will have an ambiguity (two possible solutions). By using the extra information from the two-hop communications, this ambiguity can be resolved. By applying this collaborative multilateration over a wider scope, more node positions can be determined. Further, error accumulation can be reduced more, but at the cost of increased communications and time delays.

11.3.2 Distance-Vector Hop

To overcome the drawback of the incremental method, multihop distance measurements can be employed. That is, one determines the distance from an ordinary node to anchors which are not within the radio range of the ordinary node. Once at least three such distance estimates are available, the position of the ordinary node can be determined. One such a method is the distance-vector (DV)-hop method [5]. In this method, anchors broadcast their location and other information such as their identification to their neighboring ordinary nodes, which forward the relevant information to their neighboring nodes and so on. Using the received information, each node maintains a table of the locations of the anchors and the number of hops to each of the anchors. The information originated from one anchor is finally received at all other anchors. After an anchor calculates the Euclidean distances to other anchors, it estimates an average distance of one hop. The average distance per hop for anchor i is calculated as

$$
d_i^{(h)} = \left[\sum_{\substack{j=1 \\ j \neq i}}^{N_a} \sqrt{(x_i - x_j)^2 + (y_i - y_j)^2} \right] \left(\sum_{\substack{j=1 \\ j \neq i}}^{N_a} L_{i,j} \right)^{-1} \tag{11.3}
$$

where (x_j, y_j) are the coordinates of anchor j, $L_{i,j}$ is the number of hops from anchor j to anchor i and N_a is the number of anchors. For example, as shown in Figure 11.4, there are three anchors numbered 1, 2 and 3. The average distance per hop, or average hop size, from anchor 1 to anchors 2 and 3 is calculated as $d_1^{(h)} = (70 + 130)/(3 + 6) = 22.22$ m. Similarly, the average hop sizes associated with anchors 2 and 3 are 22.14 m and 21.5 m respectively. Note that another option for calculating the average distance per hop is that for each pair of anchors an average hop size is computed based on the Euclidean distance and the number of hops between the pair of anchors.

After getting the average hop size information from the anchors, an ordinary node calculates the distance to all the anchors. Specifically, if an ordinary node is m hops away

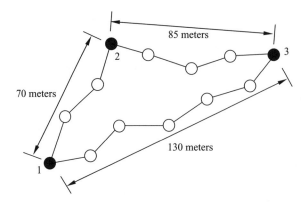

Figure 11.4 DV-hop distance determination. The anchor nodes are solid red dots, with ordinary nodes as circles. The 1–2 link has three hops and a distance of 70 m and the 1–3 link six hops and a distance of 130 m.

from an anchor, then the distance from the node to the anchor is estimated as m times the average hop size of that anchor. Based on these distance measurements, the ordinary nodes are able to localize themselves by employing a position-estimation algorithm. Alternatively, each pair of nodes that is within radio range of each other estimates the distance between them, using methods such as the signal strength or TOA measurements. The distance measurements are propagated from one node to another so that each node has the accumulative traveling distance to an anchor. This method is more accurate than the 'DV-hop' approach, but it is sensitive to measurement errors. A weighted average of these two methods can limit the effect of measurement errors.

11.3.3 Centralized Scheme

In a centralized scheme, all data are forwarded to a central point (or a location server) for processing. It is assumed that the calculations are too computation-intensive for local processing. Because of the processing power in the central server, more complex algorithms can be used for minimizing the position errors. For example optimization techniques typically require determining many possible solutions and choosing the optimum one, so that superior processing power can result in more accurate positional solutions. In typical cases, the positions of some nodes are known, whereas the locations of other nodes are to be determined by using the proximity constraints [6]. The proximity constraint is that if two nodes are in communication, then their separation must not be greater than the radio range. All nodes must communicate their connectivity information to a single computer to solve the optimization problem. Connections are used to form convex constraints. In radio communications, a radial distance constraint can be used, whereas an angular constraint can be employed in optical communications. Alternatively, Euclidean distance measurements and both the lower and upper distance bounds can be exploited to improve accuracy. Determining the optimum solution must use some metric, with the LS error being the most commonly used. However, besides the LS solution, another possible option is to minimize the sum of absolute errors to determine the unknown node positions [7]. This option tends to reduce the effect of very large errors, as the squaring operation in such cases can distort the optimization process.

In summary, the advantage of the centralized approach is that better location accuracy can be expected due to the fact that all data are processed simultaneously, allowing global minimization of errors. However, the disadvantages of the centralized approach are

- A global resource is needed challenging the ad hoc nature of the network;
- all the range, proximity and position information has to be sent back and forth to the sensor nodes, resulting in routing bottlenecks and unnecessary energy consumption.

11.3.4 Distributed Scheme

To overcome the drawbacks of the centralized approaches, localization can be performed as a distributed process. More specifically, the computation is spatially distributed across the network and each node is responsible for computing its own location estimate. This is achieved by performing local computation and communications with neighboring nodes. The common principle of distributed schemes is first to obtain an initial location estimate of each node whose position is unknown and then possibly update this estimate based on these initial estimates from neighboring nodes. Some ordinary nodes are able to get their initial position estimates earlier than others since they are directly connected to sufficient anchors. In some cases, owing to lack of sufficient numbers of directly connected anchors, local cooperation is taken among a few ordinary nodes to estimate their positions jointly. Those ordinary nodes which are not directly connected to anchors use the initial estimates of their neighbors as the reference points to produce their location estimates. This computation is repeated from node to node across the network until potentially all the nodes obtain their location estimates.

As soon as a node computes a new location estimate, it broadcasts the estimate to its neighbors and the neighbors use this new positional data to update their own position estimates. To achieve reasonable accuracy, each node may produce an updated location estimate by exploiting a minimization algorithm, provided that it has sufficiently redundant data and the necessary computational power. When making use of the updated location estimates from the neighbors, it is important to weight them based on their accuracy confidences. For instance, when the location information is provided by a node whose reference nodes are mostly anchors, its data should be weighted higher than those using only ordinary nodes. If a node is known to be stationary, and the change in each new position estimate at a node becomes sufficiently small crossing a predefined threshold, then the node stops the refinement process and reports the final position estimate.

11.3.5 Anchor Deployment

The strategy for the deployment of anchor nodes in WSNs is not trivial, since the placement of the anchors has a significant impact on localization performance. For example, when the anchors or base stations are placed on the boundary of the coverage area, the best location accuracy will be achieved provided that ordinary nodes or mobile stations have communication links with sufficient (redundant) anchors. In a large WSN, it is not feasible to deploy all the anchors around the boundary due to

- long-range communications beyond the link range, so that perhaps many hops between an ordinary node and an anchor are required, resulting in extra time delay and energy consumption;
- difficulty in obtaining accurate distance measurements.

Therefore, anchors are typically deployed in a quasi-uniform fashion, such that all locations in the coverage area can communicate with at least two (preferably three) anchors. With such a deployment of anchors, position determination is simple and has maximum accuracy. However, such a deployment may result in too many anchors from an economic or logistics point of view, so strategies using double hops might have to be considered in the design, but with the risk that there may be some locations in the intended coverage where positions cannot be obtained.

Because of the characteristics of radio propagation, it may be necessary in some circumstances to add a few extra anchors to improve location performance in some regions of the coverage area. In regions with poor localization, candidate points are selected for placing new anchors (or beacons) [8]. Further, if there is a dense deployment of anchors to achieve the required coverage, then it may be desirable to rotate functionality among anchors (by turning them on and off) to maximize lifetime. This adaptive operational density is achieved through anchor cooperation without diminishing the localization granularity. However, such a strategy is not appropriate if the ordinary nodes are mobile.

11.4 Localization with Unknown Internal Delays and Clock Offsets

In a practical ad hoc WSN, the unknown parameters are not only the coordinates of the nodes, but also the clock time synchronization and frequency offsets. This section describes methods of achieving the time and frequency synchronization using basic TOA measurements.

One possible method of range estimation is measuring the RTT delay, whereby a node transmits a signal and a second node detects this signal and then replies. Note that the reply time is not the propagation time, but rather the response time of the second node, usually software based (see Chapter 5 for a summary of the signaling protocol). For low-cost and low-complexity devices, such as ordinary sensor nodes, the clock frequency accuracy would typically be around ±5 ppm. If the ranging can be completed in 3 ms, then a TOA error of 15 ns would be incurred.

Another practical issue is the internal delays within the nodes, namely the time difference between the node clock time and the time when the radio signal arrives at (or departs from) the antenna, from which the propagation delay is measured. Depending on the transceiver structure, this internal delay can be around a few hundred nanoseconds. The variation of the internal delay is nontrivial. For instance, a variation of the delay of 3 ns will result in an error of about 1 m in distance measurement, which would make sub-meter accuracy positioning impractical.

In this section, distributed localization algorithms are studied in the following scenario. In the network there are two types of node: anchor nodes and ordinary nodes. The positions of the ordinary nodes, being static or mobile, are estimated by measuring the TOA at the nodes and their neighbors. The system is not time synchronized; therefore, there is an unknown clock offset between each pair of nodes. The system internal delay is treated as another unknown parameter. The TOA measurement error consists of three parts: measurement error due to receiver noise and channel fading, the positive bias due to the NLOS propagation and the error due to the clock frequency offset.

11.4.1 Signal and System Model

Consider a WSN in which the positions of the anchor nodes are known and the positions of other ordinary nodes are to be determined. Some ordinary nodes may have radio links with a number of anchors, while others may not have radio links with any anchor. For notational convenience there are q ordinary nodes numbered from 1 to q, and $N - q$ anchors numbered from $q + 1$ to N. The unknown parameters associated with ordinary node i are listed as follows:

- ϕ_i^c, clock time offset, which is defined as the local clock time relative to global system time
- Δ_i^{tx}, system internal delay between the clock and the transmit antenna
- Δ_i^{rx}, system internal delay between the clock and the receive antenna
- α_i, relative clock frequency, which is typically slightly smaller or greater than one
- (x_i, y_i), 2D position coordinates.

The corresponding parameters related to the anchors are assumed known *a priori*. In reality, the anchor locations are typically imperfect and their accuracy will be dependent on how the locations are determined. For instance, when the anchors are manually deployed and their locations are manually measured, the location accuracy can be as small as centimeters. On the other hand, when the locations are determined by low-cost GPS receivers, the measurement error can be several meters. The impact of anchor location errors on location accuracy of ordinary nodes can be found in Chapters 8 and 12. Here, the anchor locations are assumed to be error free. The problem is to make use of the known parameters of the anchors to determine the unknown locations of the ordinary nodes by measuring the TOA between each pair of neighboring ordinary nodes and/or between each pair of neighboring anchor and ordinary node.

The TOA measurements are the bridge for connecting the known and unknown parameters, so it is important to understand clearly the relationship among the time parameters (clock time offset, clock frequency and internal delay) and establish equations to describe the relationship. Figure 11.5 illustrates the time axes of two typical nodes and the global clock. In this figure, the clock of node i goes faster than that of the global clock ($\alpha_i < 1$), whereas the clock of node j goes slower than that of the global clock ($\alpha_j > 1$). It is assumed that the relative frequency of the global clock is equal to one; that is, it is synchronized to the global system clock.

It is worth noting that, in practical implementation, absolute or system global time is not used. The purpose of using global time is for analytical and notational convenience. Also, to be precise, ϕ is the pn-code phase.

Assume that node j and node i are within radio range. When node i transmits a signal at global time ϕ_i^c, after a delay of Δ_i^{tx}, the signal emits from the transmit antenna. The signal then travels

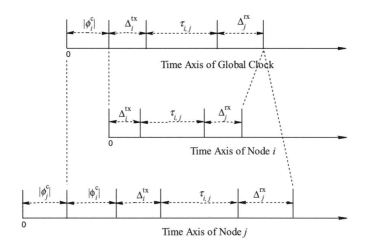

Figure 11.5 Time axes of nodes i and j and the global clock.

in the space for duration equal to

$$\tau_{i,j} = \tau_{j,i} = \frac{\sqrt{(x_i - x_j)^2 + (y_i - y_j)^2}}{c} \tag{11.4}$$

where c is the speed of radio propagation and (x_i, y_i) and (x_j, y_j) are the coordinates of node i and j respectively. The signal from node i arrives at the receiving antenna of node j and, after a delay of Δ_j^{rx} due to the radio and correlation processing, the clock is triggered and the TOA is measured. From the results in Section 5.2.4 we obtain

$$2\tau_{i,j} + \Delta_i + \Delta_j = \hat{T}_{i,j} + \hat{T}_{j,i} - (\varepsilon_{i,j} + \varepsilon_{j,i}) \tag{11.5}$$

where $\Delta_i = \Delta_i^{\text{tx}} + \Delta_i^{\text{rx}}$ is the combined internal delay of node i, $\hat{T}_{i,j}$ is the estimate of the TOA $(T_{i,j})$ of the signal transmitted from node i and received at node j, and $\varepsilon_{i,j}$ is the corresponding TOA estimation error including the measurement noise, multipath propagation, the NLOS effects and the clock frequency errors.

There are three different cases regarding the measurement equation (11.5). The first case is where both nodes i and j are to be localized such that

$$\frac{2\sqrt{(x_i - x_j)^2 + (y_i - y_j)^2}}{c} + \Delta_i + \Delta_j = \hat{T}_{i,j} + \hat{T}_{j,i} - (\varepsilon_{i,j} + \varepsilon_{j,i}) \tag{11.6}$$

where $(x_i, y_i), (x_j, y_j), \Delta_i$ and Δ_j are all unknown. The second case is where one node (say j) is an anchor and the other (say i) is an ordinary node to be localized. Then, the unknown parameters in (11.6) are (x_j, y_j) and Δ_j. Note that the last case is where both nodes are ordinary nodes but one of them has already been localized. Then, (11.6) becomes

$$\frac{2\sqrt{(x_i - \hat{x}_j)^2 + (y_i - \hat{y}_j)^2}}{c} + \Delta_i = \hat{T}_{i,j} + \hat{T}_{j,i} - \hat{\Delta}_j - (\varepsilon_{i,j} + \varepsilon_{j,i}). \tag{11.7}$$

11.4.2 Clock Frequency Offset Estimation

In Section 5.2.5, an approach was introduced for clock frequency offset estimation. In this section, we describe the estimation process in more detail. Without loss of generality, let us consider nodes i and j, whose relative clock frequencies are α_i and α_j respectively. The purpose is to estimate the frequency offset $\delta\alpha_{i,j} = \alpha_i - \alpha_j$ based on the TOA measurements, so that the TOA data can be corrected to remove the effect of the frequency offsets. The process starts when node i transmits a packet of data at the true (reference) time t_k, equivalent to the local time $\alpha_i t_k + \phi_i$. The packet is received by node j and the corresponding TOA measurement is

$$\hat{T}_{i,j}(t_k) = \tau_{i,j} + \text{mod}(\phi_i + \alpha_i t_k, T_{\text{pn}}) + \Delta_j - \text{mod}(\phi_j + \alpha_j(t_k + \tau_{i,j}), T_{\text{pn}}) + \varepsilon_{i,j}(t_k) \quad (11.8)$$

where $\varepsilon_{i,j}$ is the TOA measurement error due to noise and the NLOS effect, $\text{mod}(a, b)$ denotes the modulus operation and T_{pn} is the period of the pn-code. Like the daily clock repeating every 12 h, the pn-code clock cycles in its period, which might be in the order of microseconds. Note that, since the continuous reference time is used for notational convenience, the modulus operation is necessary. However, in reality, the modulus operation does not exist in the transceiver. At time $t_{k+1} = t_k + \delta t$, node i transmits another packet of data which is also received at node j. The TOA measurement at this time instant can be shown to be

$$\hat{T}_{i,j}(t_{k+1}) = \tau_{i,j} + \text{mod}(\phi_i + \alpha_i(t_k + \delta t), T_{\text{pn}}) + \Delta_j - \text{mod}(\phi_j + \alpha_j(t_k + \delta t))$$
$$+ 2\alpha_j \tau_{i,j}, T_{\text{pn}}) + \varepsilon_{i,j}(t_{k+1}) \quad (11.9)$$

From (11.8) and (11.9), we obtain the difference of the TOA measurements at the two neighboring time instants t_k and t_{k+1}:

$$\delta T_{i,j} = \hat{T}_{i,j}(t_{k+1}) - \hat{T}_{i,j}(t_k)$$
$$= \text{mod}(\alpha_i \delta t - \alpha_j(\delta t + \tau_{i,j}), T_{\text{pn}}) \quad (11.10)$$
$$\approx \text{mod}((\alpha_i - \alpha_j)\delta t, T_{\text{pn}})$$

where it is assumed that $\alpha_j \tau_{i,j} \ll |\alpha_i - \alpha_j|\delta t$. To obtain reliable relative frequency offset estimates from (11.10), the following inequality should be satisfied:

$$(\alpha_i - \alpha_j)\delta t < T_{\text{pn}} \quad (11.11)$$

The reason is that violating (11.11) would produce the wrong offset estimate due to the loss of the integer part of $(\alpha_i - \alpha_j)\delta t / T_{\text{pn}}$. Therefore, in system design, it is necessary to choose appropriate δt and T_{pn} so that the pn-code period is larger than the two clock time differences during the interval δt. Under the assumption, (11.10) becomes

$$\delta T_{i,j} \approx (\alpha_i - \alpha_j)\delta t \quad (11.12)$$

Thus, the raw TOA measurements must be adjusted by the TOA increment defined by (11.12). Figure 11.6 shows some typical TOA measurements between nodes i and j in the presence of noise when $T_{\text{pn}} = 20.48\ \mu s$ and $\delta t = 0.5\ s$. The first and third figures show the TOA measurements, whereas the second and fourth figures show the differential TOA measurements.

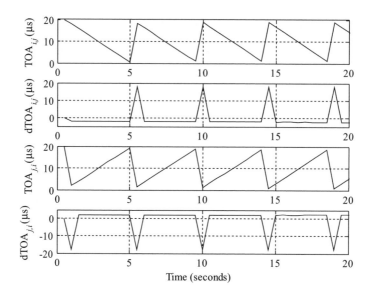

Figure 11.6 Typical TOA measurements among a pair of nodes and differential TOA measurements at node i (dTOA$_{i,j}$) and at node j (dTOA$_{j,i}$). The differential TOA measurements shown are not corrected.

When there is an abrupt jump or fall from one TOA measurement to the next due to pre-code repetition, an abnormal differential TOA measurement occurs. That is, the computed differential TOA using (11.12) does not give the right results. For instance, it can be seen that there are four and five abnormal errors in the second and fourth figure respectively. To recover the actual differential TOA values, one needs to find out when the jumps/falls happen so that necessary corrections can be made. This correction process may be summarized as follows:

- When $\delta T_{i,j} > T_{\mathrm{pn}}/2$, $\delta T_{i,j}$ is replaced by $\delta T_{i,j} - T_{\mathrm{pn}}$. Otherwise, if $\delta T_{i,j} < -T_{\mathrm{pn}}/2$, $\delta T_{i,j}$ is replaced by $\delta T_{i,j} + T_{\mathrm{pn}}$.

The thresholds $\pm T_{\mathrm{pn}}/2$ are set to detect reliably if an abrupt jump occurs. It can be shown that, when the differential TOA measurements are corrected, the results in the second and fourth figures will become straight lines in the absence of noise.

Next, dividing both sides of (11.12) by δt produces the estimate of the relative frequency offset between nodes i and j at time t_{k+1}; that is:

$$\delta \alpha_{i,j} = \alpha_i - \alpha_j \approx \frac{\delta T_{i,j}}{\delta t} \qquad (11.13)$$

Note that the receiver only knows the time interval δt based on its own clock, namely $\alpha_j \delta t$. However, when computing the relative frequency offset, the difference using the local clock is negligible and can be neglected. The frequency offset estimate is then employed to correct the corresponding TOA data. In practice, the frequency offset varies over time, so that tracking is required to follow the frequency variation using (11.13). Furthermore, the estimation performance can be improved by smoothing the raw frequency offset estimates. A variety of filtering/smoothing algorithms are available; the simple linear LS approach described in Chapter 7 is used in the simulation described below.

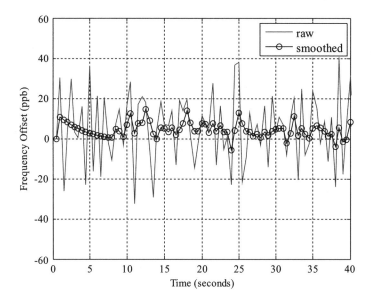

Figure 11.7   Clock frequency offset estimation errors before and after smoothing.

Figure 11.7 shows the typical results of the clock frequency offset estimation and tracking error between a pair of nodes using the synchronization scheme described above. The clock frequency offset is modeled as a summation of two parts: one is constant, randomly chosen between −5 and +5 ppm; the other is time varying, as a sinusoid of period randomly selected between 100 and 200 s and amplitude randomly chosen between 100 and 200 ppb (parts per billion). The constant part simulates the aging effects, whereas the varying component is simulating the effects caused by temperature changes in the local oscillator. The constant component and the period and amplitude of the varying component are fixed in the simulation run once they are selected. The clock offsets are randomly chosen between −0.5 and +0.5 s and fixed once selected. In addition, the TOA measurement error is modeled as a Gaussian random variable of zero mean and STD equal to 6 ns. The window length of the LS smoothing is set at 10. From Figure 11.7, it can be clearly seen that the frequency offset estimation error is mostly kept below 40 ppb without smoothing and below 10 ppb when smoothing is applied. In practice, only a few samples of frequency offset estimates may be required to produce a good frequency offset estimate, provided that the clock frequencies do not change greatly over a certain period of time.

11.4.3 Three-Stage Localization Scheme

The determination of the locations of nodes in a network is based on utilization of the known (or at least reasonably accurate) positions of anchor nodes and then, by inter-nodal ranging (or pseudo-ranging), the positions of the ordinary nodes are determined. The scheme is assumed to be based on measuring the RTT between pairs of nodes, so the pseudo-range includes a component related to the internal delays within the node, including communications between anchor nodes. Therefore, the process for position determination of ordinary nodes involves three stages: determining the anchor node delay parameters, initial local position determination of ordinary nodes (including their delay parameter) and finally the refinement of position estimates by exploiting all the data.

11.4.3.1 First Stage: Anchor Delay Determination

System internal radio delays are affected by various factors, including temperature, moisture and aging of electronic components. Consequently, real-time calibrations of the internal delays would be necessary unless the variations are negligible due to relatively constant environmental conditions over a certain period of time. In this subsection, a simple method is described to determine the internal delays of anchors based on their mutual communications and TOA estimations. RTT measurement among each pair of anchors within radio range results in

$$\Delta_i + \Delta_j = f_{i,j} + \upsilon_{i,j} \tag{11.14}$$

where

$$f_{i,j} = \hat{T}_{i,j} + \hat{T}_{j,i} - 2\tau_{i,j} \tag{11.15}$$
$$\upsilon_{i,j} = \varepsilon_{i,j} + \varepsilon_{j,i}$$

where $\tau_{i,j}$ is the known TOF of the signal from anchor i to anchor j or vice versa. Suppose that N_a anchors can communicate with each other. Then, (11.14) can be written in a compact form as

$$\mathbf{A}\Delta = \mathbf{h} + \boldsymbol{\upsilon} \tag{11.16}$$

where

$$\mathbf{A} = \begin{bmatrix} 1 & 1 & 0 & 0 & 0 & \cdots & 0 & 0 & 0 \\ \vdots & \vdots & \vdots & \vdots & \vdots & \ddots & \vdots & \vdots & \vdots \\ 1 & 0 & 0 & 0 & 0 & \cdots & 0 & 0 & 1 \\ 0 & 1 & 1 & 0 & 0 & \cdots & 0 & 0 & 0 \\ \vdots & \vdots & \vdots & \vdots & \vdots & \ddots & \vdots & \vdots & \vdots \\ 0 & 1 & 0 & 0 & 0 & \cdots & 0 & 0 & 1 \\ \vdots & \vdots & \vdots & \vdots & \vdots & \ddots & \vdots & \vdots & \vdots \\ 0 & 0 & 0 & 0 & 0 & \cdots & 0 & 1 & 1 \end{bmatrix} \in \mathbb{R}^{N_a(N_a-1)/2 \times N_a} \tag{11.17}$$

$$\Delta = \begin{bmatrix} \Delta_1 & \Delta_2 & \cdots & \Delta_{N_a} \end{bmatrix}^T \in \mathbb{R}^{N_a \times 1}$$
$$\mathbf{h} = \begin{bmatrix} f_{1,2} & \cdots & f_{1,N_a} & f_{2,3} & \cdots & f_{2,N_a} & \cdots & f_{N_a-1,N_a} \end{bmatrix}^T \in \mathbb{R}^{1 \times N_a(N_a-1)/2}$$
$$\boldsymbol{\upsilon} = \begin{bmatrix} \upsilon_{1,2} & \cdots & \upsilon_{1,N_a} & \upsilon_{2,3} & \cdots & \upsilon_{2,N_a} & \cdots & \upsilon_{N_a-1,N_a} \end{bmatrix}^T \in \mathbb{R}^{1 \times N_a(N_a-1)/2}$$

Note that $\mathbf{h}$ is the known observation vector, Δ is the unknown internal delay vector and $\mathbb{R}^{a \times b}$ denotes the real number domain of dimensional size $a \times b$. Therefore, the weighted LS solution to (11.16) is given by

$$\mathbf{A}\Delta = (\mathbf{A}^T\mathbf{W}\mathbf{A})^{-1}\mathbf{A}^T\mathbf{W}\mathbf{h} \tag{11.18}$$

where $\mathbf{W}$ is the weighting matrix, which is equal to the inverse of the covariance of $\boldsymbol{v}$; that is, $\mathbf{W} = \text{cov}^{-1}(\boldsymbol{v})$. The dimensions of $\mathbf{A}$, $\mathbf{h}$ and $\boldsymbol{v}$ will decrease if one or more anchors are out of radio range of other anchors so that some of the RTT measurements are not available. Although this is a centralized scheme to determine the internal delays of the anchors, the computational complexity would be acceptable to an anchor if the number of anchors is not large – around 10 say. In a large system, where the number of anchors may be as large as 30, the anchors may be set into different groups. The internal delays of anchors in each group are determined independently. The purpose of such grouping is to avoid long-range communications and to share the computational burden. Once the internal delays of anchors are obtained, ordinary node position determination starts.

Note that the internal radio delay of a transceiver can be calibrated when it is manufactured. Owing to the aging of electronic components and variations of environmental conditions, the internal delay varies. On the other hand, when the TOA measurement error is large due to adverse propagation conditions, the delay calculation by using (11.18) would produce large estimation errors. Therefore, it is important to choose a suitable method to determine the internal delays of anchors based on application and environmental conditions. This issue will be discussed further in Section 11.4.5.

An illustrative example is given below to show how accurate an anchor delay can be achieved. To evaluate the accuracy of the anchor delay estimation, two different models of TOA estimation errors are considered. In the first model, the TOA measurement error is a Gaussian random variable of zero mean and a variance independent of the range. This model is appropriate for open outdoor environments where LOS propagation exists and multipath is not severe. In the second model, the TOA measurement error is a mixture of a Gaussian random variable and an exponential random variable. The variances of the two random variables are dependent on the range between the transmitter and receiver. This model might be suitable for rich-multipath indoor environments. As stated in Chapter 5, a number of other models are also used to characterize the TOA or range errors. Typically, a specific environment and measurement device require a specific model. The anchor locations are randomly generated in a square area with dimensions of $100\,\text{m} \times 100\,\text{m}$ under a constraint: the distance between a pair of anchors is at least $25\,\text{m}$. It is assumed that all anchors are within radio range of each other and 2000 different anchor configurations are examined. The internal delay of each anchor is uniformly distributed between 500 and 600 ns.

Figure 11.8 shows the delay estimation results with the first model. The number of anchors ranges from 4 to 12 and the STD of the TOA errors goes from 2 to 10 ns. When there are more than six anchors and the TOA error is less than 3 ns, the internal delay estimation error can be around just 1 ns. Note that, when LOS propagation exists and multipath is minor, the TOA estimation can be very accurate, resulting in small TOA estimation errors. Figures 11.9 and 11.10 respectively show the STD and the mean of the delay estimation when using the second TOA error model. The NLOS TOA bias is modeled as an exponential random variable with mean equal to 6% of the true straight-line error-free TOA. The STD of the Gaussian component of the TOA error ranges from 1% to 6% of the error-free TOA value. Even for the Gaussian error of normalized STD 2% and 12 anchors, the STD of the delay estimation error can be as large as 4 ns. In addition, the mean of the delay estimation error is around 3 ns for all the parameters examined. Therefore, the delay estimation error can be large for severe multipath conditions, especially when the number of anchors is relatively small.

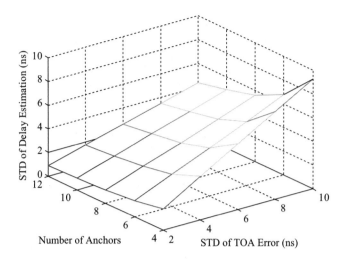

Figure 11.8   Internal delay estimation with respect to number of anchors and STD of TOA estimation error, which is a Gaussian random variable.

For indoor environments the variation of temperature is rather small and the variation of internal radio delay would be small. However, severe multipath propagation typically exists indoors, so that real-time calibration of the internal delays will produce relatively large estimation errors. Consequently, for indoor positioning under adverse propagation conditions, calibrated internal delay should be used. This method would be particularly desirable when the

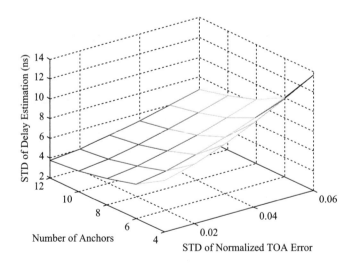

Figure 11.9   STD of internal delay estimation with respect to number of anchors and normalized STD of TOA measurement error, which is a Gaussian random variable. The NLOS bias is an exponential random variable.

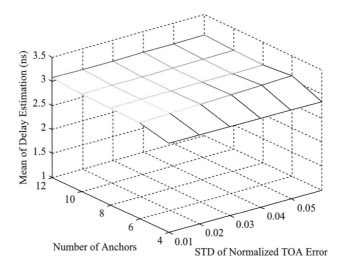

Figure 11.10   Mean of internal delay estimation with respect to number of anchors and normalized STD of TOA estimation error, which is a Gaussian random variable. The NLOS bias is an exponential random variable.

internal delays of anchors can be measured offline during system maintenance. On the other hand, for outdoor environments, the temperature, humidity and other conditions vary significantly even during a day. In this case, real-time anchor internal delay estimation is necessary, especially when offline recalibration of the delays is prohibited. In summary, by rule of thumb, calibrated anchor internal delays should be employed for indoor rich-multipath environments provided that the delay variation is relatively small, whereas real-time estimation of anchor internal delays should be carried out for outdoor environments.

In the following subsection the anchor location information and internal delay estimates are employed to determine the locations of ordinary sensor nodes in a distributed way. It is worth mentioning that centralized localization is usually impractical in a WSN due to the lack of infrastructure or a processing center. In addition, since sensor nodes are typically battery powered, it is not desirable to forward data through many hops to a central or a master node.

11.4.3.2 Second Stage: Initial Position Determination

The second stage not only generates initial location estimates, but also ensures the accuracy is as good as possible. Intuitively, the location accuracy at the second stage would certainly affect the accuracy at the refinement stage. Therefore, it is beneficial to obtain the initial location estimates as accurate as possible.

In a WSN, depending on anchor density, some ordinary nodes may have multiple neighboring anchors, whereas others may not have any neighboring anchors at all. Naturally, when localization is performed in a distributed manner, the nodes with the largest number of neighboring anchors should be localized first. Accordingly, a localization procedure with priority is described as follows:

- First, one localizes nodes that have three or more neighboring anchors; this is to make use of the accurate location information of the anchors.
- Second, one localizes nodes that have two neighboring anchors and one or more localized neighboring ordinary nodes. The purpose of this procedure is to extend the localization of more ordinary nodes, although the use of localized ordinary nodes in this process means that the quality of the positions will be slightly less than those obtained in the first step.
- Third, one localizes nodes that have one neighboring anchor and two or more localized neighboring ordinary nodes.
- Last, one localizes nodes that do not have any neighboring anchors, but each of which has three or more localized neighboring ordinary nodes.

Once a node is localized, it broadcasts its own location. When neighboring nodes receive this location information, the node will check to determine whether this new data can be used to improve its position estimate. To make location estimates as accurate as possible and to reduce error propagation, the algorithm first checks whether the three nodes with known locations form a solid triangle [9] which satisfies some conditions of the shortest edge and the smallest angle. That is

$$d_{\mathrm{m}} \sin^2 \varphi_{\mathrm{m}} > 3\sigma \qquad (11.19)$$

where φ_{m} is the smallest angle and d_{m} is the shortest edge of the triangle. In the presence of a solid triangle, the basic triangulation method is first employed, for instance with three independent equations to determine the three unknown parameters based on direct calculation. More specifically, suppose that (x_i, y_i) are the unknown coordinates of node i which has three neighbors whose positions are known and denoted by $(\hat{x}_j, \hat{y}_j)$, $j = j1, j2, j3$. In the case of anchors, $(\hat{x}_j, \hat{y}_j)$ may be treated as the true locations. It can be shown that by making round-trip TOA measurements and using the location information of the three neighbors the coordinates of node i can be determined as

$$\begin{aligned} \hat{x}_i &= h_{1,1} + h_{1,2}\hat{\Delta}_i \\ \hat{y}_i &= h_{2,1} + h_{2,2}\hat{\Delta}_i \end{aligned} \qquad (11.20)$$

By defining

$$\begin{aligned} \hat{x}_{i,j} &= \hat{x}_i - \hat{x}_j \\ \hat{y}_{i,j} &= \hat{y}_i - \hat{y}_j \\ f_{i,j} &= 0.5c(\hat{T}_{i,j} + \hat{T}_{j,i} - \hat{\Delta}_j) \\ a_{k,j} &= c(f_{i,k} - f_{i,j}) \\ g_{k,j} &= f_{i,k}^2 - f_{i,j}^2 + \hat{x}_j^2 + \hat{y}_j^2 - (\hat{x}_k^2 + \hat{y}_k^2) \\ \rho &= 4(\hat{x}_{j1,j2}\hat{y}_{j1,j3} - \hat{x}_{j1,j3}\hat{y}_{j1,j2}) \end{aligned} \qquad (11.21)$$

$\{h_{i,j}\}$ can be computed according to

$$
\begin{aligned}
h_{1,1} &= 2\rho^{-1}(\hat{y}_{j1,j3}g_{j2,j1} - \hat{y}_{j1,j2}g_{j3,j1}) \\
h_{1,2} &= 2\rho^{-1}(\hat{y}_{j1,j3}a_{j1,j2} - \hat{y}_{j1,j2}a_{j1,j3}) \\
h_{2,1} &= 2\rho^{-1}(\hat{x}_{j1,j2}g_{j3,j1} - \hat{x}_{j1,j3}g_{j2,j1}) \\
h_{2,2} &= 2\rho^{-1}(\hat{x}_{j1,j2}a_{j1,j3} - \hat{x}_{j1,j3}a_{j1,j2})
\end{aligned}
\tag{11.22}
$$

Also, $\hat{\Delta}_i$ is the estimate of the internal delay of node i given by

$$
\hat{\Delta}_i = -\frac{F}{E} \pm \sqrt{\left(\frac{F}{E}\right)^2 - \frac{G}{E}}
\tag{11.23}
$$

where

$$
\begin{aligned}
E &= h_{1,2}^2 + h_{2,2}^2 - \frac{c^2}{4} \\
F &= h_{1,2}(h_{1,1} - \hat{x}_{j1}) + h_{2,2}(h_{2,1} - \hat{y}_{j1}) + 0.5cf_{i,j1} \\
G &= (h_{1,1} - \hat{x}_{j1})^2 + (h_{2,1} - \hat{y}_{j1})^2 - f_{i,j1}^2
\end{aligned}
\tag{11.24}
$$

Apparently, there are two solutions to the location estimate; however, one of the solutions can be readily discarded by checking the residuals, which are defined as

$$
\left[\sqrt{(\hat{x}_i - \hat{x}_{j2})^2 + (\hat{y}_i - \hat{y}_{j2})^2} - f_{i,j2} + 0.5c\hat{\Delta}_i\right]^2 + \left[\sqrt{(\hat{x}_i - \hat{x}_{j3})^2 + (\hat{y}_i - \hat{y}_{j3})^2} - f_{i,j3} + 0.5c\hat{\Delta}_i\right]^2
\tag{11.25}
$$

Note that the problem can also be resolved by using the hyperbolic navigation method described in Chapter 5 and the QLS method described in Chapter 6. The direct calculation described above is just one of the methods which can be employed to solve the problem.

The solution with the smaller residual is initially chosen, while the other one with the larger residual is discarded. To further avoid abnormal estimation errors and reduce error propagation, two tests are performed. In the first test, the residual of the initially selected solution is examined. If it is less than a predefined threshold, then the solution passes the test. The second test is related to radio range, or range threshold. If the estimated location of a node is within radio range of its neighboring nodes with known locations, then the solution passes the test. To accommodate estimation errors, the range threshold may be slightly greater than the radio range, for instance setting at 1.5 times the radio range. Note that the radio range indoors can vary significantly. However, since it is intended to remove the estimated location that is too far away from its neighbors, an approximate upper bound of radio range may be used. Only when both tests are passed is the location estimate accepted.

In the case that a node has more than three (say m) neighbors whose locations are known, its location can be estimated by using the LS algorithm. Defining the parameter vector as

$$\boldsymbol{\theta} = [x_i, y_i, \Delta_i]^\mathrm{T} \tag{11.26}$$

the LS solution to $\boldsymbol{\theta}$ is given by

$$\begin{aligned}
\hat{\boldsymbol{\theta}} &= [\hat{x}_i, \ \hat{y}_i, \ \hat{\Delta}_i]^\mathrm{T} \\
&= (\mathbf{A}^\mathrm{T}\mathbf{W}\mathbf{A})^{-1}\mathbf{A}^\mathrm{T}\mathbf{W}\mathbf{g}
\end{aligned} \tag{11.27}$$

where $\mathbf{W}$ is the weighting matrix to emphasize the more reliable measurements and

$$\begin{aligned}
[\mathbf{A}]_{k,1} &= 2\hat{x}_{j_1, j_{k+1}} \qquad k = 1, 2, \ldots, m-1 \\
[\mathbf{A}]_{k,2} &= 2\hat{y}_{j_1, j_{k+1}} \\
[\mathbf{A}]_{k,3} &= c(f_{i, j_{k+1}} - f_{i, j_1}) \\
[\mathbf{g}]_{k,1} &= f_{i, j_{k+1}}^2 - f_{i, j_1}^2 + \hat{x}_{j_1}^2 + \hat{y}_{j_1}^2 - (\hat{x}_{j_{k+1}}^2 + \hat{y}_{j_{k+1}}^2)
\end{aligned} \tag{11.28}$$

The weighting matrix $\mathbf{W}$ can be based on several factors, including the type of anchor (true anchor of known position or a quasi-anchor whose position has been derived), the quality of the TOA measurement and the range (long range has lower accuracy and, hence, lower weighting). In the latter case, the SNR and the range residual can be used to define the weighting of a measurement. As mentioned previously, the hyperbolic navigation method in Chapter 5 and a number of methods in Chapter 6 can also be used to produce a position estimate.

Once a location estimate is obtained, its accuracy can be enhanced by exploiting an iterative algorithm. The estimates from the above direct calculation and the noniterative LS algorithm are used as the starting point. Two such iterative algorithms are considered: one is the SQP optimization method and the other is the TS-LS method. More details are provided in the following subsection.

11.4.3.3 Third Stage: Refinement of Position Estimates

When the first two stages are completed, all ordinary nodes should obtain their initial location estimates. In some rare circumstances, a few nodes might not be localized since they do not have three neighbors or they do not have enough neighboring anchors or localized nodes. Also, each node should have knowledge of the estimated locations and internal delay parameters of their neighbors in addition to these parameters of their neighboring anchors. In the final stage, the node location estimates are refined by employing the location information of all neighboring anchors and ordinary nodes. The simple noniterative LS algorithm can be employed to obtain refined location estimates. To achieve better location accuracy, two iterative algorithms are developed to perform the refinement of the location estimates as follows.

Taylor series method   The square root term in (11.7) can be linearized by using a Taylor series approximation, resulting in

$$\begin{aligned}
\tilde{d}_{i,j} &= \sqrt{(x_i - \hat{x}_j)^2 + (y_i - \hat{y}_j)^2} \\
&\approx \hat{d}_{i,j} + \frac{\partial \tilde{d}_{i,j}}{\partial x_i} \delta x_i + \frac{\partial \tilde{d}_{i,j}}{\partial y_i} \delta y_i
\end{aligned} \tag{11.29}$$

where $(\hat{x}_j, \hat{y}_j)$ are the known coordinates of either an anchor or a localized ordinary node and

$$
\begin{aligned}
&\hat{d}_{i,j} = \sqrt{(\hat{x}_i - \hat{x}_j)^2 + (\hat{y}_i - \hat{y}_j)^2} \\
&\frac{\partial \tilde{d}_{i,j}}{\partial x_i} = \frac{\hat{x}_i - \hat{x}_j}{\hat{d}_{i,j}} \\
&\frac{\partial \tilde{d}_{i,j}}{\partial y_i} = \frac{\hat{y}_i - \hat{y}_j}{\hat{d}_{i,j}} \\
&\delta x_i = x_i - \hat{x}_i \\
&\delta y_i = y_i - \hat{y}_i
\end{aligned}
\tag{11.30}
$$

Consequently, after ignoring the measurement noise and the linearization errors, (11.7) becomes

$$
\frac{2}{c} \frac{\partial \tilde{d}_{i,j}}{\partial x_i} \delta x_i + \frac{2}{c} \frac{\partial \tilde{d}_{i,j}}{\partial y_i} \delta y_i + \Delta_i \approx u_{i,j}
\tag{11.31}
$$

where

$$
u_{i,j} = \hat{T}_{i,j} + \hat{T}_{j,i} - \hat{\Delta}_j - \frac{2\sqrt{(\hat{x}_i - \hat{x}_j)^2 + (\hat{y}_i - \hat{y}_j)^2}}{c}
\tag{11.32}
$$

Since (11.31) is linear with respect to δx_i, δy_i and Δ_i, a linear LS solution can be readily obtained for the three unknown parameters. The initial position estimate is set at the results produced at the second stage. Then, the position estimates are updated according to

$$
\begin{aligned}
\hat{x}_i + \delta x_i &\Rightarrow \hat{x}_i \\
\hat{y}_i + \delta y_i &\Rightarrow \hat{y}_i
\end{aligned}
\tag{11.33}
$$

The updated $\hat{x}_i$ and $\hat{y}_i$ are then used to update the partial derivatives and $u_{i,j}$. Again, the linear LS estimator is employed to update the position increments. This procedure continues until a predefined threshold is crossed. This TS-LS method usually achieves superior accuracy when the initial position estimate is not far from the true value.

It is worth noting that the refined location estimates in the final stage are also examined against the residual thresholds and the radio range. If the estimates do not pass the test, then the results from the second stage are accepted. The updated location estimates can then be employed to refine the location estimates of the neighboring nodes.

Constrained optimization   Another approach suited at the final stage to obtain refined location estimates is the iterative optimization method as studied in Chapter 7. To start with, an objective/cost function is required, which is defined as the sum of the squares of the difference between the two sides of (11.7) after the measurement errors are ignored; namely:

$$
\Xi(\boldsymbol{\theta}) = \sum_j w_j \left[\frac{2\sqrt{(x_i - \hat{x}_j)^2 + (y_i - \hat{y}_j)^2}}{c} + \Delta_i - (\hat{T}_{i,j} + \hat{T}_{j,i} - \hat{\Delta}_j) \right]^2
\tag{11.34}
$$

where $\boldsymbol{\theta}$ is the parameter vector defined in (11.26) and $\{w_j\}$ are the weights. The parameter estimates are produced by minimizing the cost function; that is:

$$\hat{\boldsymbol{\theta}} = \arg \min_{\boldsymbol{\theta}} \Xi(\boldsymbol{\theta}) \qquad (11.35)$$

Note that in the TS-LS method, the solution of the coordinate increments at each iteration is produced by first linearizing the nonlinear terms in the measurement equations and then minimizing the sum of the squares of the difference between the two sides of (11.31). In contrast, we consider here a direct nonlinear minimization for the cost function (11.34). The TS-LS estimator is not optimal due to the fact that the linearization approximation causes information loss. Further, when localized ordinary nodes are used as anchors, the simple iterative LS method may not perform well, since the combination of the internal delay estimation error, the position error and the TOA measurement error can be large and the initial position estimate may not be accurate enough. Accordingly, the iterative minimization/optimization method may achieve superior performance comparatively. Certainly, in the case where the performance gain by using the complicated minimization approach is negligible, the simpler iterative LS algorithm should be used.

There are a variety of optimization/minimization algorithms which can be used to minimize the cost function in (11.34) to produce accurate position estimates. Both the unconstrained and constrained algorithms can be employed. When using constrained optimization, the cost function may be minimized under the constraints

$$\boldsymbol{\theta}^{\mathrm{L}} \leq \boldsymbol{\theta} \leq \boldsymbol{\theta}^{\mathrm{U}} \qquad (11.36)$$

where $\boldsymbol{\theta}^{\mathrm{L}}$ and $\boldsymbol{\theta}^{\mathrm{U}}$ are the lower and upper bounds of $\boldsymbol{\theta}$ respectively. The coordinate bounds may be defined according to the position information of the known nodes, or the dimensions of the location area. The internal delay bounds can be predicted based on the typical range of the internal delay of a given type of transceiver. By restricting the search space, we can reduce the number of function evaluations to find the solution. Generally speaking, more constraints and bound limitations result in fewer function evaluations because the optimization makes better decisions regarding step size and regions of feasibility than in the unconstrained case. Therefore, it is good practice to bound and constrain problems, where possible, to promote fast convergence to a solution. In the simulations in Section 11.4.5 the SQP method will be employed to solve the nonlinear minimization problem. More information about the SQP optimization method can be found in Annex 10.A. Also, the effect of the constraints will be evaluated later through simulation.

11.4.4 Cramer–Rao Lower Bound

The CRLB has been studied in Chapters 8 and 10. In this section, CRLB analysis is used on the TOA measurement errors, including the receiver noise, the random errors and positive bias due to multipath propagation, and the unknown internal delays of the ordinary nodes. The anchor positions are assumed error free and the anchor internal delays are also assumed known. Frequency offset is not taken into account, since its impact on positioning accuracy is dramatically suppressed after the frequency offset compensation is performed.

As discussed in Chapter 5, the TOA measurement error in LOS conditions is modeled as a Gaussian random variable of zero mean. The TOA error in NLOS conditions is modeled as a mixture of two parts: one is a Gaussian random variable and the other is an exponential random variable. Accordingly, in NLOS conditions, $(\varepsilon_{i,j} + \varepsilon_{j,i})$ in (11.6) consists of two Gaussian distributed components, denoted by $n_{i,j}$ and $n_{j,i}$, and two exponentially distributed components, denoted by $\eta_{i,j}$ and $\eta_{j,i}$. Suppose that $n_{i,j}$ and $n_{j,i}$ are independent and identically distributed (i.i.d.) random variables with a zero mean and a variance equal to σ_n^2, and $\eta_{i,j}$ and $\eta_{j,i}$ are also i.i.d. random variables with a mean of $\lambda_{i,j}$ and a variance of $\lambda_{i,j}^2$. Owing to the complexity of the density function of the sum of two Gaussian random variables and two exponential random variables, we apply the CLT to the sum of the four random variables. That is, we approximate $v_{i,j} = \varepsilon_{i,j} + \varepsilon_{j,i}$ as a Gaussian random variable with a mean equal to $2\lambda_{i,j}$ and a variance given by

$$\sigma_{v_{i,j}}^2 = 2\sigma_n^2 + 2\lambda_{i,j}^2 \tag{11.37}$$

This approximation would be suitable as long as $\lambda_{i,j}$ is not much greater than σ_n. The measurement equation between an anchor and an ordinary node or between two ordinary nodes can be written as

$$r_{i,j} = \frac{2d_{i,j}}{c} + \Delta_i + \Delta_j + v_{i,j} \tag{11.38}$$

where $d_{i,j}$ is not greater than the radio range. It is assumed that all nodes, either anchors or ordinary nodes, have the same radio range. Extension to the case of multiple different radio ranges is straightforward. Let us define the parameter vector as

$$\vartheta = [x_1, \ldots, x_q, y_1, \ldots, y_q, \Delta_1, \ldots, \Delta_q]^{\mathrm{T}} \in \mathbb{R}^{3q \times 1} \tag{11.39}$$

and the observation vector as

$$\mathbf{r} = [r_{1,2}, \ldots, r_{1,N}, r_{2,3}, \ldots, r_{2,N}, \ldots, r_{q,q+1}, \ldots, r_{q,N}]^{\mathrm{T}} \in \mathbb{R}^{N_r \times 1} \tag{11.40}$$

where $r_{i,j}$ is available only when node i and node j are within radio range of each other and

$$N_r \leq \frac{q(q-1)}{2} + q(N-q) \tag{11.41}$$

The equality in (11.41) holds only when all the ordinary nodes are within radio range of the anchors and each other. Typically, in a WSN, an ordinary node would be within radio range of a number of other ordinary nodes and a rather limited number of anchors or even no anchors, depending on the node density and the percentage of anchors. Then, given the parameter vector, the PDF of the observation vector $\mathbf{r}$ can be written as

$$p(\mathbf{r}|\vartheta) = \left(\prod_{\substack{i=1}}^{q-1} \prod_{\substack{j=i+1 \\ d_{i,j} \leq R_0}}^{q} p(v_{i,j}) \right) \left(\prod_{i=1}^{q} \prod_{\substack{j=q+1 \\ d_{i,j} \leq R_0}}^{N} p(v_{i,j}) \right) \tag{11.42}$$

where R_0 is the radio range, and it is assumed that all nodes including the anchors have the same radio range, and

$$p(v_{i,j}) = \frac{1}{\sqrt{2\pi}\sigma_{v_{i,j}}} \exp\left\{ -\frac{[\tilde{r}_{i,j} - (2d_{i,j}/c + \Delta_i + \Delta_j)]^2}{2\sigma_{v_{i,j}}^2} \right\} \qquad (11.43)$$

where

$$\tilde{r}_{i,j} = \begin{cases} r_{i,j} & \text{LOS condition} \\ r_{i,j} + 2\lambda_{i,j} & \text{NLOS condition} \end{cases} \qquad (11.44)$$

The CRLB for the estimate of the kth parameter of ϑ is defined by

$$\text{CRLB}(\hat{\vartheta}_k) = [\mathbf{F}^{-1}(\hat{\vartheta})]_{k,k} \qquad (11.45)$$

where $\mathbf{M}(\hat{\vartheta})$ is the FIM, whose elements are defined by

$$[\mathbf{F}(\hat{\vartheta})]_{k,\ell} = -E\left[\frac{\partial^2 \ln p(\mathbf{r}|\vartheta)}{\partial\vartheta_k \partial\vartheta_\ell}\right] \qquad k,\ell = 1,2,\ldots,3q \qquad (11.46)$$

where, after ignoring the irrelevant constants, the log likelihood function is given as

$$\Lambda(\mathbf{r}|\vartheta) = \ln p(\mathbf{r}|\vartheta)$$

$$= -\left\{ \sum_{\substack{i=1}}^{q-1} \sum_{\substack{j=i+1 \\ d_{i,j}\leq R_0}}^{q} \frac{1}{2\sigma_{v_{i,j}}^2} \left[\tilde{r}_{i,j} - \left(2\sqrt{(x_i - x_j)^2 + (y_i - y_j)^2}/c + \Delta_i + \Delta_j \right) \right]^2 \right.$$

$$\left. + \sum_{\substack{i=1}}^{q} \sum_{\substack{j=q+1 \\ d_{i,j}\leq R_0}}^{N} \frac{1}{2\sigma_{v_{i,j}}^2} \left[(\tilde{r}_{i,j} - \Delta_j) - \left(2\sqrt{(x_i - x_j)^2 + (y_i - y_j)^2}/c + \Delta_i \right) \right]^2 \right\} \qquad (11.47)$$

Then, the FIM can be derived to be

$$\mathbf{F}(\hat{\vartheta}) = \begin{bmatrix} \mathbf{A} & \mathbf{B} & \mathbf{C} \\ \mathbf{D} & \mathbf{Q} & \mathbf{S} \\ \mathbf{U} & \mathbf{V} & \mathbf{G} \end{bmatrix} \in \mathbb{R}^{3q \times 3q} \qquad (11.48)$$

where

$$[\mathbf{A}]_{i,i} = \sum_{\substack{j=1 \\ j\neq i, d_{i,j}\leq R_0}}^{q} \frac{4(x_i - x_j)^2}{c^2\sigma_{v_{i,j}}^2 d_{i,j}^2} + \sum_{\substack{j=q+1 \\ j\neq i, d_{i,j}\leq R_0}}^{N} \frac{4(x_i - x_j)^2}{c^2\sigma_{v_{i,j}}^2 d_{i,j}^2} \qquad i = 1,2,\ldots,q$$

$$[\mathbf{A}]_{i,j} = -\frac{4(x_i - x_j)^2}{c^2\sigma_{v_{i,j}}^2 d_{i,j}^2} \qquad i,j = 1,2,\ldots,q, i\neq j, d_{i,j}\leq R_0$$

$$[\mathbf{B}]_{i,i} = \sum_{\substack{j=1 \\ j\neq i, d_{i,j}\leq R_0}}^{q} \frac{4(x_i - x_j)(y_i - y_j)}{c^2 \sigma_{v_{i,j}}^2 d_{i,j}^2} + \sum_{\substack{j=q+1 \\ j\neq i, d_{i,j}\leq R_0}}^{N} \frac{4(x_i - x_j)(y_i - y_j)}{c^2 \sigma_{v_{i,j}}^2 d_{i,j}^2} \qquad i = 1, 2, \ldots, q$$

$$[\mathbf{B}]_{i,j} = -\frac{4(x_i - x_j)(y_i - y_j)}{c^2 \sigma_{v_{i,j}}^2 d_{i,j}^2} \qquad i, j = 1, 2, \ldots, q, i \neq j, d_{i,j} \leq R_0$$

$$[\mathbf{C}]_{i,i} = \sum_{\substack{j=1 \\ j\neq i, d_{i,j}\leq R_0}}^{q} \frac{2(x_i - x_j)}{c \sigma_{v_{i,j}}^2 d_{i,j}} + \sum_{\substack{j=q+1 \\ j\neq i, d_{i,j}\leq R_0}}^{N} \frac{2(x_i - x_j)}{c \sigma_{v_{i,j}}^2 d_{i,j}} \qquad i = 1, 2, \ldots, q$$

$$[\mathbf{C}]_{i,j} = -\frac{2(x_i - x_j)}{c \sigma_{v_{i,j}}^2 d_{i,j}} \qquad i, j = 1, 2, \ldots, q, i \neq j, d_{i,j} \leq R_0$$

$$[\mathbf{Q}]_{i,i} = \sum_{\substack{j=1 \\ j\neq i, d_{i,j}\leq R_0}}^{q} \frac{4(y_i - y_j)^2}{c^2 \sigma_{v_{i,j}}^2 d_{i,j}^2} + \sum_{\substack{j=q+1 \\ j\neq i, d_{i,j}\leq R_0}}^{N} \frac{4(y_i - y_j)^2}{c^2 \sigma_{v_{i,j}}^2 d_{i,j}^2} \qquad i = 1, 2, \ldots, q$$

$$[\mathbf{Q}]_{i,j} = \frac{4(y_i - y_j)^2}{c^2 \sigma_{v_{i,j}}^2 d_{i,j}^2} \qquad i, j = 1, 2, \ldots, q, i \neq j, d_{i,j} \leq R_0$$

$$[\mathbf{S}]_{i,i} = \sum_{\substack{j=1 \\ j\neq i, d_{i,j}\leq R_0}}^{q} \frac{2(y_i - y_j)}{c \sigma_{v_{i,j}}^2 d_{i,j}} + \sum_{\substack{j=q+1 \\ j\neq i, d_{i,j}\leq R_0}}^{N} \frac{2(y_i - y_j)}{c \sigma_{v_{i,j}}^2 d_{i,j}} \qquad i = 1, 2, \ldots, q$$

$$[\mathbf{S}]_{i,j} = -\frac{2(y_i - y_j)}{c \sigma_{v_{i,j}}^2 d_{i,j}} \qquad i, j = 1, 2, \ldots, q, i \neq j, d_{i,j} \leq R_0$$

$$[\mathbf{G}]_{i,i} = \sum_{\substack{j=1 \\ j\neq i, d_{i,j}\leq R_0}}^{q} \frac{1}{\sigma_{v_{i,j}}^2} + \sum_{\substack{j=q+1 \\ j\neq i, d_{i,j}\leq R_0}}^{N} \frac{1}{\sigma_{v_{i,j}}^2} \qquad i = 1, 2, \ldots, q$$

$$[\mathbf{G}]_{i,j} = -\frac{1}{\sigma_{v_{i,j}}^2} \qquad i, j = 1, 2, \ldots, q, i \neq j, d_{i,j} \leq R_0$$

$$\mathbf{D} = \mathbf{B}^{\mathrm{T}} \quad \mathbf{U} = \mathbf{C}^{\mathrm{T}} \quad \mathbf{V} = \mathbf{S}^{\mathrm{T}}$$

To provide an accuracy measure for the case of multiple node configurations, the CRLB for the position estimates

$$\hat{\mathbf{p}} = [\hat{x}_1, \ldots, \hat{x}_q, \hat{y}_1, \ldots, \hat{y}_q]^{\mathrm{T}} \in \mathbb{R}^{2q \times 1} \qquad (11.49)$$

is defined as

$$\mathrm{CRLB}(\hat{\mathbf{p}}) = \sqrt{\frac{1}{2qL} \sum_{\ell=1}^{L} \sum_{k=1}^{2q} [\mathbf{M}^{-1}(\hat{\vartheta})]_{k,k}^{(\ell)}} \qquad (11.50)$$

where ℓ indexes the node configurations and there are L node configurations in total. The above definition is actually the root-mean-square (RMS) of the STD bounds of the coordinate error of each unknown node under different node configurations.

11.4.5 Performance Comparison

In this section, simulations are conducted to evaluate the performance of the three-stage localization approach. First, the location accuracy at the initial stage and the refinement stage are examined. Then, the effect of the frequency offset and the constraints for optimization are evaluated. In the simulation, the internal delays are assumed known *a priori* based on calibration. The variation of delays is modeled as a Gaussian random variable of zero mean and STD equal to 2 ns unless it is specified otherwise.

11.4.5.1 Accuracy at the First and Second Stages

The node population is set at 40, six of which are anchors. The node locations are randomly generated in a square area with dimensions of $100\,\text{m} \times 100\,\text{m}$ under two constraints: the distance between a pair of anchors is at least 25 m and the distance between each pair of ordinary nodes is greater than 5 m. The radio range of all nodes is set at 40 m. A total of 1000 different node configurations are examined for the Monte Carlo simulation. The internal delay of all nodes is modeled as a random variable uniformly distributed between 500 and 600 ns. The TOA error due to measurement noise and channel fading is modeled as a Gaussian random variable with a zero mean and an STD set at values ranging from 1% to 7% of the true TOA values, and the TOA error bias due to the NLOS condition is modeled as an exponential random variable whose mean is 6% of the true range measured in propagation delay. Twenty percent of the TOA measurements are corrupted by NLOS propagation. The performance measure employed is the RMSE of all the position coordinate estimates. The CRLB is computed based on (11.50) to benchmark the RMSE. Note that a range of NLOS mitigation techniques are available, including, for instance, the residual weighting algorithm and the joint location and bias optimization algorithm, as presented in Chapter 10. These techniques could be readily incorporated into the developed algorithms in both the localization stages, but are not included in the simulation.

Figure 11.11 shows the RMSE of the location estimates at the first stage and the second stage. Also shown is the CRLB for performance comparison. The horizontal axis is the STD of the TOA measurement noise (excluding the NLOS bias), which is normalized by the true TOA value. The clock frequency offset estimation errors (after smoothing) are used as the frequency offsets. Three different methods are evaluated: the linear LS algorithm, the TS-LS algorithm and the SQP optimization algorithm. The first algorithm may also be used at the first stage of the later two algorithms. Also, at the first stage, the direct location calculation based on locations of three neighboring nodes may be used in all three algorithms. It can be seen that the accuracy improvement due to the second stage is not trivial. In particular, for the iterative optimization method, at the normalized STD of 0.04, the error is decreased by approximately 30%. It can be seen that the gap between the SQP algorithm and the CRLB is still relatively large at about 1 m. The gap would be reduced by better weighting and using global minimization, which does not belong to the distributed scheme. Note that, in the simulation, equal weights are simply used;

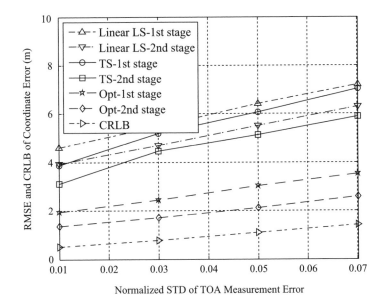

Figure 11.11 RMSE of location estimation errors of the first stage and second stage. 'Linear LS' denotes results using the linear LS algorithm; 'TS' denotes results using the TS–LS algorithm; and 'Opt' denotes results using the SQP optimization algorithm.

however, accuracy improvement can be achieved through weighting based on measurement and/or position estimation quality. The accuracy and computational complexity of the linear LS algorithm are both the lowest, whereas those of the SQP optimization algorithm are both the highest. In practice, it would be desirable to investigate the performance of other algorithms, such as those described in Chapters 6 and 7. Based on the accuracy and complexity of each of the algorithms, one could choose one or two suitable algorithms for implementation.

Figure 11.12 shows the impact of the anchor delay estimation error on the accuracy of the SQP optimization algorithm. The delay estimation error is modeled as a Gaussian random variable with a zero mean and STDs equal to 0, 2, 5, and 8 ns. When the STD of the error is 2 ns, the accuracy degradation is minor. However, when the STD of the error is 5 ns, the location accuracy degrades substantially. That is, the location error increased by between about 13% and 45% for the range of the TOA error examined. Further, when the STD of the delay error is 8 ns, the location error can be increased by up to about 80%. Therefore, it is important to obtain the estimates of the anchor internal delays as accurately as possible.

11.4.5.2 Effect of Frequency Offset

Figure 11.13 shows the effect of the clock frequency offset on the TS-LS algorithm. Four different cases are examined, corresponding to the five different frequency offsets: 0, 0.01, 0.03, 0.05, and 0.08 ppm. The clock time offset among each pair of nodes is modeled as a random variable uniformly distributed between -0.5 and $+0.5$ s. At the 0.01 ppm offset, the accuracy

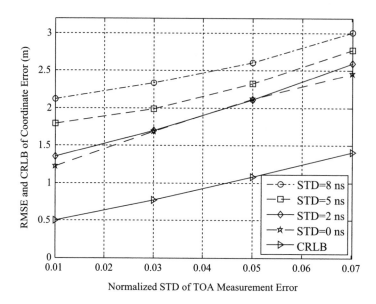

Figure 11.12 Effect of anchor internal delay errors on the accuracy of the SQP optimization algorithm. 'STD = a ns' denotes results when the STD of the delay error equals a ns.

degradation is rather negligible; however, at the 0.08 ppm offset, the location estimation error is doubled at the normalized STD of 0.4. Similar effects are observed for the other two algorithms. As a result, it would be desirable to keep the frequency offset below 0.01 ppm. Clearly, the effect of the residual frequency offset as shown in Figure 11.7 is negligible.

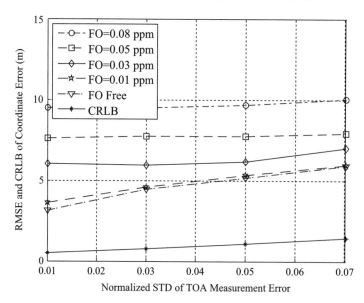

Figure 11.13 Effect of clock frequency offset on the accuracy of the TS-LS algorithm.

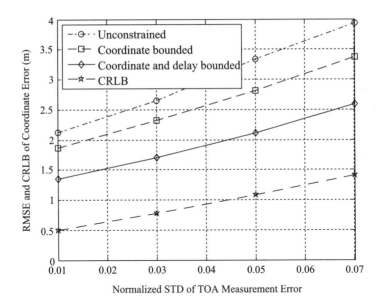

Figure 11.14   Effect of parameter constraints on the accuracy of the optimization algorithm.

11.4.5.3 Effect of Bounds on the Optimization Method

Figure 11.14 shows the effect of the parameter bounds on the optimization method. Three different cases are examined: unconstrained optimization; optimization with coordinate constraints; and optimization with both coordinate and internal delay constraints. The coordinate constraints are based on the dimensional size of the monitored area; that is, the coordinate are constrained to be less than 100 m and greater than zero in this case. The internal delays are constrained to be less than 600 ns and greater than 500 ns. In practice, the constraints can be determined based on the specific application scenarios and characteristics of the radio transceivers. The corresponding results are denoted by 'unconstrained', 'coordinate bounded', and 'coordinate and internal delay bounded'. Clearly, using more constraints achieves better accuracy.

References

[1] K. Romer and F. Mattern, 'The design space of wireless sensor networks', *IEEE Wireless Communications*, **11**(6), 2004, 54–61.

[2] N. Bulusu, J. Heidemann, and D. Estrin, 'GPS-less low-cost outdoor localization for very small devices', *IEEE Personal Communications Magazine*, **7**(5), 2000, 28–34.

[3] T. He, C. Huang, B.M. Blum, J. Stankovic and T. Abdelzaher, 'Range-free localization schemes for large scale sensor networks', in *Proceedings of the 9th Annual International Conference on Mobile Computing and Networking (MobiCom)*, San Diego, CA, USA pp. 81–95, September 2003.

[4] A. Savvides, H. Park, and M.B. Strivastava, 'The bits and flops of the N-hop multilateration primitive for node localization problems', in *Proceedings of International Workshop on Wireless Sensor Networks and Applications (WSNA)*, pp. 112–121, 2002.

[5] D. Niculescu and B. Nath, 'Ad hoc positioning system (APS)', in *Proceedings of IEEE Global Telecommunications Conference (GLOBECOM)*, San Antonio, TX, USA, pp. 2926–2931, 2001.

[6] L. Doherty, K.S.J. Pister and L.E. Ghaoui, 'Convex position estimation in wireless sensor networks', in *IEEE Conference on Computer Communications (INFOCOM)*, pp. 1655–1663, 2001.

[7] P. Biswas and Y. Ye, 'Semidefinite programming for ad hoc wireless sensor network localization', in *Proceedings of IEEE Information Processing in Sensor Networks*, pp. 46–54, April 2004.

[8] N. Bulusu, J. Heidemann, D. Estrin and T. Tran, 'Self-configuring localization systems: design and experimental evaluation', *ACM Transactions on Embedded Computing Systems*, **3**(1), 2004, 24–60.

[9] D. Moore, J. Leonard, D. Rus and S. Teller, 'Robust distributed network localization with noisy range measurements', in *Proceedings of International Conference on Embedded Networked Sensor Systems*, Baltimore, MD, USA, pp. 50–61, 2004.

12

Anchor Position Accuracy Enhancement

The localization accuracy in a WSN depends not only on the accuracy of ranging measurements, but also on the positioning accuracy of the anchors. Therefore, a WSN positioning system needs some independent method of surveying these anchor points. The simplest method is to determine the anchor positions independently using a map. This method is particularly appropriate for indoor systems, where a map of the building can be used to obtain centimeter positional accuracy. For outdoor systems, classical surveying techniques can also achieve similar accuracy. However, the logistics of surveying many anchor points in a WSN usually makes such surveying methods impractical. Ideally, the WSN infrastructure itself should perform the surveying of the anchor points.

One possible solution to self-surveying is the use of the GPS. Some advanced GPS receivers, such as differential GPS (DGPS) and kinematical GPS, can achieve accuracy ranging from a few meters down to a few decimeters [1,2]. In a DGPS system, one stationary receiver obtains corrections (based on its location being defined accurately by some other surveying method), or a separate radio receiver is used for obtaining broadcast GPS corrections applicable for the local vicinity. This master GPS node then provides the correction information to the other receivers that are either static or moving. However, such more sophisticated GPS nodes are generally not appropriate for WSN applications due to the constraints on cost and complexity. On the other hand, low-cost commercial GPS receivers may only achieve an accuracy that might be as poor as 20 m [3,4], which is typically worse than that which can be achieved within the WSN itself, so this cheap GPS option is also not appropriate in some circumstances. In addition, adverse environments, such as in urban canyons, under heavy foliage and indoor locations, may degrade the GPS measurement accuracy significantly, or result in no GPS positions being obtained.

In outdoor WSNs, it is usually the case that only a small percentage of anchor nodes are equipped with GPS receivers, whereas the other ordinary sensor nodes do not have GPS receivers. The global positions of the ordinary sensor nodes may be obtained by using anchor GPS measurements and a multilateration method directly under the global coordinate system. Alternatively, node locations can be first estimated under a local coordinate system and then node global positions are produced by transforming the local system into a global system based on the anchor global locations. Errors usually exist in ordinary sensor node location estimations due to signal parameter estimation errors even with error-free anchor locations. Erroneous anchor position estimates further decrease ordinary sensor location estimation accuracy.

In the event that the accuracy of low-cost GPS receivers is not satisfactory, it is possible to obtain better location accuracy by making use of the anchor-to-anchor signal measurements. With the wideband and UWB technologies, the accuracy of ranging can be 20–50 cm, even in multipath-dominated environments such as indoors [5–7]. The main focus of this chapter is to make use of more accurate anchor-to-anchor parameter estimates to enhance the anchor location accuracy. The accurate anchor location information will also help localize ordinary sensor nodes satisfactorily.

The chapter is organized as follows. Section 12.1 evaluates the impact of anchor location accuracy on sensor node localization. Section 12.2 describes the signal model in LOS conditions and in NLOS conditions. Section 12.3 derives the anchor position accuracy bound as a performance benchmark. Section 12.4 presents an accuracy improvement algorithm based on both distance and AOA measurements. Finally, Section 12.5 studies a distance-based approach for enhancing anchor location accuracy.

12.1 Impact of Anchor Location Accuracy on Sensor Node Localization

When evaluating the sensor node localization accuracy, the anchor locations are often assumed to be error free. In reality, however, the accuracy of the anchor location varies depending on the localization method used. In this section we evaluate how the anchor location accuracy affects the performance of the ordinary sensor node localization. It is intended to show the importance of anchor location accuracy enhancement in the case where the positional accuracy of the anchors is low. To this end, simulations are conducted under the following setup. The monitored area has a cubic shape with the edge length equal to 200 m. There are six anchors whose locations denoted by (x_i, y_i, z_i) are set at (200, 0, 0), (200, 200, 0), (0, 200, 0), (200, 0, 200), (0, 0, 200) and (0, 200, 200) (in meters). The location of the sensor node, denoted by (x, y, z), is randomly generated in the monitored area, and 5000 different sensor node locations are examined for each simulation run. The propagation between the sensor node and each anchor node is assumed to be LOS. Let $\hat{d}_i$ be the measurement of the distance between the sensor node and the ith anchor; $\hat{\phi}_i$ and $\hat{\alpha}_i$ are the azimuth and elevation angle measurements respectively made at the ith anchor. Then, the measurement equation is given by

$$\mathbf{Mp} \approx \mathbf{f} \tag{12.1}$$

where

$$\mathbf{M} = \begin{bmatrix} \mathbf{e}_N & \mathbf{0} & \mathbf{0} \\ \mathbf{0} & \mathbf{e}_N & \mathbf{0} \\ \mathbf{0} & \mathbf{0} & \mathbf{e}_N \end{bmatrix} \qquad \mathbf{p} = \begin{bmatrix} x \\ y \\ z \end{bmatrix}$$

$$\mathbf{f} = \begin{bmatrix} \hat{x}_1 + \hat{d}_1 \sin \hat{\alpha}_1 \cos \hat{\phi}_1 & \cdots & \hat{x}_N + \hat{d}_N \sin \hat{\alpha}_N \cos \hat{\phi}_N & \hat{y}_1 + \hat{d}_1 \sin \hat{\alpha}_1 \sin \hat{\phi}_1 \\ \cdots \hat{y}_N + \hat{d}_N \sin \hat{\alpha}_N \sin \hat{\phi}_N & \hat{z}_1 + \hat{d}_1 \cos \hat{\alpha}_1 & \cdots & \hat{z}_N + \hat{d}_N \cos \hat{\alpha}_N \end{bmatrix} \tag{12.2}$$

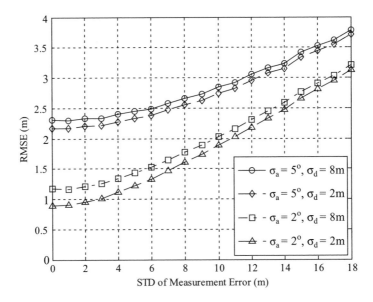

Figure 12.1 Impact of anchor location accuracy on sensor node localization. The STDs of the AOA errors are set at 2° and 5° and the STDs of the distance errors are set at 2 m and 8 m.

where $\mathbf{e}_N$ is a column vector of N ones. The LS solution for (12.1) is given by

$$\hat{\mathbf{p}} = (\mathbf{M}^T \Phi \mathbf{M})^{-1} \mathbf{M}^T \Phi \mathbf{f} \tag{12.3}$$

where Φ is a weighting matrix which may be determined based on the accuracy of the distance and angle measurements at each anchor. For simplicity, it is simply set as an identity matrix here.

Figure 12.1 shows the accuracy (RMSE) of the sensor node location estimation. The STD of the AOA (both azimuth and elevation angle) estimation errors is set at two different values, namely 2° and 5°. Further, the STDs of the distance measurement errors are set at 2 and 8 m. The STD of the anchor coordinate errors ranges from 0 to 18 m. From Figure 12.1, it can be observed that the sensor node location errors are reduced from about 1.9 m to 1.24 m under $\sigma_{AOA} = 2°$ and $\sigma_d = 2$ m when the anchor coordinate error is decreased from 10 m to 5 m. That is, the sensor location error is decreased by about 35%. When $\sigma_{AOA} = 5°$ and $\sigma_d = 8$ m, the accuracy improvement of the sensor location estimation is also considerable if the anchor location error is reduced. For instance, the sensor location error is reduced from 2.85 m to 2.45 m when the anchor location error changes from 10 m to 5 m. It can also be observed that the sensor location error is much smaller than the anchor location error. This may be due to the fact that, at each anchor, the sensor location estimate is the sum of two terms, one of which is the anchor coordinate estimate and the other is related to the distance and angle measurements. After applying the LS estimation, the effect of the anchor location errors would be largely averaged out.

In summary, the sensor node location accuracy increases substantially as the accuracy of anchor positions increase. In the remainder of this chapter, algorithms for anchor location accuracy enhancement are studied and the performance of the algorithms is evaluated.

12.2 Line-of-Sight and Non-Line-of-Sight Propagation Models

Consider a sensor network in which there are N anchors, each of which is equipped with a GPS receiver. Each anchor is able to communicate with some other anchors within the radio range to obtain distance estimates by such means as measuring the RTT. Also, each anchor is able to measure the AOA in both azimuth and elevation by using a directional antenna or an antenna array. We assume that the time and angle parameter estimation is available and they are exploited for anchor location accuracy improvement. As shown in Figure 12.2, anchors i and j are within the radio range of each other in LOS propagation conditions, whereas the radio propagation is NLOS in Figure 12.3. For clarity, let us define

- (x_i, y_i, z_i) as the global position coordinates of anchor i;
- $(x_s^{(i,j)}, y_s^{(i,j)}, z_s^{(i,j)})$ as the position coordinates of the scatterer between anchors i and j under NLOS conditions;
- $d_{i,j}$ as the LOS distance between anchors i and j in Figure 12.2 and NLOS distance in Figure 12.3;
- $\phi_{i,j}$ as the azimuth angle of the signal transmitted from anchor j and received at anchor i;
- $\alpha_{i,j}$ as the elevation angle of the signal transmitted from anchor j and received at anchor i.

In the presence of estimation error, the distance and angle estimates are modeled as

$$\begin{aligned}
\hat{d}_{i,j} &= d_{i,j} + n_{\hat{d}_{i,j}} \qquad 1 \le i, j \le N, i \ne j \\
\hat{\phi}_{i,j} &= \phi_{i,j} + n_{\hat{\phi}_{i,j}} \\
\hat{\alpha}_{i,j} &= \alpha_{i,j} + n_{\hat{\alpha}_{i,j}}
\end{aligned} \qquad (12.4)$$

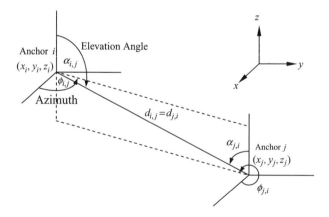

Figure 12.2 Illustration of 3D anchor positions and angles in LOS condition.

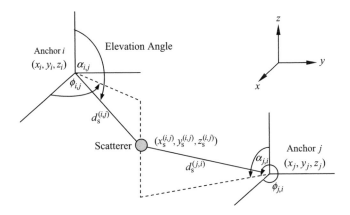

Figure 12.3　Illustration of 3D anchor positions and angles in NLOS condition.

where $n_{\hat{d}_{i,j}}$ is the distance estimation error, $n_{\hat{\phi}_{i,j}}$ is the azimuth angular estimation error and $n_{\hat{\alpha}_{i,j}}$ is the elevation angular estimation error. In LOS conditions, the distance and angles are described as

$$
\begin{aligned}
d_{i,j} &= \sqrt{(x_i - x_j)^2 + (y_i - y_j)^2 + (z_i - z_j)^2} \\
\phi_{i,j} &= \tan^{-1} \frac{y_j - y_i}{x_j - x_i} \\
\alpha_{i,j} &= \cos^{-1} \frac{z_j - z_i}{d_{i,j}}
\end{aligned}
\tag{12.5}
$$

On the other hand, under NLOS conditions, the distance and angles are defined as

$$
\begin{aligned}
d_{i,j} &= d_{\mathrm{s}}^{(i,j)} + d_{\mathrm{s}}^{(j,i)} \\
d_{\mathrm{s}}^{(i,j)} &= \sqrt{(x_{\mathrm{s}}^{(i,j)} - x_i)^2 + (y_{\mathrm{s}}^{(i,j)} - y_i)^2 + (z_{\mathrm{s}}^{(i,j)} - z_i)^2} \\
d_{\mathrm{s}}^{(j,i)} &= \sqrt{(x_{\mathrm{s}}^{(i,j)} - x_j)^2 + (y_{\mathrm{s}}^{(i,j)} - y_j)^2 + (z_{\mathrm{s}}^{(i,j)} - z_j)^2} \\
\phi_{i,j} &= \tan^{-1} \frac{y_{\mathrm{s}}^{(i,j)} - y_i}{x_{\mathrm{s}}^{(i,j)} - x_i} \\
\alpha_{i,j} &= \cos^{-1} \frac{z_{\mathrm{s}}^{(i,j)} - z_i}{d_{\mathrm{s}}^{(i,j)}}
\end{aligned}
\tag{12.6}
$$

where $d_{\mathrm{s}}^{(i,j)}$ is the LOS distance between anchor i and the scatterer that is associated with anchors i and j. Note that $(x_{\mathrm{s}}^{(i,j)}, y_{\mathrm{s}}^{(i,j)}, z_{\mathrm{s}}^{(i,j)})$ and $(x_{\mathrm{s}}^{(j,i)}, y_{\mathrm{s}}^{(j,i)}, z_{\mathrm{s}}^{(j,i)})$ are equal, whereas $d_{\mathrm{s}}^{(i,j)}$ and $d_{\mathrm{s}}^{(j,i)}$ are different. The GPS global coordinate estimates are modeled as

$$
\begin{aligned}
\hat{x}_i &= x_i + n_{\hat{x}_i} \qquad i = 1, 2, \ldots, N \\
\hat{y}_i &= y_i + n_{\hat{y}_i} \\
\hat{z}_i &= z_i + n_{\hat{z}_i}
\end{aligned}
\tag{12.7}
$$

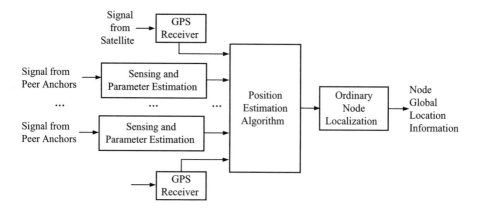

Figure 12.4   Simplified block diagram of anchor position accuracy improvement scheme.

where $n_{\hat{x}_i}$, $n_{\hat{y}_i}$ and $n_{\hat{z}_i}$ are the GPS coordinate estimation errors. Note that the GPS errors come from various sources, including the satellite geometry, satellite orbital positions, multipath effects, atmospheric effects and clock inaccuracies. The analysis can be greatly simplified by using the model in (12.7).

As shown in Figure 12.4, the aim is to re-estimate the anchor global positions by making use of anchor-to-anchor distance estimates or both distance and angle estimates, and the GPS coordinate measurements as well. Consequently the anchor positions can be improved. Also, the refined anchor position estimates are employed to determine the global locations of the ordinary sensor nodes so that the location accuracy of the ordinary sensor nodes can be improved. Note that there are different ways of improving GPS measurement accuracy. In DGPS, for example, the correction data can be forwarded from one device to others so that the accuracy of all GPS receivers in the system can be improved. This chapter just describes one alternative and potentially a way to achieve the same purpose.

12.3 Anchor Position Accuracy Bound

As mentioned previously in Chapters 8, 10 and 11, the CRLB is one of the main performance measures used in evaluating the performance of positioning algorithms. In this section the focus is on the derivation of the CRLB in LOS conditions when both the AOA and distance measurements are employed. As described in Chapters 8 and 10, the CRLB is defined as

$$\text{CRLB}(\hat{p}_k) = [\mathbf{F}^{-1}(\hat{\mathbf{p}})]_{k,k} \tag{12.8}$$

where $\mathbf{F}(\hat{\mathbf{p}})$ is the FIM, whose components are defined as

$$[\mathbf{F}(\hat{\mathbf{p}})]_{k,\ell} = -E\left[\frac{\partial^2 \ln p(\mathbf{r}|\mathbf{p})}{\partial p_k \partial p_\ell}\right] \qquad k,\ell = 1,2,\ldots,N \tag{12.9}$$

where the log likelihood function can be derived as

$$
\ln p(r|\mathbf{p}) = -\frac{1}{2}\sum_{i=1}^{N}\sum_{\substack{j=1\\j\neq i}}^{N}\left[\frac{1}{\sigma_{\hat{d}_{i,j}}^2}(\hat{d}_{i,j}-d_{i,j})^2+\frac{1}{\sigma_{\hat{\phi}_{i,j}}^2}\left(\hat{\phi}_{i,j}-\tan^{-1}\frac{y_{i,j}}{x_{i,j}}\right)^2+\frac{1}{\sigma_{\hat{\alpha}_{i,j}}^2}\left(\hat{\alpha}_{i,j}-\cos^{-1}\frac{z_{i,j}}{d_{i,j}}\right)^2\right]
$$

$$
-\frac{1}{2}\sum_{k=1}^{N}\left[\frac{1}{\sigma_{\hat{x}_k}^2}(\hat{x}_k-x_k)^2+\frac{1}{\sigma_{\hat{y}_k}^2}(\hat{y}_k-y_k)^2+\frac{1}{\sigma_{\hat{x}_k}^2}(\hat{z}_k-z_k)^2\right]
$$

$$(12.10)$$

where

$$
x_{i,j}=x_i-x_j
$$
$$
y_{i,j}=y_i-y_j
$$
$$
z_{i,j}=z_i-z_j
$$

$$(12.11)$$

From (12.9) and (12.10), the FIM can be derived as

$$
\mathbf{F}(\hat{\mathbf{p}}) = \begin{bmatrix} \mathbf{F_{xx}} & \mathbf{F_{xy}} & \mathbf{F_{xz}} \\ \mathbf{F_{yx}} & \mathbf{F_{yy}} & \mathbf{F_{yz}} \\ \mathbf{F_{zx}} & \mathbf{F_{zy}} & \mathbf{F_{zz}} \end{bmatrix} \in \mathbb{R}^{3N\times 3N}
$$

$$(12.12)$$

where the sub-blocks of the FIM are defined as follows. Define

$$
u_i = \frac{1}{2}\sum_{\substack{j=1\\j\neq i}}^{N}\left[\frac{1}{\sigma_{\hat{d}_{i,j}}^2}(\hat{d}_{i,j}-d_{i,j})^2+\frac{1}{\sigma_{\hat{d}_{j,i}}^2}(\hat{d}_{j,i}-d_{j,i})^2+\frac{1}{\sigma_{\hat{\phi}_{i,j}}^2}\left(\hat{\phi}_{i,j}-\tan^{-1}\frac{y_{j,i}}{x_{j,i}}\right)^2\right.
$$

$$
\left.+\frac{1}{\sigma_{\hat{\phi}_{j,i}}^2}\left(\hat{\phi}_{j,i}-\tan^{-1}\frac{y_{i,j}}{x_{i,j}}\right)^2+\frac{1}{\sigma_{\hat{\alpha}_{i,j}}^2}\left(\hat{\alpha}_{i,j}-\cos^{-1}\frac{z_{j,i}}{d_{j,i}}\right)^2+\frac{1}{\sigma_{\hat{\alpha}_{j,i}}^2}\left(\hat{\alpha}_{j,i}-\cos^{-1}\frac{z_{i,j}}{d_{i,j}}\right)^2\right]
$$

$$
-\frac{1}{2}\left[\frac{1}{\sigma_{\hat{x}_k}^2}(\hat{x}_k-x_k)^2+\frac{1}{\sigma_{\hat{y}_k}^2}(\hat{y}_k-y_k)^2+\frac{1}{\sigma_{\hat{x}_k}^2}(\hat{z}_k-z_k)^2\right]
$$

$$(12.13)$$

and let

$$
d_{2ij}=\sqrt{(x_{j,i})^2+(y_{j,i})^2}\quad j\neq i
$$
$$
\rho(v_{i,j})=\sigma_{v_{i,j}}^{-2}+\sigma_{v_{j,i}}^{-2}\quad v_{i,j}=d_{i,j},\phi_{i,j},\alpha_{i,j}
$$

$$(12.14)$$

Then, the components of the FIM can be derived as

$$[\mathbf{F_{xx}}]_{i,i} = E\left[\frac{\partial^2 u_i}{\partial x_i^2}\right] = \sum_{\substack{j=1 \\ j\neq i}}^{N}\left(\frac{x_{i,j}^2\rho(\hat{d}_{i,j})}{d_{i,j}^2} + \frac{y_{i,j}^2\rho(\hat{\phi}_{i,j})}{d_{2ij}^4} + \frac{x_{i,j}^2 z_{i,j}^2\rho(\hat{\alpha}_{i,j})}{d_{i,j}^4 d_{2ij}^2}\right) + \frac{1}{\sigma_{\hat{x}_i}^2}$$

$$[\mathbf{F_{xx}}]_{i,j} = E\left[\frac{\partial^2 u_i}{\partial x_i \partial x_j}\right] = -\left(\frac{x_{i,j}^2\rho(\hat{d}_{i,j})}{d_{i,j}^2} + \frac{y_{i,j}^2\rho(\hat{\phi}_{i,j})}{d_{2ij}^4} + \frac{x_{i,j}^2 z_{i,j}^2\rho(\hat{\alpha}_{i,j})}{d_{i,j}^4 d_{2ij}^2}\right)$$

$$[\mathbf{F_{xy}}]_{i,i} = E\left[\frac{\partial^2 u_i}{\partial x_i \partial y_i}\right] = \sum_{\substack{j=1 \\ j\neq i}}^{N} x_{i,j}y_{i,j}\left(\frac{\rho(\hat{d}_{i,j})}{d_{i,j}^2} - \frac{\rho(\hat{\phi}_{i,j})}{d_{2ij}^4} + \frac{z_{i,j}^2\rho(\hat{\alpha}_{i,j})}{d_{i,j}^4 d_{2ij}^2}\right)$$

$$[\mathbf{F_{xy}}]_{i,j} = E\left[\frac{\partial^2 u_i}{\partial x_i \partial y_j}\right] = -x_{i,j}y_{i,j}\left(\frac{\rho(\hat{d}_{i,j})}{d_{i,j}^2} - \frac{\rho(\hat{\phi}_{i,j})}{d_{2ij}^4} + \frac{z_{i,j}^2\rho(\hat{\alpha}_{i,j})}{d_{i,j}^4 d_{2ij}^2}\right)$$

$$[\mathbf{F_{xz}}]_{i,i} = E\left[\frac{\partial^2 u_i}{\partial x_i \partial z_i}\right] = \sum_{\substack{j=1 \\ j\neq i}}^{N} \frac{x_{i,j}z_{i,j}}{d_{i,j}^2}\left(\rho(\hat{d}_{i,j}) - \frac{\rho(\hat{\alpha}_{i,j})}{d_{i,j}^2}\right)$$

$$[\mathbf{F_{xz}}]_{i,j} = E\left[\frac{\partial^2 u_i}{\partial x_i \partial z_j}\right] = -\frac{x_{i,j}z_{i,j}}{d_{i,j}^2}\left(\rho(\hat{d}_{i,j}) - \frac{\rho(\hat{\alpha}_{i,j})}{d_{i,j}^2}\right)$$

$$[\mathbf{F_{yy}}]_{i,i} = E\left[\frac{\partial^2 u_i}{\partial y_i^2}\right] = \sum_{\substack{j=1 \\ j\neq i}}^{N}\left(\frac{y_{i,j}^2\rho(\hat{d}_{i,j})}{d_{i,j}^2} + \frac{x_{i,j}^2\rho(\hat{\phi}_{i,j})}{d_{2ij}^4} + \frac{y_{i,j}^2 z_{i,j}^2\rho(\hat{\alpha}_{i,j})}{d_{i,j}^4 d_{2ij}^2}\right) + \frac{1}{\sigma_{\hat{y}_i}^2}$$

$$[\mathbf{F_{yy}}]_{i,j} = E\left[\frac{\partial^2 u_i}{\partial y_i \partial y_j}\right] = -\left(\frac{y_{i,j}^2\rho(\hat{d}_{i,j})}{d_{i,j}^2} + \frac{x_{i,j}^2\rho(\hat{\phi}_{i,j})}{d_{2ij}^4} + \frac{y_{i,j}^2 z_{i,j}^2\rho(\hat{\alpha}_{i,j})}{d_{i,j}^4 d_{2ij}^2}\right)$$

$$[\mathbf{F_{yz}}]_{i,i} = E\left[\frac{\partial^2 u_i}{\partial y_i \partial z_i}\right] = \sum_{\substack{j=1 \\ j\neq i}}^{N} \frac{y_{i,j}z_{i,j}}{d_{i,j}^2}\left(\rho(\hat{d}_{i,j}) - \frac{\rho(\hat{\alpha}_{i,j})}{d_{i,j}^2}\right)$$

$$[\mathbf{F_{yz}}]_{i,j} = E\left[\frac{\partial^2 u_i}{\partial y_i \partial z_j}\right] = -\frac{y_{i,j}z_{i,j}}{d_{i,j}^2}\left(\rho(\hat{d}_{i,j}) - \frac{\rho(\hat{\alpha}_{i,j})}{d_{i,j}^2}\right)$$

$$[\mathbf{F_{zz}}]_{i,i} = E\left[\frac{\partial^2 u_i}{\partial z_i^2}\right] = \sum_{\substack{j=1 \\ j\neq i}}^{N}\left(\frac{z_{i,j}^2\rho(\hat{d}_{i,j})}{d_{i,j}^2} + \frac{d_{2ij}^2\rho(\hat{\alpha}_{i,j})}{d_{i,j}^4}\right) + \frac{1}{\sigma_{\hat{z}_i}^2}$$

$$[\mathbf{F_{zz}}]_{i,j} = E\left[\frac{\partial^2 u_i}{\partial z_i \partial z_j}\right] = -\left(\frac{z_{i,j}^2\rho(\hat{d}_{i,j})}{d_{i,j}^2} + \frac{d_{2ij}^2\rho(\hat{\alpha}_{i,j})}{d_{i,j}^4}\right)$$

Consequently, the variances of the coordinate estimates are bounded by

$$
\begin{aligned}
\mathrm{var}(\hat{x}_i) &\geq \mathrm{CRLB}(\hat{x}_i) = [\mathbf{F}^{-1}(\hat{\mathbf{p}})]_{j,j} & j &= i \\
\mathrm{var}(\hat{y}_i) &\geq \mathrm{CRLB}(\hat{y}_i) = [\mathbf{F}^{-1}(\hat{\mathbf{p}})]_{j,j} & j &= N+i \\
\mathrm{var}(\hat{z}_i) &\geq \mathrm{CRLB}(\hat{z}_i) = [\mathbf{F}^{-1}(\hat{\mathbf{p}})]_{j,j} & j &= 2N+i
\end{aligned}
\tag{12.15}
$$

When dropping the angle-related terms in the above CRLB, we obtain the CRLB in the presence of anchor-to-anchor distance estimates and GPS coordinate measurements. It is worth noting that, in the absence of GPS measurements, the FIM in (12.9) is associated with the localization in anchor-free sensor networks. In this case the FIM is singular, so that the standard CRLB does not exist.

Consider an example as follows. A cubic region of dimensions of 500 m × 500 m × 500 m is considered, where the anchors are randomly deployed. This dimensional size is only used to show the efficiency of the accuracy enhancement algorithms studied. In practice, the vertical dimensional size of the location area may be much smaller than the horizontal ones. A thousand anchor position configurations are examined and the performance is then averaged. That is, the average CRLB is defined as

$$
\sqrt{\frac{1}{3LN} \sum_{\ell=1}^{L} \mathrm{trace}([\mathbf{F}^{-1}(\hat{\mathbf{p}})]^{(\ell)})}
\tag{12.16}
$$

Figure 12.5 shows the CRLB with respect to the STDs of the anchor-to-anchor distance estimation errors and those of the angular estimation errors. The azimuth and elevation angle errors are assumed to have the same STDs, the STD of the GPS measurement errors is set at 10 m and there are six anchors. For the parameters examined, the CRLB is more sensitive to the

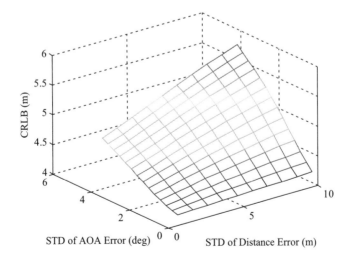

Figure 12.5 CRLB under different STDs of distance and angular estimation errors with six anchors and the STD of the GPS coordinate measurements set at 10 m.

angular estimation errors than to the distance estimation errors. This may indicate that, when both range and angle estimates are employed, the angle estimation accuracy should be higher. Note that accurate angle estimation is often difficult for the reasons described preciously. Therefore, angle-based systems may not be a solution to solve practical problems. The analysis related to AOA throughout the book mainly aims at providing theoretical insight into the positioning problem. Nevertheless, it would provide some guidance in circumstances where AOA-based positioning is considered and feasible. From Figure 12.5, it can be seen that for the parameters given the CRLB ranges roughly from 4 to 6 m, a decrease of about 50% of the original GPS measurement errors.

12.4 Accuracy Improvement Based on Distance and Angle Estimates

This section presents accuracy enhancement by fusing GPS measurements and anchor-to-anchor distance and angle measurements. Simple but efficient LS algorithms are developed for both LOS and NLOS scenarios.

12.4.1 Least-Squares Algorithm in Line-of-Sight Conditions

When both the peer-to-peer distance and angle estimates are provided, we can derive low-complexity LS algorithms to improve the anchor position accuracy under either LOS or NLOS conditions. To obtain the information of LOS and NLOS conditions, the NLOS identification methods to be presented in Chapter 14 may be applied. The identified NLOS corrupted measurements should be excluded in the event of plenty of LOS measurements. Otherwise, the NLOS measurements need to be exploited. Let us first focus on the LOS scenario. From Figure 12.2 and (12.4) and (12.5), it can be seen that the equivalent observation equations related to the anchor-to-anchor distance and angle estimates are

$$
\begin{aligned}
x_j - x_i &= (\hat{d}_{i,j} - n_{\hat{d}_{i,j}})\sin(\hat{\alpha}_{i,j} - n_{\hat{\alpha}_{i,j}})\cos(\hat{\phi}_{i,j} - n_{\hat{\phi}_{i,j}}) \\
&\approx \hat{d}_{i,j}\sin\hat{\alpha}_{i,j}\cos\hat{\phi}_{i,j} \\
y_j - y_i &= (\hat{d}_{i,j} - n_{\hat{d}_{i,j}})\sin(\hat{\alpha}_{i,j} - n_{\hat{\alpha}_{i,j}})\sin(\hat{\phi}_{i,j} - n_{\hat{\phi}_{i,j}}) \\
&\approx \hat{d}_{i,j}\sin\hat{\alpha}_{i,j}\sin\hat{\phi} \\
z_j - z_i &= (\hat{d}_{i,j} - n_{\hat{d}_{i,j}})\cos(\hat{\alpha}_{i,j} - n_{\hat{\alpha}_{i,j}}) \\
&\approx \hat{d}_{i,j}\cos\hat{\alpha}_{i,j} \qquad i, j = 1, 2, \ldots, N, i \neq j
\end{aligned} \tag{12.17}
$$

where the approximations result from dropping the estimation errors. Since (12.7) and (12.17) are linear in terms of the anchor coordinates, we can write them in a compact form as

$$
\mathbf{Ap} \approx \mathbf{r} \tag{12.18}
$$

where $\mathbf{p}$ is the position vector defined as

$$
\mathbf{p} = \begin{bmatrix} x_1, x_2, \ldots, x_N, & y_1, y_2, \ldots, y_N, & z_1, z_2, \ldots, z_N \end{bmatrix}^{\mathrm{T}} \in \mathbb{R}^{3N \times 1} \tag{12.19}
$$

A is a constant matrix with elements equal to 0, $+1$ or -1 and **r** is the constant observation vector. Both **A** and **r** are defined in Annex 12.A. Applying the weighted LS (WLS) estimator, the refined anchor location estimates become

$$\hat{\hat{\mathbf{p}}} = (\mathbf{A}^T\mathbf{W}\mathbf{A})^{-1}\mathbf{A}^T\mathbf{W}\mathbf{r} \tag{12.20}$$

where we use double hats to distinguish it from the original GPS coordinate measurements and **W** is the weighting matrix to emphasize the more reliable estimates. For example, the received SNR may be used to choose the weights. The matrix inverse is assumed to exist, and this is usually true for an overdetermined system. It would be feasible to run this algorithm at any anchor due to its low complexity, although an anchor with the most computational power might be chosen to run the algorithm. In the event that the anchors are sparsely deployed in a large sensor network, the anchors may be divided into a number of groups and the algorithm run at each group independently in order to avoid undesirable long-distance communications.

In practice, angular measurements will suffer from multipath propagation. Also, the orientation measurements will be affected by magnetic anomalies. Therefore, caution must be taken to ensure the accuracy of the angular measurements. Inaccurate AOA measurements should not be considered for position determination.

The efficiency of the method is evaluated through an illustrative example as follows. The simulation setup is the same as in Section 12.3 and the performance measure is the RMSE of the anchor location estimation, which is computed according to

$$\sqrt{\frac{1}{3LN}\sum_{\ell=1}^{L}\sum_{k=1}^{N}\left[(\hat{\hat{x}}_k^{(\ell)} - x_k^{(\ell)})^2 + (\hat{\hat{y}}_k^{(\ell)} - y_k^{(\ell)})^2 + (\hat{\hat{z}}_k^{(\ell)} - z_k^{(\ell)})^2\right]} \tag{12.21}$$

where ℓ indexes the anchor configurations and k indexes the anchors. Note that the RMSE is equivalent to the STD of the estimation error when the mean of the error is equal to zero.

Figure 12.6 shows the accuracy of the linear LS algorithms with respect to the distance estimation errors, ranging from 1 to 10 m. Two different STDs of the original (GPS) measurement errors ($\sigma = 10$ m and $\sigma = 18$ m) are examined, the STD of the angular estimation errors is set at 2.1° and there are six anchors. It can be observed that the anchor location error can be decreased by around 50% by employing the simple LS algorithm.

Figure 12.7 shows the accuracy of the linear LS algorithms versus the accuracy of the angle (both azimuth and elevation) estimation, ranging from 0.3° to 4°. The STD of the distance estimation errors is set at 6 m (this is just an example). Also, two different STDs of the original (GPS) measurements are examined (10 and 18 m). Clearly, over the AOA error range, the location accuracy gain is substantial. Comparatively, the accuracy is more sensitive to the angular estimation errors than the distance errors. The reason is probably that, when the distance is large, a small angular error would produce a larger distance error.

12.4.2 Least-Squares Algorithm in Non-Line-of-Sight Conditions

Let us discuss the case in which all anchor-to-anchor radio propagations are under NLOS conditions. As shown in Figure 12.3, the radio propagation between anchors i and j is NLOS and the signal is reflected only once before arriving at the receiver. $\phi_{i,j}$ is the azimuth angle of

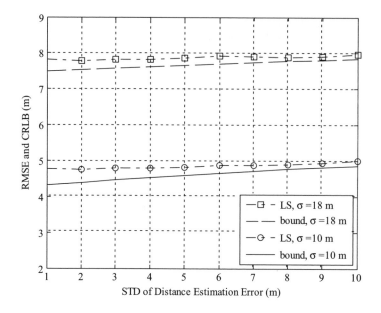

Figure 12.6 Impact of distance accuracy on the LS algorithm in LOS conditions. The STD of original coordinate estimates equals either 10 or 18 m and there are six anchors.

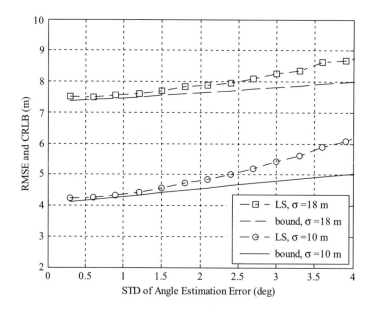

Figure 12.7 Impact of AOA accuracy on the LS algorithms in LOS conditions. The STD of GPS coordinate estimates equals either 10 or 18 m and there are six anchors.

the signal transmitted from anchor j, reflected by the scatterer, and received at anchor i; $\alpha_{i,j}$ is the elevation angle of the signal transmitted from anchor j, reflected by the scatterer, and received at anchor i. Note that the problem is simplified for analytical convenience. In reality, there are multiple scattering points and the scattering is not from points but from surfaces of certain dimensions, depending on the scatterers.

The purpose is to introduce the scatterer locations to reduce the NLOS effect. From Figure 12.3 and (12.4) and (12.6), we can readily obtain the equivalent observation equations related to the anchor-to-anchor distance and angle estimates as

$$
\begin{aligned}
x_s^{(i,j)} - x_i &\approx d_s^{(i,j)} \sin \hat{\alpha}_{i,j} \cos \hat{\phi}_{i,j} \qquad i = 1, 2, \ldots, N-1, j = i+1, \ldots, N \\
y_s^{(i,j)} - y_i &\approx d_s^{(i,j)} \sin \hat{\alpha}_{i,j} \sin \hat{\phi}_{i,j} \\
z_s^{(i,j)} - z_i &\approx d_s^{(i,j)} \cos \hat{\alpha}_{i,j} \\
x_s^{(i,j)} - x_j &\approx (\hat{d}_{i,j} - d_s^{(i,j)}) \sin \hat{\alpha}_{j,i} \cos \hat{\phi}_{j,i} \\
y_s^{(i,j)} - y_j &\approx (\hat{d}_{i,j} - d_s^{(i,j)}) \sin \hat{\alpha}_{j,i} \sin \hat{\phi}_{j,i} \\
z_s^{(i,j)} - z_j &\approx (\hat{d}_{i,j} - d_s^{(i,j)}) \cos \hat{\alpha}_{j,i}
\end{aligned}
\tag{12.22}
$$

Define the parameter vector as

$$
\boldsymbol{\theta} = \begin{bmatrix} \mathbf{x}^T & \mathbf{y}^T & \mathbf{z}^T & \mathbf{x}_s^T & \mathbf{y}_s^T & \mathbf{z}_s^T & \mathbf{d}_s \end{bmatrix}^T \in \mathbb{R}^{N(2N+1) \times 1}
\tag{12.23}
$$

where

$$
\begin{aligned}
\mathbf{x} &= [x_1, x_2, \ldots, x_N]^T \\
\mathbf{y} &= [y_1, y_2, \ldots, z_N]^T \\
\mathbf{z} &= [z_1, z_2, \ldots, z_N]^T \\
\mathbf{x}_s &= [x_s^{(1,2)} \ldots, x_s^{(1,N)}, x_s^{(2,3)} \ldots, x_s^{(2,N)} \ldots, x_s^{(N-1,N)}] \in \mathbb{R}^{N_1 \times 1} \\
\mathbf{y}_s &= [y_s^{(1,2)} \ldots, y_s^{(1,N)}, y_s^{(2,3)} \ldots, y_s^{(2,N)} \ldots, y_s^{(N-1,N)}] \in \mathbb{R}^{N_1 \times 1} \\
\mathbf{z}_s &= [z_s^{(1,2)} \ldots, z_s^{(1,N)}, z_s^{(2,3)} \ldots, z_s^{(2,N)} \ldots, z_s^{(N-1,N)}] \in \mathbb{R}^{N_1 \times 1} \\
\mathbf{d}_s &= [d_s^{(1,2)} \ldots, d_s^{(1,N)}, d_s^{(2,3)} \ldots, d_s^{(2,N)} \ldots, d_s^{(N-1,N)}] \in \mathbb{R}^{N_1 \times 1}
\end{aligned}
\tag{12.24}
$$

Then, (12.7) and (12.22) can be written in a compact form as

$$
\mathbf{B}\boldsymbol{\theta} \approx \mathbf{h}
\tag{12.25}
$$

where the matrix $\mathbf{B}$ and the observation vector $\mathbf{h}$ are defined in Annex 12.B. The WLS solution of (12.25) is given by

$$
\hat{\boldsymbol{\theta}} = (\mathbf{B}^T \mathbf{Q} \mathbf{B})^{-1} \mathbf{B}^T \mathbf{Q} \mathbf{h}
\tag{12.26}
$$

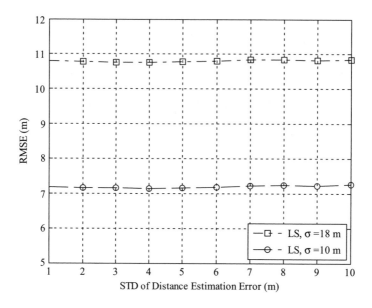

Figure 12.8 Impact of distance accuracy on the LS algorithm in NLOS conditions. The STD of original coordinate estimates equals either 10 or 18 m and there are six anchors.

where $\mathbf{Q}$ is the weighting matrix. Owing to the fact that the first $3N_1$ components of $\mathbf{h}$ are zero, which can be seen from (12.22), the refined anchor position estimates can be computed according to

$$\hat{\mathbf{p}} = \left[\hat{\mathbf{x}}^{\mathrm{T}}, \hat{\mathbf{y}}^{\mathrm{T}}, \hat{\mathbf{z}}^{\mathrm{T}}\right]^{\mathrm{T}}$$
$$= \mathbf{V}\tilde{\mathbf{h}} \tag{12.27}$$

where $\mathbf{V} \in \mathbb{R}^{3N \times 3N_1}$ is the top-right sub-block matrix of $(\mathbf{B}^{\mathrm{T}}\mathbf{Q}\mathbf{B})^{-1}\mathbf{B}^{\mathrm{T}}\mathbf{Q}$ and $\tilde{\mathbf{h}} \in \mathbb{R}^{3N_1 \times 1}$ is the column vector that is subset of vector $\hat{\mathbf{h}}$ starting from the $(3N_1 + 1)$th component to the last one.

The performance of the accuracy enhancement algorithm in NLOS conditions can be evaluated in a similar way to that in LOS conditions. The scatterers' locations are randomly produced in the location area for simplicity. Figures 12.8 and 12.9 show the simulated results of the LS algorithm with respect to the distance error and the AOA error respectively. The STD of the original (GPS) coordinate measurement error is set at two different values, 10 m and 18 m. Compared with the accuracy improvement in LOS conditions, there is considerable accuracy degradation under the NLOS condition. However, the accuracy improvement in NLOS conditions is still substantial. It can be seen that the accuracy gain drops sharply as the AOA error increases. Therefore, when the AOA error is large due to severe multipath and NLOS condition, the accuracy enhancement may be marginal. In the event that there are both NLOS and LOS distance and angular estimates, the accuracy improvement of the LS

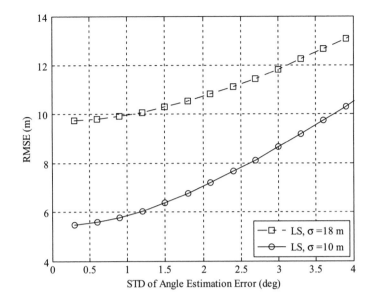

Figure 12.9 Impact of AOA accuracy on the LS algorithm in NLOS conditions. The STD of original coordinate estimates equals either 10 or 18 m and there are six anchors.

algorithm will be between the best (all LOS conditions) and the worst (all NLOS conditions) results.

12.5 Accuracy Improvement Based on Distance Estimates

In most situations, anchors do not have antenna arrays or directional antennas, so angle estimates are not available. In this case, the LS algorithms described in the previous section are no longer applicable. In this section, we can employ the iterative optimization method to improve the anchor location accuracy.

Making use of all the anchor-to-anchor distance estimates, the cost function is defined as

$$\varepsilon(\mathbf{p}) = \sum_{i=1}^{N} \sum_{\substack{j=1 \\ j \neq i}}^{N} w_{i,j} (\hat{d}_{i,j} - d_{i,j})^2 \qquad (12.28)$$

where $\{w_{i,j}\}$ are the weights. The anchor global positions can be refined by minimizing the above cost function with respect to the anchor coordinates; that is:

$$\hat{\mathbf{p}} = \arg\min_{\mathbf{p}} \varepsilon(\mathbf{p}) \qquad (12.29)$$

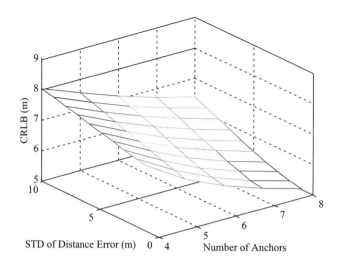

Figure 12.10   Impact of distance estimation accuracy and number of anchors on the CRLB when STD of GPS coordinates estimates equals 10 m.

The GPS measurements are used as the initial position coordinate estimates. Intuitively, accurate distance measurements can achieve more accurate relative positions of the anchors. Since the GPS measurements are employed as the starting point, the iterative process would result in more accurate global positions. A range of optimization methods can be considered and a couple of examples of using optimization in positioning have already been discussed in Chapters 7, 8 and 10. In this simulation, the LM method [8,9] will be employed. Please refer to Chapter 7 for more information about the iterative optimization algorithms.

In the same way as evaluating the enhancement algorithm based on both distance and AOA estimation in the preceding section, the performance of the distance-based algorithm can be examined. Figure 12.10 illustrates the impact of the distance estimation error and the number of anchors on the CRLB when the STD of the original GPS measurement error is set at 10 m in the absence of AOA estimates. It is seen that the accuracy improves as the number of anchors increases. Perhaps, this is due to the fact that the addition of one anchor introduces three more unknown position parameters; however, $6(N-1)$ more peer-to-peer measurements and three more GPS measurements are produced. Generally speaking, more independent measurements could result in a redundancy/diversity gain. Figure 12.11 shows the performance of the LM-based iterative optimization algorithm under the same conditions. Basically, there is a good match between Figures 12.10 and 12.11. Over the range of distance estimation errors and the numbers of the anchors, the resulting RMSE of the enhanced location estimation ranges from 5 to 9 m. Clearly, the location accuracy improvement is substantial.

Figure 12.12 shows the CRLB versus the accuracy of the distance estimation and that of the GPS measurements when there are six anchors. Roughly, there exists a linear relation between the CRLB and the accuracy of the distance estimation or the GPS measurements. For the range of the distance estimation and GPS measurement errors, the accuracy of the

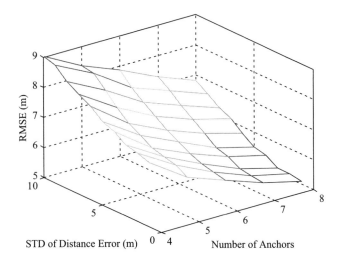

Figure 12.11 Impact of distance estimation accuracy and number of anchors on the accuracy of the LS algorithm when STD of GPS coordinates estimates equals 10 m.

refined anchor position estimation is around between 3 and 12 m. Figure 12.13 shows the accuracy of the iterative optimization algorithm under the same setup as that in Figure 12.12. Clearly, the accuracy of the algorithm closely approaches the CRLB. Thus, the distance-based algorithm can also provide a good accuracy gain, which is also observed in Figure 12.11.

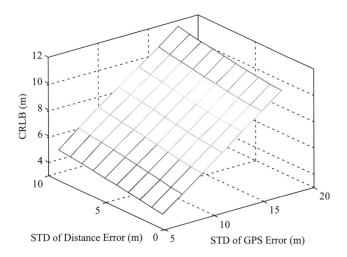

Figure 12.12 Impact of distance estimation accuracy and GPS measurement error on the CRLB when there are six anchors.

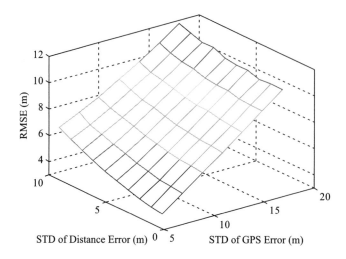

Figure 12.13 Impact of distance estimation accuracy and GPS measurement error on the LM algorithm when there are six anchors.

Annex 12.A: Definition of Matrix and Vector in Line-of-Sight Conditions

In Section 12.4.1 an LS estimator was used to determine the refined anchor location estimates. For convenience, the constant matrix and vector in (12.18) are presented in this annex. The constant matrix $\mathbf{A}$ in (12.18) is defined as

$$\mathbf{A} = \begin{bmatrix} \mathbf{C}^T & \mathbf{C}^T & \mathbf{0} & \mathbf{0} & \mathbf{0} & \mathbf{0} & \mathbf{I}_N & \mathbf{0} & \mathbf{0} \\ \mathbf{0} & \mathbf{0} & \mathbf{C}^T & \mathbf{C}^T & \mathbf{0} & \mathbf{0} & \mathbf{0} & \mathbf{I}_N & \mathbf{0} \\ \mathbf{0} & \mathbf{0} & \mathbf{0} & \mathbf{0} & \mathbf{C}^T & \mathbf{C}^T & \mathbf{0} & \mathbf{0} & \mathbf{I}_N \end{bmatrix}^T \in \mathbb{R}^{3(2N_1+N)\times 3N} \tag{12.30}$$

where

$$N_1 = \frac{1}{2}N(N-1) \tag{12.31}$$

$\mathbf{I}_N$ is an identity matrix of dimensions of $N \times N$ and

$$\mathbf{C} = \begin{bmatrix} 1 & -1 & 0 & 0 & 0 & \cdots & 0 & 0 & 0 \\ 1 & 0 & -1 & 0 & 0 & \cdots & 0 & 0 & 0 \\ \vdots & \vdots & \vdots & \vdots & \vdots & \ddots & \vdots & \vdots & \vdots \\ 1 & 0 & 0 & 0 & 0 & \cdots & 0 & 0 & -1 \\ 0 & 1 & -1 & 0 & 0 & \cdots & 0 & 0 & 0 \\ 0 & 1 & 0 & -1 & 0 & \cdots & 0 & 0 & 0 \\ \vdots & \vdots & \vdots & \vdots & \vdots & \ddots & \vdots & \vdots & \vdots \\ 0 & 1 & 0 & 0 & 0 & \cdots & 0 & 0 & -1 \\ \vdots & \vdots & \vdots & \vdots & \vdots & \ddots & \vdots & \vdots & \vdots \\ 0 & 0 & 0 & 0 & 0 & \cdots & 0 & 1 & -1 \end{bmatrix} \in \mathbb{R}^{N_1\times N} \tag{12.32}$$

The constant observation vector $\mathbf{r}$ in (12.18) is defined as

$$\mathbf{r} = [\mathbf{r}_{1a}^T, \mathbf{r}_{1b}^T, \mathbf{r}_{2a}^T, \mathbf{r}_{2b}^T, \mathbf{r}_{3a}^T, \mathbf{r}_{3b}^T, \hat{\mathbf{x}}^T, \hat{\mathbf{y}}^T, \hat{\mathbf{z}}^T]^T \in \mathbb{R}^{3(2N_1+N)\times 1} \tag{12.33}$$

where

$$\hat{\mathbf{x}} = [\hat{x}_1, \hat{x}_2, \ldots, \hat{x}_N]^T$$
$$\hat{\mathbf{y}} = [\hat{y}_1, \hat{y}_2, \ldots, \hat{z}_N]^T \tag{12.34}$$
$$\hat{\mathbf{z}} = [\hat{z}_1, \hat{z}_2, \ldots, \hat{z}_N]^T$$

and

$$\mathbf{r}_{1a} = \Big[\hat{d}_{1,2}\sin\hat{\alpha}_{1,2}\cos\hat{\phi}_{1,2} \;\cdots\; \hat{d}_{1,N}\sin\hat{\alpha}_{1,N}\cos\hat{\phi}_{1,N} \; \hat{d}_{2,3}\sin\hat{\alpha}_{2,3}\cos\hat{\phi}_{2,3} \;\cdots\; \hat{d}_{2,N}\sin\hat{\alpha}_{2,N}\cos\hat{\phi}_{2,N}$$
$$\cdots \hat{d}_{N-1,N}\sin\hat{\alpha}_{N-1,N}\cos\hat{\phi}_{N-1,N}\Big]^T \in \mathbb{R}^{N_1\times 1}$$

$$\mathbf{r}_{1b} = \Big[\hat{d}_{2,1}\sin\hat{\alpha}_{2,1}\cos\hat{\phi}_{2,1} \;\cdots\; \hat{d}_{N,1}\sin\hat{\alpha}_{N,1}\cos\hat{\phi}_{N,1} \; \hat{d}_{3,2}\sin\hat{\alpha}_{3,2}\cos\hat{\phi}_{3,2} \;\cdots\; \hat{d}_{N,2}\sin\hat{\alpha}_{N,2}\cos\hat{\phi}_{N,2}$$
$$\cdots \hat{d}_{N,N-1}\sin\hat{\alpha}_{N,N-1}\cos\hat{\phi}_{N,N-1}\Big]^T \in \mathbb{R}^{N_1\times 1}$$

$$\mathbf{r}_{2a} = \Big[\hat{d}_{1,2}\sin\hat{\alpha}_{1,2}\sin\hat{\phi}_{1,2} \;\cdots\; \hat{d}_{1,N}\sin\hat{\alpha}_{1,N}\sin\hat{\phi}_{1,N} \; \hat{d}_{2,3}\sin\hat{\alpha}_{2,3}\sin\hat{\phi}_{2,3} \;\cdots\; \hat{d}_{2,N}\sin\hat{\alpha}_{2,N}\sin\hat{\phi}_{2,N}$$
$$\cdots \hat{d}_{N-1,N}\sin\hat{\alpha}_{N-1,N}\sin\hat{\phi}_{N-1,N}\Big]^T \in \mathbb{R}^{N_1\times 1}$$

$$\mathbf{r}_{2b} = \Big[\hat{d}_{2,1}\sin\hat{\alpha}_{2,1}\sin\hat{\phi}_{2,1} \;\cdots\; \hat{d}_{N,1}\sin\hat{\alpha}_{N,1}\sin\hat{\phi}_{N,1} \; \hat{d}_{3,2}\sin\hat{\alpha}_{3,2}\sin\hat{\phi}_{3,2} \;\cdots\; \hat{d}_{N,2}\sin\hat{\alpha}_{N,2}\sin\hat{\phi}_{N,2}$$
$$\cdots \hat{d}_{N,N-1}\sin\hat{\alpha}_{N,N-1}\sin\hat{\phi}_{N,N-1}\Big]^T \in \mathbb{R}^{N_1\times 1}$$

$$\mathbf{r}_{3a} = \Big[\hat{d}_{1,2}\cos\hat{\alpha}_{1,2} \;\cdots\; \hat{d}_{1,N}\cos\hat{\alpha}_{1,N} \; \hat{d}_{2,3}\cos\hat{\alpha}_{2,3} \;\cdots\; \hat{d}_{2,N}\cos\hat{\alpha}_{2,N} \;\cdots$$
$$\hat{d}_{N-1,N}\cos\hat{\alpha}_{N-1,N}\Big]^T \in \mathbb{R}^{N_1\times 1}$$

$$\mathbf{r}_{3b} = \Big[\hat{d}_{2,1}\cos\hat{\alpha}_{2,1} \;\cdots\; \hat{d}_{N,1}\cos\hat{\alpha}_{N,1} \; \hat{d}_{3,2}\cos\hat{\alpha}_{3,2} \;\cdots\; \hat{d}_{N,2}\cos\hat{\alpha}_{N,2}$$
$$\cdots \hat{d}_{N,N-1}\cos\hat{\alpha}_{N,N-1}\Big]^T \in \mathbb{R}^{N_1\times 1} \tag{12.35}$$

Note that in the above definitions of $\mathbf{A}$ and $\mathbf{r}$ it is assumed that each anchor is within radio range of other anchors. In the event that a pair of anchors is out of the radio range, the corresponding components in the matrices and vectors disappear.

Annex 12.B: Definition of Matrix and Vector in Non-Line-of-Sight Conditions

Section 12.4.2 presented anchor position accuracy enhancement in NLOS scenarios. The anchor positions are refined by solving the linear equation (12.25). The constant matrix and the observation vector in this equation are provided in this annex. The matrix $\mathbf{B}$ in (12.25) is defined as

$$
\mathbf{B} = \begin{bmatrix}
\mathbf{A}_1 & \mathbf{0} & \mathbf{0} & -\mathbf{I}_{N_1} & \mathbf{0} & \mathbf{0} & \mathbf{A}_{sc} \\
\mathbf{0} & \mathbf{A}_1 & \mathbf{0} & \mathbf{0} & -\mathbf{I}_{N_1} & \mathbf{0} & \mathbf{A}_{ss} \\
\mathbf{0} & \mathbf{0} & \mathbf{A}_1 & \mathbf{0} & \mathbf{0} & -\mathbf{I}_{N_1} & \mathbf{A}_s \\
\mathbf{B}_1 & \mathbf{0} & \mathbf{0} & -\mathbf{I}_{N_1} & \mathbf{0} & \mathbf{0} & \mathbf{B}_{sc} \\
\mathbf{0} & \mathbf{B}_1 & \mathbf{0} & \mathbf{0} & -\mathbf{I}_{N_1} & \mathbf{0} & \mathbf{B}_{ss} \\
\mathbf{0} & \mathbf{0} & \mathbf{B}_1 & \mathbf{0} & \mathbf{0} & -\mathbf{I}_{N_1} & \mathbf{B}_s \\
\mathbf{I}_N & \mathbf{0} & \mathbf{0} & \mathbf{0} & \mathbf{0} & \mathbf{0} & \mathbf{0} \\
\mathbf{0} & \mathbf{I}_N & \mathbf{0} & \mathbf{0} & \mathbf{0} & \mathbf{0} & \mathbf{0} \\
\mathbf{0} & \mathbf{0} & \mathbf{I}_N & \mathbf{0} & \mathbf{0} & \mathbf{0} & \mathbf{0}
\end{bmatrix} \in \mathbb{R}^{3N^2 \times N(2N+1)} \qquad (12.36)
$$

where

$$
\mathbf{A}_1 = \begin{bmatrix}
\mathbf{e}_{N-1} & \mathbf{0} & \mathbf{0} & \cdots & \mathbf{0} & \mathbf{0} & \mathbf{0} \\
\mathbf{0} & \mathbf{e}_{N-1} & \mathbf{0} & \cdots & \mathbf{0} & \mathbf{0} & \mathbf{0} \\
\vdots & \vdots & \vdots & \ddots & \vdots & \vdots & \vdots \\
\mathbf{0} & \mathbf{0} & \mathbf{0} & \mathbf{0} & \mathbf{e}_2 & \mathbf{0} & \mathbf{0} \\
\mathbf{0} & \mathbf{0} & \mathbf{0} & \mathbf{0} & \mathbf{0} & 1 & \mathbf{0}
\end{bmatrix} \in \mathbb{R}^{N_1 \times N} \qquad (12.37)
$$

where $\mathbf{e}_m$ is a column vector of m ones and

$$
\mathbf{B}_1 = \begin{bmatrix}
\mathbf{0} & \mathbf{0} & \cdots & \mathbf{0} & 0 \\
\mathbf{I}_{N-1} & \mathbf{I}_{N-2} & \cdots & \mathbf{I}_2 & 1
\end{bmatrix}^{\mathrm{T}} \in \mathbb{R}^{N_1 \times N} \qquad (12.38)
$$

Also:

$$
\begin{aligned}
\mathbf{A}_{sc} &= \mathrm{diag}\{\sin \hat{\alpha}_{1,2} \cos \hat{\phi}_{1,2} \quad \cdots \quad \sin \hat{\alpha}_{1,N} \cos \hat{\phi}_{1,N} \quad \sin \hat{\alpha}_{2,3} \cos \hat{\phi}_{2,3} \quad \cdots \quad \sin \hat{\alpha}_{2,N} \cos \hat{\phi}_{2,N} \\
&\quad \cdots \quad \sin \hat{\alpha}_{N-1,N} \cos \hat{\phi}_{N-1,N}\} \in \mathbb{R}^{N_1 \times N_1} \\
\mathbf{A}_{ss} &= \mathrm{diag}\{\sin \hat{\alpha}_{1,2} \sin \hat{\phi}_{1,2} \quad \cdots \quad \sin \hat{\alpha}_{1,N} \sin \hat{\phi}_{1,N} \quad \sin \hat{\alpha}_{2,3} \sin \hat{\phi}_{2,3} \quad \cdots \quad \sin \hat{\alpha}_{2,N} \\
&\quad \sin \hat{\phi}_{2,N} \cdots \sin \hat{\alpha}_{N-1,N} \sin \hat{\phi}_{N-1,N}\} \in \mathbb{R}^{N_1 \times N_1} \\
\mathbf{A}_c &= \mathrm{diag}\{\cos \hat{\alpha}_{1,2} \quad \cdots \quad \cos \hat{\alpha}_{1,N} \quad \cos \hat{\alpha}_{2,3} \quad \cdots \quad \cos \hat{\alpha}_{2,N} \quad \cdots \quad \cos \hat{\alpha}_{N-1,N}\} \in \mathbb{R}^{N_1 \times N_1}
\end{aligned} \qquad (12.39)
$$

and

$$
\begin{aligned}
\mathbf{B}_{sc} &= -\mathrm{diag}\{\sin \hat{\alpha}_{2,1} \cos \hat{\phi}_{2,1} \quad \cdots \quad \sin \hat{\alpha}_{N,1} \cos \hat{\phi}_{N,1} \quad \sin \hat{\alpha}_{3,2} \cos \hat{\phi}_{3,2} \quad \cdots \\
&\quad \sin \hat{\alpha}_{N,2} \cos \hat{\phi}_{N,2} \cdots \sin \hat{\alpha}_{N,N-1} \cos \hat{\phi}_{N,N-1}\} \in \mathbb{R}^{N_1 \times N_1} \\
\mathbf{B}_{ss} &= -\mathrm{diag}\{\sin \hat{\alpha}_{2,1} \sin \hat{\phi}_{2,1} \quad \cdots \quad \sin \hat{\alpha}_{N,1} \sin \hat{\phi}_{N,1} \quad \sin \hat{\alpha}_{3,2} \sin \hat{\phi}_{3,2} \quad \cdots \\
&\quad \sin \hat{\alpha}_{N,2} \sin \hat{\phi}_{N,2} \cdots \sin \hat{\alpha}_{N,N-1} \sin \hat{\phi}_{N,N-1}\} \in \mathbb{R}^{N_1 \times N_1} \\
\mathbf{B}_c &= -\mathrm{diag}\{\cos \hat{\alpha}_{2,1} \quad \cdots \quad \cos \hat{\alpha}_{N,1} \quad \cos \hat{\alpha}_{3,2} \quad \cdots \\
&\quad \cos \hat{\alpha}_{N,2} \cdots \cos \hat{\alpha}_{N,N-1}\} \in \mathbb{R}^{N_1 \times N_1}
\end{aligned} \qquad (12.40)
$$

The observation vector **h** in (12.25) is given by

$$\mathbf{h} = \left[\mathbf{0}_{3N_1}^T, \quad \mathbf{h}_{sc}^T, \quad \mathbf{h}_{ss}^T, \quad \mathbf{h}_c^T, \quad \hat{\mathbf{x}}^T, \quad \hat{\mathbf{y}}^T, \quad \hat{\mathbf{z}}^T \right]^T \in \mathbb{R}^{3N^2 \times 1} \qquad (12.41)$$

where $\mathbf{0}_m$ is a column vector of m zeros, $\hat{\mathbf{x}}, \hat{\mathbf{y}}$ and $\hat{\mathbf{z}}$ are the GPS measurements defined in (12.34) and

$$\mathbf{h}_{sc} = -\{\hat{d}_{2,1} \sin \hat{\alpha}_{2,1} \cos \hat{\phi}_{2,1} \quad \cdots \quad \hat{d}_{N,1} \sin \hat{\alpha}_{N,1} \cos \hat{\phi}_{N,1} \quad \hat{d}_{3,2} \sin \hat{\alpha}_{3,2} \cos \hat{\phi}_{3,2} \quad \cdots$$
$$\hat{d}_{N,2} \sin \hat{\alpha}_{N,2} \cos \hat{\phi}_{N,2} \cdots \hat{d}_{N,N-1} \sin \hat{\alpha}_{N,N-1} \cos \hat{\phi}_{N,N-1}\}^T \in \mathbb{R}^{N_1 \times 1}$$
$$\mathbf{h}_{ss} = -\{\hat{d}_{2,1} \sin \hat{\alpha}_{2,1} \sin \hat{\phi}_{2,1} \quad \cdots \quad \hat{d}_{N,1} \sin \hat{\alpha}_{N,1} \sin \hat{\phi}_{N,1} \quad \hat{d}_{3,2} \sin \hat{\alpha}_{3,2} \sin \hat{\phi}_{3,2} \quad \cdots$$
$$\hat{d}_{N,2} \sin \hat{\alpha}_{N,2} \sin \hat{\phi}_{N,2} \cdots \hat{d}_{N,N-1} \sin \hat{\alpha}_{N,N-1} \sin \hat{\phi}_{N,N-1}\}^T \in \mathbb{R}^{N_1 \times 1}$$
$$\mathbf{h}_c = -\{\hat{d}_{2,1} \cos \hat{\alpha}_{2,1} \quad \cdots \quad \hat{d}_{N,1} \cos \hat{\alpha}_{N,1} \quad \hat{d}_{3,2} \cos \hat{\alpha}_{3,2} \quad \cdots \quad \hat{d}_{N,2} \cos \hat{\alpha}_{N,2} \quad \cdots$$
$$\hat{d}_{N,N-1} \cos \hat{\alpha}_{N,N-1}\}^T \in \mathbb{R}^{N_1 \times 1} \qquad (12.42)$$

References

[1] P. Misra, B.P. Burke and M.M. Pratt, 'GPS performance in navigation', *Proceedings of the IEEE*, **87**(1), 1999, 65–85.
[2] A.S. Zaidi and M.R. Suddle, 'Global navigation satellite systems: a survey', in *Proceedings of International Conference on Advances in Space Technologies*, pp. 84–87, September 2006.
[3] D. Accardo, F. Esposito and A. Moccia, 'Low-cost avionics for autonomous navigation software and hardware testing', in *Proceedings of IEEE Aerospace Conference*, pp. 3016–3024, 2004.
[4] C. Wang, Z. Hu, S. Kusuhara and K. Uchimura, 'Vehicle localization with global probability density function for road navigation', in *Proceedings of IEEE Intelligent Vehicles Symposium*, pp. 1033–1038, June 2007.
[5] R.J. Fontana, 'Recent system applications of short-pulse ultra-wideband (UWB) technology', *IEEE Transactions on Microwave Theory and Technology*, **52**(9), 2004, 2087–2104.
[6] I. Oppermann, M. Hamalainen and J. Iinatti, *UWB Theory and Applications*, John Wiley & Sons, Ltd, Chichester, 2004.
[7] X. Li and K. Pahlavan, 'Super-resolution TOA estimation with diversity for indoor geolocation', *IEEE Transactions on Wireless Communications*, **3**(1), 2004, 224–234.
[8] D. Marquardt, 'Algorithm for least-squares estimation of nonlinear parameters', *SIAM Journal on Applied Mathematics*, **11**(2), 1963, 431–441.
[9] K. Yu and I. Oppermann, 'UWB positioning for wireless embedded networks', in *Proceedings of IEEE RAWCON*, Atlanta, USA, pp. 459–462, 2004.

13

Anchor-free Localization

Node localization is studied in Chapter 11 for scenarios where some nodes are anchors whose absolute positions are known *a priori*. In some circumstances, however, there is no anchor nodes in the positioning network so that no absolute location information is available at any node. For such anchor-free networks, the key issue of localization is to determine the relative positions of the nodes accurately and to accurately determine the graph of the node configuration. In the case where a very limited number of anchors are available, it may be still better to treat the network as an anchor-free one and localize all the nodes under a local coordinate system first. Then, one transforms the local coordinate system into the global one based on the known global locations of the anchors to achieve better location accuracy.

In the absence of anchor nodes and, thus, an absolute coordinate system, a local coordinate system needs to be established. In a 2D environment, one can select three nodes first under the following conditions:

- the nodes are within radio range each other;
- the nodes do not lie in a straight line;
- and the shortest side and the smallest angle of the triangle formed by the three nodes should be greater than some predefined values.

The coordinates of the three nodes are assigned and computed by using distance measurements between each pair of nodes within the radio range. Then, the coordinates of nodes are incrementally calculated using the distances to the three nodes with already calculated coordinates. This incremental scheme is simple and easy to implement; however, the disadvantage is that it incurs error propagation, especially for a large multihop network. To overcome the drawback of the basic incremental approach, different techniques and algorithms must be employed. In this chapter, a number of such anchor-free localization algorithms are studied. Also, localization accuracy measures are developed for anchor-free localization, including the CRLB and the approximate distance error lower bound.

Ground-Based Wireless Positioning   Kegen Yu, Ian Sharp and Jay Guo
© 2009 John Wiley & Sons, Ltd

13.1 Robust Quads

Flip ambiguity is often the dominant factor that affects generating the precise graph of the nodes in multihop wireless networks such as WSNs. Since quadrilaterals (quads) are the smallest possible sub-graph that can be unambiguously localized in isolation, the network nodes can be grouped into quads. The nodes in each quad are within the radio range of each other, so six distance measurements are available. When no three nodes are collinear, the distance constraints make the quad globally rigid. That is, the relative positions of the four nodes are uniquely defined subject to a global rotation, translation and reflection. Also, any number of quads chained together in the way that two chained quads share three nodes forms a globally rigid graph. However, global rigidity is not sufficient to guarantee a unique graph realization in the presence of measurement noise. Therefore, the robustness of each quad must be examined and only the robust quads (RQs) should be exploited. A quad is robust if all the four triangles formed by the nodes are robust. Once the locations of three nodes of a quad are available, the other node can be localized. This procedure continues until all nodes in the chained graph are localized. This is called the RQ method [1].

The RQ method defines a parameter of robustness, which may be set to three times the STD σ of the distance measurement errors. When the smallest angle φ_m and the shortest edge d_m of a triangle satisfy

$$d_m \sin^2\varphi_m > 3\sigma \tag{13.1}$$

the triangle is regarded as robust. Figure 13.1 illustrates an RQ and Figure 13.2 illustrates a non-RQ. When the positions of the three nodes of an RQ are known, the fourth node can be robustly localized. How to estimate the unknown location by using the known locations of three nodes has been discussed in previous chapters; for instance, by using three measurement equations to determine the two unknown coordinates uniquely or by using the linear LS algorithm.

13.2 Multidimensional Scaling Method

The multidimensional scaling (MDS) approach makes use of a distance-like matrix as input to output a coordinate configuration of a set of objects. MDS has been widely used in the physical,

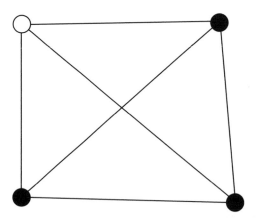

Figure 13.1   Illustration of an RQ.

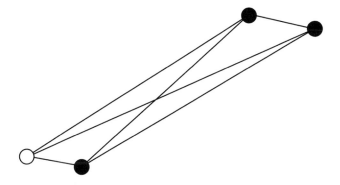

Figure 13.2   Illustration of a non-RQ.

biological and behavioral sciences. The MDS technique has also been used to determine the locations of nodes [2]. Two different approaches can be implemented based on the MDS technique: centralized and distributed. In the centralized scheme, the distance between a pair of nodes which are out of radio range is estimated as the shortest path distance. Basically, the shortest path distance can be determined in a number of different ways. One way is to use the average hop distance and the number of hops so that the distance between two nodes is approximated by

$$\hat{d}_{i,j} = N_{\mathrm{h}}^{(i,j)} d_{\mathrm{h}}$$ (13.2)

where $N_{\mathrm{h}}^{(i,j)}$ is the minimum number of hops between the two nodes and d_{h} is the estimated distance per hop. Alternatively, the distance can be estimated as

$$\hat{d}_{i,j} = \sum_{k=1}^{N_{\mathrm{h}}^{(i,j)}} \hat{d}_{k,k+1}^{(i,j)}$$ (13.3)

where $\hat{d}_{k,k+1}^{(i,j)}$ is the measured distance between the kth node and the $(k+1)$th node on the shortest path from node i to node j. In some cases, to improve the location estimation accuracy, the Euclidian distance can be directly estimated by, for instance, making use of both the distance and AOA measurements between each pair of nodes which are within radio range. This technique is already studied in anchor-based localization. Another way of producing the Euclidean distance estimate is to establish the relation between the Euclidean distance and the shortest path distance in a uniformly deployed sensor network.

Once the pairwise distance estimates are obtained, the squared-distance matrix is constructed as

$$\mathbf{D} = \begin{bmatrix} 0 & \hat{d}_{1,2}^2 & \hat{d}_{1,3}^2 & \cdots & \hat{d}_{1,N-1}^2 & \hat{d}_{1,N}^2 \\ \hat{d}_{2,1}^2 & 0 & \hat{d}_{2,3}^2 & \cdots & \hat{d}_{2,N-1}^2 & \hat{d}_{2,N}^2 \\ \vdots & \vdots & \vdots & \ddots & \vdots & \vdots \\ \hat{d}_{N,1}^2 & \hat{d}_{N,2}^2 & \hat{d}_{N,3}^2 & \vdots & \hat{d}_{N,N-1}^2 & 0 \end{bmatrix}$$ (13.4)

where N is the number of nodes in the network. Clearly, the distance matrix is symmetric matrix whose elements are nonnegative real numbers. Define

$$\mathbf{C} = \mathbf{I}_N - \frac{1}{N}\mathbf{e}_N\mathbf{e}_N^T \tag{13.5}$$

where $\mathbf{I}_N$ is the identity matrix of dimensions of $N \times N$ and $\mathbf{e}_N$ is a column vector of length N whose elements are all ones. Define another matrix as

$$\mathbf{A} = -\frac{1}{2}\mathbf{CDC} \tag{13.6}$$

Next, the matrix $\mathbf{A}$ is decomposed according to singular value decomposition (SVD)[1] as

$$\mathbf{A} = \mathbf{U}\Lambda\mathbf{U}^T \tag{13.7}$$

where

$$\Lambda = \text{diag}\{\lambda_1, \lambda_2, \ldots, \lambda_N\} \tag{13.8}$$

is a diagonal matrix with the eigenvalues of $\mathbf{A}$ as its diagonal elements and $\mathbf{U}$ is a unitary matrix whose column vectors are the corresponding eigenvectors. The eigenvectors are then normalized to be unit eigenvectors and the eigenvalues are sorted in a nonincreasing order to form Λ'. The unit-vector matrix is rearranged accordingly, resulting in $\mathbf{U}'$. Finally, the coordinates of all the nodes are estimated as the first two columns of the matrix

$$\mathbf{V} = \mathbf{U}'\sqrt{\Lambda'} \tag{13.9}$$

That is:

$$\hat{\mathbf{P}} = \begin{bmatrix} \hat{x}_1 & \hat{y}_1 \\ \hat{x}_2 & \hat{y}_2 \\ \vdots & \vdots \\ \hat{x}_N & \hat{y}_N \end{bmatrix} = \begin{bmatrix} v_{1,1} & v_{1,2} \\ v_{2,1} & v_{2,2} \\ \vdots & \vdots \\ v_{N,1} & v_{N,2} \end{bmatrix} \tag{13.10}$$

where $v_{i,j} = \{\mathbf{V}\}_{i,j}$. This centralized scheme can produce the location estimates of all the nodes simultaneously. On the other hand, it involves long-range communications which may result in long delays and unnecessary power dissipation. When new nodes are added, they need to establish distance estimates, either shortest path or Euclidean, to all other nodes. The MDS algorithm has to be run again to produce locations of all existing nodes and the newly added nodes. In addition, the location accuracy would degrade considerably when the shortest path distance estimates are used to form the squared-distance matrix and the sensor deployment is heterogeneous, as shown in Figure 13.3.

To cope with the problems of the centralized scheme, a distributed MDS method can be employed. One way to implement a distributed algorithm is to divide all the sensor nodes into clusters based on their neighborhoods. The clusters are pairwise partially overlapped and a local coordinate system is established in each cluster. The MDS algorithm is applied to localize

---

[1] It is known that SVD and eigenvalue decomposition are equivalent in the case of a full rank square matrix.

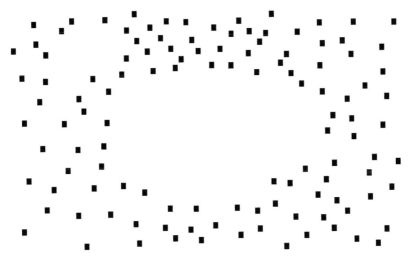

Figure 13.3 An illustration of heterogeneous deployment of sensor nodes.

the nodes within each cluster. The size of a cluster will certainly affect the final location accuracy, and how to select the cluster size may not be an easy task. One extreme case is that each node sets up its own local coordinate system and the other extreme is just one cluster, which corresponds to the centralized scheme. Since each cluster uses its own coordinate system, a process called stitching is required to form a global coordinate system. In general, one cluster is treated as a base and its coordinate system is chosen as the global one. This cluster is usually the one with the largest number of nodes, and it would be preferable that the cluster is close to the center of the location area.

Clearly, there are different ways of stitching all the clusters to form one under the chosen coordinate system. The simplest way is to stitch the clusters one by one starting from the clusters neighboring the base cluster. When stitching two neighboring clusters, coordinate transformation is performed. How to perform the transform is not trivial. One such method is as follows. Suppose that there are N_c nodes whose locations are already determined under both coordinate systems. Let the location estimates of the common nodes be

$$
\begin{aligned}
\hat{\mathbf{p}}_i^{(a)} &= [\hat{x}_i^{(a)}, \hat{x}_i^{(a)}]^{\mathrm{T}} \qquad i = 1, 2, \ldots, N_c \\
\hat{\mathbf{p}}_i^{(b)} &= [\hat{x}_i^{(b)}, \hat{x}_i^{(b)}]^{\mathrm{T}}
\end{aligned}
\tag{13.11}
$$

under coordinate system **a** and **b** respectively. Assume that the purpose is to transform the node location estimates $\hat{\mathbf{p}}_i^{(a)}$, $i > N_c$, into the corresponding values under the coordinate system **b**. The centroids of the location estimates of the common nodes are computed according to

$$
\begin{aligned}
\bar{\mathbf{p}}^{(a)} &= \frac{1}{N_c} \sum_{i=1}^{N_c} \hat{\mathbf{p}}_i^{(a)} \\
\bar{\mathbf{p}}^{(b)} &= \frac{1}{N_c} \sum_{i=1}^{N_c} \hat{\mathbf{p}}_i^{(b)}
\end{aligned}
\tag{13.12}
$$

Then, in the 2D case, the transform matrix

$$\mathbf{Q} = \begin{bmatrix} q_{1,1} & q_{1,2} \\ q_{2,1} & q_{2,2} \end{bmatrix} \in \mathbb{R}^{2\times 2} \tag{13.13}$$

is determined by minimizing the following cost function with respect to the transform matrix:

$$\varepsilon(\mathbf{Q}) = \sum_{i=1}^{N_c} ||\mathbf{Q}(\hat{\mathbf{p}}_i^{(a)} - \bar{\mathbf{p}}^{(a)}) - (\hat{\mathbf{p}}_i^{(b)} - \bar{\mathbf{p}}^{(b)})||^2 \tag{13.14}$$

That is:

$$\hat{\mathbf{Q}} = \arg \min_{\mathbf{Q}} \varepsilon(\mathbf{Q}) \tag{13.15}$$

Once $\hat{\mathbf{Q}}$ is produced, the transform is performed according to

$$\hat{\mathbf{Q}}\hat{\mathbf{p}}_i^{(a)} + \bar{\mathbf{p}}^{(a)} - \bar{\mathbf{p}}^{(b)} \qquad i > N_c \tag{13.16}$$

It is worth noting that the cluster-based scheme overcomes some problems associated with the centralized scheme and the location estimates in each cluster would be more accurate than with the centralized scheme. However, accuracy degradation will also occur when stitching the clusters together.

13.3 Mass–Spring Model

Mass–spring optimization is widely used in the field of force-directed drawing. Howard et al. [3] applied the concept to robot localization and Priyantha et al. [4] proposed a technique to reduce the probability of converging to local minima. The fundamental principle of the mass–spring optimization-based localization can be described as follows. First, an initial position estimate is obtained for each node. This can be achieved in many different ways, but it is crucial that the initial location estimates roughly represent the original network configuration without severe folding. There are a number of approaches to generate the initial graph structurally similar to the original one. For instance, in robot localization, each robot is equipped with odometric equipment and moves around to provide initial coordinates. The second approach is to apply the MDS method to produce the initial location estimates based on the shortest path distance estimates, as described earlier. Another approach is to establish a referent coordinate system by choosing a number of landmarks, as discussed later in this section. Each node is then able to use the shortest path distance estimates to the landmarks and the locations of the landmarks to obtain their initial location estimates.

At any given time, each node has a current estimate of its position. It also periodically sends this position estimate to all its neighbors. Once knowing its own estimated position and the estimated positions of all its neighbors, each node calculates the estimated distance to each neighbor as

$$\hat{d}_{i,j} = \sqrt{(\hat{x}_i - \hat{x}_j)^2 + (\hat{y}_i - \hat{y}_j)^2} \tag{13.17}$$

The distance measurement between two neighboring nodes i and j is known and described as

$$r_{i,j} = \sqrt{(x_i - x_j)^2 + (y_i - y_j)^2} + n_{i,j} \qquad (13.18)$$

where $n_{i,j}$ is the distance measurement error. A force between two nodes (say i and j) is defined as

$$\vec{F}_{i,j} = \vec{u}_{i,j}(\hat{d}_{i,j} - r_{i,j}) \qquad (13.19)$$

where $\vec{u}_{i,j}$ is the unit vector in the direction from $(\hat{x}_i, \hat{y}_i)$ to $(\hat{x}_j, \hat{y}_{ji})$. Accordingly, the resultant force exerted on a node is defined as the sum of the scaled unit vectors, with the scaling factors as the differences between the measured and estimated distances; that is:

$$\vec{F}_i = \sum_j \vec{F}_{i,j} \qquad (13.20)$$

where j indexes the neighboring nodes of node i. In the same way, the total energy of node i is defined as the sum of the squares of the difference between the measured and estimated distances as

$$E_i = \sum_j E_{i,j} = \sum_j (\hat{d}_{i,j} - r_{i,j})^2 \qquad (13.21)$$

It is worth mentioning that, in data fusion, when distance measurements are made from different approaches, different spring constants are needed to scale distance measurements of different degrees of accuracy [5]. The energy of each node reduces when it moves by an infinitesimal amount in the direction of the resultant force. The amount of movement needs to ensure that the new position has a lower energy than the original position and that such movement does not result in local minima. That is, the total energy of the system

$$E = \sum_i E_i \qquad (13.22)$$

should be decreased when the location of a node is changed. Note that (13.22) is typically the cost function for a number of distance-based optimization algorithms. The main difference between the algorithms is that each algorithm uses a different searching method to approach the global minima. One of the main advantages of this mass–spring model-based method lies in its robustness.

13.4 Hybrid Approach

In the preceding two sections, the MDS method and the RQ method are studied. The advantage of the MDS method is that each node can be localized if the one-hop or multihop distance between the node and any other node can be estimated. However, the accuracy of the MDS method may not be satisfactory due to the use of multihop distances, which usually have large errors, especially when the distance is over many hops. On the other hand, the RQ method can achieve good accuracy by decreasing the probability of flipping, but the number of nodes that cannot be localized due to lack of RQs may be large. The essence of the hybrid algorithm is to exploit the advantages of both the MDS and the RQ methods and to compensate for their drawbacks.

It is assumed that distance measurements between each pair of nodes that are within radio range are available. That is:

$$\hat{d}_{i,j} = d_{i,j} + \varepsilon_{i,j} \qquad 1 \leq i,j \leq N, i \neq j \qquad (13.23)$$

where it is assumed that there are N nodes in the network, $\varepsilon_{i,j}$ is the distance estimation error and $d_{i,j} = \sqrt{(x_i - x_j)^2 + (y_i - y_j)^2}$, where (x_i, y_i) and (x_j, y_j) are the coordinates of node i and j respectively. It is worth mentioning that the coordinates of each node depend on the selection of a coordinate system.

To begin with, we need to select a group of nodes in the network that can communicate with each other and the number of the nodes in the group should be as large as possible. The reason for just choosing nodes within radio range of each other is that the MDS algorithm performs well when using one-hop distance estimates. On the other hand, the MDS algorithm tends to produce large errors when a few multihop/shortest path distance estimates are used. Then, we choose three nodes in the group to set up a local coordinate system. Without loss of generality, the three nodes are numbered 1, 2 and 3. It is required that the three chosen nodes do not lie in a straight line. One of the three (say node 1) is set at the origin $(x_1, y_1) = (0,0)$ and the other (say node 2) is on the positive x-axis $(x_2, y_2) = (d_{1,2}, 0)$, where $d_{1,2}$ is the distance between nodes 1 and 2. The third node (say node 3) has a positive y-axis coordinate; that is:

$$
\begin{aligned}
x_3 &= \frac{x_2^2 + d_{1,3}^2 - d_{2,3}^2}{2x_2} \\
y_3 &= \sqrt{d_{1,3}^2 - x_3^2}
\end{aligned}
\qquad (13.24)
$$

Once the three nodes are chosen and their coordinates are assigned, the coordinates of other nodes are uniquely established. Note that, since the distance estimates instead of the true values are available, we can only obtain the estimates of the coordinates x_2, x_3 and y_3.

The selected group of nodes is localized by applying the MDS method that is described earlier in this chapter to produce the location estimates of the base nodes as

$$\hat{P} = \begin{bmatrix} \hat{x}_1 & \hat{y}_1 \\ \hat{x}_2 & \hat{y}_2 \\ \vdots & \vdots \\ \hat{x}_{N_B} & \hat{y}_{N_B} \end{bmatrix} \in \mathbb{R}^{N_B \times 2} \qquad (13.25)$$

where N_B is the number of nodes in the base. Since the position estimates are produced purely based on distance measurements, the resulting coordinate system can be significantly different from the established one. Therefore, if the coordinate system is already established, such as defined by nodes 1, 2 and 3, a coordinate transformation is required. In the

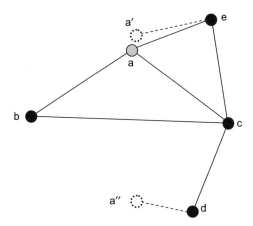

Figure 13.4   An example of localizing a node based on the RTRR approach.

case that there is no coordinate system developed, the generated coordinate system from the MDS may be adopted. In this case, the origin of the coordinate system is not set at the location of a node in general. The advantage of the adopted coordinate system is that it avoids the errors caused by coordinate transformation. The position estimation accuracy may be enhanced by using optimization techniques such as those discussed in Chapter 7. This requires a node in the group to have the computational power to run an iterative minimization algorithm. After the n selected nodes are localized, we make use of the RQ approach to localize the other nodes in the network. An unknown node can be localized if it and three other nodes can form an RQ.

In the event that there are nodes whose locations cannot be determined by the RQ method due to the absence of RQs, the robust triangle plus radio range (RTRR) approach can be exploited to localize the remainder of the nodes, which can be described using Figure 13.4. Node **a** is the neighbor of both nodes **b** and **c** which have been localized, but not necessarily in radio connection to each other. Node **d** has been localized and it is the neighbor of node **c**, but not the neighbor of **a**. In the case that the triangle **abc** is robust, two solutions exist for the position estimate of node **a**; that is, the two locations indicated by **a′** and **a″**. Node **e** is the neighbor of **a** and **a′**, while node **d** is the neighbor of **a′**, but not the neighbor of **a**. We can eliminate the solution (**a″**), since **a′** is not the neighbor of node **d**, but the neighbor of **e**, whereas **a″** is the neighbor of **d**, but not the neighbor of **e**. Therefore, by making use of the neighbors of nodes **a**, **b** and **c**, which have already been localized, the location of node **a** can be successfully determined. More similar information would result in a more accurate judgment. Also, multiple robust triangles may produce multiple position estimates which can be further processed, such as by averaging to obtain more accurate position information. The RTRR method is applied to each unlocalized node and then the RQ approach is applied if there are still nodes to be localized. The RQ and the RTRR algorithms are performed alternately until no more nodes can be localized by the two methods. Figure 13.5 shows the block diagram of the proposed hybrid localization scheme.

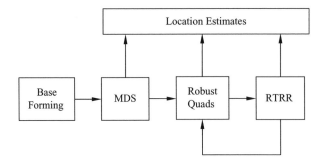

Figure 13.5 Block diagram of the hybrid localization scheme.

It is nontrivial to point out that once a node is localized, its position estimate can be refined by making use of the position information and distance measurements of all its localized neighbors, and by running a minimization algorithm, provided that the node has the computational power to do so.

When a location estimate is obtained, residual checking is applied to ensure the reliability of the estimate and, thus, to reduce the effect of abnormal errors and error propagation. In the presence of multiple location estimates of the same node due to the node being in multiple RQs, averaging is used to produce improved location estimate. Therefore, it is important to take any necessary measures to achieve satisfactory localization performance.

13.5 Graphical Model

A graphical model is a collection of random variables whose joint distribution is specified by a set of potentials. Potentials are real-valued functions of realizations of these random variables, and they determine the nature and the strength of correlations [6]. In a sensor network, each node of the graph denotes a random quantity that pertains to a sensor measurement, and the graph structure connecting the nodes reflects first-order dependencies between the measurements of various sensors. Discussion of application of the graphical models to sensor networks, including self-localization, can be found in [7]. The key ideas of a number of graphical-model-based localization are summarized below, and more details, such as the mathematical derivations, can be found in the references given.

When statistical information of the measurement errors is known, but probably imperfectly, the factor graph and sum-product algorithm can be utilized to solve the location problem in sensor networks [8]. Using a factor graph, the system complexity can be reduced such as by dividing the 2D problem (in the case of 2D positioning) into a 1D problem (x-coordinate and y-coordinate groups). In a factor graph, two types of nodes are defined, namely a variable node and an agent node, which are connected by an edge. Soft information is passed among neighboring nodes and exchanged among local processing units. An initial node location is obtained first and then the location estimate is purified iteratively.

Nonparametric belief propagation (BP) can be used for localization [9]. Like other graphical-model-based methods, the BP method exploits the local nature of the problem; that is, the sensor node location estimate depends primarily on the information about nearby nodes and allowing for a distributed estimation procedure. It is not restricted to Gaussian measurement models and produces both location estimate and information on location uncertainties. BP is aimed at estimating the posterior marginal distribution of each variable. However, neither the discrete nor Gaussian BP is well suited for the sensor self-localization due to the large amount of computation involved and the presence of nonlinear relationships and potentially highly non-Gaussian uncertainties.

Distributed particle filtering can be employed for localization in sensor networks [10]. One such approach is based on factorizing the likelihood and forming parametric approximations to products of likelihood factors (using the particles and their associated likelihoods as training data). The model parameters are then exchanged between sensor nodes, instead of the data or exact particle information. This approach places restrictions on the structure of the problem, because it must be possible both to factorize the likelihood and develop reasonably accurate, low-dimensional parametric models to describe the factors. The approach results in substantial communication savings when the data dimension is much higher than the dimension of the parameter space of the models which approximate the likelihood factors. Another such distributed algorithm makes use of an adaptive data encoding approach. It involves the training of predictive linear quantizers at every time-step based on a common PF maintained at all nodes. Sensor nodes transmit the quantized data to one another, resulting in substantial compression because the PF can provide a good indication of where a sensor measurement is likely to be. This approach places no restrictions on the nature of the likelihood function, but the training of optimal linear quantizers is computationally expensive. Consequently, the amount of data that can be processed at each time-step is limited by the computational power of the sensor nodes.

When using a distributed state-space approach for tracking multiple targets [11], the state-centric multi-modal mechanism systematically decomposes a high-dimensional estimation problem into more manageable sub-problems. The basic idea is to factor a joint state space of all targets of interest, whenever possible, into lower dimensional subspaces, each of which can be represented and processed in a distributed way.

13.6 Clustering and Stitching

In a large WSN with hundreds or even thousands of nodes, both the accuracy and the speed of localization require attention. To achieve both good accuracy and quickness, clustering-based localization can be considered. As usual, clustering involves dividing the network into overlapping sub-graphs and local coordinate establishment at each sub-graph. For instance, in Figure 13.6 the network is divided into four clusters, one of which is connected to at least one of the others and all clusters are connected together. Stitching deals with merging the individual coordinate systems to form a global coordinate system and all the sub-graphs are stitched to form a global graph. There are different ways of forming clusters and stitching, and a number of approaches are described below.

The simplest way of clustering is that every node establishes its local coordinate system and forms a cluster with its neighbors, which can be detected by using beacons. After the absence of a certain number of successive beacons, the node is considered no longer a neighbor. The node of interest becomes the center of its own local coordinate system. Two other nodes are randomly chosen under the condition that the three nodes do not lie in a straight line. It is defined that one of the two nodes lies on the positive x-axis and the other has a positive y-axis

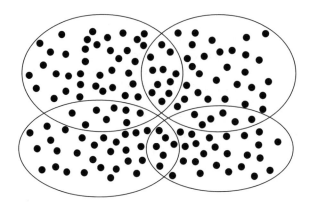

Figure 13.6   Illustration of network clustering.

component such that the local coordinate system is uniquely defined. The position of any other node can be determined if the distances between the node to three nodes with known locations can be measured. The location estimates can be improved by further processing, such as minimization. Clearly, this method is simple, but it requires a large amount of stitching to merge all clusters into one cluster. Also, it would be a challenge for this method to achieve both good accuracy and quickness.

Clustering may also be realized in the following way. Once the nodes are deployed, each node starts to decrement a random timer. If the timer of a node expires before it is contacted by any other nodes, then the node becomes a master node and broadcasts a message establishing itself as a master. The nodes that are within the radio range of the 'master node' and receive this message stop decrementing their timers and become slave nodes. This set of nodes is referred as the domain of the master node and, thus, a cluster is formed. Also, some nodes in the domain/cluster hear from other master nodes. These nodes are called border nodes, which are essential for stitching clusters.

Another possible way of forming clusters is to set up a base cluster and then generate other clusters. More specifically, a node with a certain ID coordinates to form a cluster called a base cluster. Then, four nodes in the base cluster are chosen, which are around the boundary of the base cluster and are approximately evenly spaced. The purpose is to determine four directions to form other clusters. One way to choose the four nodes is to localize the base cluster first. Then, it is easy to choose the four nodes based on their locations. The other way is to apply method two in Section 13.7.2. Once the four nodes are chosen, they coordinate to form four different clusters. The procedure continues until no more clusters can be formed. Certainly, some nodes may not belong to any cluster; however, they can usually be localized after the neighboring clusters are localized.

Once a cluster is localized, it is stitched with other clusters. First, it is required to establish a global coordinate system. For instance, in the case of a master node in each cluster, the coordinate system of the cluster whose master node has the lowest ID is simply chosen as the global coordinate system. This cluster may be called the central cluster. Another way is to choose the coordinate system of the cluster that has the largest number of nodes. Alternatively, the cluster in the center of the network may be selected and the corresponding coordinate system becomes the global one. Then, the stitching order needs to be determined. One way to perform the stitching is to start from the central cluster. All clusters neighboring the central cluster are stitched simultaneously. Then, the clusters next to the stitched clusters are merged to the stitched clusters. The process repeats until all clusters are merged together. Another way to choose the order is based on their respective master node IDs. For instance, a coordinate system whose master node (or coordinating node) has a higher ID changes to the system of the master node with the lower ID. Also, if a master node shares border nodes with more than one master node, then it changes its coordinates to those of the master node with the smallest ID.

13.7 Referent Coordinate System Establishment

In an anchor-free network there is no absolute coordinate system. However, a referent coordinate system needs to be established so that node locations can be described by means of coordinates. Four different approaches to developing a referent coordinate system are briefly described below.

13.7.1 Method One

When the network is divided into overlapping clusters and the nodes in each cluster are localized under the local coordinate system of the cluster, the clusters need to be stitched together to form a global graph. It is convenient to choose one of the local coordinate systems as the referent system and the other local systems are transformed to the referent one. The corresponding chosen cluster should have the largest number of nodes and it is preferable that the selected cluster is close to the center of the network so that long-range stitching can be avoided.

13.7.2 Method Two

In this method [4], five reference nodes are first selected. Four of the five nodes are selected such that they are on the periphery of the graph (node configuration) and a pair of them is roughly perpendicular to the other pair. The fifth node is selected such that it is about in the center of the graph. These five nodes are selected in the following five steps:

Step 1   Select an arbitrary node v_0; a simple way to do so is to pick the node with the smallest ID. Then, select reference node v_1 to maximize the hop count $h_{0,1}$ (along the shortest path between nodes v_0 and v_1). Consequently, node v_1 is near the boundary of the graph.

Step 2   Select reference node v_2 to maximize $h_{1,2}$. In the case of ties, the node with the smaller ID is selected. The selected node v_2 is the maximum hop count away from v_1.

Step 3   Select reference node v_3 to minimize $|h_{1,3} - h_{2,3}|$. In the case of ties, pick the node that maximizes $h_{1,3} + h_{2,3}$ from the contenders.

Step 4   Select reference node v_4 to minimize $|h_{1,4} - h_{2,4}|$ and, in the event of ties, pick the node that maximizes $h_{3,4}$. As a result, all four selected nodes are close to the boundary and the lines connecting v_1 and v_2 and connecting v_3 and v_4 are approximated as the two coordinate axes.

Step 5   Select reference node v_5 to minimize $|h_{1,5} - h_{2,5}|$; the tie-breaking rule is to pick the node that minimizes $|h_{3,5} - h_{4,5}|$. This step selects a node representing the rough center of the graph.

13.7.3 Method Three

A set of nodes are chosen to form a so-called location reference group (LRG) which is stable and is less likely to disappear from the network. One way to choose the LRG is that the density of the nodes in the LRG is the highest in the network. The network center is not a particular node, but is set at the center of the LRG. When the nodes are moving, the LRG center is recomputed. It is expected that the moving speed of the LRG center is smaller than the average speed of the nodes. The direction of the network coordinate system is computed as the mean value of the directions of the local coordinate systems of the nodes in the LRG.

13.7.4 Method Four

In this method, a number of nodes are selected to serve as landmarks. In selecting the landmarks, initially, any node is a candidate as a landmark if its stability and computing power are higher than a predefined threshold. Owing to the fact that node location accuracy is highest when the landmarks are deployed around the boundary of the location area, the landmarks should be close to the edge of the area as much as possible. One of the indexes for determining whether a selected landmark is close to the boundary is the sum of the reciprocals of the squared distance between the examined landmark and the other selected landmarks. The node with the highest index is probably located at the center and should be removed from the set of landmarks. Then, all the indexes are computed again and the process repeats until the desired number of landmarks are left.

Every anchor estimates its distance to other anchors and exchanges the distance information obtained with each other. As a result, an anchor has accumulated a full set of distances between any two anchors in the network. Then, the coordinate system is established by minimizing the sum of the errors of the distances between any two anchors. That is:

$$\hat{\mathbf{p}} = \arg\min_{\mathbf{p}} \sum_{i=1}^{N_a-1} \sum_{j=i+1}^{N_a} \left[\sqrt{(x_i - x_j)^2 + (y_i - y_j)^2} - \hat{d}_{i,j} \right]^2 \qquad (13.26)$$

where $\mathbf{p}$ is the coordinate vector of the selected anchors, $\hat{d}_{i,j}$ is the estimate of the distance between anchors i and j and there are N_a selected anchors. To produce a unique solution to the coordinate vector, three anchors are assigned their coordinates in advance. For instance, an anchor is set at the origin, another anchor has a positive x-coordinate value and the third anchor has a positive y-coordinate value.

13.8 Cramer–Rao Lower Bound

The CRLB is widely used to evaluate the accuracy of any unbiased estimators and provides a performance benchmark. In the case of positioning in cellular networks and anchor-based localization in WSNs, the CRLB has been studied in early chapters. Here, the focus is on deriving the CRLB to benchmark localization accuracy in an anchor-free WSN.

Define the parameter vector as

$$\boldsymbol{\theta} = [\,x_1 \quad \cdots \quad x_N \quad y_1 \quad \cdots \quad y_N\,]^T \in \mathbb{R}^{2N \times 1} \tag{13.27}$$

and the measurement vector as

$$\hat{\mathbf{r}} = [\,\hat{d}_{1,2} \quad \cdots \quad \hat{d}_{1,N} \quad \hat{d}_{2,3} \quad \cdots \quad \hat{d}_{2,N} \quad \cdots \quad \hat{d}_{N-1,N}\,]^T \in \mathbb{R}^{N_r \times 1} \tag{13.28}$$

where $\hat{d}_{i,j}$ is the distance measurement between nodes i and j which are within the radio range of each other as defined in (13.23) and N_r is the length of the distance measurement vector $\hat{\mathbf{r}}$. In general, $N_r \le N(N-1)/2$, since some pairs of nodes are out of the radio range. The equality holds only when all nodes are within radio range of each other.

When the distance measurement error $\varepsilon_{i,j}$ is a Gaussian random variable of zero mean and variance $\sigma_{\hat{d}_{i,j}}^2$, giving the node locations, the PDF of the distance measurement vector is

$$p(\hat{\mathbf{r}}|\boldsymbol{\theta}) = (\sqrt{2\pi})^{-N_r} \prod_{\substack{i=1}}^{N-1} \prod_{\substack{j=i+1 \\ d_{i,j}<R_0}}^{N} \sigma_{\hat{d}_{i,j}}^{-1} \exp\left[-\sum_{i=1}^{N-1} \sum_{\substack{j=i+1 \\ d_{i,j}<R_0}}^{N} \frac{(\hat{d}_{i,j} - d_{i,j})^2}{2\sigma_{\hat{d}_{i,j}}^2} \right] \tag{13.29}$$

where R_0 is the radio range. Let $\hat{\boldsymbol{\theta}}$ be the estimate of $\boldsymbol{\theta}$. Then, as described earlier, the FIM of the estimated parameter vector $\hat{\boldsymbol{\theta}}$ is defined as

$$[\mathbf{F}(\hat{\boldsymbol{\theta}})]_{i,j} = -E\left[\frac{\partial^2 \ln p(\hat{\mathbf{r}}|\boldsymbol{\theta})}{\partial \theta_i \partial \theta_j} \right] \qquad i,j = 1,2,\ldots,N \tag{13.30}$$

where θ_i is the ith element of $\boldsymbol{\theta}$ and the log likelihood function $\ln p(\hat{\mathbf{r}}|\boldsymbol{\theta})$ is given by

$$\ln p(\hat{\mathbf{r}}|\boldsymbol{\theta}) = -\sum_{i=1}^{N-1} \sum_{\substack{j=i+1 \\ d_{i,j}<R_0}}^{N} \frac{(\hat{d}_{i,j} - d_{i,j})^2}{2\sigma_{\hat{d}_{i,j}}^2} \tag{13.31}$$

where the irrelevant constants are dropped. The FIM can be written in a block matrix form as

$$\mathbf{F}(\hat{\boldsymbol{\theta}}) = \begin{bmatrix} \mathbf{F_{xx}} & \mathbf{F_{xy}} \\ \mathbf{F_{yx}} & \mathbf{F_{yy}} \end{bmatrix} \in \mathbb{R}^{2N_r \times 2N_r} \tag{13.32}$$

where by defining

$$u_i = \sum_{\substack{j=1 \\ j \ne i, d_{i,j}<R_0}}^{N} \frac{(\hat{d}_{i,j} - d_{i,j})^2}{2\sigma_{\hat{d}_{i,j}}^2} \tag{13.33}$$

the components of the FIM can be derived as

$$[\mathbf{F_{xx}}]_{i,i} = E\left[\frac{\partial^2 u_i}{\partial x_i^2}\right] = \sum_{\substack{j=1 \\ j \neq i, d_{i,j} < R_0}}^{N} \frac{(x_i - x_j)^2}{\sigma_{\hat{d}_{i,j}}^2 \, d_{i,j}^2}$$

$$[\mathbf{F_{xx}}]_{i,j} = E\left[\frac{\partial^2 u_i}{\partial x_i \partial x_j}\right] = -\frac{(x_i - x_j)^2}{\sigma_{\hat{d}_{i,j}}^2 \, d_{i,j}^2} \qquad j \neq i, d_{i,j} \leq R_0$$

$$[\mathbf{F_{xy}}]_{i,i} = E\left[\frac{\partial^2 u_i}{\partial x_i y_i}\right] = \sum_{\substack{j=1 \\ j \neq i, d_{i,j} < R_0}}^{N} \frac{(x_i - x_j)(y_i - y_j)}{\sigma_{\hat{d}_{i,j}}^2 \, d_{i,j}^2}$$

$$[\mathbf{F_{xy}}]_{i,j} = E\left[\frac{\partial^2 u_i}{\partial x_i \partial y_j}\right] = -\frac{(x_i - x_j)(y_i - y_j)}{\sigma_{\hat{d}_{i,j}}^2 \, d_{i,j}^2} \qquad j \neq i, d_{i,j} \leq R_0$$

$$[\mathbf{F_{yy}}]_{i,i} = E\left[\frac{\partial^2 u_i}{\partial y_i^2}\right] = \sum_{\substack{j=1 \\ j \neq i, d_{i,j} < R_0}}^{N} \frac{(y_i - y_j)^2}{\sigma_{\hat{d}_{i,j}}^2 \, d_{i,j}^2}$$

$$[\mathbf{F_{yy}}]_{i,j} = E\left[\frac{\partial^2 u_i}{\partial y_i \partial y_j}\right] = -\frac{(y_i - y_j)^2}{\sigma_{\hat{d}_{i,j}}^2 \, d_{i,j}^2} \qquad j \neq i, d_{i,j} \leq R_0$$

It can be seen that the standard CRLB defined as the inverse of the FIM does not exist for anchor-free localization since the FIM is singular. The rank of the FIM can be up to $2N - 3$, instead of $2N$. This is due to the fact that the node configuration remains the same after translation, rotation or reflection. To handle the complication of the singular FIM, we make use of the results in [12]. That is, given

$$\mathbf{r} = [d_{1,2}, \quad \ldots, \quad d_{1,N}, \quad d_{2,3}, \quad \ldots, \quad d_{2,N}, \quad \ldots, \quad d_{N-1,N}]^T \in \mathbb{R}^{N_r \times 1} \tag{13.34}$$

the covariance matrix of $\hat{\boldsymbol{\theta}}$

$$\Phi = E[(\hat{\boldsymbol{\theta}} - E[\hat{\boldsymbol{\theta}}])(\hat{\boldsymbol{\theta}} - E[\hat{\boldsymbol{\theta}}])^T] \tag{13.35}$$

satisfies

$$\Phi \geq \Phi_b = \mathbf{H} \mathbf{F}^+ \mathbf{H}^T \tag{13.36}$$

where

$$\mathbf{H} = \frac{\partial (E[\hat{\boldsymbol{\theta}}] - \boldsymbol{\theta})}{\partial \boldsymbol{\theta}^T} + \frac{\partial \mathbf{r}}{\partial \boldsymbol{\theta}^T} \tag{13.37}$$

and $\mathbf{F}^+$ is the Moore–Penrose pseudo-inverse of the FIM $\mathbf{F}$, which is defined as

$$\mathbf{F}^+ = (\mathbf{F}^T\mathbf{F})^{-1}\mathbf{F}^T \tag{13.38}$$

In the event that $\hat{\boldsymbol{\theta}}$ is an unbiased estimate of $\boldsymbol{\theta}$, $\mathbf{H}$ is given as

$$
\begin{aligned}
\mathbf{H} &= \frac{\partial \mathbf{r}}{\partial \boldsymbol{\theta}^T} \\
&= \begin{bmatrix}
\dfrac{\partial r_1}{\partial x_1} & \cdots & \dfrac{\partial r_1}{\partial x_N} & \dfrac{\partial r_1}{\partial y_1} & \cdots & \dfrac{\partial r_1}{\partial y_N} \\
\dfrac{\partial r_2}{\partial x_1} & \cdots & \dfrac{\partial r_2}{\partial x_N} & \dfrac{\partial r_2}{\partial y_1} & \cdots & \dfrac{\partial r_2}{\partial y_N} \\
\vdots & \ddots & \vdots & \vdots & \ddots & \vdots \\
\dfrac{\partial r_{N_r}}{\partial x_1} & \cdots & \dfrac{\partial r_{N_r}}{\partial x_N} & \dfrac{\partial r_{N_r}}{\partial y_1} & \cdots & \dfrac{\partial r_{N_r}}{\partial y_N}
\end{bmatrix} \in \mathbb{R}^{N_r \times 2N}
\end{aligned} \tag{13.39}
$$

As mentioned in the previous chapters, the accuracy of an algorithm is often calculated as the RMSE of the location estimates. To benchmark the accuracy in terms of the RMSE of the localization algorithms, the averaged CRLB can be used, which is defined as

$$
\begin{aligned}
\mathrm{CRLB}_{\mathrm{av}} &= \sqrt{\frac{1}{2N}\sum_{i=1}^{2N}[\Phi_b]_{i,i}} \\
&= \sqrt{\frac{1}{2N}\mathrm{trace}(\Phi_b)}
\end{aligned} \tag{13.40}
$$

13.9 Accuracy of Location Estimates

The CRLB is the lower bound that any unbiased estimators cannot surpass. The accuracy of a specific localization algorithm is often evaluated by calculating the RMSE of the estimated location coordinates and comparing with the CRLB. It is noted that the true positions of the nodes are expressed in the coordinate system established when forming the base. Owing to estimation errors, the true positions of the second node and the third node when choosing three nodes to set up a coordinate system may not be the same as the estimated ones. The difference between the true positions of the three chosen nodes or the base nodes described in the hybrid localization algorithm and the estimated ones means that, even when the relative positions of all other nodes can be calculated precisely, the RMSE could still be significant. Since we are dealing with anchor-free sensor networks, the effect of the coordinate system shift must be removed by using a coordinate transformation in order to access the accuracy of the location estimates fairly.

Here, an efficient method to perform the mapping is described as follows. Let

$$\mathbf{p}_i = [\, x_i \quad y_i \,]^T \qquad i = 1, 2, \ldots, N \tag{13.41}$$

be the predefined node locations and let

$$\hat{\mathbf{p}}_i = [\hat{x}_i \quad \hat{y}_i]^{\mathrm{T}} \qquad i = 1, 2, \ldots, N \tag{13.42}$$

be the location estimates. The centroid of the predefined error-free locations of all the localized nodes is defined as

$$\mathbf{c}^o = \frac{1}{N} \sum_{i=1}^{N} \mathbf{p}_i \tag{13.43}$$

and the centroid of the estimated locations is given as

$$\mathbf{c}^e = \frac{1}{N} \sum_{i=1}^{N} \hat{\mathbf{p}}_i \tag{13.44}$$

The first step in the mapping process is to translate the centroid of the estimated locations to the centroid of the predefined node locations. That is, the estimated locations are changed as

$$\hat{\mathbf{p}}'_i = \hat{\mathbf{p}} + (\mathbf{c}^o - \mathbf{c}^e) \tag{13.45}$$

Then, the estimated locations are transformed according to

$$\hat{\mathbf{p}}''_i = \mathbf{T}\hat{\mathbf{p}}'_i \tag{13.46}$$

where

$$\mathbf{T} = \begin{bmatrix} t_{1,1} & t_{1,2} \\ t_{2,1} & t_{2,2} \end{bmatrix} \in \mathbb{R}^{2 \times 2} \tag{13.47}$$

is the transform matrix, which is determined by

$$\mathbf{T} = \arg\min_{\mathbf{T}} \left\{ \sum_{i=1}^{N} \|\mathbf{T}\hat{\mathbf{p}}'_i - \mathbf{p}_i\|^2 \right\} \tag{13.48}$$

where $\|.\|$ is the vector norm. That is, the problem is to determine the matrix $\mathbf{T}$ that minimizes

$$\sum_{i=1}^{N} \|\mathbf{T}\hat{\mathbf{p}}'_i - \mathbf{p}_i\|^2 = \sum_{i=1}^{N} [(t_{1,1}x'_i + t_{1,2}y'_i - x_i)^2 + (t_{2,1}x'_i + t_{2,2}y'_i - y_i)^2] \tag{13.49}$$

where $x'_i = [\hat{\mathbf{p}}'_i]_{1,1}$ and $y'_i = [\hat{\mathbf{p}}'_i]_{2,1}$. Differentiating the right-hand side of (13.49) with respect to $t_{1,1}$, $t_{1,2}$, $t_{2,1}$ and $t_{2,2}$ and then setting them to zero, we obtain

$$\begin{cases} t_{1,1} \sum_{i=1}^{N} (x'_i)^2 + t_{1,2} \sum_{i=1}^{N} x'_i y'_i = \sum_{i=1}^{N} x_i x'_i \\ t_{1,1} \sum_{i=1}^{N} x'_i y'_i + t_{1,2} \sum_{i=1}^{N} (y'_i)^2 = \sum_{i=1}^{N} x_i y'_i \end{cases} \tag{13.50}$$

and

$$
\begin{cases}
t_{2,1}\sum_{i=1}^{N}(x_i')^2 + t_{2,2}\sum_{i=1}^{N}x_i'y_i' = \sum_{i=1}^{N}x_i'y_i \\
t_{2,1}\sum_{i=1}^{N}x_i'y_i' + t_{2,2}\sum_{i=1}^{N}(y_i')^2 = \sum_{i=1}^{N}y_i'y_i
\end{cases}
\tag{13.51}
$$

Accordingly, the components of the transform matrix are solved as

$$
\begin{aligned}
t_{1,1} &= \frac{1}{\Delta}\left[\sum_{i=1}^{N}(x_i x_i')\sum_{j=1}^{N}(y_j')^2 - \sum_{i=1}^{N}(x_i y_i')\sum_{j=1}^{N}(x_j' y_j')\right] \\
t_{1,2} &= \frac{1}{\Delta}\left[\sum_{i=1}^{N}(x_i')^2\sum_{j=1}^{N}(x_j y_j') - \sum_{i=1}^{N}(x_i x_i')\sum_{j=1}^{N}(x_j' y_j')\right] \\
t_{2,1} &= \frac{1}{\Delta}\left[\sum_{i=1}^{N}(x_i' y_i)\sum_{j=1}^{N}(y_j')^2 - \sum_{i=1}^{N}(x_i' y_i')\sum_{j=1}^{N}(y_j' y_j)\right] \\
t_{2,2} &= \frac{1}{\Delta}\left[\sum_{i=1}^{N}(x_i')^2\sum_{j=1}^{N}(y_j' y_j) - \sum_{i=1}^{N}(x_i' y_i')\sum_{j=1}^{N}(x_j' y_j)\right]
\end{aligned}
\tag{13.52}
$$

where

$$
\Delta = \sum_{i=1}^{N}(x_i')^2 \sum_{j=1}^{N}(y_j')^2 - \left[\sum_{i=1}^{N}(x_i' y_i')\right]^2
\tag{13.53}
$$

Therefore, after the mapping, the accuracy of the coordinate estimates in terms of RMSE can be fairly evaluated with respect to the predefined node locations. The RMSE is now computed according to

$$
\sqrt{\frac{1}{2N}\sum_{i=1}^{N}[(\hat{x}_i'' - x_i)^2 + (\hat{y}_i'' - y_i)^2]}
\tag{13.54}
$$

where $\hat{x}_i''$ and $\hat{y}_i''$ are the adjusted location estimates determined by (13.46).

Let us show that the transformation of estimated locations is necessary through an example. Figure 13.7 shows the localization result with no nodes based on the hybrid algorithm described earlier under one network configuration. The dots denote the true node configuration under a predefined coordinate system and the circles denote the configuration of the estimated node locations. The two locations of the same node are connected by a solid line. Figure 13.8 shows the corresponding results after the location estimates, which are already shown in Figure 13.7, are transformed to better match the predefined node locations. Based on the length of the connection lines, we can roughly judge that the estimated locations in Figure 13.8 are more accurate than those in Figure 13.7. More precisely, it is found that the RMSE of the coordinate estimation is 10.9 m and 26.7 m with and without transformation respectively. Clearly, it is crucial to transform the estimated locations before calculating the coordinate

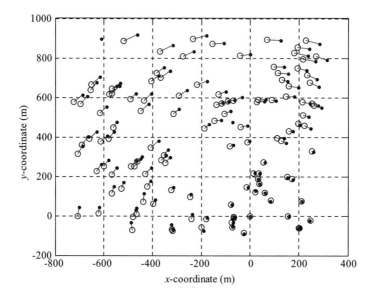

Figure 13.7 Position estimation results for one network configuration. The true and estimated positions are denoted by dot and circle respectively and connected by a solid line.

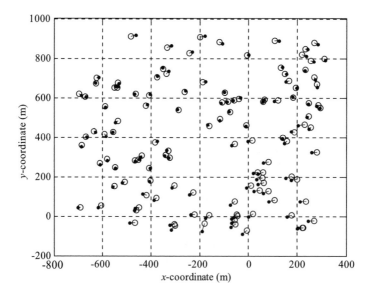

Figure 13.8 The predefined locations and the transformed location estimates.

estimation errors, otherwise the location accuracy would be underestimated. Since the distance between each pair of estimated locations does not change with the transformation, the RMSE of the distance estimates remains the same and equal to 8.3 m. Therefore, the RMSE of the distance errors might provide a more accurate measure of the anchor-free location algorithms.

13.10 Distance-Error-based Accuracy Measure

Another performance measure that can be used to evaluate the accuracy of anchor-free localization algorithms is the RMS of the difference between the true distance and the estimated one of each pair of nodes; that is:

$$\sqrt{\frac{2}{N(N-1)} \sum_{i=1}^{N-1} \sum_{j=i+1}^{N} (\tilde{d}_{i,j} - d_{i,j})^2} \tag{13.55}$$

where $d_{i,j}$ is the true distance between node i and j and $\tilde{d}_{i,j}$ is the distance between the estimated locations of node i and j, which is defined as

$$\tilde{d}_{i,j} = \sqrt{(\hat{x}_i - \hat{x}_j)^2 + (\hat{y}_i - \hat{y}_j)^2} \tag{13.56}$$

Note that the distance measurement $\hat{d}_{i,j}$ is only available when the two nodes are within radio range of each other, whereas the distance $d_{i,j}$ can be computed for any pair of nodes which are localized. It is obvious that for a given node configuration the distance between a pair of nodes remains the same when any coordinate transformation is performed on the node locations. This is the advantage of using the distance error

$$\delta d_{i,j} = \tilde{d}_{i,j} - d_{i,j} \tag{13.57}$$

as an accuracy measure. As the CRLB described earlier is used as a lower bound on the accuracy of the coordinate estimates, one is motivated to derive an approximate lower bound on the distance errors. Expanding the estimated distance $\tilde{d}_{i,j}$ in a Taylor series at the true distance and retaining the first two terms below the second-order term, it can be seen that

$$\tilde{d}_{i,j} \approx d_{i,j} + \frac{\partial d_{i,j}}{\partial x_i} \Delta x_i + \frac{\partial d_{i,j}}{\partial x_j} \Delta x_j + \frac{\partial d_{i,j}}{\partial y_i} \Delta y_i + \frac{\partial d_{i,j}}{\partial y_j} \Delta y_j \tag{13.58}$$

where

$$
\begin{aligned}
\Delta x_i &= \hat{x}_i - x_i \qquad 1 \le i \le N \\
\Delta y_i &= \hat{y}_i - y_i \\
\frac{\partial d_{i,j}}{\partial x_i} &= \frac{1}{d_{i,j}}(x_i - x_j) \qquad i = 1, 2, \dots, N-1, j = i+1, \dots, N \\
\frac{\partial d_{i,j}}{\partial y_i} &= \frac{1}{d_{i,j}}(y_i - y_j)
\end{aligned}
\tag{13.59}
$$

Assuming that Δx_i, Δx_j, Δy_i and Δy_j are mutually independent, we obtain the variance of $\delta d_{i,j}$ as

$$
\sigma_{\delta d_{i,j}}^2 \approx \left(\frac{\partial d_{i,j}}{\partial x_i}\right)^2 \sigma_{\hat{x}_i}^2 + \left(\frac{\partial d_{i,j}}{\partial x_j}\right)^2 \sigma_{\hat{x}_j}^2 + \left(\frac{\partial d_{i,j}}{\partial y_i}\right)^2 \sigma_{\hat{y}_i}^2 + \left(\frac{\partial d_{i,j}}{\partial y_j}\right)^2 \sigma_{\hat{y}_j}^2 \tag{13.60}
$$

where $\sigma_{\hat{x}_i}^2$ and $\sigma_{\hat{y}_i}^2$ are variances of the coordinate estimation errors of node i. Replacing the variances of the coordinate estimates in the above equation by their corresponding CRLBs defined in (13.36), we obtain the approximate lower bound on the distance error $\delta d_{i,j}$ as

$$
\mathbf{LB}_{\delta d_{i,j}} \approx \left(\frac{\partial d_{i,j}}{\partial x_i}\right)^2 [\Phi_b]_{i,i} + \left(\frac{\partial d_{i,j}}{\partial x_j}\right)^2 [\Phi_b]_{j,j} + \left(\frac{\partial d_{i,j}}{\partial y_i}\right)^2 [\Phi_b]_{N+i,N+i} + \left(\frac{\partial d_{i,j}}{\partial y_j}\right)^2 [\Phi_b]_{N+j,N+j}
$$

$$
\tag{13.61}
$$

so that

$$
\sigma_{\delta d_{i,j}}^2 \geq \mathbf{LB}_{\delta d_{i,j}} \tag{13.62}
$$

Accordingly, the distance error lower bound can be expressed in an average form as

$$
\mathbf{LB}_{\delta d} = \sqrt{\frac{2}{N(N-1)} \sum_{i=1}^{N-1} \sum_{j=i+1}^{N} \mathbf{LB}_{\delta d_{i,j}}} \tag{13.63}
$$

13.11 Accuracy Evaluation

In this section, three algorithms, namely the hybrid algorithm, the RQ method and the MDS algorithm, are evaluated through simulation. Also, the accuracy measures discussed in previous sections are examined. It is assumed that the nodes are randomly deployed in a square area with dimensions of $1\,\text{km} \times 1\,\text{km}$. The radio/communication range of each node is set to 300 m. The distance measurement error is modeled as a Gaussian random variable of zero mean and a STD proportional to the true distance, ranging from 1% to 8%.

13.11.1 Impact of Distance Measurement Accuracy and Network Size

First, the hybrid algorithm is examined under different network size and distance measurement error. Figures 13.9–13.11 show the RMSE of the coordinate estimation and that of the distance errors with respect to the STD of the distance measurement errors when the node populations are 90, 120 and 150. Also shown are the CRLB and the approximate lower bound. When the STD is set at 4% of the true distance, the RMSEs of both the coordinate estimates and the distance estimates are less than 8.5 m under the three different node populations. As expected, the location accuracy degrades as the STD of the distance measurement error increases. The gap between the accuracy of the location estimates and the CRLB and the approximate lower bound is relatively large, especially when the distance measurement error is large. One possible

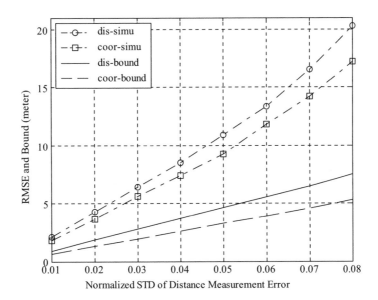

Figure 13.9 Lower bounds and the accuracy of the hybrid algorithm with 90 nodes in the network: 'dis-simu' denotes the RMSE of the distance errors calculated by (13.55); 'coor-simu' denotes the RMSE of the coordinate estimation errors computed by (13.54); 'dis-bound' denotes the approximate lower bound determined by (13.63); and 'coor-bound' denotes the CRLB determined by (13.40).

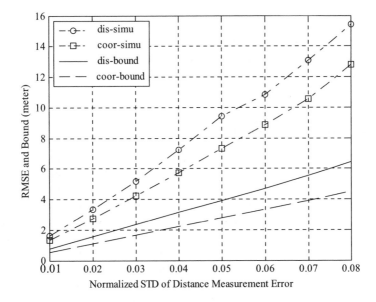

Figure 13.10 Lower bounds and the accuracy of the hybrid algorithm in the event of 120 nodes in the network. The legend items have the same definitions as in Figure 13.9.

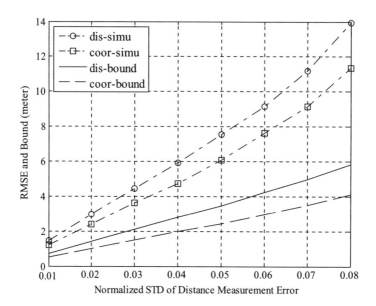

Figure 13.11   Lower bounds and the accuracy of the hybrid algorithm with 150 nodes in the network. The legend items have the same definitions as in Figure 13.9.

way to reduce the gap is to perform the global optimization over all the location estimates. This will increase the computational complexity tremendously and it is not a solution for a distributed network. Another way to narrow down the gap is to exploit the error statistics of the distance measurement. It would be interesting to continue the work to see how close the accuracy of the algorithm will approach the bounds.

Figure 13.12 compares the RMSE of the coordinate estimation under three node populations. As observed, the location accuracy improves as the node population increases. This is due to the fact that a higher node density results in more neighboring nodes, such that more information can be employed to localize the unknown nodes.

13.11.2 Performance Comparison

This subsection provides performance comparison among three different algorithms, the hybrid algorithm, the RQ algorithm and the MDS method. Figure 13.13 shows the cumulative distribution of the absolute values of the distance errors when there are 150 and 90 nodes in the network. The accuracy of the three algorithms, namely the hybrid algorithm, the RQ algorithm and the MDS method, is plotted for comparison. For each node population (150 and 90), 100 different node deployment realizations are examined. As expected, more nodes in the network produce better position estimation accuracy, as already observed earlier. It can be seen that the accuracy of the hybrid algorithm is considerably higher than that of the MDS method and the RQ method. Compared with the RQ method, the accuracy improvement of the hybrid algorithm comes from the good accuracy of the location estimates of the base nodes, the residual checking to avoid abnormal errors and the averaging of estimates from multiple RQs.

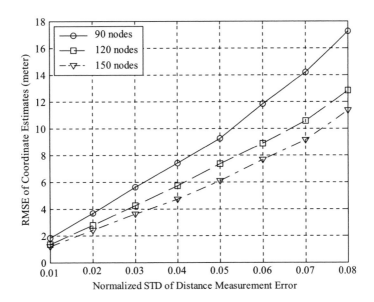

Figure 13.12 STD of the coordinate estimation errors (simulated results) of the hybrid localization algorithm when there are 90, 120 and 150 nodes in the network.

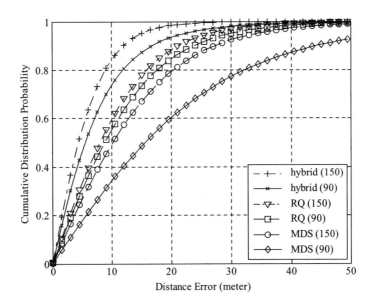

Figure 13.13 The CDP of the distance error of the hybrid algorithm, the RQ method and the MDS method when there are 150 and 90 nodes in the network. 'hybrid (150)' denotes results for the hybrid algorithm with 150 nodes in the network.

Table 13.1 Success rate of localization of the hybrid algorithm and the RQ method when there are either 150 or 90 nodes in the network.

Nodes	Success rate (%)	
	Hybrid method	RQ method
150	99.47	95.24
90	96.48	53.13

Table 13.1 shows the average success rate of localization of the hybrid algorithm and the RQ method. The success rate of localization is defined as the ratio of the number of nodes that have been localized to the number of all nodes in the network. The unlocalized nodes result from the condition that localizing a node is not possible. For instance, a node cannot communicate with at least two other nodes so that its location cannot be determined based on distance measurements.[2] Another example is that a node cannot form an RQ with other three nodes so that its location cannot be determined if only the RQ method is used. Clearly, the success rate of the hybrid algorithm is higher than that of the RQ method, especially when the node density is relatively low. This is because of the fact that, even when the RQ algorithm stops due to the lack of RQs, the RTRR approach can resume the localization process. This is attributed to the conditions of the RTRR approach usually being satisfied more easily than those of the RQ method. On the other hand, all nodes can be localized by using the MDS method provided that any pair of nodes is connected via a one-hop or multihop path, indicating an advantage of the MDS method. The distance between two nodes that are out of radio range is computed based on the shortest path distance in the MDS method.

References

[1] D. Moore, J. Leonard, D. Rus and S. Teller, 'Robust distributed network localization with noisy range measurements', in *Proceedings of International Conference on Embedded Networked Sensor Systems*, Baltimore, MD, USA, pp. 50–61, 2004.

[2] Y. Shang, W. Rumi, Y. Zhang and M. Fromherz, 'Localization from connectivity in sensor netwroks', *IEEE Transactions on Parallel and Distributed Systems*, **15**(11), 2004, 961–974.

[3] A. Howard, M. Mataric and G. Sukhatme, 'Relaxation on a mesh: a formalism for generalized localization', in *Proceedings of IEEE International Conference on Intelligent Robots and Systems*, October 2001.

[4] N.B. Priyantha, H. Balakrishnan, E. Demaine and S. Teller, 'Anchor-free distributed localization in sensor networks', MIT LCS Technical Report, No. 892, April 2003.

[5] S. Pandey, P. Prasad, P. Sinha and P. Agrawal, 'Localization of sensor networks considering energy accuracy tradeoffs', in *Proceedings of International Conference on Collaborative Computing: Networking, Applications and Worksharing*, pp. 1–10, December 2005.

[6] E.B. Ermis, M. Alanyali and V. Saligrama, 'Search and discovery in an uncertain networked world', *IEEE Signal Processing Magazine*, **23**(4), 2006, 107–118.

[7] M. Cetin, L. Chen, J.W. Fisher III, A.T. Ihler, R.L. Moses, M.J. Wainwright and A.S. Willsky, 'Distributed fusion in sensor networks', *IEEE Signal Processing Magazine*, **23**(4), 2006, 42–55.

[8] J.-C. Chen, Y.-C. Wang, C.-S. Maa and J.-T. Chen, 'Network-side mobile position location using factor graphs', *IEEE Transactions on Wireless Communications*, **5**(10), 2006, 2696–2704.

---

[2] When a node is within radio range of another node, its location is constrained to a circular region with a radius equal to the radio range.

[9] A.T. Ihler, J.W. Fisher III, R.L. Moses and A.S. Willsky, 'Nonparametric belief propagation for self-calibration in sensor networks', in *Proceedings of IEEE International Symposium on Information Processing in Sensor Networks (IPSN)*, pp. 225–233, April 2004.

[10] M. Coates, 'Distributed particle filters for sensor networks', in *Proceedings of IEEE International Symposium on Information Processing in Sensor Networks (IPSN)*, pp. 99–107, Apr. 2004.

[11] J. Liu, M. Chu, J. Liu, J. Reich, and F. Zhao, 'Distributed state representation for tracking problems in sensor networks', *Proceedings of IEEE International Symposium on Information Processing in Sensor Networks (IPSN)*, pp. 234–242, April 2004.

[12] P. Stoica and T.L. Marzetta, 'Parameter estimation problems with singular information matrices', *IEEE Transactions on Signal Processing*, **49**(1), 2001, 87–90.

14

Non-Line-of-Sight Identification

In radio positioning, one of the dominant factors that affect the positioning accuracy is the NLOS radio propagation that happens when the direct, straight radio path between the transmitter and receiver is blocked. NLOS propagation exists in a variety of scenarios, such as in dense urban environments, inside buildings and in forests. Compared with the LOS condition, the signal travels an extra distance and propagation time under the NLOS condition. NLOS propagation also results in an extra power loss and an AOA bias. To achieve satisfactory location estimation accuracy, it is desirable to determine whether the measurements come from LOS propagation or NLOS propagation. The identified NLOS-corrupted measurements can usually be discarded when there are enough measurements identified as from LOS propagation. Otherwise, the NLOS measurements may be exploited, but the contribution of LOS measurements should be emphasized, such as by larger weights. Alternatively, the NLOS-corrupted measurements are reconstructed to form estimated LOS measurements. Although it is a challenge to distinguish between LOS and NLOS measurements in general, a range of NLOS identification methods and techniques have been proposed.

This chapter studies the main methods and techniques in the identification of NLOS propagation from the point view of signal detection. Although some methods are developed under idealized conditions, the results of those methods may serve as a benchmark for more practical methods, and may also produce the effect of stimulating the development of more practical approaches. Section 14.1 presents identification based on calculating error variance through data smoothing. Section 14.2 studies a number of well-known distribution tests, which can be considered for NLOS identification. Section 14.3 uses level crossing rate and fade duration of the received signal envelope in fading channels. Section 14.4 studies a method based on calculating the Rician factor of the received signal envelope. The generalized likelihood ratio test is studied in Section 14.5. A nonparametric method is given in Section 14.6. A location estimates-based method is given in Section 14.7. The Neyman–Pearson test is employed for an idealized case in Section 14.8. A joint TOA and RSS-based method is presented in Section 14.9. Finally, the AOA-based method is described in Section 14.10.

Ground-Based Wireless Positioning   Kegen Yu, Ian Sharp and Jay Guo
© 2009 John Wiley & Sons, Ltd

14.1 Data Smoothing

In cellular networks, both distance and TDOA measurements are usually employed to determine the locations of mobile stations. In particular, the distance measurements are widely obtained by determining the propagation time for the signal traveling from the mobile station to the base station or vice versa. As discussed in Chapters 5 and 6, the propagation time can be obtained by estimating the TOA at both the mobile station and base station to produce the RTT, which is twice the propagation time, or by estimating the difference of TOA measurements of an ultrasonic signal and a radio signal.

For positioning in cellular systems or sensor networks, clock frequency synchronization between two nodes in communication may be necessary depending on how quickly the communications are completed and how stable the crystal oscillators are. For instance, if the round trip of the radio signal can be realized in 100 ms and the frequency offset is 1 ppm, then the corresponding ranging error can be 30 m. On the other hand, if the round trip can be completed in 10 ms and the frequency offset is just 0.1 ppm, then the incurred ranging error is just 0.3 m, which can be ignored when the location accuracy is required only to a few meters or even tens of meters. Another practical consideration is the internal radio delay, which makes the measured TOA greater than the true TOA value by a quantity that depends on the specific transceiver. The internal delay can be around hundreds of nanoseconds and can be calibrated when the transceiver is built. The variation of the internal delay may be completely ignored for outdoor cellular positioning.

One of the basic NLOS identification methods is based on the prior knowledge of statistics of the measurement errors in LOS conditions. For example, the distance measurement error in LOS conditions is approximately a Gaussian random variable with a STD equal to $\sigma^{(m)}$ for a particular monitored area, which is known in advance based on field measurements. During real-time operations, measurements are taken and the statistics of the measurements are calculated. When the estimated STD of the measured distances is sufficiently larger than $\sigma^{(m)}$, the distance measurements are assumed corrupted by NLOS errors.

Specifically, consider a cellular network in which a mobile station has radio links with M fixed base stations and the mobile station is moving. The distance between the mobile station and each base station is estimated at time instants t_i, $i = 1, 2, \ldots, L$, producing L distance measurements $\{r_i^{(m)}\}$, where m indexes the base station. To calculate the variance of the distance measurements taken over different temporal and spatial points, the true distances are required. However, the true distances are not known, so only approximate true distances can be determined. One way to obtain the approximate true distances is to smooth the raw distance measurements. One such smoothing technique is proposed in [1]. At each base station, the distance measurements are smoothed according to a polynomial fit; that is:

$$r_i^{(m)} = \sum_{j=0}^{J-1} a_j^{(m)} t_i^j \qquad i = 1, 2, \ldots, L \tag{14.1}$$

where $J - 1$ is the order of the polynomial. Suppose that L is greater than J; this assumption can be readily satisfied, since J is typically a small integer. Then, the polynomial coefficients $\{a_j^{(m)}\}$ can be determined by the LS technique, resulting in $\{\hat{a}_j^{(m)}\}$. Accordingly, the smoothed measurements are represented as

$$\tilde{r}_i^{(m)} = \sum_{j=0}^{J-1} \hat{a}_j^{(m)} t_i^j \qquad i = 1, 2, \ldots, L \tag{14.2}$$

Taking $\tilde{r}_i^{(m)}$ as the reference, the sample STD of $r_i^{(m)}$ can be computed as

$$\hat{\sigma}^{(m)} = \sqrt{\frac{1}{L} \sum_{i=1}^{L} (r_i^{(m)} - \tilde{r}_i^{(m)})^2} \tag{14.3}$$

When the measurements come from LOS propagation, $\hat{\sigma}^{(m)}$ may be close to the known STD $\sigma^{(m)}$ of the measurement errors in LOS condition. Therefore, the difference between $\hat{\sigma}^{(m)}$ and $\sigma^{(m)}$ can be exploited to judge the LOS/NLOS condition. The LOS hypothesis is accepted when $\hat{\sigma}^{(m)} < \sigma^{(m)}$, whereas the NLOS hypothesis is accepted when $\hat{\sigma}^{(m)} > \kappa \sigma^{(m)}$, $\kappa > 1$, where κ is a predefined constant.

It is worth mentioning that selection of parameters such as $\sigma^{(m)}$, J (for polynomial fitting) and κ will be dependent on the particular environments and the positioning systems. Field trials must be undertaken to ensure that the selected parameters are best suited to achieve reliable NLOS identification. In addition to the polynomial fitting, other smoothing techniques can also be employed. For instance, Kalman filtering has been widely used in many fields for parameter filtering and object tracking. In Chapters 7 and 10, Kalman filtering is used for several different purposes, including data smoothing. Therefore, we do not repeat the description of Kalman filtering here.

14.2 Distribution Tests

It is known that TOA and distance measurements manifest different distributions in LOS conditions from those in NLOS conditions. For instance, the envelope of the received complex low-pass signal can be modeled as a random variable with a Rayleigh distribution for NLOS propagation, whereas the envelope has a Rician distribution in the case of LOS propagation [2]. Therefore, the NLOS condition can be detected by testing if the signal envelope has a Rayleigh distribution. Another example is that the TOA and, hence, distance measurements are typically Gaussian distributed under LOS propagation, whereas the NLOS TOA measurement errors might be exponentially distributed. Accordingly, LOS propagation can be identified by testing whether the TOA measurements have a Gaussian distribution. A variety of test approaches exist to test whether a group of measurements belongs to a specific distribution so that the LOS or NLOS condition can be identified. Here, a number of well-known tests are briefly studied.

14.2.1 Kolmogorov–Smirnov Test

Suppose that there are L measurements which are ordered as $r_1, r_2, \ldots, r_L$. Then, the empirical cumulative distribution function (or cumulative frequency) of the L observations is defined as

$$F_O(i) = \frac{\nu(i)}{L} \qquad i = 1, 2, \ldots, L \tag{14.4}$$

where $\nu(i)$ is the number of observation points that are not larger than r_i. Clearly, this is a step function that increases by $1/L$ at each ordered data point. Then, one calculates the cumulative distribution function of the expected theoretical distribution, denoted by F_E. The D-statistic is defined as the greatest discrepancy between the observed and the expected cumulative distribution functions as

$$\hat{D} = \max_{1 \le i \le L} |F_O(i) - F_E(i)| \qquad (14.5)$$

where $F_E(i)$ is the theoretical cumulative distribution function value at data point r_i. The computed D-statistic is compared against the critical D-statistic for the same sample size. The critical D-statistic can be computed based on the given level of significance. For instance, when the level of significance equals 0.05, the critical D-statistic is

$$\hat{D} = \frac{1.358}{\sqrt{L}} \qquad (14.6)$$

If the calculated D-statistic is greater than the critical one, then the empirical distribution is not considered as the expected distribution and, thus, the hypothesis is rejected. Otherwise, the hypothesis is accepted.

The Kolmogorov–Smirnov (K–S) test is an exact and robust test and does not depend on the underlying cumulative distribution function being tested. The disadvantage of the test is that it only applies to continuous distributions and is more sensitive near the center of the distribution than at the tails.

14.2.2 Anderson–Darling Test

The Anderson–Darling (A–D) test is also used to judge whether a sample of data came from a population with a specific distribution. It is a modification of the K–S test and gives more weight to the tails than the K–S test does to reduce the sensitivity near the center of the distribution. Unlike the K–S test, the A–D test makes use of the specific distribution in calculating critical values. This has the advantage of allowing a more sensitive test and the disadvantage that critical values must be calculated for the specific distribution that is being tested. The A–D test statistic is defined as

$$A^2 = -L - \zeta \qquad (14.7)$$

where

$$\zeta = \sum_{i=1}^{L} \frac{2i-1}{L} \{\ln F(i) + \ln[1 - F(L+1-i)]\} \qquad (14.8)$$

where $F(i)$ is the specified cumulative distribution function value at data point r_i. As for the critical values, tabulated values and formulas are available for some selected distributions, such as the normal, log-normal, exponential and Weibull distributions [3]. If A^2 is greater than the computed critical value, then it is assumed that the measurements are not from the specific distribution.

14.2.3 Shapiro–Wilk Test

The Shapiro–Wilk test aims at testing normality [4]. In this test the W-statistic is defined as

$$W = \frac{\left[\sum_{i=1}^{L/2} a_i(r_{L-i+1} - r_i)\right]^2}{\sum_{i=1}^{L} (r_i - \bar{r})^2} \tag{14.9}$$

where $\{r_i\}$ are the ordered sample values (measurements), with r_1 being the smallest, and $\{a_i\}$ are the tabulated coefficients derived from the means, variances and covariances of the order statistics of a sample of size L from a normal distribution. Table 14.1 shows the coefficients for the number of samples ranging from 10 to 25. For samples from a normal distribution, the numerator and the denominator of W are both estimating the same quantity, whereas for a non-normal population of samples the denominator of W tends to be larger, resulting in small values of W. Accordingly, when the W computed from measurements is smaller than that from normal distribution, a non-normal distribution is assumed.

14.2.4 Chi-Square Test

The chi-square test is an alternative to the A–D and K–S goodness-of-fit tests. Unlike the A–D test and the K–S test, the chi-square goodness-of-fit test can be applied to discrete distributions, such as the binomial and the Poisson distributions. In the test, the measurements are divided into κ bins and the test statistic is defined as

$$\chi^2 = \sum_{i=1}^{\kappa} \frac{1}{F_E(i)} (F_O(i) - F_E(i))^2 \tag{14.10}$$

where $F_O(i)$ and $F_E(i)$ are respectively the observed frequency and the expected frequency for bin i. The expected frequency is defined as

$$F_E(i) = L(F(t_U) - F(t_L)) \tag{14.11}$$

where $F(t)$ is the cumulative distribution function for the distribution being tested, t_U and t_L are respectively the upper limit and the lower limit of the measurements for bin i and L is the sample size. Let $\chi^2_{a_1,a_2}$ be the chi-square percentage point function with a_2 degrees of freedom and a significance level a_1. Then, the hypothesis that the measurements come from a population with the specified distribution is rejected if

$$\chi^2 > \chi^2_{a_1,a_2} \tag{14.12}$$

Table 14.1 Tabulated coefficients for the Shapiro–Wilk test.

i	10	11	12	13	14	15	16	17	18	19	20	21	22	23	24	25
										L						
1	0.574	0.560	0.548	0.536	0.525	0.515	0.506	0.497	0.489	0.481	0.473	0.464	0.459	0.454	0.449	0.445
2	0.329	0.332	0.333	0.333	0.332	0.331	0.329	0.327	0.325	0.323	0.321	0.319	0.316	0.313	0.310	0.307
3	0.214	0.226	0.235	0.241	0.246	0.250	0.252	0.254	0.255	0.256	0.257	0.258	0.257	0.256	0.255	0.254
4	0.122	0.143	0.159	0.171	0.180	0.188	0.194	0.199	0.203	0.206	0.209	0.220	0.213	0.214	0.215	0.215
5	0.040	0.070	0.092	0.110	0.124	0.135	0.145	0.152	0.159	0.164	0.169	0.174	0.176	0.179	0.181	0.182
6		0.000	0.030	0.054	0.073	0.088	0.101	0.111	0.120	0.127	0.133	0.140	0.144	0.148	0.151	0.154
7				0.000	0.024	0.043	0.059	0.073	0.084	0.093	0.101	0.109	0.115	0.120	0.125	0.128
8						0.000	0.020	0.036	0.050	0.061	0.071	0.080	0.088	0.094	0.100	0.105
9								0.000	0.013	0.030	0.042	0.053	0.062	0.070	0.076	0.082
10										0.000	0.014	0.026	0.037	0.046	0.054	0.061
11												0.000	0.012	0.023	0.032	0.040
12														0.000	0.011	0.020

One disadvantage of the chi-square test is that the test statistic is dependent on how the measurements are grouped. Another disadvantage of the test is that it requires sufficient samples in order for the chi-square approximation to be valid.

14.2.5 Grubbs Test

The Grubbs test is used to detect outliers in a univariate data set. It is also known as the maximum normalized residual test. This test assumes normality; that is, the TOA or distance measurements can be reasonably approximated by a normal distribution, which is only approximately true under LOS conditions. There are two hypotheses in the test: there are no outliers in the measurements and there is at least one outlier in the data. The Grubbs test detects one outlier at a time. In the case of multiple outliers, once an outlier is detected, it is removed from the data. Then, the test is used again to detect the next outlier. The procedure continues until no more outliers are detected. The Grubbs test statistic is defined as

$$G = \frac{\max_i(|r_i - \bar{r}|)}{\sigma} \tag{14.13}$$

where $\bar{r}$ and σ are respectively the sample mean and STD of the data. That is, the Grubbs test statistic is the largest absolute deviation from the sample mean in units of the sample STD. Given the level of significance as α, the hypothesis that there is at least one outlier is accepted provided that

$$G > \frac{L-1}{\sqrt{L}} \sqrt{\frac{t^2_{(\alpha/(2L),L-2)}}{L-2+t^2_{(\alpha/(2L),L-2)}}} \tag{14.14}$$

where $t_{(\alpha/(2L),L-2)}$ is the critical value of the t-distribution with $(L-2)$ degrees of freedom and a significance level of $\alpha/(2L)$. Some critical values of the t-distribution with 15 different degrees of freedom and five different significance levels are listed in Table 14.2.

14.2.6 Skewness and Kurtosis Tests

Consider the case where a time series of range measurements is made to estimate the distance between a base station and a moving mobile station and only a small fraction of measurements might be corrupted by NLOS propagation. In the absence of NLOS propagation, the measurements can be well approximated by a Gaussian distribution. However, in the presence of NLOS errors, the measurements would have a skewed distribution. The test statistic for skewness test based on the third moment is defined as

$$b_3 = \frac{\frac{1}{L}\sum_{i=1}^{L}(r_i - \bar{r})^3}{\sigma^3} \tag{14.15}$$

Table 14.2 Critical values for t-distribution of α_t degrees of freedom and significance level β.

α_t	$\beta = 0.05$	$\beta = 0.02$	$\beta = 0.01$	$\beta = 0.005$	$\beta = 0.001$
1	12.71	31.82	63.66	127.32	318.31
2	4.30	6.97	9.93	14.09	22.33
3	3.18	4.54	5.84	7.45	10.22
4	2.78	3.75	4.60	5.60	7.17
5	2.57	3.37	4.03	4.77	5.89
6	2.45	3.14	3.71	4.32	5.21
7	2.37	3.00	3.50	4.03	4.78
8	2.31	2.90	3.36	3.83	4.50
9	2.26	2.82	3.25	3.69	4.30
10	2.23	2.76	3.17	3.58	4.14
11	2.20	2.72	3.11	3.50	4.02
12	2.18	2.68	3.06	3.43	3.93
13	2.16	2.65	3.01	3.37	3.85
14	2.15	2.62	2.98	3.33	3.79
15	2.13	2.60	2.95	3.29	3.77

where $\bar{r}$ and σ are the sample mean and STD of the data respectively. The kurtosis test is based on the fourth moment and is given by

$$b_4 = \frac{\frac{1}{L}\sum_{i=1}^{L}(r_i - \bar{r})^4}{\sigma^4} \tag{14.16}$$

All normal distributions are symmetrical, so their skewness is equal to zero. They also have a kurtosis equal to three. For a finite number of samples of a normal distribution, the skewness and kurtosis would be very close to zero and three respectively. When $b_3 > 0$, the data are skewed to the right of the mean, whereas the left side of the distribution has a longer tail when $b_3 < 0$. Further, a symmetric distribution with $b_4 > 3$ has heavier tails than a normal distribution, while $b_4 < 3$ indicates thinner tails than normal. Critical-value tables exist for skewness and kurtosis. When using the tables, one needs to pay attention to the formulas, since there are different formulas to calculate the skewness and kurtosis. For instance, the skewness and kurtosis may also be computed according to

$$b_3 = \frac{\sqrt{L(L-1)}}{L-2}\frac{m_3}{m_2^{3/2}}$$

$$b_4 = \frac{L^2 - 1}{(L-2)(L-3)}\left(\frac{m_4}{m_2^2} - 3\right) \tag{14.17}$$

where

$$m_\kappa = \frac{1}{L}\sum_{i=1}^{L}(r_i - \bar{r})^\kappa \tag{14.18}$$

Some of the tests have already been considered for improving positioning accuracy [5]. Generally speaking, it is difficult to determine which test is the best, although one of them may be suited for more circumstances than the others. In practice, the selection of the tests will depend on the characteristics of measurements, which mainly rely on the environmental conditions and the transceivers being used. Certainly, the performance of the different tests can be compared in advance based on an established database or by making field measurements. As a result, the appropriate test can be selected for a specific application.

14.3 Calculating Level Crossing Rate and Fade Duration

In fading channels, the level crossing rate and average fade duration are also useful statistics which can be used to distinguish different fading distributions [6]. This method is only valid for fading channels when either the mobile is moving or the environment produces fading. The rates depend on the wavelength and the speed; hence, the Doppler frequency given by

$$f_D = \frac{f_o v}{c} \tag{14.19}$$

where f_o is the carrier frequency of the radio signal, c is the speed of propagation of the radio wave and v is the velocity of the transmitter relative to the receiver. Thus, the speed must be known (from the positions). This is probably done using a Kalman filter. Alternatively, the Doppler frequency must be estimated by other techniques.

The level crossing rate is defined as

$$\xi_{LCR} = \frac{\eta_{NC}}{T} \tag{14.20}$$

where η_{NC} is the number of crossings of the specified signal level and T is the measurement duration. The average fade duration is defined by

$$t_{FD} = \frac{1}{\eta_{NC}} \sum_{i=1}^{\eta_{NC}} t_i \tag{14.21}$$

where t_i is the time duration of the ith individual fade. As indicated in the preceding section, the received signal envelope is Rice distributed in the LOS condition and has a Rayleigh distribution in the NLOS condition. For a Rice distribution, the level crossing rate is given by

$$\xi_{LCR} = \sqrt{2\pi(K+1)} f_D \rho \exp\left[-K - (K+1)\rho^2\right] I_0\left[2\rho\sqrt{K(K+1)}\right] \tag{14.22}$$

where $I_0(.)$ is the modified Bessel function of the first kind, K is the Rice factor, f_D is the Doppler frequency, and

$$\rho = \frac{A_{LT}}{A_{RMS}} \tag{14.23}$$

where A_{LT} is the envelope level threshold and A_{RMS} is the root-mean-square of the envelope. For a Rayleigh distribution, (14.22) reduces to

$$\xi_{LCR} = \sqrt{2\pi}f_D\rho\exp(-\rho^2) \tag{14.24}$$

Figure 14.1 shows the level crossing rate versus the ratio parameter ρ under four different Rician factors. Clearly, as K increases, the difference between the two rates (Rayleigh and Rician), defined in (14.22) and (14.24), becomes greater and, hence, the two distributions can be better distinguished. In the hypothesis test, one calculates the level crossing rate based on the measurements for a given level threshold. Then, one compares it to the theoretical ones defined by (14.22) and (14.24). If the measured rate is closer to that in (14.24), then the LOS hypothesis is rejected.

The average fade duration for Rice distribution is defined as

$$t_{FD} = \frac{1 - Q(\sqrt{2K}, \sqrt{2(K+1)\rho^2})}{\sqrt{2\pi(K+1)}f_D\rho\exp[-K - (K+1)\rho^2]I_0[2\rho\sqrt{K(K+1)}]} \tag{14.25}$$

where $Q(a, b)$ is the Marcum Q function [7]. For Rayleigh fading, (14.25) reduces to

$$t_{FD} = \frac{\exp(\rho^2) - 1}{\sqrt{2\pi}f_D\rho} \tag{14.26}$$

In the hypothesis test, the measured average fade duration is compared with the theoretical ones given by (14.25) and (14.26). The distribution with a closer match is selected and the

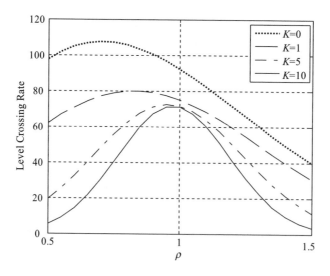

Figure 14.1 Level crossing rate with respect to the ratio of the level threshold over the RMS of the envelope when the Rician factor is 0, 1, 5, and 10. The Doppler frequency is set at 100 Hz.

corresponding LOS or NLOS hypothesis is accepted. Certainly, the results of the test based on the level crossing rate and those based on the average fade duration can be combined to produce more-accurate NLOS identification. Probably the key issues of the above method is to obtain an accurate estimate of the Doppler frequency and choosing the envelope crossing level.

14.4 Estimating the Rician Factor

An alternative way of distinguishing between Rice fading and Rayleigh fading is to estimate the Rice factor. Therefore, the LOS/NLOS hypothesis test can be formulated based on the Rician K factor estimation [8]. When the estimated Rician K factor in decibels is close to zero, Rayleigh fading is assumed and, hence, the NLOS hypothesis is accepted. Otherwise, Rice fading and the LOS hypothesis are accepted. When the moments of the received signal envelope are employed, the estimation of the Rician K factor can be performed as follows [9]. First, estimate the ratio of the first two moments of the envelope according to

$$\xi_{1/2} = \frac{\bar{r}}{\sqrt{\sigma_r^2 + \bar{r}^2}} \tag{14.27}$$

where $\bar{r}$ and σ_r^2 are the sample mean and variance of the measurements respectively. Then, the estimate of the Rician K factor is obtained by solving

$$\xi_{1/2} = \frac{\sqrt{\pi}\exp\left(-\frac{K}{2}\right)}{2\sqrt{K+1}}\left[(K+1)I_0\left(\frac{K}{2}\right) + KI_1\left(\frac{K}{2}\right)\right] \tag{14.28}$$

where $I_0(.)$ and $I_1(.)$ are modified Bessel functions of the first and second kind respectively. A closed-form expression for K cannot be obtained from the above equation, so a numerical solution is produced. Once the estimate of the Rician K factor is obtained, the hypothesis test can be readily performed to choose either the LOS or NLOS hypothesis. The Rician K factor can also be estimated by other approaches; for instance, the ML estimation-based method [10], the method using the covariance of the received signal power [11] and the method using two estimated moments of the received power [12].

14.5 Generalized Likelihood Ratio Test

It is observed that distance measurement error in NLOS conditions is significantly different from that in LOS conditions. Naturally, the distinct distributions of the LOS or NLOS errors with respective prior probabilities can be exploited to identify the LOS/NLOS conditions. The LOS distance measurement noise is typically Gaussian distributed of mean and variance known (the assumption here is that the measurement errors are related to the SNR and other fixed processes, such as digital signal processing-related errors and variations in the analog components, particularly the radio) *a priori*, while the NLOS error is also Gaussian (the basis here is that the measurement is the sum of many processes of unknown statistics, but due to the CLT the resulting combined statistics is Gaussian) but with unknown mean and variance. In this section, the generalized likelihood ratio test (GLRT) for identifying LOS/NLOS conditions is studied when the NLOS error mean and variance are deterministic and the NLOS/LOS probabilities are either known or unknown [13].

Suppose that there are L distance measurement samples to form an observation vector as

$$\mathbf{r} = [r_1 \quad r_2 \quad \cdots \quad r_L]^{\mathrm{T}} \tag{14.29}$$

whose elements are defined as

$$r_i = d + \varepsilon_i \tag{14.30}$$

where d is the true distance and ε_i is the distance measurement error. Define

- $P(H_\ell)$ as the known prior probability of LOS condition;
- $P(H_n)$ as the known prior probability of NLOS condition;
- μ_{los} and μ_{nlos} as the means of ε_i in the LOS and NLOS condition respectively;
- σ_{los}^2 and σ_{nlos}^2 as the variance of ε_i in the LOS and NLOS condition respectively;
- $p(\mathbf{r}|d, H_\ell)$ as the conditional PDF of $\mathbf{r}$ under hypothesis H_ℓ, given by

$$p(\mathbf{r}|d, H_\ell) = \frac{1}{\sqrt{2\pi}\sigma_{\mathrm{los}}^L} \exp\left\{ -\frac{1}{2\sigma_{\mathrm{los}}^2} \sum_{i=1}^{L} [r_i - (\mu_{\mathrm{los}} + d)]^2 \right\} \tag{14.31}$$

- $p(\mathbf{r}|d, H_n)$ as the conditional PDF of $\mathbf{r}$ under hypothesis H_n, given by

$$p(\mathbf{r}|d, \mu_{\mathrm{nlos}}, \sigma_{\mathrm{nlos}}, H_n) = \frac{1}{\sqrt{2\pi}\sigma_{\mathrm{nlos}}^L} \exp\left\{ -\frac{1}{2\sigma_{\mathrm{nlos}}^2} \sum_{i=1}^{L} [r_i - (\mu_{\mathrm{nlos}} + d)]^2 \right\} \tag{14.32}$$

Then, according to the GLRT, we decide H_n if

$$\Lambda(\mathbf{r}) = \frac{p(\mathbf{r}|\hat{d}_{\mathrm{nlos}}, \hat{\mu}_{\mathrm{nlos}}, \hat{\sigma}_{\mathrm{nlos}}, H_n)}{p(\mathbf{r}|\hat{d}_{\mathrm{los}}, H_\ell)} > \frac{P(H_\ell)}{P(H_n)} \tag{14.33}$$

where $\hat{d}_{\mathrm{los}}, \hat{d}_{\mathrm{nlos}}, \hat{\mu}_{\mathrm{nlos}}$ and $\hat{\sigma}_{\mathrm{nlos}}$ are the ML estimates of the unknown LOS and NLOS distances, unknown noise mean and unknown noise STD respectively. To determine $\hat{d}_{\mathrm{nlos}}, \hat{\mu}_{\mathrm{nlos}}$ and $\hat{\sigma}_{\mathrm{nlos}}$, taking logarithms on the right-hand side of (14.32), differentiating the resulting logarithm with respect to $\eta = \mu_{\mathrm{nlos}} + d$ and σ_{nlos} and setting them to zero produce

$$\hat{\eta} = \hat{\mu}_{\mathrm{nlos}} + \hat{d}_{\mathrm{nlos}} = \frac{1}{L} \sum_{i=1}^{L} r_i$$

$$\hat{\sigma}_{\mathrm{nlos}}^2 = \frac{1}{L} \sum_{i=1}^{L} (r_i - \hat{\eta})^2 \tag{14.34}$$

Owing to the fact that the variance of the measurement error in the NLOS condition is typically greater than that in the LOS condition, one constraint can be applied to the variance estimate:

$$\hat{\sigma}_{\text{nlos}}^2 > \kappa \sigma_{\text{los}}^2 \qquad \kappa > 1 \tag{14.35}$$

In the same way, the ML estimate of the distance under hypothesis H_ℓ is obtained as

$$\hat{d}_{\text{los}} = \frac{1}{L} \sum_{i=1}^{L} r_i - \mu_{\text{los}} \tag{14.36}$$

By substituting the above ML estimates into the likelihood functions in (14.31) and (14.32), the generalized likelihood ratio in (14.33) can be computed as

$$\Lambda(\mathbf{r}) = \left(\frac{\sigma_{\text{los}}}{\hat{\sigma}_{\text{nlos}}} \right)^L \exp \left[\left(\frac{\hat{\sigma}_{\text{nlos}}^2 - \sigma_{\text{los}}^2}{2\hat{\sigma}_{\text{nlos}}^2 \sigma_{\text{los}}^2} \right) \sum_{i=1}^{L} (r_i - \hat{\eta})^2 \right] \tag{14.37}$$

Consequently, the decision can be made to choose one of the two hypotheses. The above hypothesis testing is based on the known prior LOS and NLOS probabilities. In practice, the prior probabilities of either the LOS or NLOS condition of a transmission may be unknown due to either lack of historical data or the complex dynamics of the environment. In the absence of the prior probabilities, other tests should be considered.

14.6 Nonparametric Method

As mentioned in the previous sections, the probability distribution of the distance or TOA measurements in the LOS condition is usually known except for its mean. On the other hand, the probability distribution of the measurement errors and, hence, that of the measurements in the NLOS condition may not be known in most circumstances. One way to handle the case of unknown NLOS error statistics is to adopt a nonparametric approach [14]. That is, the PDF of the measurements in the NLOS condition is approximated by some certain functions. For instance, given L i.i.d. distance measurements as defined in (14.29), the PDF can be approximated using the Parzen window density function:

$$\hat{p}_{\text{r}}(t) = \frac{1}{L} \sum_{i=1}^{L} \frac{1}{h_L} f \left(\frac{t - r_i}{h_L} \right) \tag{14.38}$$

where $f(.)$ is the window function and h_L is a scaling parameter. The window function is a PDF, so that it is nonnegative and integrates to one. Many functions can be used as a window function; however, Gaussian and rectangular windows are among the commonly used window functions.

The decision of LOS or NLOS condition is based on testing whether the distance measurements are suited to one of the two distributions. When $\hat{p}_{\text{r}}(t)$ and the PDF in the LOS condition are sufficiently close, the LOS hypothesis is accepted. Otherwise the NLOS hypothesis is accepted. The problem now is how to determine the distance between the two distributions and how to choose the distance threshold for testing. For given PDFs p_1 and p_2, the distance between the two distributions can be computed according to the Kullback–Leibler distance:

$$D(p_1\|p_2) = \int p_1(t)\log\frac{p_1(t)}{p_2(t)}\,dt \tag{14.39}$$

To compute the above distribution distance, the true distance between the transmitter and the receiver is required. For a zero-mean Gaussian measurement error and a symmetric window function – that is, $f(t) = f(-t)$ for all t – the sample mean of the distance measurements

$$\hat{d} = \frac{1}{L}\sum_{i=1}^{L} r_i \tag{14.40}$$

is the distance estimate which minimizes the distribution distance. The threshold is usually set to satisfy the probability of false alarm (PFA), which, in this case, is the probability that the distribution distance exceeds the threshold given that the LOS is the true hypothesis. It can be proved that, for a zero mean Gaussian measurement error and a symmetric window function, the PFA can be set independently of the true distance between the transmitter and the receiver.

Another nonparametric method is to jointly estimate the probability density and parameters iteratively [15]. As discussed in Chapter 6, after introducing a new variable, the distance measurement equations can be written in a linear form as

$$\mathbf{u} = \mathbf{A}\boldsymbol{\theta} + \mathbf{v} \tag{14.41}$$

The ordinary LS solution of $\boldsymbol{\theta}$ is

$$\hat{\boldsymbol{\theta}} = (\mathbf{A}^T\mathbf{A})^{-1}\mathbf{A}^T\mathbf{u} \tag{14.42}$$

Suppose that the distribution of the noise $\mathbf{v}$ has an unknown density function denoted by $f_{\mathbf{v}}$, the ML estimate of $\boldsymbol{\theta}$ is given by

$$\hat{\boldsymbol{\theta}}_{\mathrm{ML}} = \arg\min_{\boldsymbol{\theta}}\{-\log f_{\mathbf{v}}(\mathbf{u} - \mathbf{A}\boldsymbol{\theta})\} \tag{14.43}$$

If the log-likelihood function is a convex function, then a unique ML solution can be obtained by solving

$$\mathbf{A}\psi(\mathbf{u} - \mathbf{A}\boldsymbol{\theta}) = 0 \tag{14.44}$$

where ψ is the location score function of $f_{\mathbf{v}}(\mathbf{v})$, which is defined as

$$\psi = -\frac{1}{f_{\mathbf{v}}}\frac{\partial f_{\mathbf{v}}}{\partial\boldsymbol{\theta}} \tag{14.45}$$

After getting the density estimate, the score function estimate is obtained as

$$\hat{\psi} = -\frac{1}{\hat{f}_{\mathbf{v}}}\frac{\partial\hat{f}_{\mathbf{v}}}{\partial\boldsymbol{\theta}} \tag{14.46}$$

Next, the parameter estimation and the core function estimation are performed alternatively and iteratively. Specifically, an initial parameter estimate is first obtained, such as from (14.42). Then, the residual is computed according to

$$\hat{\mathbf{v}} = \mathbf{u} - \mathbf{A}\hat{\boldsymbol{\theta}}^{(i)} \tag{14.47}$$

The parameter estimate and the residual are used to obtain the score function estimate $\hat{\psi}^{(i)}$ from (14.46). Finally, the parameter estimate is updated, such as by the M-estimator, as

$$\hat{\boldsymbol{\theta}}^{(i+1)} = \hat{\boldsymbol{\theta}}^{(i)} + \mu(\mathbf{A}^\mathrm{T}\mathbf{A})^{-1}\mathbf{A}^\mathrm{T}\hat{\psi}^{(i)} \tag{14.48}$$

where μ is the step size. The procedure continues until $\|\hat{\boldsymbol{\theta}}^{(i+1)} - \hat{\boldsymbol{\theta}}^{(i)}\|$ is sufficiently small. The advantage of this nonparametric method is that it does not require any training data. On the other hand, it requires the estimation of the distribution function.

14.7 Using Intermediate Location Estimation

The purpose of NLOS identification is to remove or mitigate the effect of the NLOS-corrupted measurements and, hence, to improve location estimation accuracy. Note that the multipath effect on the first-path signal detection can often be reduced first at the receiver, such as by using the leading-edge detection techniques studied in Chapter 4. However, it is very difficult, if not impossible, to mitigate the NLOS impact at the receiver. On the other hand, it is feasible to identify the NLOS-corrupted measurements from a mixture of LOS and NLOS measurements. The measurements can be the time or distance measurements directly. Also, the position estimation results, either from all LOS measurements or from both LOS and NLOS measurements, can be employed to assist NLOS identification and mitigation.

Is it possible to obtain a location estimate first and then perform the NLOS detection? The answer is positive, since the estimation residuals of NLOS-corrupted measurements are usually greater than those of measurements in LOS conditions. Also, the location estimation errors in the LOS condition may have a different distribution than the location errors using NLOS-corrupted measurements.

In macrocellular networks, where the scatterers are usually around the mobile station, the NLOS errors in AOA measurements are usually small when the AOA is measured at the base stations, where it is usually feasible to set up antenna arrays [16]. Note that, in a macrocellular network, if the AOA is measured at the mobile site, then the NLOS-incurred AOA errors can be very large. An initial position estimate can be obtained by using all the AOA measurements. For instance, when the statistics of the angular measurement errors are known, the ML estimator can be employed. Then, the angular measurement errors can be computed according to

$$\delta\phi_i = \frac{|(\hat{x}_{\mathrm{ML}} - x_i)\sin\hat{\phi}_i - (\hat{y}_{\mathrm{ML}} - y_i)\cos\hat{\phi}_i|}{\sqrt{(\hat{x}_{\mathrm{ML}} - x_i)^2 + (\hat{y}_{\mathrm{ML}} - y_i)^2}} \tag{14.49}$$

where $(\hat{x}_{\mathrm{ML}}, \hat{y}_{\mathrm{ML}})$ is the ML estimate of the mobile station location, (x_i, y_i) are the known locations of the base stations and $\{\hat{\phi}_i\}$ are the AOA measurements. Once the angular measurement errors of all base stations are estimated, their RMS is computed. Then, one picks up the measurement whose error deviates from the RMS the greatest and is greater than, for example, 1.5 times the RMS and excludes the measurement. With the remaining measurements, the mobile station location is estimated again. This would be an improved estimation. The procedure may be repeated, especially when multiple NLOS base stations likely exist. The process would automatically stop when there are no more base stations whose angular measurement error is much bigger than normal. This method is suited for situations where there are sufficient base stations and a few measurements are NLOS corrupted.

In some circumstances, only one or two base stations may suffer from NLOS propagation, resulting in one or two NLOS-corrupted measurements which are either TOA, TDOA or distance. To identify the NLOS propagation, the measurements of base stations can be arranged into combinations/sets. For example, consider 3D positioning with three unknown position coordinates and in the presence of six base stations. To generate a unique position estimate, measurements from at least four base stations are required. When the six measurements are grouped into combinations, each of which has four measurements, there are 15 different combinations. Also, there are six combinations, each of which has five measurements. In addition, the six measurements can be formed into one combination, resulting in a total of 22 combinations.

First, a reference mobile station location is produced by using each of the combinations of measurements. Then, the reference mobile station locations are used to calculate the measurement residuals. For each base station/measurement, one sums up the residuals of the base station sets that a specific base station belongs to. Then, by ranking each base station according to its total residual, one picks up the base station with the largest residual and the related measurements are excluded.

As indicated in [17], the false detection rate largely depends on the combination set size in addition to the mobile station location estimate reference. When all the measurements are employed to generate the mobile station location reference, the false detection rate increases as the set size gets larger. On the other hand, when the intermediate location estimate based on the measurements of the combination is used as the reference, the false detection rate first decreases with the set size and then starts to increase when the set size is greater than a certain value. Therefore, in practice, it is necessary to tune the set size based on the choice of mobile station location reference and the total number of base stations involved in the presence of database or field measurements.

Another way to deal with the NLOS identification is to determine the LOS dimension first [18]; that is, the number of LOS base station measurements. To start with, the base stations or measurements are grouped into different combinations. For instance, in the case of seven base stations and 2D positioning, each group has three, four, five, six or seven base stations, resulting in a total of $35 + 35 + 21 + 7 + 1 = 99$ combinations. An intermediate mobile station location estimate is yielded based on each combination of measurements, denoted by $(\hat{x}(k), \hat{y}(k))$. Suppose that the LOS dimension is seven; that is, all base stations are in the LOS condition. The squares of the normalized residuals are defined as

$$
\begin{aligned}
\chi_x^2(k) &= \frac{|\hat{x}(k) - \hat{x}(99)|^2}{B_x(k)} \qquad k = 1, 2, \ldots, 98 \\
\chi_y^2(k) &= \frac{|\hat{y}(k) - \hat{y}(99)|^2}{B_y(k)}
\end{aligned}
\tag{14.50}
$$

where $(\hat{x}(99), \hat{y}(99))$ are the coordinate estimates using all measurements, which are the best under the assumption of LOS dimension of seven. $B_x(k)$ and $B_y(k)$ are the respective approximations of the x-coordinate and y-coordinate CRLBs for the kth combination of measurements. When the distance measurement errors are i.i.d. random variables, the normalized residuals in (14.50) have an approximate central chi-square PDF of one degree of freedom. The approximation comes from the fact that $B_x(k)$ and $B_y(k)$ are only approximate CRLBs; the true mobile station location is approximated by $(\hat{x}(99), \hat{y}(99))$; and $\hat{x}(k)$ and $\hat{y}(k)$ may be partially correlated. However, when there are one or more NLOS-corrupted measurements, the residuals in (14.50) will have a noncentral chi-square distribution. Clearly, a noncentral chi-square distribution has heavier tails than the central chi-square distribution. Therefore, a threshold can be chosen to judge whether the distribution is central chi-square. For instance, when the number of the residuals in (14.50) that are above the threshold is less than 10% of the total number of the residuals, the central chi-square distribution hypothesis is accepted. Otherwise, the LOS dimension is assumed less than seven. In the latter case, one checks if the LOS dimension is six. The measurements are grouped into seven combinations, each of which has six measurements. Then, one follows the same procedure to check whether each of the seven sets of measurements has an LOS dimension of six. In the case that more than one set has a dimension of six, one chooses the set with the smaller number of residuals above the threshold. If all sets have dimensions below six, then one checks whether the LOS dimension equals five by following the same procedure. Relatively speaking, underestimating the LOS dimension is better than overestimating it, since, when overestimating, NLOS-corrupted measurements will be used and large location estimation errors might be produced. However, falsely reducing the dimension will result in larger errors. Also, this approach does not take into account the effect of GDOP, which is studied in Chapter 9. Bad geometry with fairly good data can be worse than including one measurement with moderate errors but good GDOP. Thus, in practice, this method should be used with caution.

14.8 Neyman–Pearson Test

In Sections 14.5 and 14.6, two different identification methods have been proposed based on multiple TOA-based range measurements. The measurement noise statistics in LOS condition are assumed known; for example, from an established database or from field measurements in the monitored area. This assumption may also be based on the SNR performance of the receiver and, thus, is based on knowledge of the SNR from the receiver. Note that even with LOS conditions there will at least be reflections from the ground, which result in signal strength variations, and bias errors. Indoors, reflection from walls can occur in LOS conditions. On the other hand, the error statistics in the NLOS condition are not known. In this section we consider the idealized case in which the true distance and bias are known so that the optimal Neyman–Pearson (NP) test can be applied. Although the NP test is not realizable due to the unknown true distance and bias, the performance of the NP test can be used as a bound for performance comparison. This is analogous to the use of the CRLB on unbiased estimator variance.

It is assumed that N TOA-based distance measurements between a mobile station and a base station are taken. Also, the change in the location of the mobile station during these measurements are ignored, so that the distance between the mobile station and the base station can be considered approximately constant for the identification purpose. Then, the identification problem becomes

$$
\begin{aligned}
H_\ell : \quad & \hat{d}_i = d + w_{\text{los},i} & i = 1, 2, \ldots, N & \quad \text{LOS condition} \\
H_{\text{n}} : \quad & \hat{d}_i = d + b + w_{\text{nlos},i} & i = 1, 2, \ldots, N & \quad \text{NLOS condition}
\end{aligned}
\tag{14.51}
$$

where d is the true straight-line range between the two nodes, such as a mobile station and a base station, $\hat{d}_i$ is the ith measurement of d, b is the extra measured distance (positive bias) due to the blockage of the direct path, and $w_{los,i}$ and $w_{nlos,i}$ are the measurement noises under the LOS and NLOS conditions respectively. $w_{los,i}$ and $w_{nlos,i}$ are modeled as white Gaussian random variables with zero means and variances equal to $\sigma^2_{w_{los}}$ and $\sigma^2_{w_{nlos}}$ respectively. The NLOS bias b in both indoor and outdoor environments is modeled as an exponential random variable with a mean λ and a variance λ^2.

The sample mean of the N measurements is defined as

$$\hat{d} = \frac{1}{N}\sum_{i=1}^{N}\hat{d}_i = \begin{cases} d + w_{los} & \text{LOS condition} \\ d + b + w_{nlos} & \text{NLOS condition} \end{cases} \tag{14.52}$$

where

$$w_{los} = \frac{1}{N}\sum_{i=1}^{N}w_{los,i}$$
$$w_{nlos} = \frac{1}{N}\sum_{i=1}^{N}w_{nlos,i} \tag{14.53}$$

are the sample means of the measurement noise in the LOS and NLOS conditions respectively. Clearly, w_{los} and w_{nlos} are Gaussian distributed with zero mean and variances $\sigma^2_{w_{los}}/N$ and $\sigma^2_{w_{nlos}}/N$ respectively. Note that the measurements are taken when the mobile is moving, so d is actually the average of the distances at the measurement points. When the mobile moves over a short distance during the measurements, we may consider the distance from the mobile to the base station as constant. This also applies to the NLOS bias b.

Then, given the true distance and bias, the sample mean $\hat{d}$ is also Gaussian distributed with a mean equal to the true distance and the variance $\sigma^2_{w_{los}}/N$ in the LOS condition and $\sigma^2_{w_{nlos}}/N$ in the NLOS condition, respectively. According to the NP theorem, the detector decides the hypothesis H_n if

$$\frac{p(\hat{d}|d,b,H_n)}{p(\hat{d}|d,H_\ell)} = \frac{\sigma_{w_{los}}}{\sigma_{w_{nlos}}}\exp\left\{\frac{(\hat{d}-d)^2}{2\sigma^2_{w_{los}}/N} - \frac{[\hat{d}-(d+b)]^2}{2\sigma^2_{w_{nlos}}/N}\right\} > \kappa \tag{14.54}$$

The decision rule in (14.54) is equivalent to deciding H_n if

$$\hat{d} > \gamma \tag{14.55}$$

where, giving the PFA ε, the threshold γ is determined by

$$\varepsilon = \int_{\gamma}^{\infty}\frac{1}{\sqrt{2\pi/N}\sigma_{w_{los}}}\exp\left[-\frac{(x-d)^2}{2\sigma^2_{w_{los}}/N}\right]dx$$
$$= Q\left(\frac{\gamma - d}{\sigma_{w_{los}}/\sqrt{N}}\right) \tag{14.56}$$

Accordingly, we can compute the theoretical probability of detection (POD) by

$$P_{\mathrm{D}} = \int_{\gamma}^{\infty} p(\hat{d}|d,b,H_{\mathrm{n}})\,\mathrm{d}\hat{d}$$

$$= \int_{\gamma}^{\infty} \frac{1}{\sqrt{2\pi/N}\sigma_{w_{\mathrm{nlos}}}} \exp\left\{-\frac{[x-(d+b)]^2}{2\sigma_{w_{\mathrm{nlos}}}^2/N}\right\}\mathrm{d}x \qquad (14.57)$$

$$= Q\left[\frac{\gamma-(d+b)}{\sigma_{w_{\mathrm{nlos}}}/\sqrt{N}}\right]$$

In the simulation we will compare the performance between the TOA-based approaches and the joint TOA and RSS method, which will be discussed in the following section.

The following example illustrates the performance of the NP-test-based identification. The distance measurement noise is modeled as a Gaussian random variable of mean zero and an STD that equals 4% and $1.3 \times 4\%$ of the true distance in the LOS and NLOS conditions respectively. The NLOS bias is modeled as an exponential random variable with a mean that equals 8% of the true distance. Ten thousand different distance samples (from 20 m to 1000 m) are examined for each simulation run and the performance is then averaged.

Figure 14.2 shows the POD versus the PFA of the three TOA-based methods when four distance samples are used for each decision-making. The three curves are produced by using the GLRT method, the nonparametric method and the idealized case in Section 14.8. Clearly, there is a relatively large gap between the performance of the idealized case and that of the other two methods. Therefore, it would be interesting to develop practical estimation approaches to narrow the gap.

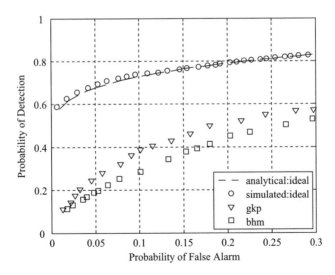

Figure 14.2 POD versus PFA of the TOA-based identification with four measurements for a given distance.

14.9 Joint Time-of-Arrival and Received Signal Strength-based Approaches

In nearly all wireless applications the RSS measurements are readily available. If the TOA measurements are also made by the system, then the TOA and RSS measurements can be jointly exploited to improve the NLOS identification performance. In this section, one such joint TOA and RSS-based identification method is described.

14.9.1 Algorithm Derivations

The RSS method uses an empirically developed path loss model to determine the propagation distance between the transmitter and receiver. There are a number of well-known path loss models for describing the radio signal propagation in different scenarios. The Walfisch–Ikegami path loss model [19] is suitable for urban areas, whereas the log-distance model is suited for indoor environments.

When using a path loss model for determining the propagation distance, it is crucial to tune the model parameters well so that there is a good match between the model and the field measurements. Owing to multipath fading, the path loss computation is based on the mean received signal power of multiple measurements and the known transmitted signal power. Note that the receiver power must be measured over a considerable area in outdoor environments to obtain the large-scale signal strength. The variation has three scales: short (up to a wavelength); medium (10–100 wavelengths with log-normal distribution with typical STD 6–8 dB; large scale (>100 wavelengths). This greatly reduces the effectiveness. Clearly, the accuracy can only be of the order of a few hundred wavelengths. Here, we exploit the Walfisch–Ikegami model for study.[1] In this model, the path loss (in decibels) is computed according to

$$L_{\mathrm{p}} = \begin{cases} A_{\mathrm{los}} + 26 \log_{10} d & \text{LOS condition} \\ A_{\mathrm{nlos}} + 38 \log_{10} d & \text{NLOS condition} \end{cases} \quad (14.58)$$

where A_{los} and A_{nlos} are parameters that are dependent on the signal carrier frequency, transmitter and receiver antenna heights, structure of buildings and roads and street orientation relative to the direct radio path; d is the LOS distance between the transmitter and receiver. Note that, in reality, both A_{los} and A_{nlos} are random variables, not constants. For example, they vary with orientation of the person when the mobile is on a person. The power-law parameters are also random variables, depending on the propagation environment. Therefore, one should be cautious in choosing the parameters in practice In the presence of measurement noise and modeling error, the path loss can be expressed as

$$\hat{L}_{\mathrm{p}} = \begin{cases} A_{\mathrm{los}} + 26 \log_{10} d + \nu_{\mathrm{los}} & \text{LOS condition} \\ A_{\mathrm{nlos}} + 38 \log_{10} d + \nu_{\mathrm{nlos}} & \text{NLOS condition} \end{cases} \quad (14.59)$$

where ν_{los} and ν_{nlos} are the path loss model errors.

The purpose is to identify the NLOS condition by jointly using the TOA-based distance estimates and the path loss measurements. The question is how to combine both the

---

[1] In practice, better models may be developed so that they are well suited to particular environments.

measurements/estimates effectively. Our approach is described as follows. Let

$$
\begin{aligned}
\log_{10}(\hat{d} - w_{\text{los}}) &= \log_{10}\hat{d} + \rho_{\ell} \\
\log_{10}[\hat{d} - (b + w_{\text{los}})] &= \log_{10}\hat{d} + \rho_{n}
\end{aligned}
\tag{14.60}
$$

where $\hat{d}$ is the sample mean of the N TOA-based distance estimates and w_{los} and w_{nlos} are the corresponding sample means of the measurement noise in LOS condition and NLOS condition respectively, as defined in (14.53). It can be shown that under the assumption of $w_{\text{los}} \ll \hat{d}$ and $w_{\text{nlos}} + b \ll \hat{d}$, and using the approximation of $\ln(1 + x) \approx x$ for small x, ρ_{ℓ} and ρ_{n} can be approximated respectively as

$$
\begin{aligned}
\rho_{\ell} &\approx -\frac{w_{\text{los}}}{2.3\hat{d}} \\
\rho_{n} &\approx -\frac{b + w_{\text{nlos}}}{2.3\hat{d}}
\end{aligned}
\tag{14.61}
$$

Then, replacing the true distance in (14.59) by *both* the distance estimate *and* the distance error in (14.51), and using the approximations in (14.61), the joint TOA and RSS-based detection problem becomes

$$
\begin{aligned}
H_{\ell} : \quad & \hat{L}_{\text{p}} = A_{\text{los}} + 26\log_{10}d + v_{\text{los}} \approx A'_{\text{los}} + v'_{\text{los}} \\
H_{n} : \quad & \hat{L}_{\text{p}} = A_{\text{nlos}} + 38\log_{10}\hat{d}
\end{aligned}
\tag{14.62}
$$

where

$$
\begin{aligned}
A'_{\text{los}} &= A_{\text{los}} + 26\log_{10}\hat{d} \\
A'_{\text{nlos}} &= A_{\text{nlos}} + 38\log_{10}d + v_{\text{nlos}} \approx A'_{\text{nlos}} + v'_{\text{nlos}} \\
v'_{\text{los}} &= v_{\text{los}} - \frac{26}{2.3\hat{d}}w_{\text{los}} \\
v'_{\text{nlos}} &= \left(v_{\text{nlos}} - \frac{38}{2.3\hat{d}}w_{\text{nlos}}\right) - \frac{38}{2.3\hat{d}}b
\end{aligned}
\tag{14.63}
$$

Let v_{los} and v_{nlos} be Gaussian random variables with means $\bar{v}_{\text{los}}$ and $\bar{v}_{\text{nlos}}$ and variances $\sigma^2_{v_{\text{los}}}$ and $\sigma^2_{v_{\text{nlos}}}$ respectively. Also, let w_{los} and w_{nlos} be Gaussian random variables with zero means and variances equal to $\sigma^2_{w_{\text{los}}}$ and $\sigma^2_{w_{\text{nlos}}}$ respectively, and b be an exponential random variable with mean λ and variance λ^2 as assumed earlier. Under the assumption that the five random variables are mutually independent and given the TOA-based distance measurement $\hat{d}_i$, it is seen that v'_{los} is a Gaussian random variable with mean $\bar{v}_{\text{los}}$ and variance

$$
\sigma^2_{v'_{\text{los}}} = \sigma^2_{v_{\text{los}}} + \left(\frac{26}{2.3\hat{d}}\right)^2 \sigma^2_{w_{\text{los}}}.
\tag{14.64}
$$

Define

$$
\begin{aligned}
u &= v_{\text{nlos}} - \frac{38}{2.3\hat{d}}w_{\text{nlos}} \\
s &= \frac{38}{2.3\hat{d}}b
\end{aligned}
\tag{14.65}
$$

Apparently, u is a Gaussian random variable with mean $\bar{u} = \bar{v}_{\text{nlos}}$ and variance

$$\sigma_u^2 = \sigma_{v_{\text{nlos}}}^2 + \left(\frac{38}{2.3\hat{d}}\right)^2 \sigma_{w_{\text{nlos}}}^2 \tag{14.66}$$

and s is an exponential random variable with mean given by

$$\lambda_s = \frac{38}{2.3\hat{d}}\lambda \tag{14.67}$$

Based on the fact that the PDF of the sum of two random variables equals the convolution of each of their distributions, the PDF of v'_{nlos} is determined as follows:

$$\begin{aligned} P_{v'_{\text{nlos}}}(v'_{\text{nlos}}) &= \int_{-\infty}^{\infty} p_u(u)p_s(u - v'_{\text{nlos}})du \\ &= \int_{v'_{\text{nlos}}}^{\infty} \frac{1}{\sqrt{2\pi}\sigma_u}\exp\left[-\frac{(u-\bar{u})^2}{2\sigma_u^2}\right]\frac{1}{\lambda_s}\exp\left(-\frac{u - v'_{\text{nlos}}}{\lambda_s}\right)du \\ &= \frac{1}{\lambda_s}\exp\left(\frac{v'_{\text{nlos}}}{\lambda_s} + \beta_1\right)Q\left(\frac{v'_{\text{nlos}}}{\sigma_u} + \beta_2\right) \end{aligned} \tag{14.68}$$

where

$$\begin{aligned} \beta_1 &= \frac{\sigma_u^2}{2\lambda_s^2} - \frac{\bar{u}}{\lambda_s} \\ \beta_2 &= \frac{\sigma_u}{\lambda_s} - \frac{\bar{u}}{\sigma_u} \end{aligned} \tag{14.69}$$

Therefore, the PDFs under H_ℓ and H_n are

$$p(\hat{L}_p|\hat{d}, H_\ell) = \frac{1}{\sqrt{2\pi}\sigma_{v'_{\text{los}}}}\exp\left\{-\frac{[\hat{L}_p - (A'_{\text{los}} + \bar{v}_{\text{los}})]^2}{2\sigma_{v'_{\text{los}}}^2}\right\}$$

$$p(\hat{L}_p|\hat{d}, H_n) = \frac{1}{\lambda_s}\exp\left(\frac{\hat{L}_p - A'_{\text{nlos}}}{\lambda_s} + \beta_1\right)Q\left(\frac{\hat{L}_p - A'_{\text{nlos}}}{\sigma_u} + \beta_2\right) \tag{14.70}$$

According to the NP test, we decide the hypothesis H_n (that is, the NLOS condition), when the likelihood ratio satisfies

$$\frac{p(\hat{L}_p|\hat{d}, H_n)}{p(\hat{L}_p|\hat{d}, H_\ell)} > \kappa \tag{14.71}$$

where κ is the threshold that depends on the pre-assigned PFA. If (14.71) is not satisfied, then we decide H_ℓ is true. It is seen that the inequality in (14.71) is equivalent to

$$\hat{L}_p > \gamma \tag{14.72}$$

where γ, when assigning a small value (ε) to the PFD, is determined by solving

$$\int_{\gamma}^{\infty} p(\hat{L}_p | \hat{d}, H_\ell) d\hat{L}_p = Q\left[\frac{\gamma - (A'_{los} + \bar{v}_{los})}{\sigma_{v'_{los}}} \right] = \varepsilon \tag{14.73}$$

When γ is given, we can determine the POD as

$$\begin{aligned}
P_D &= \int_{\gamma}^{\infty} p(\hat{L}_p | \hat{d}, H_n) d\hat{L}_p \\
&= \int_{\gamma}^{\infty} \frac{1}{\lambda_s} \exp\left(\frac{\hat{L}_p - A'_{nlos}}{\lambda_s} + \beta_1 \right) Q\left(\frac{\hat{L}_p - A'_{nlos}}{\sigma_u} + \beta_2 \right) d\hat{L}_p \\
&= \frac{1}{\lambda_s} e^{\beta_1} \int_{\gamma - A'_{nlos}}^{\infty} \exp\left(\frac{x}{\lambda_s} \right) Q\left(\frac{x}{\sigma_u} + \beta_2 \right) dx \\
&= Q\left(\frac{\gamma - A'_{nlos} - \bar{u}}{\sigma_u} \right) - \exp\left(\frac{\gamma - A'_{nlos} - \bar{u}}{\lambda_s} + \frac{\sigma_u^2}{2\lambda_s^2} \right) Q\left(\frac{\gamma - A'_{nlos} - \bar{u}}{\sigma_u} + \frac{\sigma_u}{\lambda_s} \right)
\end{aligned} \tag{14.74}$$

where details about how to get the last equality in (14.74) are given in Annex 13.B. In reality, there always exist some model parameter errors, especially in NLOS conditions. In the following section, the impact of the model errors on the performance of the method will be evaluated through simulation.

14.9.2 Identification Accuracy Evaluation

The accuracy of the joint TOA and RSS method is illustrated by the following example. The path loss model parameter A_{los} is set at 30.2 at frequency 1.9 GHz and A_{nlos} is set at 31.0 based on some typical building and road parameters. The other parameters are the same as in evaluating the TOA-based method in Section 14.8. Figure 14.3 shows the POD versus the PFA under three different STD values (6, 9, and 12 dB) of the path loss error in LOS conditions. The STD of the path loss error in NLOS conditions is set at 30% larger than that in LOS conditions. Each decision is made based on one TOA-based distance measurement and one path loss sample. For comparisons, the results of the TOA-based method in idealized conditions and with $N = 4$ samples are also plotted. Both the analytical (denoted by lines) and simulated (denoted by symbols) results are presented. The analytical results are computed using the last equation in (14.74), while the simulated results are obtained based on (14.72). To achieve a POD of 90%, the PFA needs to be set at about 0.5, 5 and 17.5% under the three STD values. Clearly, the method performs well and it outperforms the TOA-based methods considerably, even in presence of relatively large path loss measurement errors.

Figure 14.4 shows the impact of the distribution parameter mismatch on the POD and PFA. The STD estimate of the path loss error in LOS and NLOS conditions is set to 9 dB and 1.3×9 dB respectively. Similar to Figure 14.3, three different levels of the model parameter errors are examined. The STDs of the path loss and the TOA-based distance measurement noise in both LOS and NLOS conditions are assumed uniformly distributed between $(1 - q) \times$ (estimated value) and $(1 + q) \times$ (estimated value). It can be seen that the impact

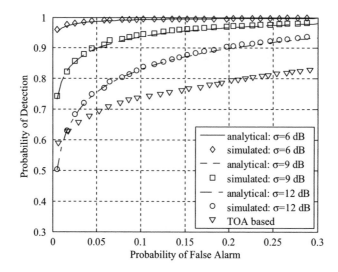

Figure 14.3 POD versus PFA of the joint TOA and signal-strength-based NLOS identification. Three different path loss error STDs (6, 9, and 12 dB) are examined. 'TOA based' denotes results of the TOA-based method in idealized conditions.

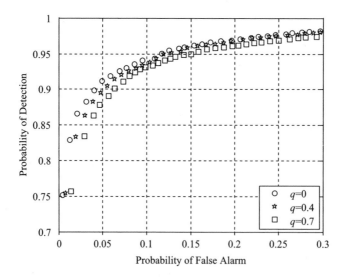

Figure 14.4 Impact of distribution parameter errors when the STD estimates of the path loss error in LOS and NLOS conditions are set to 9 dB and 1.3 × 9 dB respectively, whereas the true STD is uniformly distributed between $(1 - q)$ × (estimated value) and $(1 + q)$ × (estimated value).

of the model parameter mismatch on the identification method is marginal. It is worth noting that, when the distribution parameters are fixed, the relation between the true PFA and POD remains the same regardless of the parameters' estimation errors. This is a property of the NP detection rule. However, modeling error will result in overestimating or underestimating the PFA and POD when using the analytical formulae. Therefore, when designing the identification parameters in the presence of model mismatch, caution should be taken to ensure the desired PFA and POD.

14.10 Angle-of-Arrival-based Methods

When both ends of the communication channel are equipped with a directional antenna or an antenna array, AOA estimates can be used for NLOS identification. The AOA estimates may be obtained by using the high-resolution approaches, including the MUSIC and ESPRIT algorithms. In this section, we show how to identify NLOS conditions based on angular measurements in 3D physical environments. Two theorems and one corollary are first presented and the corresponding proofs provided. Then, the hypothesis test is developed when either 1D or 2D angular measurements are employed.[2]

14.10.1 Line-of-Sight Judgment in Ideal Conditions

Theorem 1   Assume that node A and node B are in communication. Let α_a be the elevation angle[3] of the incoming signal at node A and α_b be the elevation angle of the incoming signal at node B, as shown in Figure 14.5. In the absence of angular and orientation measurement error, the condition for the existence of LOS propagation is

$$\alpha_a + \alpha_b = \pi \tag{14.75}$$

and that the received signal has not been reflected by any scatterer on a 3D surface defined by

$$\frac{z_s - z_a}{\sqrt{(x_s - x_a)^2 + (y_s - y_a)^2}} + \frac{z_s - z_b}{\sqrt{(x_s - x_b)^2 + (y_s - y_b)^2}} = 0 \tag{14.76}$$

where (x_a, y_a, z_a) and (x_b, y_b, z_b) are the coordinates of the two nodes, and (x_s, y_s, z_s) are the coordinates of the scatterer.

Theorem 2   Assume that node A and node B are in communication. Let ϕ_a be the azimuth angle[4] of the incoming signal at node A and ϕ_b be the azimuth angle of the incoming signal at node B, as shown in Figure 14.6. In the absence of angular and orientation measurement error, the LOS condition is guaranteed if the two azimuth angles satisfy

$$|\phi_a + \phi_b| = \pi \tag{14.77}$$

---

[2] Note that, in outdoor (especially macro-) cellular environments, the elevation angle may not vary hugely across the cell compared with the azimuth angle, although it will vary more for indoor deployments.

[3] The elevation angle is defined as the angle of the beam with respect to the positive z-axis and it ranges from zero to π.

[4] The azimuth angle is defined as the angle of the incoming signal beam on the xy-plane with respect to the positive x-axis and it can range from zero and 2π.

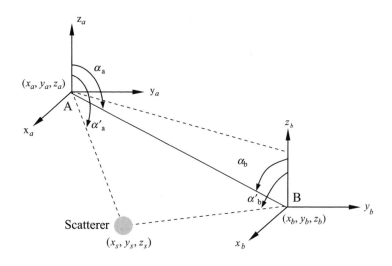

Figure 14.5 LOS/NLOS identification using elevation angles.

where $|.|$ is the operation of taking the absolute value, and that the first received signal has not been reflected from any scatterer on the planar area enclosed by the two z-axes.

Corollary 1   Assume that node A and node B are in communication. In the absence of angular measurement and orientation measurement error, the LOS condition can be uniquely determined if both (14.75) and (14.77) are satisfied.

The corresponding proofs of the above results are presented in the Annex 13.A. In the remainder of this section we apply hypothesis testing using a statistical decision theory to identify the NLOS condition.

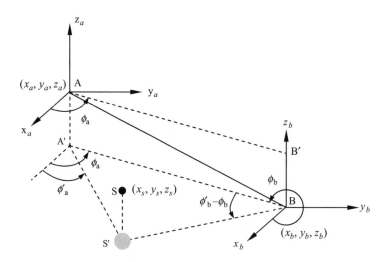

Figure 14.6 LOS/NLOS identification using azimuth angles.

14.10.2 *Identification Based on Azimuth and Elevation Angles*

First, we deal with the identification problem based on either azimuth or elevation angular measurements. Then, the problem is handled based on the 2D angular measurements. Later, the performance is analyzed for identification based on the 1D azimuth angular measurements.

It is assumed that the received signal is not reflected by scatterers on the 3D surface defined by (14.76), so that satisfying (14.75) means the LOS condition. Also assume that the distributions of the estimation errors are available, which may be determined *a priori* from field measurements [20]. For notational simplicity, let

$$\alpha = \alpha_a + \alpha_b - \pi \tag{14.78}$$

Note that α is actually equal to zero when α_a and α_b are the true values of the elevation angles in the LOS condition. Then, the two competing hypotheses are

$$
\begin{array}{lll}
H_\ell : & \hat{\alpha} = n_\alpha & \text{LOS condition} \\
H_n : & \hat{\alpha} = \xi + n'_\alpha & \text{NLOS condition}
\end{array}
\tag{14.79}
$$

where $\hat{\alpha}$ is the estimate of α, n_α and n'_α are the corresponding estimation errors under the LOS and NLOS conditions respectively and ξ is the extra angle resulting from the NLOS propagation. Since $0 \le \alpha_a, \alpha_b \le \pi, \xi$ is constrained by $-\pi \le \xi \le \pi$. For analytical simplicity, it is assumed that ξ is uniformly distributed in $[-\pi, \pi]$. For instance, this assumption would be approximately true for macrocellular environments. In other circumstances, more accurate distributions must be established in advance in order to use the identification methods presented in this section. Otherwise, alternative AOA-based techniques should be considered. Therefore, the results presented may be approximated as a lower bound or a reference for practical applications. n_α and n'_α are Gaussian distributed with means $\bar{n}_\alpha$ and $\bar{n}'_\alpha$ and variances $\sigma^2_{n_\alpha}$ and $\sigma^2_{n'_\alpha}$ respectively, and that ξ, n_α and n'_α are mutually independent. It is known that the PDF of the sum of the two random variables $\hat{\alpha} = \xi + n'_\alpha$ is equal to the convolution of the PDFs, $p_\xi(\xi)$ and $p_{n'_\alpha}(n'_\alpha)$, of the two random variables. That is:

$$
\begin{aligned}
p(\hat{\alpha}|H_n) &= \int_{-\infty}^{\infty} p_{n'_\alpha}(n'_\alpha) p_{\xi_\alpha}(\hat{\alpha} - n'_\alpha) dn'_\alpha \\
&= \frac{1}{2\pi} \int_{\hat{\alpha}-\pi}^{\hat{\alpha}+\pi} p_{n'_\alpha}(n'_\alpha) dn'_\alpha \\
&= \frac{1}{2\pi} \left[Q\left(\frac{\hat{\alpha} - \bar{n}'_\alpha - \pi}{\sigma_{n'_\alpha}} \right) - Q\left(\frac{\hat{\alpha} - \bar{n}'_\alpha + \pi}{\sigma_{n'_\alpha}} \right) \right]
\end{aligned}
\tag{14.80}
$$

where $Q(.)$ is the standard Q function. In this binary hypothesis test, two types of misjudgment may be made. One is that we decide H_ℓ but H_n is true with the probability denoted by $P(H_\ell, H_n)$, whereas the other is that we decide H_n but H_ℓ is true with the probability denoted by $P(H_n, H_\ell)$. The latter is referred to as the PFA,[5] denoted by P_{FA}. Note

---

[5] Here, we focus on the detection of the NLOS condition, so we call $P(H_n, H_\ell)$ the PFA and $P(H_n, H_n)$ the POD. In the event that identification of the LOS condition is the main issue, then $P(H_\ell, H_n)$ would be called the PFA and $P(H_\ell, H_\ell)$ the POD.

that these two probabilities cannot be reduced simultaneously. However, we can assign a small value to P_{FA} and then maximize the POD: $P(H_\mathrm{n}, H_\mathrm{n})$; that is, the probability of choosing H_n when H_n is true. According to the NP theorem, the POD is maximized for a given $P_{\mathrm{FA}} = \varepsilon$ when we decide H_n is true provided that

$$\frac{p(\hat{\alpha}|H_\mathrm{n})}{p(\hat{\alpha}|H_\ell)} > \kappa \tag{14.81}$$

where $p(\hat{\alpha}|H_\ell)$ is the PDF of n_α, which is Gaussian, and κ is a quantity that is dependent on the PFA. When the inequality in (14.81) is not valid, we decide H_ℓ is true. It may be impractical to derive a simple mathematical expression to describe the relation between κ and $\hat{\alpha}$ due to the complexity of the PDF in (14.80). However, we may use the fact that the PDF in (14.80) is rather flat in $[-\pi, \pi]$, whereas the PDF of n_α resembles a spike in the same region, provided that the angle errors n_α and n'_α are much smaller than π. Figure 14.7 shows an example of the two PDFs, $p(\hat{\alpha}|H_\ell)$ and $p(\hat{\alpha}|H_\mathrm{n})$, when $\bar{n}_\alpha = \bar{n}'_\alpha = 0$ and $\sigma_{n_\alpha} = \sigma_{n'_\alpha} = 10°$.

Note that in reality, the angle error would be a few degrees. Therefore, (14.81) could be approximated by

$$\begin{cases} \hat{\alpha} \leq \bar{n}_\alpha - \gamma \\ \hat{\alpha} \geq \bar{n}_\alpha + \gamma \end{cases} \tag{14.82}$$

where γ is determined by

$$\int_{-\infty}^{\bar{n}_\alpha - \gamma} p(\hat{\alpha}|H_\ell)\mathrm{d}\hat{\alpha} + \int_{\bar{n}_\alpha + \gamma}^{\infty} p(\hat{\alpha}|H_\ell)\mathrm{d}\hat{\alpha} = 2Q\left(\frac{\gamma}{\sigma_{n_\alpha}}\right) = \varepsilon \tag{14.83}$$

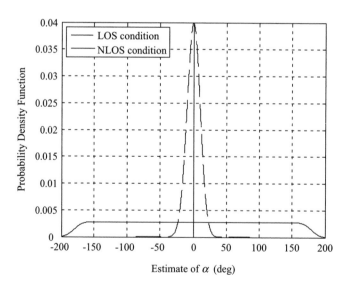

Figure 14.7 An example of $p(\hat{\alpha}|H_\ell)$ (dashed line) and $p(\hat{\alpha}|H_\mathrm{n})$ (solid line).

Since there is no closed-form solution to the Q function, γ can be determined numerically. For example, one may assign a group of possible values to γ, compute the corresponding Q function values and find out the Q function value that is closest to 0.5ε. Then, the corresponding γ value is chosen as the solution to (14.83). Note that the Q function can be computed using numerical integration and truncation. Accordingly, given γ, we can determine the POD as

$$P_{\text{D1}} = P(H_{\text{n}}, H_{\text{n}}) = \int_{-\infty}^{\bar{n}_\alpha - \gamma} p(\hat{\alpha}|H_{\text{n}})\mathrm{d}\hat{\alpha} + \int_{\bar{n}_\alpha + \gamma}^{\infty} p(\hat{\alpha}|H_{\text{n}})\mathrm{d}\hat{\alpha} \qquad (14.84)$$

Applying the properties of the Q function (see equation (3.58) in [21])

$$\int_0^{t_1} Q(t)\mathrm{d}t = t_1 Q(t_1) + \frac{1}{\sqrt{2\pi}}\left[1 - \exp\left(-\frac{t_1^2}{2}\right)\right] \qquad (14.85)$$

the POD can be derived as

$$P_{\text{D1}} = 1 - \frac{\sigma_{n'_\alpha}}{2\pi}\Bigg\{ g_1 Q(g_1) - g_2 Q(g_2) - g_3 Q(g_3) + g_4 Q(g_4)$$

$$- \frac{1}{\sqrt{2\pi}}\left[\exp\left(-\frac{g_1^2}{2}\right) - \exp\left(-\frac{g_2^2}{2}\right) - \exp\left(-\frac{g_3^2}{2}\right) + \exp\left(-\frac{g_4^2}{2}\right)\right] \Bigg\}$$

$$(14.86)$$

where

$$g_1 = \frac{\bar{n}_\alpha - \bar{n}'_\alpha - \pi + \gamma}{\sigma_{n'_\alpha}}$$

$$g_2 = \frac{\bar{n}_\alpha - \bar{n}'_\alpha - \pi - \gamma}{\sigma_{n'_\alpha}}$$

$$g_3 = \frac{\bar{n}_\alpha - \bar{n}'_\alpha + \pi + \gamma}{\sigma_{n'_\alpha}} \qquad (14.87)$$

$$g_4 = \frac{\bar{n}_\alpha - \bar{n}'_\alpha + \pi - \gamma}{\sigma_{n'_\alpha}}$$

Let us make use of Theorem 2 to deal with the identification problem. Also assume that the first received signal is not reflected by scatterers on the 2D surface enclosed by the two z-axes, so that satisfying (14.77) means the LOS condition. Similar to (14.78), let

$$\phi = \begin{cases} \phi_{\text{a}} - \phi_{\text{b}} - \pi & \phi_{\text{a}} > \phi_{\text{b}} \\ \phi_{\text{b}} - \phi_{\text{a}} - \pi & \phi_{\text{b}} > \phi_{\text{a}} \end{cases} \qquad (14.88)$$

The two hypotheses are now defined as

$$\begin{aligned} H_\ell : & \quad \hat{\phi} = n_\phi & \text{LOS condition} \\ H_{\text{n}} : & \quad \hat{\phi} = \eta + n'_\phi & \text{NLOS condition} \end{aligned} \qquad (14.89)$$

where $\hat{\phi}$ is the estimate of ϕ and n_ϕ and n'_ϕ are the corresponding estimation errors under the LOS and NLOS conditions respectively. η is the angle bias caused by the scatterer and is constrained by $-\pi \leq \eta \leq \pi$. The azimuth angle estimation errors are modeled as Laplacian distributed random variables based on measurements in both indoor and urban environments [22]. That is:

$$
\begin{aligned}
p_{n_\phi}(n_\phi) &= \frac{1}{\sqrt{2}\sigma_{n_\phi}} \exp\left(-\frac{\sqrt{2}|n_\phi|}{\sigma_{n_\phi}}\right) \\
p_{n'_\phi}(n'_\phi) &= \frac{1}{\sqrt{2}\sigma_{n'_\phi}} \exp\left(-\frac{\sqrt{2}|n'_\phi|}{\sigma_{n'_\phi}}\right)
\end{aligned}
\tag{14.90}
$$

Assuming that η is uniformly distributed in $[-\pi, \pi]$, similar to (14.80), the PDF of $\hat{\phi} = \eta + n'_\phi$ can be derived as

$$
\begin{aligned}
p(\hat{\phi}|H_n) = \frac{1}{4\pi} \Bigg\{ &\mathrm{sgn}(\hat{\phi}+\pi)\left[1-\exp\left(-\frac{\sqrt{2}|\hat{\phi}+\pi|}{\sigma_{n'_\phi}}\right)\right] \\
&- \mathrm{sgn}(\hat{\phi}-\pi)\left[1-\exp\left(-\frac{\sqrt{2}|\hat{\phi}-\pi|}{\sigma_{n'_\phi}}\right)\right] \Bigg\}
\end{aligned}
\tag{14.91}
$$

where $\mathrm{sgn}(.)$ is the sign function

$$
\mathrm{sgn}(t) = \begin{cases} 1 & t > 0 \\ 0 & t = 0 \\ -1 & t < 0 \end{cases}
\tag{14.92}
$$

Also, making use of the NP theorem, we decide H_n when

$$
\frac{p(\hat{\phi}|H_n)}{p(\hat{\phi}|H_\ell)} > \kappa
\tag{14.93}
$$

is satisfied. Similarly, (14.93) can be equivalent to

$$
\begin{cases} \hat{\phi} \leq -\gamma \\ \hat{\phi} \geq \gamma \end{cases}
\tag{14.94}
$$

where γ is determined by solving

$$
\begin{aligned}
\varepsilon &= \int_{-\infty}^{-\gamma} p(\hat{\phi}|H_\ell)\,d\hat{\phi} + \int_{\gamma}^{\infty} p(\hat{\phi}|H_\ell)\,d\hat{\phi} \\
&= 2\int_{\gamma}^{\infty} p(\hat{\phi}|H_\ell)\,d\hat{\phi} \\
&= \exp\left(-\frac{\sqrt{2}\gamma}{\sigma_{n_\phi}}\right)
\end{aligned}
\tag{14.95}
$$

That is

$$\gamma = -\frac{\sigma_{n_\phi}}{\sqrt{2}} \ln \varepsilon \tag{14.96}$$

Therefore, when given γ, the POD can be computed by

$$
\begin{aligned}
P_{D2} &= \int_{-\infty}^{-\gamma} p(\hat{\phi}|H_n)d\hat{\phi} + \int_{\gamma}^{\infty} p(\hat{\phi}|H_n)d\hat{\phi} \\
&= 2\int_{\gamma}^{\infty} p(\hat{\phi}|H_n)d\hat{\phi} \\
&= [\mathrm{sgn}(\gamma + \pi) - \mathrm{sgn}(\gamma - \pi)]\frac{\pi - \gamma}{2\pi} + \frac{\sigma_{n'_\phi}}{2\sqrt{2\pi}}\left[\exp\left(-\frac{\sqrt{2}|\gamma - \pi|}{n'_\phi}\right)\right. \\
&\quad \left. - \exp\left(-\frac{\sqrt{2}|\gamma + \pi|}{n'_\phi}\right)\right]
\end{aligned} \tag{14.97}
$$

Now, let us relax the restrictions that the first received signal is not reflected by scatterers either on the 3D surface defined by (14.76) or on the 2D surface enclosed by the two z-axes. As a result, the above derived PFA and POD results are for deciding whether (14.75) or (14.77) is satisfied or not. According to the Corollary 1, the NLOS condition is decided when both (14.75) and (14.77) are satisfied. Otherwise, the LOS condition is assumed. Therefore, the joint POD is given by

$$P_D = P_{D1}P_{D2} \tag{14.98}$$

and the resulting PFA is

$$P_{FA} = \varepsilon^2 \tag{14.99}$$

Let us consider an example to show the performance of the 2D angle-based identification. The means of the azimuth and elevation angular errors in both the LOS and NLOS conditions are set to zero. Four different STDs ($2°, 5°, 8°$ and $11°$) of the angular errors in LOS conditions are examined, and the STD of the angular errors in NLOS conditions is set at 30% larger than that in the LOS conditions. One hundred thousand samples of $\hat{\alpha}$ and $\hat{\phi}$ are generated. Figure 14.8 shows the POD with respect to the PFA, in which the analytical POD denoted by a solid line is generated from (14.98), whereas the simulated results denoted by circles are produced by making decisions based on (14.82) and (14.94). From the figure, it can be seen that, when setting the PFA to 2.5%, the POD is about 85%, 88%, 92.5% and 97% under the four STDs. To achieve a POD of 90%, the PFA is required to be greater than about 0.1%, 0.8%, 5% and 12.5%, respectively.

Figure 14.9 shows the impact of the model parameter errors on the accuracy of the azimuth and elevation angle-based method. To include the model errors, the noise STDs in both LOS and NLOS conditions are assumed to be time varying and uniformly distributed between $(1 - q) \times$ (estimated value) and $(1 + q) \times$ (estimated value). The constant q is set to three different values: 0, 0.4 and 0.7. Also, a Gaussian random variable is added to the uniformly

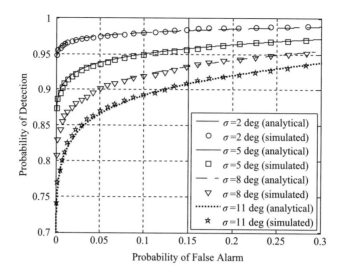

Figure 14.8 POD versus PFA of the azimuth and elevation angle-based NLOS identification. Four different angle error STDs ($\sigma_{n_\alpha} = \sigma_{n_\phi} = \sigma = 2°, 5°, 8°$ and $11°$) are examined.

distributed angle biases, with zero mean and STDs $0°$, $8°$ and $14°$, corresponding to the three q values respectively. Clearly, even when there are relatively large errors in the distribution parameters, the impact of the errors on the identification accuracy is marginal. However, as mentioned previously, in the presence of modeling errors or statistic parameter estimation

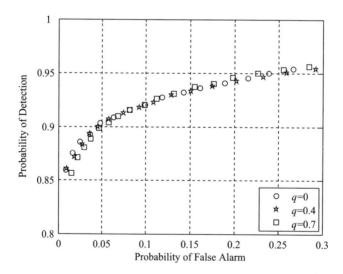

Figure 14.9 Impact of the model parameter errors when the true noise STD ranges from $(1 - q) \times$ (estimated value) to $(1 + q) \times$ (estimated value). The estimated STDs of the angular measurement noise in LOS and NLOS conditions are set to $8°$ and $1.3 \times 8°$ respectively.

error, the theoretical PFA and POD would be either overestimated or underestimated. Therefore, uncertainty should be taken into account to make sure the desired PFA and POD can be achieved.

14.10.3 Identification Based on Azimuth Angle

In reality some nodes may produce both the azimuth and elevation angular measurements, whereas the others may only be able to provide azimuth angular measurements. In the event that there are only azimuth angular measurements at both communication nodes, we may solely rely on (14.77) to judge the NLOS conditions. In this case the performance will depend on the probability that the first received signal has been reflected by a scatterer located in the plane enclosed between the two z-axes. Let us define

- $P(H_\ell)$: probability of the LOS condition
- $P(H_n)$: probability of the NLOS condition
- $P(Z|H_\ell)$: probability that (14.77) is satisfied under the LOS condition
- $P(Z|H_n)$: probability that (14.77) is satisfied under the NLOS condition

where the probabilities satisfy

$$P(H_\ell) + P(H_n) = 1 \qquad P(Z|H_\ell) = P(H_\ell) \tag{14.100}$$

and typically

$$P(Z|H_\ell) \gg P(Z|H_n) \tag{14.101}$$

Since NLOS conditions may also satisfy (14.77), the decision threshold would be different from that given by (14.96). Giving the probabilities $P(H_\ell)$ and $P(Z|H_n)$ and the PFA ε, we have

$$\begin{aligned} \varepsilon &= \frac{P(H_\ell)}{P(H_\ell) + P(Z|H_n)} \left[\int_{-\infty}^{-\gamma} p(\hat{\phi}|H_\ell)\mathrm{d}\hat{\phi} + \int_{\gamma}^{\infty} p(\hat{\phi}|H_\ell)\mathrm{d}\hat{\phi} \right] \\ &= \frac{P(H_\ell)}{P(H_\ell) + P(Z|H_n)} \exp\left(-\frac{\sqrt{2}\gamma}{\sigma_{n_\phi}} \right) \end{aligned} \tag{14.102}$$

resulting in

$$\gamma = -\frac{\sigma_{n_\phi}}{\sqrt{2}} \ln\left[1 + \frac{P(Z|H_n)}{P(H_\ell)} \right] \varepsilon \tag{14.103}$$

Accordingly, the POD becomes

$$P''_{\mathrm{D2}} = \frac{P(H_n) - P(Z|H_n)}{P(H_n)} P'_{\mathrm{D2}} + \Delta \tag{14.104}$$

where P'_{D2} results from substituting γ in (14.103) into (14.97) and

$$
\begin{aligned}
\Delta &= \frac{P(Z|H_n)}{P(H_n)} \left\{ \int_{-\infty}^{-\gamma} p(\hat{\phi}|H_\ell)\mathrm{d}\hat{\phi} + \int_{\gamma}^{\infty} p(\hat{\phi}|H_\ell)\mathrm{d}\hat{\phi} \right\} \\
&= \frac{P(Z|H_n)}{P(H_n)} \exp\left(-\frac{\sqrt{2}\gamma}{\sigma_{n_\phi}} \right)
\end{aligned}
\tag{14.105}
$$

Another practical issue is that there exists a mismatch between the noise distribution model and the true measurement noise in general. Typically the distribution of the measurement noise or the scatterers is approximate, not strictly fitting to a specific distribution. Another possibility is that the variances of the angular measurement noise are either overestimated or underestimated. Accordingly, the decision threshold based on the estimated variance would not produce the predefined PFA and POD. Therefore, it is necessary to pay attention to the uncertainties of the variances to achieve both the desirable POD and PFA.

The performance of the 1D angle-based identification are illustrated by Figure 14.10. Since (14.77) is not a sufficient condition for the LOS condition, the accuracy of the method relies on how often the NLOS propagation satisfies (14.77). In this plot, four cases are examined. That is, the probability $P(Z|H_n)$ is equal to 0%, 1%, 3% and 5%. For comparison, the results using both azimuth and elevation angle measurements are also shown. It can be seen that when the probability $P(Z|H_n)$ is less than 1%, the identification accuracy would be very similar when using either azimuth angle or both azimuth and elevation angle measurements. In this circumstance the azimuth angle-based method would be more desirable, since it is more feasible to make 1D angular measurements in practice.

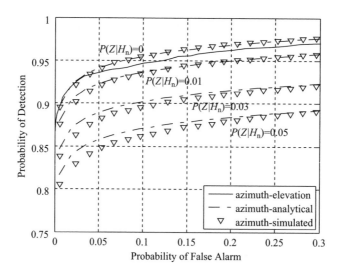

Figure 14.10 POD versus PFA of the azimuth angle-based method. 'azimuth-elevation' denotes results based on both azimuth and elevation angle measurements. 'azimuth-analytical' stands for the analytical results based on (14.104). 'azimuth-simulated' denotes the simulated results based on whether the threshold is crossed. The STD σ_{n_ϕ} of the angle measurement errors is 5°.

Annex 14.A: Proofs of Theorems and Corollary

Proof of Theorem 1 As shown in Figure 14.5, when the direct path between A and B is free of blockage, it is clear that the sum of the two elevation angles is equal to π due to the fact that the pair of angles is supplementary; that is, (14.75) is satisfied. In the NLOS situation, for instance, the scatterer located at (x_s, y_s, z_s) results in the elevation angles α'_a and α'_b for nodes A and B respectively, which are determined by

$$
\begin{aligned}
\tan \alpha'_a &= \frac{z_s - z_a}{\sqrt{(x_s - x_a)^2 + (y_s - y_a)^2}} \\
\tan \alpha'_b &= \frac{z_s - z_b}{\sqrt{(x_s - x_b)^2 + (y_s - y_b)^2}}
\end{aligned}
\tag{14.106}
$$

When $\alpha'_a + \alpha'_b = \pi$, we obtain $\tan \alpha'_a = -\tan \alpha'_b$. Consequently, we obtain (14.76) and Theorem 1 is proved. Geometrically, by rotating line AB around axis Z_a, we obtain a cone-shaped surface, called cone a. Any scatterer on cone a will produce an elevation angle equal to α_a. Similarly, we can have cone b on which any scatterer will yield an elevation angle equal to α_b. Clearly, the intersection of the two cones is the line AB under the LOS condition. Widening the two cones simultaneously and keeping (14.75) satisfied, they intersect each other until the two surfaces become flat and parallel to each other. The intersection is determined by (14.76). In the event of $z_a = z_b$, (14.76) defines the whole $X_a Y_a$ or $X_b Y_b$ plane.

Proof of Theorem 2 As shown in Figure 14.6, there are two nodes located at points A and B. Under the LOS condition, the azimuth angles are ϕ_a and ϕ_b respectively and they satisfy (14.77) because ϕ_a and $2\pi - \phi_b$ are supplementary angles. Under the NLOS condition, will the pair of azimuth angles also satisfy (14.77)? When the signal is reflected at the scatterer S before arriving at points A or B, the corresponding azimuth angles are ϕ'_a and ϕ'_b respectively and

$$
\begin{aligned}
\tan \phi'_a &= \frac{z_s - z_a}{x_s - x_a} \\
\tan \phi'_b &= \frac{z_s - z_b}{x_s - x_b}
\end{aligned}
\tag{14.107}
$$

Using the trigonometric relations

$$
\tan^{-1}\kappa_1 - \tan^{-1}\kappa_2 = \{0, \pi, -\pi\} + \tan^{-1}\frac{\kappa_1 - \kappa_2}{1 + \kappa_1\kappa_2}
\tag{14.108}
$$

where choosing which of the three constants depends on κ_1 and κ_2, and assuming

$$
\phi'_a - \phi'_b = \pm\pi
\tag{14.109}
$$

we obtain

$$
(x_s - x_a)(y_s - y_b) - (x_s - x_b)(y_s - y_a) = 0
\tag{14.110}
$$

which can be written as

$$
(y_b - y_a)x_s - (x_b - x_a)y_s + x_a y_b - x_b y_a = 0
\tag{14.111}
$$

Clearly, (14.111) is a linear equation with respect to the scatterer coordinates, so that the propagation is along a straight line. On the X_aY_a plane, this line includes the segment AB', while on the X_bY_b plane this line includes the segment A'B. Clearly, when the projections of S on the X_aY_a and X_bY_b planes are on the segments AB' and A'B respectively, this is equivalent to the LOS condition. However, when the projections of S on the two planes are on the extensions from either end of the segments AB' and A'B, then the resulting two azimuth angles are equal, since they are corresponding angles, so that (14.77) is not satisfied. That is, in the 3D space, the azimuth angles caused by any point/scatterer on the plane only enclosed by the Z_a and Z_b axes satisfy (14.77) or (14.109). This proves Theorem 2.

Proof of Corollary 1 Clearly, Corollary 1 is the direct result from Theorem 1 and Theorem 2. The common part of the uncertain regions in Theorem 1 and Theorem 2 is the line segment AB, so that satisfying (14.75) and (14.77) means the LOS condition.

Annex 14.B: Derivation of the Probability of Detection

In Section 14.9.1, the POD when using the joint TOA and RSS-based identification is presented. Here, let us show how to derive the POD. For notational convenience, let us deal with the integral given by

$$I = \int_{v'_{nlos}}^{\infty} e^{a_1 t} Q(a_2 t + a_3) dt \tag{14.112}$$

which can be determined as follows:

$$I = \int_{v'_{nlos}}^{\infty} e^{a_1 t} Q(a_2 t + a_3) d\left(\frac{1}{a_1} e^{a_1 t}\right)$$
$$= \frac{1}{a_1} e^{a_1 t} Q(a_2 t + a_3)\Big|_{\gamma}^{\infty} - \frac{1}{a_1} \int_{v'_{nlos}}^{\infty} e^{a_1 t} d(Q(a_2 t + a_3)) \tag{14.113}$$

Making use of the formulae of differentiation under the integral sign:

$$\frac{d}{dt} F(t) = f(t, g_u(t)) \frac{dg_u(t)}{dt} - f(t, g_\ell(t)) \frac{dg_\ell(t)}{dt} + \int_{g_\ell(t)}^{g_u(t)} \frac{\partial}{\partial t} f(t, s) ds \tag{14.114}$$

where

$$F(t) = \int_{g_\ell(t)}^{g_u(t)} f(t, s) ds \tag{14.115}$$

we obtain

$$d(Q(a_2t + a_3)) = -\frac{a_2}{\sqrt{2\pi}} \exp\left[-\frac{(a_2t + a_3)^2}{2}\right] dt \tag{14.116}$$

Then:

$$
\begin{aligned}
I &= \frac{1}{a_1} e^{a_1\gamma} Q(a_2\gamma + a_3) + \frac{a_2}{a_1\sqrt{2\pi}} \int_{v'_{nlos}}^{\infty} \exp\left(-\frac{a_2t + a_3}{2} + a_1t\right) dt \\
&= \frac{1}{a_1} \left\{ -e^{a_1\gamma} Q(a_2\gamma + a_3) + \exp\left[-\frac{a_1}{a_2}\left(a_3 - \frac{a_1}{2a_2}\right)\right] Q\left(a_2\gamma + a_3 - \frac{a_1}{a_2}\right) \right\}
\end{aligned}
\tag{14.117}
$$

Letting

$$a_1 = \frac{1}{\lambda_s} \quad a_2 = \frac{1}{\sigma_u} \quad a_3 = \beta_2 \tag{14.118}$$

we can readily obtain the fourth equation in (14.74).

References

[1] M.P. Wylie and J. Holtzmann, 'The non-line of sight problem in mobile location estimation', in *Proceedings of IEEE Conference on Universal Personal Communications*, pp. 827–831, 1996.

[2] J.G. Proakis, *Digital Communications*, McGraw-Hill, New York, 1995.

[3] M.A. Stephens, 'EDF statistics for goodness of fit and some comparisons', *Journal of the American Statistical Association*, **69**(347), 1974, 730–737.

[4] S. Shapiro and M. Wilk, 'An analysis of variance test for normality (complete samples)', *Biometrika*, **52**(3–4), 1965, 591–611.

[5] S. Venkatraman and J. Caffery Jr, 'A statistical approach to non-line-of-sight BS identification', in *Proceedings of International Symposium on Wireless Personal Multimedia Communications*, pp. 296–300, October 2002.

[6] S. Al-Jazzar and J. Caffery Jr, 'ML and Bayesian TOA location estimators for NLOS environments', in *Proceedings of IEEE Vehicular Technology Conference (VTC)*, pp. 1178–1181, September 2002.

[7] M. Schwartz, W.R. Bennett and S. Stein, *Communication Systems and Techniques*, new edition, IEEE Press, 1995.

[8] F. Benedetto, G. Giunta, A. Toscano and L. Vegni, 'Dynamic LOS/NLOS statistical discrimination of wireless mobile channels', in *Proceedings of IEEE Vehicular Technology Conference (VTC)*, pp. 3071–3075, April 2007.

[9] C. Tepedelenlioglu, A. Abdi and G.B. Giannakis, 'The Rician K factor: estimation and performance analysis', *IEEE Transactions on Wireless Communications*, **2**(4), 2003, 799–810.

[10] T.L. Marzetta, 'EM algorithm for estimating the parameters of a multivariate complex Rician density for polarimetric SAR', in *Proceedings of IEEE International Conference on Acoustics, Speech, and Signal Processing (ICASSP)*, pp. 3651–3654, May 1995.

[11] C. Tepedelenlioglu and G.B. Giannakis, 'On velocity estimation and correlation properties of narrow-band mobile communication channels', *IEEE Transactions on Vehicular Technology*, **50**(4), 2001, 1039–1052.

[12] L.J. Greenstein, D.G. Michelson and V. Erceg, 'Moment-method estimation of the Rician K factor', *IEEE Communications Letters*, **3**(6), 1999, 175–176.

[13] J. Borras, P. Hatrack and N.B. Mandayam, 'Decision theoretic framework for NLOS identification', in *Proc. IEEE Vehicular Technology Conf. (VTC)*, pp. 1583–1587, May 1998.

[14] S. Gezici, H. Kobayashi, and H.V. Poor, 'Non-parametric non-line-of-sight identification', in *Proceedings of IEEE Vehicular Technology Conference (VTC)*, pp. 2544–2548, October 2003.

[15] C.-H. Lim, A.M. Zoubir, C.-M.S. See and B.-P. Ng, 'A robust statistical approach to non-line-of-sight mitigation', *Proc. IEEE Workshop on Statistical Signal Processing*, pp. 428–432, Aug. 2007.

[16] L. Xiong, 'A selective model to suppress NLOS signals in angle-of-arrival (AOA) location estimation', in *Proceedings of IEEE International Symposium on Personal, Indoor and Mobile Radio Communications (PIMRC)*, pp. 461–465, September 1998.

[17] L. Cong and W. Zhuang, 'Non-line-of-sight error mitigation in TDOA mobile location', in *Proceedings of IEEE Global Telecommunications Conference (GLOBECOM)*, pp. 680-684, November 2001.

[18] Y.-T. Chan, W.-Y. Tsui, H.-C. So and P.-C. Ching, 'Time-of-arrival based localization under NLOS conditions', *IEEE Transactions on Vehicular Technology*, **55**(1), 2006, 17–24.

[19] E. Damosso (ed.), Digital mobile radio towards future generation systems. COST 231 Final Report, 1996.

[20] K.I. Pedersen, P.E. Mogensen and B.H. Fleury, 'A stochastic model of the temporal and azimuthal dispersion seen at the base station in outdoor propagation environments', *IEEE Transactions on Vehicular Technology*, **49**(2), 2000, 437–447.

[21] S. Verdu, *Multiuser Detection*. Cambridge University Press, Cambridge, UK, 1998.

[22] Q. Spencer, M. Rice, B. Jeffs and M. Jensen, 'A statistical model for angle of arrival in indoor multipath propagation', in *Proceedings of IEEE Vehicular Technology Conference (VTC)*, pp. 1415–1419, May 1997.

Appendix A: Hyperbolic Navigation

This appendix provides an analytical analysis of the hyperbolic navigation problem for the 2D case with three base stations. The problem may be defined either as a navigation problem (receiver in the mobile device) or as a tracking problem (transmitter in the mobile device). In either case, it is assumed that the receiver measures the pseudo-range to the three base stations and from these data the position of the mobile device is sought.

From the definition of the pseudo-ranges, the differential pseudo-ranges are equal to the differential ranges to the base stations. Thus, for any two base stations (a and b), the differential range can be estimated from the measured pseudo-ranges by

$$R_a - R_b = P_a - P_b \tag{A.1}$$

From the properties of a hyperbola, the locus of positions defined by (A.1) is a hyperbola. Similarly, using another pair of base stations (say a and c), another hyperbola is defined. The position of the mobile node is thus defined by the intersection of the two hyperbolas. Thus, the problem of solving the hyperbolic navigation problem is equivalent to solving the general case of the intersection of hyperbolas.

The intersection of two hyperbolas can potentially result in three outcomes. In general, there can be two, one or no intersection points; the latter case could occur if there are measurement errors. Thus, if analytical geometry is used, it is expected that the solution will involve a quadratic equation with two, one or no (real) solutions.

A.1 Analytical Equations of a Hyperbola

The following is a brief review of the properties of hyperbolas. In classical geometry the hyperbola can be defined by a slanted plane cutting a cone. (The circle, ellipse and parabola are three other conic sections.) However, the following analysis uses analytical geometry techniques.

The usual description of a hyperbola in Cartesian coordinates is

$$\left(\frac{x}{a}\right)^2 - \left(\frac{y}{b}\right)^2 = 1 \tag{A.2}$$

The geometry and the axes are shown in Figure A.1. The hyperbola cuts the x-axis at $x = \pm a$, so that there are two branches: one to the right of the y-axis and one to the left (mirror image). In the following analysis, only the branch with $x > 0$ is considered. At large distances from the

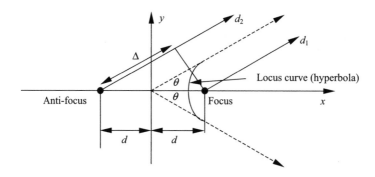

Figure A.1   Geometry of the differential definition of a hyperbola. The asymptotic straight lines are shown above and below the x-axis.

focus point the hyperbola is asymptotic to straight lines at angles relative to the x-axis of

$$\theta = \pm\cos^{-1}\left(\frac{\Delta}{2d}\right) \qquad (0 \le \Delta \le 2d) \tag{A.3}$$

where Δ is the differential distance as measured from the two foci and $2d$ is the separation of the foci. If the differential distance is zero, then the hyperbola is reduced to the straight line $x = 0$ (y-axis); and when the differential distance is the maximum $2d$, the hyperbola is reduced to the straight line $y = 0$ (x-axis). For other differential distances the hyperbola is defined by the equation $d_1 - d_2 = \Delta$.

Noting that using Cartesian coordinates the two distances (d_1 and d_2) involve square roots, the equation $d_1 - d_2 = \Delta$ must be squared to transform it into a standard form. The resulting expression $d_1^2 + d_2^2 - \Delta^2 = 2d_1 d_2$ also involves square roots and, thus, must also be squared, resulting in

$$d_1^4 + d_2^4 + \Delta^4 = 2[d_1^2 d_2^2 + \Delta^2(d_1^2 + d_2^2)] \tag{A.4}$$

Upon substituting $d_1^2 = (x+d)^2 + y^2$ and $d_2^2 = (x-d)^2 + y^2$ and simplifying the resulting expression, (A.4) becomes the standard form

$$\left(\frac{x}{a}\right)^2 - \left(\frac{y}{b}\right)^2 = 1$$
$$a = \frac{\Delta}{2} \qquad b = \sqrt{d^2 - a^2} \qquad (a \le d) \tag{A.5}$$

The eccentricity e of the hyperbola is related to the hyperbola parameters by the expressions

$$e = \frac{2d}{\Delta} \ge 1$$
$$b = a\sqrt{e^2 - 1} \tag{A.6}$$

Thus the differential distance definition of the hyperbola has been reduced to the standard Cartesian coordinate form, with the parameters related to the separation of the foci and the differential distance parameter.

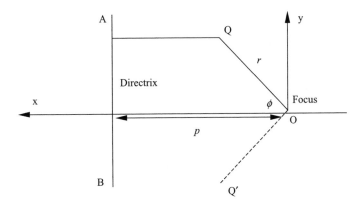

Figure A.2 Geometry of arbitrary conic section, showing the focus and directrix.

Another method of defining an arbitrary conic section (and a hyperbola in particular) is illustrated in Figure A.2. The conic section can be defined using polar coordinates, such that the ratio of the radial distance r to a point Q from the focus to the perpendicular distance from Q to the directrix AB is a constant, namely the eccentricity. Thus, the equation defining the locus is

$$e = \frac{r}{AQ} = \frac{r}{p - r\cos\phi} \tag{A.7}$$

Note that the point Q′ is also defined by the above ratio; thus, the locus is symmetrical about the x-axis. Equation (A.7) can be solved for the radial distance, resulting in

$$r = \frac{p}{1/e + \cos\phi} \tag{A.8}$$

Alternatively, this polar equation can be converted to Cartesian coordinates (with $e > 1$), which results in the equation of a hyperbola:

$$\left(\frac{x'}{\alpha}\right)^2 - \left(\frac{y}{\beta}\right)^2 = 1$$
$$\alpha = \frac{ep}{e^2 - 1} \qquad \beta = \alpha(\sqrt{e^2 - 1}) \tag{A.9}$$
$$x' = x - \frac{pe^2}{e^2 - 1}$$

Thus, the polar definition reduces to the standard Cartesian coordinate representation, with the parameters defined in terms of the eccentricity and the directrix distance. The x-axis for the standard representation is shifted by a distance $pe^2/(e^2 - 1)$. By comparing (A.5) and (A.6), the directrix distance p is given by

$$p = d(1 - 1/e^2) = \frac{\Delta}{2}e(1 - 1/e^2) \tag{A.10}$$

As the directrix distance can be expressed in terms of either the distance between the foci or the differential distance (plus the eccentricity), the polar form can be derived from the Cartesian coordinate representation parameters, and vice versa.

A.2 Solution to Hyperbolic Navigation

If the system consists of three base stations, then the differential ranges defined by (A.1) will result in two hyperbolas. The solution of the hyperbolic navigation problem basically requires the determination of the location where these two hyperbolas intersect. An analytical method based on the Cartesian coordinate form is straightforward, but results in a fourth-order equation. While an analytical solution to fourth-order polynomial equations exists, the resulting expression is complex and, thus, difficult to use in practice. However, as two hyperbolas intersect at only two locations, there is a potential that another analytical approach will result in a second-order equation. The following analysis is such a method.

The following analysis uses the polar definitions for hyperbolas, as it avoids the troublesome square root of the Cartesian solution. In particular, if the base station common to both pairs is used, then an arbitrary hyperbola can be defined by a simple polar rotation. (The corresponding rectangular coordinate system results in much more complex expressions.) With reference to Figure A.3, if the line joining the base stations is rotated an angle α relative to the x-axis, then the hyperbola can be expressed in the polar form as

$$r = \frac{p}{1/e - \cos(\phi - \alpha)} = \frac{D}{2} \frac{1 - 1/e^2}{1/e - \cos(\phi - \alpha)}$$
$$p = \frac{D}{2}(1 - 1/e^2)$$
(A.11)

In (A.11), e is the eccentricity of the hyperbola and is given by $e = D/\Delta \geq 1$. (See also (A.6) and (A.10).) Thus, the hyperbola is completely defined by three parameters (D, α, e). The location of the base stations defines the separation D and the orientation α, while the differential distance Δ defines the shape (eccentricity). As shown by (A.1), this latter parameter can be estimated from the differential pseudo-ranges.

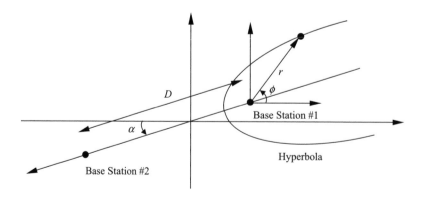

Figure A.3 Geometry of the hyperbola, showing the polar coordinates and the rotation by angle α.

As the radial distance r becomes large ($r \gg D$), then, from (A.11), the angle ϕ approaches a constant limit given by

$$\phi \to \alpha + \cos^{-1}(1/e) \tag{A.12}$$

(See also (A.3).) Thus, as required for a hyperbola, the curve becomes asymptotic to a straight line, the slope of which is governed by the eccentricity. If the differential distance Δ approaches zero, then the eccentricity approaches infinity and the equation approaches

$$r \to \frac{-D}{2\cos(\phi - \alpha)} \tag{A.13}$$

This expression is the equation of a straight line bisecting the line joining the base stations at right angles.

The location is determined by the intersection of two hyperbolas. Expressions in the form of (A.11) apply to the two hyperbolas. At their intersection, clearly $r_1 = r_2$ and $\phi_1 = \phi_2$. By equating the radial distances at the intersection point, the following equation as a function of the angle θ is obtained:

$$\frac{p_1}{1/e_1 - \cos(\phi - \alpha_1)} = \frac{p_2}{1/e_2 - \cos(\phi - \alpha_2)} \tag{A.14}$$

Rearranging results in

$$c = \frac{p_1}{e_2} - \frac{p_2}{e_1} = p_1\cos(\phi - \alpha_2) - p_2\cos(\phi - \alpha_1) \tag{A.15}$$

Expanding the cosine terms, and setting $u = \cos\phi$, the following quadratic equation results:

$$\begin{aligned} a &= p_1\cos\alpha_2 - p_2\cos\alpha_1 \\ b &= p_1\sin\alpha_2 - p_2\sin\alpha_1 \\ c &= au \pm b\sqrt{1 - u^2} \end{aligned} \tag{A.16}$$

The solution to this quadratic equation (A.16) in u is

$$u = \frac{ac \pm b\sqrt{a^2 + b^2 - c^2}}{a^2 + b^2} = \cos\phi_1 = \cos\phi_2 \tag{A.17}$$

In general, there are two solutions for $\cos\phi$, but in most cases one of these solutions can be eliminated. The sine of the angle will have an ambiguity in the sign, so that $s = \sin\phi = \pm\sqrt{1 - u^2}$. The radius r is given by

$$r = \frac{p}{1/e - s\cos\alpha \pm \sqrt{1 - s^2}\sin\alpha} \tag{A.18}$$

The radius r as given be (A.18) will have four-way ambiguity for both r_1 and r_2. To resolve these ambiguities, the following tests are made to eliminate all the false solutions

$$\begin{aligned} |s_1| &\leq 1 \quad |s_2| \leq 1 \\ r_1 &> 0 \quad\quad r_2 > 0 \\ r_1 &= r_2 \end{aligned} \tag{A.19}$$

Using the constraints defined in (A.19), the four-way ambiguity can be reduced to the single solution (r, ϕ), or at most two solutions. In the latter case, no further elimination of ambiguity is possible, as the two hyperbolas intersect at two points.

A.3 Solution to Example Problem

To illustrate the application of the above solution to the hyperbolic navigation problem, consider the example as summarized in Figure A.4. Note that the distances (dotted lines) from the mobile point P to the base stations are pseudo-ranges, not the physical ranges. The base stations form an equilateral triangle of side 10 m.

The solution described in Section A.2 can now be applied to the example problem. For this particular case, the coordinates of the fixed points (base stations) are $(0,0)$, $(10,0)$ and $(5, 5\sqrt{3})$. The differential distances are $AP - BP = 3$, and $CP - PB = -6$. The separation D is 10 in both cases, namely AB and BC. As B is the common point, this is the origin for this solution, with the x-axis in the direction BA.

Thus, the corresponding eccentricities $(e = D/|\Delta|)$ are 10/3 and 10/6 and, from (A.11), the directrix distances p are given by

$$p_1 = \frac{10}{2}\left[1 - \left(\frac{3}{10}\right)^2\right] = \frac{91}{20} \qquad p_2 = \frac{10}{2}\left[1 - \left(\frac{6}{10}\right)^2\right] = \frac{16}{5} \qquad \text{(A.20)}$$

Next the rotations of the two hyperbolas are calculated. With B as the origin, the vector BA has an angle π and the vector BC has the angle $2\pi/3$. Thus, the parameters in (A.16) are

$$a = \frac{-p_1}{2} + p_2 = \frac{37}{40} \qquad b = \frac{\sqrt{3}}{2}p_1 = \frac{91\sqrt{3}}{40} \qquad c = -\left(\frac{6p_1 + 3p_2}{10}\right) = -\frac{369}{100} \quad \text{(A.21)}$$

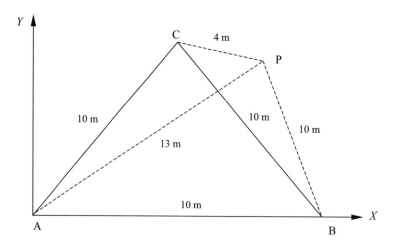

Figure A.4 Geometry of the problem. Note the (dashed) distances AP, BP and CP are the measured pseudo-ranges, not the actual distances. The other distances shown are true distances.

The solution to the quadratic equation (see (A.17)) is

$$u = \frac{ac \pm d}{a^2 + b^2}$$

$$d = b\sqrt{a^2 + b^2 - c^2} = \frac{91}{1000}\sqrt{5187}$$

$$a^2 + b^2 = \frac{6553}{400} \qquad ac = \frac{13\,653}{4000}$$

$$u = \pm\frac{182}{32\,765}\sqrt{5187} - \frac{13\,653}{65\,530}$$

(A.22)

In (A.22) there are two possible solutions for u, but only one is the solution to this particular problem, namely

$$u = \cos\phi = \frac{182}{32\,765}\sqrt{5187} - \frac{13\,653}{65\,530} \approx 0.192 \qquad (A.22)$$

The radial distance to the required point P can now be evaluated using (A.18). Using the AB pair of points, the radial distance from B is given by

$$r = \frac{p_1}{1/e_1 + s} = \frac{6553}{4(33 + 2\sqrt{5187})} = 9.253 = 10 - 0.747 \qquad (A.23)$$

The final expression for the radial distance is in the form $r = P_b - R_c$, so that the pseudo-range constant is given by

$$R_c = \frac{80\sqrt{5187} - 5233}{4(33 + 2\sqrt{5187})} = \frac{51}{4} - \frac{\sqrt{5187}}{6} \approx 0.747 \qquad (A.24)$$

Finally, the position of the point P can be determined from the polar coordinates (r, ϕ). As the required origin is at A (not B), the P(X, Y) is given by

$$X = P_b - r\cos\phi = 10 - ru = \frac{26\,853 + 436\sqrt{5187}}{40(33 + 2\sqrt{5187})} = 8.226$$

$$Y = r\sin\phi = r\sqrt{1 - u^2} = \frac{33\,579\sqrt{3} + 148\sqrt{1729}}{40(33 + 2\sqrt{5187})} = 9.082$$

(A.25)

Appendix B: Radio Propagation Measurement Techniques

The development of mathematical models of radio propagation in a severely scattering environment as described in Chapter 2 relies heavily on measurements. This appendix describes some measurement techniques which allow both the loss and the delay excess characteristics to be determined.

There are fundamentally two approaches to measuring the radio propagation characteristics: in the time domain or the frequency domain. In theory, as measurement in one domain can be converted to the other domain using a Fourier transform (or its inverse), the choice of method is more related to practical considerations associated with the equipment. The most flexible method is to use a network analyzer to measure the frequency response of the environment. This method can determine the transfer function at each frequency across a wide band of frequencies and, thus, determine the propagation losses and the propagation delays by an inverse Fourier transform of the spectral data. However, as will be explained, the method is somewhat limited by the need for a cable from the transmitter to the receiver, and the sensitivity can be less than required for long propagation paths. The alternative time-domain approach is more sensitive and does not need a cable for relative delay measurements, but absolute delays with this method cannot be measured. Thus, there is no one perfect method for all situations, so both techniques will be described.

Because of the requirement to measure the scattering from major objects along the path from the transmitter to the receiver, and the desire for high measurement resolution (ideally sub-meter range resolution), very wide bandwidths are necessary. If a bandwidth of 1 GHz is used, then the time resolution is about 2 ns (60 cm). These bandwidths are associated with UWB systems; but, from the wideband data, the performance of narrower band systems can be derived by the appropriate mathematical processing of the raw data.

B.1 Measurements with a Network Analyzer

A network analyzer is a general instrument designed to measure the frequency characteristics of a two-port network. While its use is usually associated with physical hardware components, the propagation characteristics between a transmitting antenna and a receiving antenna can also be interpreted as a two-port impedance which varies as a function of frequency. Thus, by attaching antennas (via suitable cables) to the two ports of the network analyzer, the complex spectrum as defined by (2.2) can be directly measured. In practice, network analyzers usually measure the 'scattering parameters' (or S parameters), which, for a general two-port, is represented by a 2×2 **S** matrix. Of particular interest for propagation measurements is the parameter S_{21}, which is the scattering parameter from port 1 to port 2.

The measurement technique requires the system to be calibrated. The calibration involves determining the characteristics of the antennas (including the associated cables) used in the experiment. In particular, for absolute power measurements, the antenna gains and matching must be determined. The network analyzer itself has a calibration procedure so that the effect of the cables used to connect the antennas to the network analyzer does not contribute to the measured S parameters.

B.1.1 Measurements with a Reference Cable

To verify the measurement technique, the system is first tested using a length of cable. The resulting impulse response should closely approximate a pure delay and the impulse shape should approach the ideal shape. Figure B.1 shows an example of the computed impulse response of the cable from measured data, which ideally should approach a delta function for wide bandwidths. The delay associated with the cable can be determined from the phase response across the band. In particular, the delay τ is given by

$$\tau = \frac{1}{2\pi} \frac{d\phi(f)}{df} \tag{B.1}$$

By performing a linear LS fit to the phase data, the delay was determined to be 9.7 ns. As the cable was 2.1 m in length, the propagation constant for the cable is calculated to be 0.72.

The calculation of the impulse response without weighting results in the impulse response being convolved by a sinc function, resulting in the 'skirts' of the impulse function. The exact shape of the skirt depends on the position of the delay peak relative to the sampling; but, in

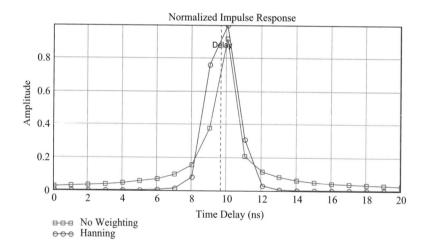

Figure B.1 Computed impulse response from the measured frequency band data. Both the raw unweighted impulse and the Hanning weighted response are shown. The 'Delay' is based on the measured phase response. The measurement bandwidth is 1 GHz.

general, the skirt amplitude decays inversely as the time separation from the peak. To improve the response characteristics, the measured frequency data can be weighted by an appropriate function. The resulting impulse shape will be wider that the unweighted pulse, but the skirt will fall off much more rapidly. For these measurements, a Hanning[1] weighting function was used; this weighting function has a reasonable tradeoff between the pulse width and the skirt fall-off rate. For large time offsets it can be shown that the Hanning impulse response falls off as the inverse cube of the time offset, compared with the inverse of time offset for the rectangular impulse response. The resulting impulse is then confined to about ± 2 ns, and the 50% amplitude (-6 dB) width to about ± 1 ns, compared with the original pulse width (-3 dB) of ± 0.5 ns, but with wide skirts. As the impulse response skirts interfere with the ability to measure the small multipath signals, some reduction in resolution is necessary to allow the environment impulse response to be measured for delays in excess of 2 ns relative to the direct path.

B.1.2 Calibration of Antennas

The calibration of the antennas uses the same technique as described for the cable, but the signal path includes the transmitter antenna, the propagation path and the received signal. The network analyzer measures the return loss (S_{11} and S_{22}) of each antenna and the propagation loss (S_{21}). Ideally, these measurements are performed in an anechoic chamber, but a room with about 1 m clearance to the obstacles will suffice if the bandwidth is 1 GHz or greater. However, to minimize reflections from the floor, some absorbent material should be placed on the floor. As the resolution is of the order of 2 ns (0.6 m), reflections from objects in the room should not greatly affect the measurement. The total measured transmission loss S_{21} in decibels is given by

$$
\begin{aligned}
S_{21} &= L_1 + G_1 + G_{\text{LOS}} + G_2 + L_2 \\
G_{\text{LOS}} &= -20 \log\left(4\pi \frac{D}{\lambda}\right) \\
L_1 &= -10 \log(1 - |S_{11}|^2) \\
L_2 &= -10 \log(1 - |S_{22}|^2)
\end{aligned}
\tag{B.2}
$$

From (B.2), the total antenna gains G_1 and G_2 can be estimated, as the other parameters can be measured or estimated. In the calibration, the LOS loss is assumed to be the free-space loss; but after the calibration, (B.2) can be used to estimate the propagation loss (not necessarily LOS) from the measured S_{21} and the other parameters measured or determined during calibration. Normally, the same antenna is used for both the transmitter and the receiver so that they should have the same gain, which can be determined from (B.2).

Figure B.2 shows an example of the calibrations impulse response. Reflections from the objects in the room result in the small delayed versions of the main impulse. These scattered signals decay approximately exponentially with delay. The 'Delay' parameter is based on the 1 m separation of the antennas and agrees with measurement within the resolution of the measurement system (1–2 ns). Note that the technique has both measured the propagation loss (from S_{21}) and the impulse response, including the propagation delay between the antennas, with the effects of the antennas and cables calibrated out.

---

[1] The Hanning weighting function is given by $H(f) = 0.5[1 + \cos(2\pi f/B)]$, where the signal is constrained to a band defined by $-B/2 \le f \le B/2$.

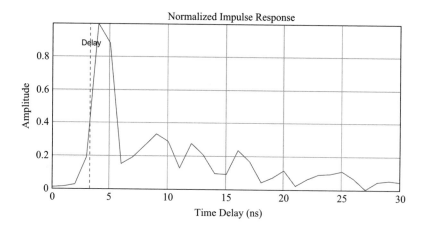

Figure B.2 Computed impulse response for the two antenna separated by 1 m. The 'Delay' indication is the true propagation delay (about 3 ns), which agrees well with the expected value.

B.1.3 Propagation Measurements

Once the system has been calibrated, the environmental propagation characteristics can be measured. The measurement procedure allows the determination of the frequency response across the RF band, the associated propagation loss at each frequency and the impulse response and the associated propagation delay. For positioning systems, the propagation delay is of particular interest. From Figure B.2 it can be observed that the propagation delay can be determined by appropriate processing of the leading edge of the impulse response. Appropriate techniques are discussed in Chapter 4, so no details are given here. However, note that the number of sample points on the leading edge will typically be limited to just two and that the peak of the impulse response is not particularly accurate. For maximum accuracy, the delay measurements are calibrated using a cable of known electrical length, thus allowing the small delays associated with the antennas to be determined and calibrated out. Using this method, the measurement accuracy is on the order of one-tenth the impulse width (or better) in a multipath environment, or about 0.2 ns for a 1 GHz system.

The measurements using the network analyzer will require a long cable connecting the receiving antenna to the input port of the network analyzer. At high frequencies (3–10 GHz for UWB measurements), this cable will be quite lossy, so that a low-noise amplifier is required at the output of the receiving antenna and connecting to the cable. The gain of this amplifier should be at least 10 dB greater than the cable loss, so that the overall noise figure of the measurement system will be essentially that of the low-noise amplifier. For long-range measurements up to 50 m, the gain typically required, even with low-loss cables, is 50–60 dB. Note also that the cable calibration procedures described in Section B.1.1 must include the low-noise amplifier. By applying (B.2), this time with the known calibration gains and the measured S_{21}, the propagation loss can be calculated in absolute terms at each frequency. The impulse response measurements are the same as discussed previously in Section B.1.2.

B.2 Time-Domain Measurements

The time-domain method of measuring the propagation characteristics of an indoor environment is based on using direct-sequence spread-spectrum signals. The method uses the correlation properties of the signal, which approximate a delta function; thus, the received signal after correlation is a direct measure of the impulse response of the environment. For more details on the correlation of direct-sequence spread-spectrum signal, refer to Chapter 3.

The test equipment consists essentially of a specialized radio transmitter and receiver. To provide good time resolution, the radios should have a bandwidth of at least 1 GHz and, thus, can be considered a UWB radio system which is constrained by the requirements for UWB transmissions (low transmitter power, operating in the 3–10.7 GHz band). A simplified block diagram of a typical system is shown in Figure B.3. This system is based on purpose-designed equipment, with the signal processing and display in a PC.

For propagation experiments, the transmitted signal is a direct-sequence spread-spectrum signal, typically with a chip rate of 1 Gchip/s, for a band limited to 1 GHz, or proportional chip rates at other bandwidths. The potential time resolution of such a system is of the order of 2 ns, so that indoor scattering can be resolved in good detail. To achieve the maximum sensitivity, a long integration time (say 0.5 s) means that the signal must be coherent over this time period, which implies accurate control of the RF frequency to about 1 Hz or better. Thus, to provide accurate frequency control, the transmitter and receiver local oscillators must be locked to very accurate sources. One possible design is to use atomic reference oscillators (such as rubidium oscillators) in both the transmitter and the receiver to obtain the necessary frequency accuracy, with no need to interconnect the transmitter and receiver by a cable. Alternatively, but more restricting, an interconnecting cable can be used to supply the reference signal to both the transmitter and the receiver.

An example of a PC control interface is shown in Figure B.4. The controls allow the type of transmission signal to be defined, as well as specifying the output parameters. Also shown are typical output displays, with the received signal spectrum on the left and the corresponding correlation output on the right.

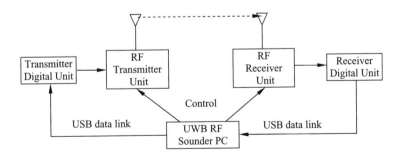

Figure B.3 Simplified block diagram of the UWB RF sounder hardware. The USB data link on the transmitter is only required during the setup, after which the transmitter continuously transmits in spread-spectrum signal. The USB link on the receiver allows the downloading of data from the receiver unit for signal processing.

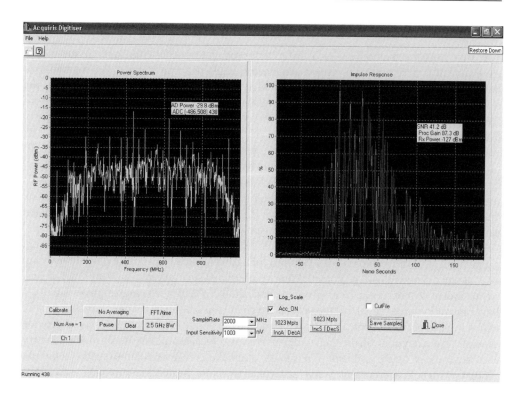

Figure B.4   Picture of the user interface screen for the RF sounder mode of operation. The spectrum is on the left and the environment impulse response on the right. The operating controls are at the bottom, below the displays.

The key to obtaining superior performance from the equipment is obtaining very high process gains.[2] In fact, it will be shown for the above-described system that process gains up to 87 dB can be achieved, provided the equipment is used in an appropriate manner. The high process gains are enabled by a very high sample-rate data logger with a large data buffer of at least 1 Gsamples. The large size of the buffer means that even at a transmission rate of 2 Gsamples/s (Nyquist rate for 1 GHz bandwidth) the transmission period that can be logged is 0.5 s. To maintain coherence over the period Δt of 0.5 s, the transmitter and receiver frequencies must be closely similar so that the phase error is less than (say) 90°. In particular, the phase error $\Delta\phi$ is given by

$$\Delta\phi = 2\pi\Delta f \Delta t \tag{B.3}$$

With the above constraint, the restriction on the frequency error is thus

$$\Delta f \le \frac{1}{4\Delta t} \tag{B.4}$$

---

[2] See Chapter 3 for details on process gain.

Thus, applying (B.4) the frequency error should be less than 0.5 Hz. As the RF is of the order of 10^{10} Hz, the frequency accuracy required is about 0.05 ppb. This accuracy cannot be achieved by normal oscillators, but can be achieved with rubidium or other similar oscillators.

If the frequency can be accurately controlled, then the bandwidth of the correlator is of the order of the reciprocal of the integration period, namely 1/0.5 s or 2 Hz in this case. Thus, the effective receiver bandwidth is 2 Hz. Based on this bandwidth, the link budget summary is as follows:

UWB transmitter power density[3] (dBm/MHz)	-41
Transmitter bandwidth B (MHz)	1000
Transmitter power (dBm)	-11
Receiver effective bandwidth (Hz)	2
Receiver noise figure F (typical) (dB)	3
Receiver noise $kTBF/G_p$ (dBm)	-168
Required SNR (dB)	10
Available loss (dB)	147

Thus, if the frequency can be accurately controlled, then the available loss, even with the low power constraint of UWB, is 147 dB, with an output SNR of 10 dB.

To measure the propagation loss the received signal strength must be measured. As the RSSI radio hardware will be based on the IF bandwidth of 1 GHz, the receiver noise floor ($kTBF$) is -81 dBm. However, most of the measurements of interest are at much lower signal levels, for which the hardware RSSI measurement is too limiting. An alternative approach is required for weaker signal levels, as will now be described.

The receiver automatic gain control system, like the hardware RSSI, functions on the total power in the receiver IF. This operation is defined by

$$\text{RSSI} = S + N = S\left(1 + \frac{1}{\text{SNR}_{\text{in}}}\right) \tag{B.5}$$

If the signal is much less than the receiver noise ($\text{SNR}_{\text{in}} \ll 1$), then the receiver gain is essentially fixed based on the receiver noise only. While the input SNR cannot be measured, the receiver output SNR can be measured, as there is a very large process gain G_p. Thus, from (B.5), the signal level is given by

$$S = \frac{\text{RSSI}}{1 + G_p/\text{SNR}_{\text{out}}} = \frac{N}{1 + G_p/\text{SNR}_{\text{out}}} = \frac{kTBF}{1 + G_p/\text{SNR}_{\text{out}}} \quad (N \gg S) \tag{B.6}$$

As the receiver noise is a known constant, the process gain of the correlation process can be calculated from the signal processing parameters and the output SNR can be estimated from the output correlation function, the receiver signal level can thus be estimated even when the signal level is much less than the receiver noise level. Further, as the process gain is much greater than the output SNR at the limit of the receiver performance, (B.6) can be further simplified to

$$S = \frac{(kTBF)\text{SNR}_{\text{out}}}{G_p}$$
$$S^* = N^* + \text{SNR}_{\text{out}}^* - G_p^* \tag{B.7}$$

---

[3] dBm: power in decibels relative to a milliwatt.

where the asterisk signifies the measurement in decibels. With the noise floor being -81 dBm, the maximum process gain of 87 dB, the minimum signal level with an output SNR of 10 dB is -157 dBm. With the transmitter power limited to -11 dBm, the available loss is 146 dB, which is in close agreement with the theoretical link budget given previously.

Thus, the RF sounder with the above specification is capable of measuring the impulse response to a resolution of 2 ns, and with a propagation loss of 146 dB even with the severe power restrictions applicable to UWB transmissions. The receiver sensitivity is much greater than that achievable using the frequency-domain method with a network analyzer, and no cable is required between the transmitter and the receiver. This method cannot measure the absolute propagation delay, as the transmitter and receiver are not time synchronized. However, by using a cable between the transmitter and the receiver and a power splitter so that both the cable and air transmissions can be simultaneously measured, absolute measurement can be achieved. With this configuration, the delay through the cable will exceed that through the air because the propagation speed in the cable is less than through the air. In this case, the cable signal will appear as a delayed multipath whose delay is known from the cable length. Using this as a reference, the propagation delays through the air can be measured in an absolute sense by using the cable delay as a reference.

Index